D0522085

DEVELOPMENTS IN FOOD ENGINEERING

Proceedings of the 6th International Congress on Engineering and Food

Proceedings of the Sixth International Congress on Engineering and Food
23-27 May 1993, Makuhari Messe, Chiba, Japan.

ORGANIZING COMMITTEE

Chairman	YANO, Toshimasa	Prof. Emeritus, Univ. of Tokyo
General Secretary	NAKAMURA, Kozo	Prof., Univ. of Tokyo
Program Chairman	MATSUNO, Ryuichi	Prof., Kyoto Univ.
Advisor	TOEI, Ryozo	Prof. Emeritus, Kyoto Univ.
Advisor	FUJIMAKI, Masao	Prof. Emeritus, Univ. of Tokyo

ADACHI, Shuji	Kyoto Univ.
ARAI, Soichi	Univ. of Tokyo
FUJIO, Yusaku	Kyushu Univ.
FUKUSHIMA, Masayoshi	Snow Brand Milk Products Co., Ltd.
IMAMURA, Kazuo	FAIS
KAWADE, Hiroyuki	Fuji Oil Co., Ltd.
KIMURA, Shoji	Univ. of Tokyo
KOBAYASHI, Takeshi	Nagoya Univ.
KUBOTA, Kiyoshi	Hiroshima Univ.
MIYAWAKI, Osato	Univ. of Tokyo
NAKANISHI, Kazuhiro	Okayama Univ.
NOGUCHI, Akinori	National Food Research Institute
OHKUBO, Yukima	Ajinomoto Co., Inc.
OHSAKI, Katsumichi	Kikkoman Corp.
SAGARA, Yasuyuki	Univ. of Tokyo
WATANABE, Atsuo	TOTO Ltd.
WATANABE, Hisahiko	Tokyo Univ. of Fisheries

INTERNATIONAL ADVISORY COMMITTEE

Bimbenet, J.J.	(France)
Bruin, S.	(The Netherlands)
Earl, R.L.	(New Zealand)
Farkas, D.	(USA)
Filka, P.	(Czech)
Hallström, B.	(Sweden)
Jowitt, R.	(UK)
LeMaguer, M.	(Canada)
Lewicki, P.	(Poland)
Linko, P.	(Finland)
Lund, D.	(USA)
Martin, A.M.	(Canada)
McKenna, B.	(Ireland)
Schubert, H.	(Germany)
Spiess, W.E.L.	(Germany)
Yano, T.	(Japan, Chairman)

DEVELOPMENTS IN FOOD ENGINEERING

Proceedings of the 6th International Congress on Engineering and Food

Edited by

Professor Toshimasa Yano (Professor Emeritus, University of Tokyo),
Department of Bioengineering, Faculty of Engineering,
Yokohama National University,
Tokiwadai, Hodogaya-ku, Yokohama 240, Japan

Professor Ryuichi Matsuno,
Department of Food Science and Technology,
Faculty of Agriculture, Kyoto University,
Sakyo-ku, Kyoto 606-01, Japan

and

Professor Kozo Nakamura
Department of Agricultural Chemistry,
Faculty of Agriculture, University of Tokyo,
Yayoi, Bunkyo-ku, Tokyo 113, Japan

BLACKIE ACADEMIC & PROFESSIONAL
An Imprint of Chapman & Hall
London · Glasgow · Weinheim · New York · Tokyo · Melbourne · Madras

Published by Blackie Academic & Professional, an imprint of Chapman & Hall, Wester Cleddens Road, Bishopbriggs, Glasgow G64 2NZ

Chapman & Hall, 2-6 Boundary Row, London SE1 8HN, UK

Blackie Academic & Professional, Wester Cleddens Road, Bishopbriggs, Glasgow G64 2NZ, UK

Chapman & Hall GmbH, Pappelallee 3, 69469 Weinheim, Germany

Chapman & Hall USA, One Penn Plaza, 41st Floor, NY 10119, USA

Chapman & Hall Japan, ITP-Japan, Kyowa Building, 3F, 2-2-1 Hirakawacho, Chiyoda-ku, Tokyo 102, Japan

DA Book (Aust.) Pty Ltd, 648 Whitehorse Road, Mitcham 3132, Victoria, Australia

Chapman & Hall India, R. Seshadri, 32 Second Main Road, CIT East, Madras 600 035, India

First edition 1994

Printed in Great Britain by Hartnolls Ltd., Bodmin, Cornwall

ISBN 0 7514 0224 9
1 work in 2 parts, not available separately

A catalogue record for this book is available from the British Library

Library of Congress Catalog Card Number: 94-70986

∞ Printed on acid-free text paper, manufactured in accordance with ANSI/NISO Z39.48-1992 (Permanence of Paper).

PREFACE

The necessity of prediction and fine control in the food manufacturing process is becoming more important than ever before, and food researchers and engineers must confront difficulties arising from the specificity of food materials and the sensitivity of human beings to taste. Fortunately, an overview of world research reveals that the mechanisms of the many complex phenomena found in the food manufacturing process have been gradually elucidated by skilful experiments using new analytical tools, methods and theoretical analyses.

This book, the proceedings of the 6th International Congress on Engineering and Food (ICEF6), held for the first time in Asia – in Chiba, Japan May 23 - 27, 1993 – summarizes the frontiers of world food engineering in 1993. Congress was joined by the 4th International Conference on Fouling and Cleaning. There were 476 active members from 31 countries participating in the Congress. The editors hope that readers will find this book to be a useful review of the current state of food engineering, and will consider future developments in this research field.

The editors extend thanks to the members of the organizing committee of ICEF6, and the advisors, Dr. Ryozo Toei, Professor Emeritus of Kyoto University and Dr. Masao Fujimaki, Professor Emeritus of the University of Tokyo. They also acknowledge the international advisory board members who helped the organizing committee in many ways, and the 10 foundations and 66 companies that financially supported the ICEF6. Finally, the editors are indebted to the reviewers of the manuscripts of these proceedings.

September, 1993

Toshimasa Yano
Ryuichi Matsuno
Kozo Nakamura

ACKNOWLEDGEMENTS

The Sixth International Congress on Engineering and Food has been financially supported by the following foundations and companies.

The Commemorative Association for the Japan World Exposition (1970)
Chiba Convention Bureau
Hosokawa Powder Technology Foundation
Nestlé Science Promotion Committee
The Kajima Foundation
The Iijima Memorial Foundation for the Promotion of Food Science and Technology
The Iwatani Naoji Foundation
The Naito Foundation
TOKYO OHKA Foundation for the Promotion of Science and Technology
The Asahi Glass Foundation

Advance Co., Ltd.
Ajikan Co., Ltd.
Ajinomoto Co., Inc.
APV Crepaco Far East, Inc.
Asahi Breweries Ltd.
Calbee Foods Co., Ltd.
Dai Nippon Printing Co., Ltd.
Daido Steel Co., Ltd.
Dainippon Ink & Chemicals Co., Ltd.
The Federation of Electric Power Companies
Fuji Electric Co., Ltd.
Fuji Oil Co., Ltd.
Gekkeikan Sake Co., Ltd.
Godo Steel Ltd.
Hayashibara Co., Ltd.
Hokkaido Sugar Co., Ltd.
House Food Industry Corp., Ltd.
Izumi Food Machinary Co., Ltd.
Japan Automobile Manufacturer's Association, Inc.
The Japan Steel, Ltd.
Japan Tobacco Inc.
K F Engineering Co., Ltd.
Kagome Co., Ltd.
Kamewada Technical Consulting Engineers Office
Kaneka Corporation
Kawasaki Steel Corporation
Kikkoman Corporation
Kirin Brewery Co., Ltd.
Kobe Steel Ltd.
Kohjin Co., Ltd.
Kubota Corporation
Kurimoto, Ltd.
Kyowa Hakko Kogyo Co., Ltd.
Matsushita Electric Industrial Co., Ltd.
Meiji Milk Products Co., Ltd.
Meiji Seika Kaisha, Ltd.
Membrance Researcn Circle of Food
Mitsubishi Steel Manufacturing Co., Ltd.
Miura Co., Ltd.
Momoya Co., Ltd.
Morinaga Engineering Co., Ltd.
Morinaga Milk Industry Co., Ltd.
Morishita Jintan Co., Ltd.
Nakayama Steel Works Ltd.
Nippon Dairy Co., Ltd.
Nippon Kokan K. K.
Nippon Lever B. V.
Nippon Steel Corporation

Nippon Suisan Kaisha, Ltd.
Nisshin Flour Milling Co., Ltd.
Nisshin Steel Co., Ltd.
NOF Corporation
Research Conference of Membrane Application
Sapporo Brewery Co., Ltd.
Satake Corporation
Senbatoka Industry Co., Ltd.
Snow Brand Milk Products Co., Ltd.
Sumitomo Metal Industries Ltd.
Suntory Ltd.
Tokyo Banker's Association, Inc.
Topy Industries, Ltd.
Toshiba Corporation
TOTO Ltd.
Toyo Seikan Kaisha, Ltd.
Wado Doctor's Group
Yamazaki Baking Co., Ltd.
Yodogawa Steel Works Ltd.

CONTENTS

Keynote Lecture

Plenary Lectures

Physical and Physicochemical Properties of Food

Application of NMR to Food Engineering

Mechanical Processing of Food

Thermal and Mass Transfer Operations of Food

Phase Change Operation

Concentration and Dehydration Processes

Reaction Kinetics in Food Processing

Fermentation Processes

Bioreactors Using Enzymes and Cells

Separation and Purification Processes

Membrane Processes

Pasteurization and Sterilization Processes

Aseptic Processes

Packaging Science and Technology

Fouling and Cleaning

Processing under Unusual Conditions

Transportation and Preservation of Food

Sensors, Process Control, and Factory Automation

Innovation in Equipment Design and Plant Operation

Environmental Problems in Food Industry

Innovation in Traditional Food Processing

Design of Physiological Functions of Foods from Engineering Viewpoints

MODELLING OF TRANSPORT PHENOMENA DURING THE CULTIVATION OF *Bacillus thuringiensis* FOR THE PRODUCTION OF BIOINSECTICIDES.

Pérez Galindo, A., Velázquez de la Cruz, G., Medrano-Roldán, H., Robles-Cárdenas, M., Rodríguez-Padilla, C.(*), Tamez-Guerra, R.(*), and J.L. Galán-Wong (*). Instituto Tecnológico de Durango, Unidad de Biotecnología Industrial. Av. Felipe Pescador 1830 Ote. 34080 Durango, Dgo. México. (*) Fac. de Ciencias Biológicas U.A.N.L. Monterrey, N.L. México.

ABSTRACT

This paper reports the preliminary design and application of a mathematical model which seeks to explain observed growth rates during propagation of a native strain of *Bacillus thuringiensis* var. *kumamotoensis* for the production of a bioinsecticide. The model, based on the mass transfer coefficient, has the following assumptions: 1) Mass transport in the total of cells may be described by a single balance equation. 2) Each cell behaves as the average of the whole population. 3) The substrate used by the microorganism is homogeneous.

INTRODUCTION

In recent years, there has been considerable interest in the application of mathematical models to microbial processes (1-3). Biological models try to predict the growth rate as a function of substrate, cell and inhibitory products concentrations.

In this work we studied the kinetics of a native strain of *Bacillus thuringiensis* characterized as *kumamotoensis* during its exponential and stationary phases. In the latter, we are interested in getting information about production of an insecticide. Measurements were used to design a mathematical model, based on mass transfer phenomena, to be applied to the automatic control of the process. The model could be made more general by using parameters directly related with process fermentation variables such as pH, temperature, aeration and agitation rates, etc.

MATERIALS AND METHODS

The model was tested with experimental data obtained from the literature and from our 20 liter semi-pilot fermentation plant Lh Engineering Series 2000. Equations for substrate consumption and growth rate were posed with Monod's model and our model. Model parameters were calculated from the data by means of a least squares subroutine of EUREKA. They were used in the differential equations to perform simulations with the aid of ISIM. Comparisons of models' results against experimental data, by graphical and numerical means, were effected with QUATTRO PRO.

RESULTS AND DISCUSSION

The model is based on the general equation of growth which states that the rate of change of the population is proportional to the population itself. In this model, an expression for the specific growth rate was derived by assuming that it depends on the flow of substrate to the cell, which is a function of the mass transfer coefficient and microbial physiology.

Solving the mass tranfer equations for the microbe-substrate system, under assumptions mentioned in the abstract, the specific growth rate is:

$$\mu = ASX\ (1-(B(1-e^{-ct}))) \qquad (1)$$

where A,B and C are the model constants.

Substitution of Eqn.(1) into the general equations for microbial concentration, X, and substrate consumption, S, with a yield coefficient Y results in the following equations:

$$dX/dt = ASX^2\ (1-(B(1-e^{-ct}))) \quad ; \quad dS/dt = -\ (dX/dt)/Y \qquad (2)$$

Figure 1 shows experimental data (**R** points), along with predictions from our model (**P** points) and Monod's model (**M** points). As seen there, the growth and substrate consumption curves (on legend as **Log cfu** and **Subs**, respectively) are correctly predicted by both models.

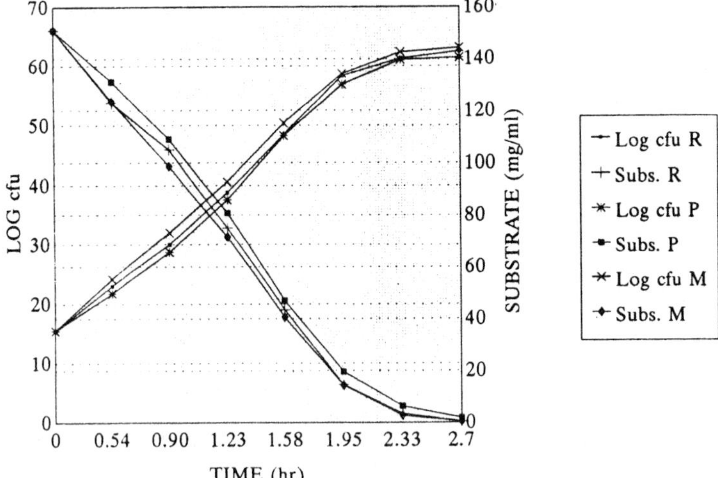

Figure 1. Comparison of model predictions with yeast kinetics

Data from our experiments are presented in the first two columns of Table 1. Model predictions are shown in columns third to sixth of the table, as ratios of predicted to measured values. A noteworthy fact is the abnormal behavior in the growth and substrate evolutions, typical of our strain. A visual comparison of the numbers in columns third and fourth (our model) against those of Monod's model (columns fifth and sixth), shows the former to yield better predictions, with numbers always closer to one. This is due to the basic formulation of Monod's model, which fails to predict these type of kinetics.

TABLE 1

Comparison of Model Results with Our Experimental Measurements

Time (hr)	Experimental data Log(cfu/ml)	Subst. (mg/ml)	This Model/Exp. fraction		Monod's/Exp. fraction	
0	6.00	1.72	1.000	1.000	1.000	1.000
2	6.68	1.70	0.997	0.826	0.975	0.883
4	7.32	1.20	0.984	0.956	0.955	1.079
6	7.92	0.74	0.962	1.279	0.939	1.494
8	8.48	0.50	0.936	1.589	0.924	1.868
10	8.24	0.50	0.993	1.362	0.994	1.565
12	7.92	0.50	1.055	1.194	1.073	1.302
18	8.12	0.50	1.065	0.912	1.130	0.721
25	8.64	0.51	1.015	0.784	1.112	0.339
35	8.92	0.52	0.982	0.781	1.108	0.111

CONCLUSIONS

A model was presented which is able to predict the kinetics of *Bacillus thuringiensis* var. *kumamotoensis*. To evaluate the model, its predictions were compared with experimental measurements of yeast kinetics with good results. This model proved to be better than Monod's model due to the fact that its formulation is based on mass tranfer. Monod's model, based on the Michaelis-Menten equation for enzyme kinetics, can not predict the abnormal behavior of our strain.

REFERENCES

1. Rolz,C., Engineering of biological reactions and processes. Center for Scientific and Technological Studies (ICAITI). Guatemala. 1989.

2. Korte,G., Rinas,U., Krake-Helm,A. and Schurgel,K., Structured model for cell growth and enzyme production by recombinant *Escherichia coli*. Appl. Microbiol. Biotechnol. 1991, **35**, 185-188.

3. Bovee,J.P., Strehaiano, P., Goma, G. and Sevely, Y., Alcoholic fermentation: Modelling based on sole substrate and product measurement. Biotechnol. and Bioeng. 1984, **24**, 328-334.

MEASUREMENT OF THE MAIN INDUSTRIAL FERMENTATION PARAMETERS GOVERNING THE PRODUCTION OF BIOINSECTICIDES BY *Bacillus thuringiensis* var. *kumamotoensis*

MEDRANO-ROLDAN,H., VELAZQUEZ,C.G., ROBLES,C.M., PEREZ-GALINDO,A., CORREA,C. P., RODRIGUEZ-PADILLA,C.(*), TAMEZ-GUERRA,R.(*) AND J.L. GALAN-WONG(*). Instituto Tecnológico de Durango, Centro de Graduados e Investigación, Sección de Biotecnología. Av. Felipe Pescador 1830 Ote. 34080 Durango, Dgo. México. (*) Facultad de Ciencias Biológicas U.A.N.L. Monterrey N.L. México.

ABSTRACT

We decided to study the independent influences of impeller dimensions, pumping impellers capacity, impeller speed and bulk dissolved oxygen concentration on the oxygen uptake kinetics of *Bacillus thuringiensis* var. *kumamotoensis* propagated at 20 liter semi-pilot fermentation plant level and some aspects on rheology *in situ*. The mixing times were shown to quantitatively influence the substrate transport into cell and the impeller parameters were shown to have strong and independent influences on the significant oxygen uptake rate. Important data was found about differences between rheology *in situ* and *in vitro*.

INTRODUCTION

Many industrial fermentations utilize several kinds of substrates similar to those used in bioinsecticides production from *Bacillus thuringiensis* strain (1,2,3). When Non-Newtonian viscous behavior occurs, the local cell environment is only indirectly related to conditions in the bulk medium.

A logical starting point for analyzing physical influences on biological activity in the fermenter would be to define the role of broth rheology to fermenter performance, specially with regard to the oxygen supply to the cells. On the other hand, very little is known about the *in situ* relationships between shear, mixing, and biological activity in systems such as this. It is intended that this research contribute to the ultimate objective of relating the pertinent independent variables, such as vessel dimensions, impeller dimensions, impeller speed, etc.

MATERIALS AND METHODS

Microorganism: A native strain of *Bacillus thuringiensis* characterized as *kumamotoensis* isolated from dead larvae of *Spodoptera frugiperda* was used in this research.

Growth Medium: For preservation Nutritive Agar (Merck) was used and for propagation at flask and semi-pilot plant level the composition of the medium was as follows (g/l): Cane Sugar Molasses, 20; Soybean Meal, 20; Corn Steep Liquor, 30; $CaCO_3$, 0.1; Tap Water, 1000 ml; pH, 7.0 (3). All the nutrients were industrial grade.

Rheology in vitro and in situ: Rheology *in vitro* was measured by using a viscometer Brookfield LVT and *in situ* was based on measurements of impeller speed and power consumption (3).

Mass Transfer from the Bulk Liquid to the Cells: Using the body of the data describing the influence of the dissolved oxygen concentration and the impeller parameters on oxygen uptake kinetics, our observations were related to more fundamental principles. One approach is to interpret the results in terms of impeller shear and flow. Both of these can be estimated in terms of impeller speed, diameter, and blade height (3).

RESULTS AND DISCUSSION

Process Fermentation for the Production of Bioinsecticides: Table 1 shows the kinetics of *Bacillus thuringiensis* var. *kumamotoensis*. Important data is the value on oxygen demand (0.70 $gO_2/lxHR$) which is low compared with others in the literature (1.2-1.8 $gO_2/lxHR$) (3).

Rheological Aspects on the Production of Bioinsecticides: The design of the impeller which leads to different levels of pumping capacity is related with the mass transfer coefficient in Figure 1. Physically, impeller pumping could be related to broth velocity, circulation time, or to mixing time. The impeller pumping capacity has been postulated to be an important parameter in the fluid mechanics of fermentation broths.

Results from experiments *in vitro* (Brookfield LVT on degassed samples removed from the fermenter) and *in situ* (Impeller Power)are showed in Table 2. This Table contains values for the flow consistency index, the flow behavior index, and the calculated apparent viscosity projected from the shear stress vs. shear rate diagrams using both methods. In all cases, the flow consistency index, is higher for the *in situ* measurements. Comparing the values of the flow behavior index, in this Table, the *in situ* broth also appears to be much more sensitive to shear thinning than the *in vitro* broth.

As a final note, the results in Table 2 show that the apparent viscosity as calculated using the *in vitro* technique increases with cell concentration whereas the *in situ* measurement shows the reverse relationship so it is our opinion that the *in situ* measurement depicts the rheological characteristics of a broth more realistically.

CONCLUSIONS

1).- *Bacillus thuringiensis* var. *kumamotoensis* shows lower oxygen demand values than others aerobic microorganisms.

2).- *In situ* measurements depict the rheological characteristics of a broth more realistically.

REFERENCES

1.- Abdel-Hameed, A. Studies on *Bacillus thuringiensis* H-14 strain isolated in Egypt-II. Ultrastructure Studies. World Jour. of Microbiol. and Biotechnol. 1990. **6**, 302-312.

2.- Blokhina, T.P., Variability in *Bacillus thuringiensis* under various growth conditions. Mikrobiologiya 53 (3), 427-431.

3.- Medrano-Roldán, H., Estudio sobre los parámetros de fermentación de importancia industrial durante la propagación de *Bacillus thuringiensis* var. *kumamotoensis* C-4 para la producción de bioinsecticidas. Tesis Doctoral. Fac. de Ciencias Biológicas U.A.N.L. Monterrey, N.L. México. 1992.

TABLE 1.
Effect of agitation rate on microbial kinetics of Bacillus thuringiensis var. kumamotoensis at pilot plant level

Agitation Rate (RPM)	X (gDC/l)	$Y_{(x/s)}$ (gDC/gS)	μ (HR^{-1})	Na (gO_2/lxHR)	Q_{O2} (gO_2/gDCxHR)	K_{LA} (HR^{-1})	Hp/ 1000 l
200	2.7	0.45	0.20	0.54	0.21	179	0.015
300	2.7	0.45	0.20	0.56	0.22	156	0.015
400	2.7	0.45	0.25	0.60	0.24	199	0.016
500	2.8	0.48	0.25	0.65	0.26	216	0.017
600	2.8	0.48	0.30	0.65	0.27	226	0.019
700	3.0	0.50	0.30	0.70	0.28	223	0.019
800	3.0	0.50	0.30	0.70	0.29	235	0.019

TABLE 2.
Comparison of rheological properties of the fermentation broth for Bacillus thuringiensis var. kumamotoensis at pilot plant level

Concentration of biomass (gDC/l)	Flow consistency index (g/cm x sec^{1-n})		Flow behavior index		Apparent viscosity (pseudopoises)	
	in vitro[a]	*in situ*[b]	*in vitro*[a]	*in situ*[b]	*in vitro*[a]	*in situ*[b]
2.0	2	18	0.90	0.60	0.43	1.8
5.0	3.5	45	0.70	0.40	0.68	1.7
8.0	4.5	62	0.65	0.30	0.77	1.60

a) Determined with the viscosimeter Brookfield on degassed samples
b) Determined by measurement of agitation power consumption

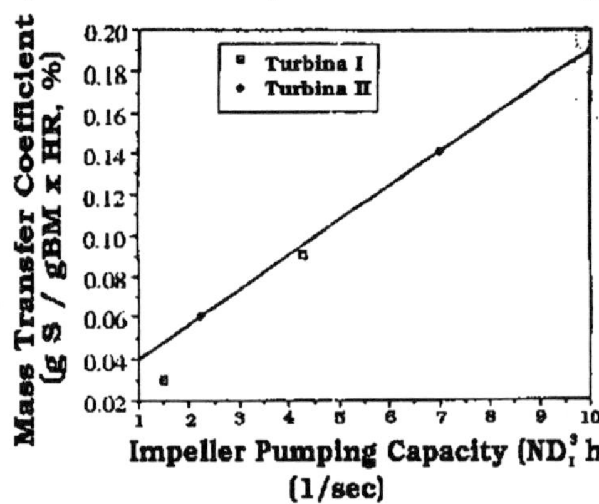

Figure 1. Effect of impeller pumping capacity on the liquid-solid mass transfer coefficient

USE OF SWEET POTATO PROCESSING WASTEWATER AS SUBSTRATE FOR FERMENTATIONS

C. A. A. RIBEIRO; M. E. CAVENAGUI; A. E. R. DE PONTES; G. L. BARROS;
A. L. PESSOA; A. C. BARANA; F. R. GUIDOLIN; V. L. DEL BIANCHI; I. O. MORAES
Departamento de Engenharia e Tecnologia de Alimentos
UNESP – Universidade Estadual Paulista
Rua Cristovao Colombo, 2265 – S. J. Rio Preto, S. P. 15054–000, BR

ABSTRACT

The aim of this work is to use, at laboratory scale, wastewaters resulting from sweet potato processing to obtain different bioproducts. The wastewater is treated in a bubble column to a COD reduction of 92%. This wastewater, rich in sugars and nitrogen, with different concentrations of total sugars, was added to enrich a semi solid process with soybean bran to produce mold glucoamylase, in both flask and column fermentation, and as substrate to produce *Bacillus thuringiensis* cells by submerged fermentation. Except in solid state column fermentation, satisfactory results have demonstrated that these wastewaters have a high potential as a substrate for fermentations.

INTRODUCTION

The food industry, besides its products, is a potential producer of wastes with high organic loads, which must be treated appropriately to reduce this load to admissible levels, before being discarded into aquatic bodies, to reduce the environmental impact originated by the wastes.

However, these wastes may be useful. This is the case for molasses, corn steep liquor and manipueira's wastes in submerged processes (1,2) and manipueira and paper processing wastes in solid state fermentation (2). In this work, the sweet potato processing wastewater was studied as a substrate for both solid state and submerged fermentation processes to obtain some bioproducts and for wastewater treatment.

MATERIALS AND METHODS

For the wastewater treatment a bubble column reactor of 5 liters (9 × 9 × 62 cm) was used, inoculated with activated sludge from the milk processing industry. After a sludge acclimatization time, different residence time (18, 24 h), cell concentration (1500, 2600, 3330 mg MLVSS/l) and COD affluent concentration (1200, 1500, 1850 mg COD/l) were used. The aeration rate was about 1 v.v.m. The COD reduction was measured by potassium dicromate method.

To the glucoamylase production by flask fermentation, 5 g of soybean meal in 250 ml Erlenmeyers flasks were used, and then humidified with tap water and sweet potato processing wastewater (100, 500, 1000, 3000 mg COD/l) at 1:1 (w/v), 37°C for 60 h. The enzymatic activity was measured by the Somogyi method.

To the glucoamylase production by column fermentation 16 g of soybean bran in a packed column of 5 cm diameter and 10 cm height were used and then were humidified with tap water and sweet potato processing wastewater (1000 mg COD/1) at 1:1 (w/v), ambient temperatures for 48 h. The enzymatic activity was measured by the Somogyi method.

To the *Bacillus thuringiensis* production 100 ml of molasses/urea medium and sweet potato processing wastewater (500, 1000, 2000, 3000 mg COD/l) in 500 ml Erlenmeyer flasks were used at 30°C for 48 h with constant agitation. The cellular growth was measured by optical density.

RESULTS

Table 1 shows that, at same fermentation conditions, the cell concentration and residence time are the prior factors influencing the COD removal. Over 24 hours, this gives an average value of 92%.

Table 2 shows that sweet potato processing wastewater can be used as a substrate to produce glucoamylase instead of water at a COD concentration of 1000 mg/1 after 36 h. At this point, wastewater was 10% better than water as humidifier.

Table 3 shows that, in solid state fermentation column reactor, sweet potato processing wastewater did not achieve good results. Water humidification was at least 37% better than the wastewater.

Table 4 shows that sweet potato processing wastewater can be used as substrate to produce *Bacillus thuringiensis* cells instead of water at a 3000 mg COD/l at the end of 60 h of fermentation. At this point, the wastewater was 68.5% better than molasses/urea as substrate.

Table 1 Average COD affluent, COD effluent and COD removal data for each cell concentration at 18/24 h

	[X] (average value) (mg MLVSS/1)			
Average value	*1522*	*2500*	*2597*	*3314*
COD affluent (mg/l)	1300	1534	1805	1862
COD effluent (mg/1)	300	217	143	159
COD removal (%)	81	86	91	92
Residence time (hours)	18	18	24	24

Table 2 Enzymatic activities variation at different COD concentrations

	Fermentation time (h)						
	18	*24*	*30*	*36*	*48*	*54*	*60*
Water	0.897	0.541	0.327	0.359	0.669	0.602	0.774
COD 100	0.863	0.607	0.316	0.534	0.692	0.263	0.592
COD 500	0.703	0.722	0.186	0.916	0.617	0.300	0.502
COD 1000	0.802	0.723	0.163	0.991	0.653	0.035	0.215
COD 3000	0.705	0.785	0.028	0.802	0.647	0.143	0.223

Table 3 Enzymatic activities variation with water and sweet potato process wastewater as a humidifier

	Fermentation time (h)			
	12	*24*	*36*	*48*
Water	0.273	0.308	0.325	0.285
Wastewater	0.109	0.162	0.179	0.105

Table 4 Cellular growth at different COD concentration

	Fermentation time (h)						
	18	*24*	*30*	*36*	*48*	*54*	*60*
Molasses/urea	0.102	0.213	0.254	0.325	0.368	0.424	0.597
COD 500	0.049	0.110	0.140	0.140	0.145	0.150	0.213
COD 1000	0.117	0.130	0.168	0.188	0.274	0.287	0.319
COD 2000	0.175	0.260	0.287	0.304	0.332	0.340	0.454
COD 3000	0.143	0.451	0.552	0.601	0.725	0.753	0.871

CONCLUSIONS

It was observed that sweet potato processing wastewater has a good potential as a fermentation substrate in both solid state and submerged processes.

REFERENCES

1. Moraes, I. O., Studies of submerged fermentation to the production of biological insecticides in lab scale. *PhD thesis*, FEEA – UNICAMP, Brazil, 1976.
2. Capalbo, D. M. F.; Moraes, I. O., A simple culture medium for *Bt* production. 8th World Congress of Food Science and Technology, Toronto, Canada, p. 326, 1991.

FERMENTATIVE PRODUCTION OF L-LACTATE FROM XYLOSE

Tomoko Ueda, Kenji Tanaka and Ayaaki Ishizaki.
Department of Food Science and Technology,
Faculty of Agriculture, Kyushu University,
Hakozaki 6-10-1, Higashi-ku, Fukuoka-shi 812, Japan

ABSTRACT

L-lactic acid fermentation employing *Lactococcus lactis* strain IO-1 (JCM7638) was examined on xylose, and glucose and xylose media. The yield of lactate on xylose was 0.47g lactate/g xylose at an initial xylose concentration of 51.2g/*l* and the μ_{max} was 1.02 1/h. Xylose culture was more susceptible to lactate inhibition than were glucose cultures but showed similar kinetic behavior. The microorganism was capable of complete sugar utilization when grown on a mixture of 20g/*l* xylose and 20g/*l* glucose and synthesized 0.66g lactate/g sugar.

INTRODUCTION

Biological conversion of xylose into simple chemicals creates an interest in order to utilize lignocellulosic biomass. *L. lactis* IO-1 is a homolactic L-lactic acid producing bacterium isolated and characterized in our laboratory. We have demonstrated that IO-1 is capable of high conversion rates of glucose to L-lactic acid, with only slight production of volatile fatty acids, but is subject to end-product uncompetitive inhibition (3). *L. lactis* IO-1 utilizes xylose as carbon source too, and so we have studied the fermentative production of L-lactic acid from xylose (1,2).

MATERIALS AND METHODS

Medium

The basal medium consisted of (per litre) 5.0g sodium chloride, 5.0g polypepton (Nihon Seiyaku, Co. Ltd., Japan) and 5.0g yeast extract (Difco Laboratories, U.S.A.). The medium was supplemented with up to 51.2g/*l* xylose and 20g/*l* glucose. The same basal medium was used for inoculum preparation.

Culture conditions

Batch cultures were carried out in a 1-L fermenter (working volume; 400cm^3), agitated at 400rpm without gas feed. Temperature was controlled at 37°C by a circulating water bath and the pH was controlled at 6.0 by automatic addition of 2N NaOH.

Analysis

Cell growth was monitored by absorbance at 562nm and converted to dry cell weight. Residual glucose and L-lactate concentrations in cell-free broth were assayed with a YSI model 23A glucose analyzer and a YSI model 23L lactate analyzer, respectively (Y. S. I. Co. Ltd., U.S.A.). Residual xylose in cell-free broth was assayed by using the method of Somogyi-Nelson to estimate the total reducing sugar content and compensating for the concentration of glucose in mixed substrate fermentation.

RESULTS AND DISCUSSION

Figure 1 shows the fermentation profiles of *L. lactis* IO-1 on 51.2g/*l* initial xylose. The yield of lactate in xylose culture was 0.47g lactate/g xylose in comparison with a yield of 0.88g lactate/g glucose in glucose culture (3).

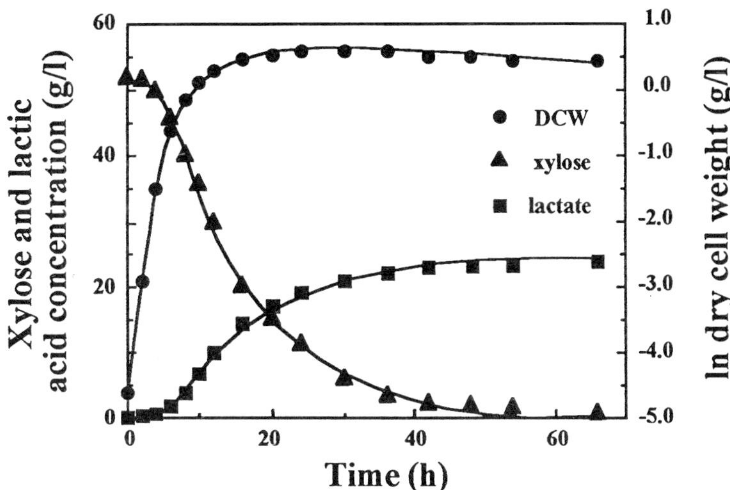

Figure 1. Fermentation profile of *L. lactis* IO-1 on 51.2g/*l* initial xylose.

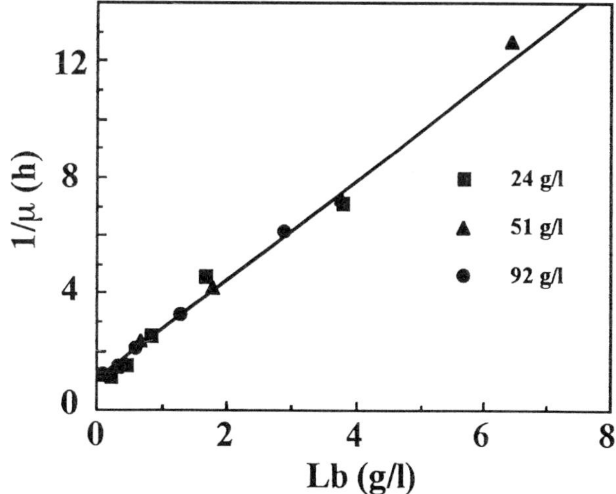

Figure 2. The effect of L-lactate concentration on $1/\mu$ in xylose batch culture. These data were obtained from batch cultures in which the initial xylose concentrations were 24, 51 and 92g/*l*.

L. lactis IO-1 was grown in batch culture on initial xylose concentrations of 24, 52 and 92g/*l* (1). Viable cells were counted to determine the specific growth rates during lactic acid accumulation phase in the xylose cultures and Figure 2 represents the relationship between the reciprocal of these specific growth rates and L-lactate concentration. The straight line shown in Figure 2 conforms with the relationship predicted by equation [1] with μ_{max} of 1.05 1/h and Ki of 1.86 *l*/g.

$$\mu = 1.05 / (1 + 1.86\, Lb) \qquad [1]$$

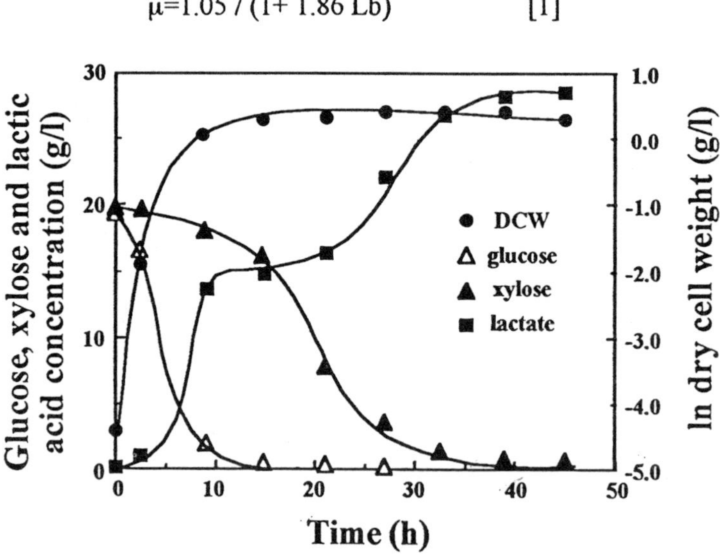

Figure 3. The growth, L-lactate production and substrate consumption of *L. lactis* IO-1 on mixed medium of 20g/l glucose and 20 g/l xylose.

Tyree *et al.* (4) demonstrated that the ability of *Lactobacillus xylosus* to utilize xylose was markedly inhibited by only 2g/*l* glucose and was completely inhibited at glucose concentrations greater than 5g/*l*. The ability of *L. lactis* IO-1 to utilize a mixture of 20g/*l* xylose and 20g/*l* glucose is illustrated in Figure 3. Although glucose was used at a higher rate than was xylose in the early stages of the culture, both sugars were utilized simultaneously. After the glucose concentration declined below 0.49g/*l*, the utilization rate of xylose increased significantly and there was a suggestion of a diauxic relationship in the data. The overall yield of lactate was 0.67g lactate/g sugar, intermediate between the values obtained for the single sugar fermentations.

REFERENCES

1. Ishizaki, A., Ueda, T., Tanaka, K. and P. F. Stanbury, L-Lactate production from xylose employing *Lactococcus lactis* IO-1. Biotechnol. Lett.,1992,**14**,**7**,599-604.

2. Ishizaki, A., Ueda, T., Tanaka, K. and P. F. Stanbury., The kinetics of end-product inhibition of L-lactate production from xylose and glucose by *Lactococcus lactis* IO-1. Biotechnol. Lett., 1993,**15**,**5**,489-494.

3. Ishizaki, A., Ohta, T. and Kobayashi, G., Batch culture growth model for lactate fermentation. J. Ferment. Bioeng.,1989,**68**,**2**,123-139.

4. Tyree, R. W., Clausen, E. C. and Gaddy, J. L. The fermentative characteristics of *Lactobacillus xylosus* on glucose and xylose. Biotechnol. Lett.,1990,**12**,**1**,51-56.

OPTIMIZATION STUDIES OF A LACTIC ACID FERMENTED BEVERAGE PROCESS

TINA MOE & JENS ADLER-NISSEN
Center for Food Research at DTH, Dept. of Biotechnology,
Technical University of Denmark, B-221, 2800 Lyngby, Denmark

ABSTRACT

The present work is concerned with a process for preparation of a lactic acid fermented beverage based on soymilk and cereals. Considerations on re-design and optimization of the process is made using new approaches. The intention is to develop an optimization strategy as a supplement to traditional design strategies known from the chemical industry. The principal idea is that the process should be optimized as a whole by optimizing key unit operations in priority order upstream.

INTRODUCTION

Lactic acid fermentation of cereals, legumes and green vegetables is applied nearly all over the world in traditional food processes based on experience and good craftsmanship. In the recent years there has, however, been a strong trend to develop some of these processes into modern, large-scale food biotechnology processes. Due to the often poor understanding of the actual chemical changes - or biotransformations - taking place and their inherently complex nature the traditional process design methodology known from the chemical industry is much too rigid to be sufficient for the optimization of these food biotechnology processes.

The aim is therefore, to develop a tool which can facilitate the decision making on the optimization and re-design aspects as well as on which further investigations to carry out on the process. The strategy is based on the proposed concept of a principal process step. The most significant parameters for e.g. product quality are identified and an order of priority for optimization of the individual parameters is established. The parameter of most importance for the product e.g. taste is focused upon at first. The optimization is then initiated at the process step which has the highest impact on this parameter - the principal process step. If output of critical components, such as organic acids, carbohydrates and soluble proteins, can be related successfully to input, a cut can be made in the flow sheet and each half of it can be redesigned separately.

OPTIMIZATION STUDIES

The Fermented Beverage Process

The strategy for optimization is currently tested on a non-commercial process for production of fermented soymilk. The process begins with a soymilk preparation. Rice flour and barley malt are thereafter added and a mashing process is carried out to give fermentable carbohydrates, sweetness and nutritional value. To ensure a high content of soluble proteins in the beverage after the lactic fermentation an enzymatic proteolysis step is included concomitant with the mashing. Finally the lactic fermentation brought about by a selected *Leuconostoc mesenteroides* strain [1] gives the beverage a pleasant fresh or apple-juice-like flavour. Being a combination - or rather addition upon addition - of separate technologies, the process from the outset presented a challenge with respect to optimization.

The Fermentation as Principal Process Step

Probably the most important parameter in food processing is the palatability of the product. In this process the palatability depends to a great extent on the fermentation step which was therefore selected as the first principal step [2]. From studies by Chung [3] it was known that a balanced formation of lactic and acetic acid during fermentation combined with a correct level of residual sugars was a prerequisite for an acceptable taste. The taste is therefore influenced by the process steps upstream, especially the mashing step which is essential for both the supply of fermentable sugars to the fermentation and for the supply of maltose to give sweetness. The mashing was therefore the natural choice as the next principal step.

The Mashing Step as Principal Process Step

The significant part of the mashing step output is a certain combination of fermentable sugars. The question was in what way this was related to the mashing step input. At a fixed mashing temperature-time profile the most crucial parameters in the input were found to be the malt:rice ratio and the initial pH at mashing.

Mashing is a process well known from the brewing industry where the process conditions malt:unmalted cereal ratio of about 60:40 (% w/w) and an initial pH of 5.3 are typical. However, one may prefer to keep the amount of malt used low (due to availability, cost etc.) and a pH of 5.3 could not be achieved naturally in this process.

The consequences by moving away from the well known optimal conditions were investigated by response surface experiments. The effect of malt:rice ratio and initial pH on the amount of fermentable sugars is shown in figure 1. Similar figures for the specific mono- and disaccharides were achieved (results not shown). The results showed that the concentration of glucose varied with varying process conditions while the concentration of maltose did not. The results gave the correlations how to adjust the sugar concentration (and combination) to the demands of the fermentation by changing pH or malt concentrations. Or the other way around, what consequences restrictions in pH or malt use will have on the sugar combination and then further on the residual sweetness of the beverage.

Typical yields of fermentable sugars in the brewers mashing are around 75 %. The figure shows that the losses derived by mashing at non-conventional process conditions are relatively low. It can be concluded that these small losses should be without significant influence on the process optimization especially as the non fermentable carbohydrates are carried through the process and give nutritional value to the final product. The advantage of the partial up-stream optimization is obvious by a change in starch source. Only a few extra experiments of mashing will be needed to know the sugar spectrum and the impact upon the product will then be predictable.

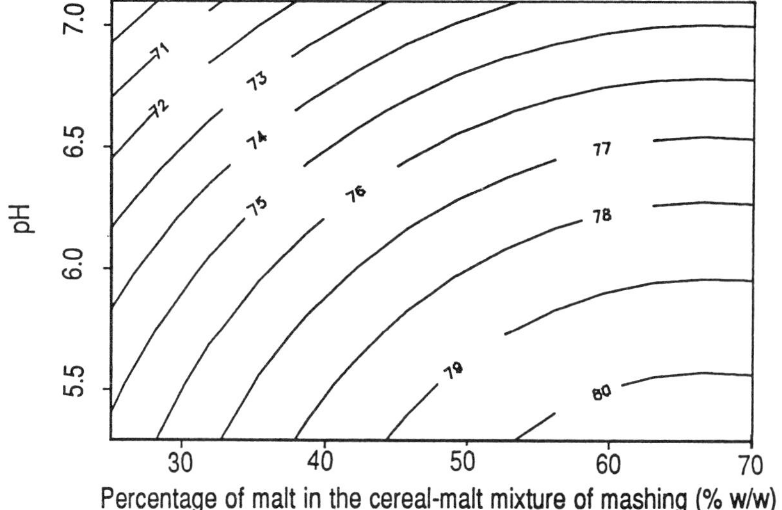

Figure 1. Response surface plot for yield of fermentable sugars (% of total available carbohydrates) as function of malt concentration and pH.

The pH of the beverage changes during the mashing step and the change is found to depend on the initial mashing pH. The correlation between the two pH-values was determined, since it is of importance for the choice of optimal pH and pH adjustment.

Energy considerations based on the temperature-time profile of the process led to a number of considerations of optimization through re-design. The evaluation of the effect of the change in process flow sheet was much facilitated by the knowledge achieved by the above drafted approach.

CONCLUSION

The development and test of an up-stream strategy for optimization of a food biotechnology process by focusing on the crucial product parameters is still in its innovative phase. The mashing example is given as an illustration of the advantage of the partial up-stream optimization achieved by the concept of a principal step.

REFERENCES

1. Lee, C.H., Min, K.C., Souane, M., Chung, M.-J., Mathiasen, T. and Adler-Nissen, J., Fermentation of prefermented and extruded rice flour by the lactic acid bacteria from Sikhae. Food Biotechnology, 1992, **6**, 239-255.
2. Adler-Nissen, J. and Moe, T., Scale-up studies for the production of high protein content lactic beverage. In book of proceedings from UNIDO International Workshop on Lactic Acid Fermentation of Non-Dairy Food and Beverages, Korea Food Research Institute, Songnam, June 25-27, 1992, pp. 149-161.
3. Chung, M.-J., Flavor formation during lactic acid fermentation of rice-soymilk mixture. In book of proceedings from UNIDO International Workshop on Lactic Acid Fermentation of Non-Dairy Food and Beverages, Korea Food Research Institute, Songnam, June 25-27, 1992, pp. 48-57.

GENERAL METHOD FOR LACTIC ACID BATCH FERMENTATION PROCESSES MONITORING

ERIC LATRILLE, DANIEL PICQUE and GEORGES CORRIEU
Laboratoire de Génie des Procédés Biotechnologiques Agro-Alimentaires
Centre de Biotechnologies Agro-Industrielles
I.N.R.A.
F-78850 Thiverval-Grignon, France

ABSTRACT

The determination of kinetics by measuring pH and electrical conductivity changes in fermented milk permits to definite feature points. These points are very useful to characterize the fermentation state and its well behaviour. Predictions of the future pH values over a long time are obtained with a good accuracy by using neural networks models and geometrical methods.

INTRODUCTION

It is of value to determine the changes in the kinetics of certain physicochemical parameters during a thermophilic lactic acid fermentation, e.g. pH or the electrical conductivity of the medium.

Yuguchi et al.[1] established correlations between the rate of pH changes and the viscoelastic properties of yogurt. The same authors [2] proposed a positive correlation between the mean particle size of yogurt gel and the fermentation rate, determined by measuring pH changes in the milk. In the case of lactic acid fermentations using *L. plantarum*, Lievense et al. [3] showed a positive linear correlation between inoculation size (or activity related to the physiological state of the bacteria) and the maximal rate of pH decrease. Using *Streptococcus thermophilus*, Spinnler and Corrieu [4] reported a linear relationship between the time at which the maximal acidification rate appeared and the logarithm of the inoculum size. In addition, the absence of simple and identifiable models in the case of batch fermentations for accessing biological parameters (biomass, substrate and metabolite concentrations) requires a different approach when interpreting the measurements.

The shapes of the pH change and electrical conductivity curves can be compared and classed by determining the characteristic points of the recorded curves [5]. These characteristics are most often points of inflection and changes in the curvature of the curve, which correspond to local minima and maxima of first derivatives vs. time.

Moreover, in batch processes, it is often very important and useful to know and to predict on-line the evolution of key data characteristics versus time. However, the attainment of this objective is not always possible due to the lack of reproducibility of the process. During the production of fermented milk, the initial pH of the medium is in the vicinity of 6.4 and it decreases to pH values as low as 4.5. Each type of fermented milk, available on the market, has its own characteristic final pH. The final pH is a feature of the dairy product and it is determined by the manufacturer. In a plant where many fermentations are performed simultaneously, the prediction, in real time, of the final time of each fermentation is very important to schedule the use of the heat exchangers and of the fermentors.

Some trials to use neural networks for on-line prediction in fermentation were done recently [6,7].

First, we will describe a method for the real-time determination of the fermentation kinetics of pH and conductivity [8].

In second part, we propose different methods, based on neural networks and on geometrical approaches applied to lactic acid batch fermentation. The purpose is to enhance the quality of the prediction of future pH values and to determine, in real time, the final time at which a predetermined value of pH is obtained [9].

RESULTS AND DISCUSSION

Kinetics and interpretation of feature points in terms of stages of fermentation

The on-line measurements of pH and electrical conductivity data have made it possible to access time and rate feature points of thermophilic lactic acid fermentations. The first derivative $(dx/dt)_{tn}$ at instant t_n used the central difference method [10] defined as :

$$\left(\frac{dx}{dt}\right)_{t_n} = \frac{x_{t_{n+1}} - x_{t_{n-1}}}{t_{n+1} - t_{n-1}}$$

where x is the measurement of pH or conductivity, t the time and *n* an index of time sampling. Ten feature points characterize curves of acidification and of conductivity changes using the main points of inflection observed (see figures 1 and 2). These feature points have shown the excellent reproducibility of fermentations conducted in standard conditions, with coefficients of variations lower than 5.1% for nine feature points.

The curve dG/dt = f(t) is always positive, with two local maxima and one local minimum, shown by the presence of three points of inflection on the curve of conductivity changes with time. The first maximum tG1 reflects the urease activity of *S.thermophilus* when the second maximum tG3 reflects the increasing production of lactic acid by *S.thermophilus* and *L.bulgaricus*. These feature points can also been used to compare the effect of type of starter, temperature and culture medium. The temperature effect results in different urease activity and acidification optima. The presence of fat (32 g/l) in the medium does not change any feature point, while the presence of sucrose (90 g/l) results in a decrease in the acidification rate and a longer fermentation time.

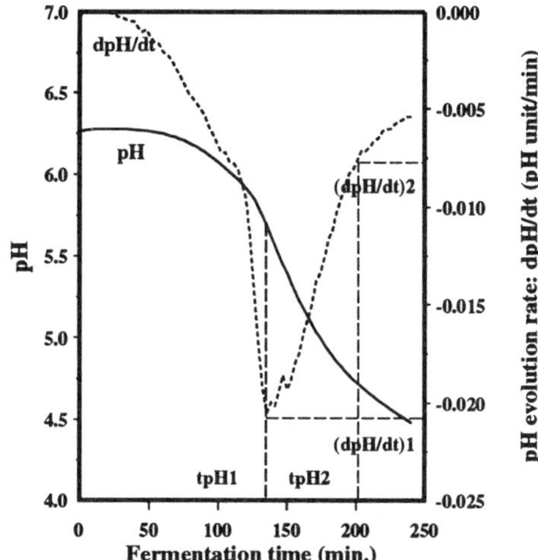

FIGURE 1 : pH and pH evolution rate (dpH/dt) changes versus fermentation time and determination of feature points. Skim milk (13% w/v), temperature 44 C, starter (S.thermophilus & L.bulgaricus)

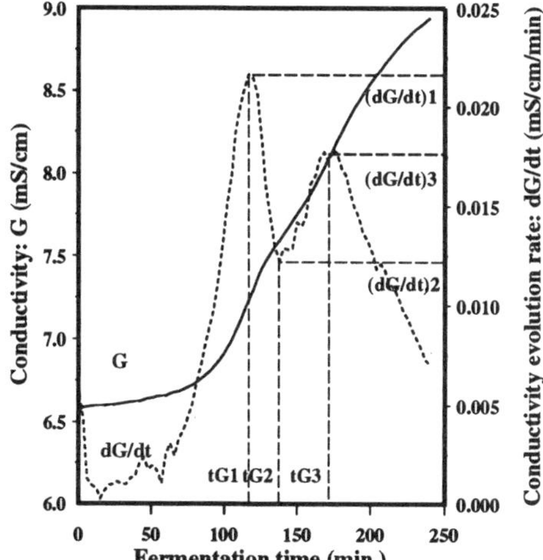

FIGURE 2 : Conductivity G and conductivity evolution rate dG/dt changes versus fermentation time and determination of feature points. Skim milk (13% w/v), temperature 44 C, starter (S.thermophilus & L.bulgaricus)

pH prediction and final fermentation time determination

Lactic acid batch fermentations, with uncontrolled pH on skim milk, can be modelled with a feedforward neural network to generate the model (pH versus time) of a reference fermentation (figure 3). In essence, the feedforward neural network is used as a general nonlinear model to store the information of a series of well-behaved fermentations. This neural network stores the information of previous fermentations and defines a reference fermentation. The distinct advantage of neural networks over other class of models is the plasticity of its structure which allows to easily capture the shape of the fermentation curve. This reference fermentation is used to perform a comparison with the actual fermentation for the on-line prediction of future pH values and of the final fermentation time. This time, which occurs at a predetermined pH (pH=4.6 for instance), is predicted with an accuracy of less than 20% at the onset of fermentation and with a much better accuracy as the fermentation proceeds. The future pH values are predicted with a mean error of 0.05 pH for a 3 hour prediction horizon. The prediction is obtained with four geometrical methods by sliding the curve of the reference fermentation along the curve of the actual fermentation.

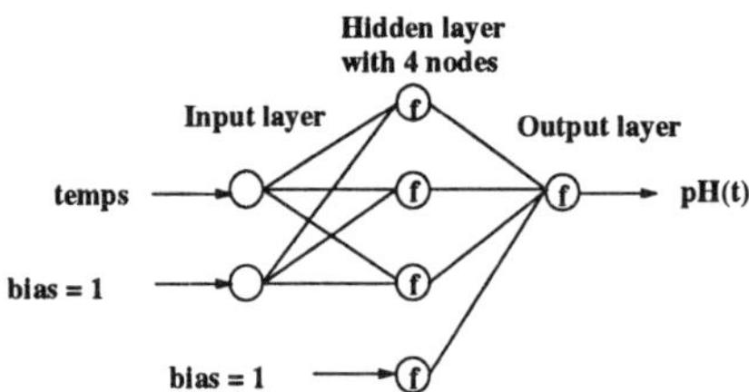

FIGURE 3 : Example of neural network structure used to generate a reference fermentation curve.

CONCLUSIONS

During batch lactic acid fermentation, the measures of the conductivity changes permit to clearly distinguish between the urease activity and the acidification activity of the starter. Feature points are very useful to characterize the fermentation state and its well behaviour. Predictions of the future pH values over a long time are obtained with a good accuracy by using neural networks models and geometrical methods. We could hope to derive a more general fermentation reference by including the effect of temperature, for instance.

REFERENCES

1. Yuguchi H., Tanai S., Iwatsuki K. and Okonogi S.: The influence of fermentation velocity on the properties of yogurt. Jpn. J. Zootech. Sci., 1989, **60**, 742-746.
2. Yuguchi H., Tanai S., Iwatsuki K. and Okonogi S.: The influence of fermentation velocity on the curd particle size and properties of yoghurt. Jpn. J. Zootech. Sci., 1989, **60**, 619-926.
3. Lievense, L.C., Van't Riet, K., and Noomen, A.: Measuring and modelling the glucose-fermenting activity of *Lactobacillus plantarum*. Appl. Microbiol. Biotechnol., 1990, **32**, 669-673.
4. Spinnler, H.E., and Corrieu, G.: Automatic method to quantify starter activity based on pH measurement. J. Dairy Res., 1989, **56**, 755-764.
5. Konstantinov K.B., and Yoshida T. : Real-Time Qualitative Analysis of the Temporal Shapes of (Bio)process Variables. AIChE Journal, 1992, **38**, 1703-1715.
6. Cléran, Y., Thibault, J., Chéruy, A. and Corrieu, G. : Comparison of prediction performances between models obtained by the Group Method of Data Handling and Neural Networks for the alcoholic fermentation rate in enology. J. Ferm. Bioeng., 1991, **71**, 356-362.
7. Van Breusegem, V., Thibault, J. and Chéruy, A. : Adaptive neural models for on-line prediction in fermentation. Can. J. of Chem. Eng., 1991, **69**, 481-487.
8. Latrille, E., Picque, D., Perret, B. and Corrieu, G. : Characterizing acidification kinetics by measuring pH and electrical conductivity in batch thermophilic lactic fermentations. J. Ferm. Bioeng., 1992, **74** (1), 32-38.
9. Latrille E., Corrieu G. and Thibault J. : pH prediction and final fermentation time determination in lactic acid batch fermentations. Suppl. to Computers & Chemical Engineering, 1993, **17**, S423-S428.
10. Siano, S.A., and Mutharasan, R.: NADH fluorescence and oxygen uptake responses of hybridoma cultures to substrate pulse and step changes. Biotechnol. Bioeng., 1991, **37**, 141-159.

ANAEROBIC CONTINUOUS ETHANOL FERMENTATION USING A COMPUTER-COUPLED MEDIUM FEEDING SYSTEM WHICH HAS DDC CONTROL pH OF THE CULTURE BROTH

SUDARUT TRIPETCHKUL[1], MICHIO TONOKAWA[1], AYAAKI ISHIZAKI[1], ZHONGPING SHI[2] AND KAZUYUKI SHIMIZU[2]
1. Department of Food Science and Technology, Faculty of Agriculture, Kyushu University, Fukuoka, Japan 812
2. Department of Biochemical Engineering and Science, Kyushu Institute of Technology, Iizuka, Fukuoka, Japan 820

ABSTRACT

The utilization of pH changes to set the rate of addition of fresh medium to a continuous ethanol fermentation has been investigated. The dilution rate and cell mass concentration depended on the pH set point and the control interval. The dilution rate, however, is independent of the difference in pH between the culture broth and feed medium.

INTRODUCTION

The pH-stat control used the pH change caused by growth to set the rate of addition of medium. This method was introduced for the mass cultivation of bacteria (1,2,3). Larrson et al.(4) also used the pH-stat as a tool for studying microbial dynamic in continuous fermentation.

This report aims to investigate how pH-stat can be used to control feed rate of medium. The parameters that affected automated substrate supply in pH-stat modal feeding technique in continuous ethanol fermentation using *Zymomonas mobilis* have also investigated.

MATERIAL AND METHODS

Microorganism and media

Zymomonas mobilis NRRL-B-14023 was used in this study. The growth medium was composed of 100 g/l glucose, 10 g/l yeast extract, 1 g/l KH_2PO_4, 1 g/l $(NH_4)_2SO_4$ and 0.5 g/l $MgSO_4.7H_2O$. Composition of feed medium is the same as the growth medium.

Continuous culture

Continuous culture was performed in 1 l jar fermentor with a working volume of 500 ml at 30°C and pH 5.50. When the continuous procedure was switched on, an additional glucose was intermittently supplied by pH-stat with separate inflows of alkali and feed medium. The fermentor was connected to a computer where the medium flow rate and the control interval were registered.

Analytical methods

Ethanol was estimated with a gas chromatography GC-8 APE . Glucose concentration was determined with a glucose analyzer . Biomass was expressed as dry cell weight calibrated to an optical density of 562 nm using a spectrophotometer.

RESULTS AND DISCUSSION

The pH-stat function by using the pH of medium to maintain growth at a constant rate and its principle is based on a growth associated production or consumption of acid or based (4). The acid production rate can be related to the specific growth rate under a balanced growth of Z. mobilis in batch culture of ethanol fermentation (Data is not shown).

Larrson and Enfors(4) reported that the steady-state biomass concentration could be controlled by the flow quotient between the separated alkali/total flow to the fermentor. Therefore the pH-stat cultivation with separating inflows of medium and alkali was operated. By means of this method, flow rate of medium and alkali is controlled by metabolic activity of microorganism. Considerable improvements of techniques are need for the system to prove of practical value. Thus the parameters that affected automated supply in pH-stat for the control of pH growth rate by regulating the feed rate have been also investigated.

The results suggest that pH set point affected cell mass concentration and dilution rate (Table 1). The shorter control interval achieved, the higher medium flow rate and the bigger dilution rate (Table 2). Increasing in biomass yield occurred with increasing pH in feed medium. However, the dilution rate was not effect at all (Table 3).

From all of these results the use of pH mediated computer in continuous ethanol fermentation can be considered as a tool to control the feed rate of the medium. To developed this method much remains in the optimization of the parameters involved in the method such as the buffer capacity of the inflow medium should be investigated.

REFERENCES

1. Martin, G.A. and Hemfling, W.P., Method for the regulation of microbial population density during continuous culture at high growth rate. Arch. Microbial., 1976, **107**, 41-47.
2. Stouthamer, A.H. and Bettenhaussen, C.W., Energetic aspects of anaerobic growth of Aerobacter aerogens in complex

medium. Arch. Microbiol., 1976, **11**, 21-23.
3. Oltmanm, L.F., Schoenmaker, G.S., Reijnders, W.N.M. and Stouthmer, A.H., Modification of the pH-stat culture method for the mass cultivation of bacteria. Biotechnol. Bioeng., 1978, **20**, 921-925.
4. Larrson, G. and Enfors, S.o., The pH-auxostat as a tool for studying microbial dynamics in continuous fermentation. Biotechnol. Bioeng., 1990, **36**, 224-239.

TABLE 1
Effect of pH set point for feed medium into the fermentor (pH 5.50) on the kinetic parameters of the pH-stat cultivation

	pH set point		
	5.50	5.52	5.55
Specific glucose consumption rate(g/g/h)	7.27	4.21	0.81
Specific ethanol production rate (g/g/h)	3.67	2.06	0.008
Dilution rate (h^{-1})	0.16	0.08	0.025
Cell concentration (g/l)	2.18	1.93	2.90
Biomass yield (g/g)	0.022	0.019	0.033
Ethanol yield (g/g)	0.49	0.48	0.49
Ethanol productivity (g/l/h)	8.0	4.0	1.25

TABLE 2
Effect of the control interval of pH value measurement on the kinetic parameters of the pH-stat cultivation

	Control interval	
	60 s	5 s
Specific glucose consumption rate (g/g/h)	7.27	8.07
Specific ethanol production rate (g/g/h)	3.67	4.27
Dilution rate (h^{-1})	0.16	0.21
Cell concentration (g/l)	2.18	2.53
Biomass yield (g/g)	0.022	0.026
Ethanol yield (g/g)	0.49	0.49
Ethanol productivity (g/l/h)	7.3	10.8

TABLE 3
Effect of pH difference between the culture and the inflow medium on the kinetic parameters of continuous ethanol fermentation using *Z. mobilis* in 100 g/l glucose medium

	pH of feed medium		
	5.50	5.55	6.0
Specific glucose consumption rate(g/g/h)	9.61	8.92	8.59
Specific ethanol production rate (g/g/h)	4.49	4.24	4.08
Dilution rate (h^{-1})	0.25	0.25	0.25
Cell concentration (g/l)	2.27	2.50	2.76
Biomass yield (g/g)	0.026	0.028	0.030
Ethanol yield (g/g)	0.48	0.48	0.49
Ethanol productivity (g/l/h)	10.2	10.6	11.3

METHOD FOR ON-LINE PREDICTION OF THE ALCOHOLIC FERMENTATION RATE IN WINE-MAKING

B. PERRET and G. CORRIEU
INRA - LGPBA - F-78850-THIVERVAL-GRIGNON

ABSTRACT

In oenology, the alcoholic fermentation rate is proportional to the outlet CO_2 gas flow rate. In isothermal conditions, this process is characterized at the onset of the fermentation, by a constant increase ($dCO_2 / dt = K1.t$), and then by an exponential decrease ($dCO_2 / dt = e^{-K2.t}$) of the fermentation rate with the time. The intercept of these two phases determines the maximum fermentation rate (Vm).

The method for the on-line prediction of the alcoholic fermentation rate consists in :

1st : determining the acceleration coefficient (K1) at the beginning of the process (five hours after the CO_2 emission).

2nd : using K1 in simple mathematical models, to determine Vm, K2 and tf (end time of fermentation).

This method can be improved if the experimental measurement of Vm is taken into account. In this case, the accuracy of K2 and tf determination felt respectively to 4% and 7% instead of 7% and 8%.

I DESCRIPTION OF THE FERMENTATION KINETIC

The alcoholic fermentation rate is computed from the emitted CO_2 volume measurement (1,2). In isothermal conditions, (usual temperatures within 15 and 30°C), the fermentation kinetic is characterized by three successive phases (Figure 1) :

- onset of fermentation : lag phase of cell growth and progressive saturation of the medium with CO_2. The end of this phase is characterized by the time (td) of onset of gas release.
- a phase of linear increase of the fermentation rate characterized by the acceleration coefficient K1 ($dCO2 / dt = K1.t$). This phase goes on until the maximum fermentation rate Vm ($Vm = (dCO_2 / dt)_{max}$) is obtained. tvm is the corresponding fermentation time..
- a phase where the fermentation rate decreases according to an exponential rule. It is characterized by a decrease constant K2 and it leads to $dCO_2 / dt = Vm.e^{-K2.(t-tVm)}$.

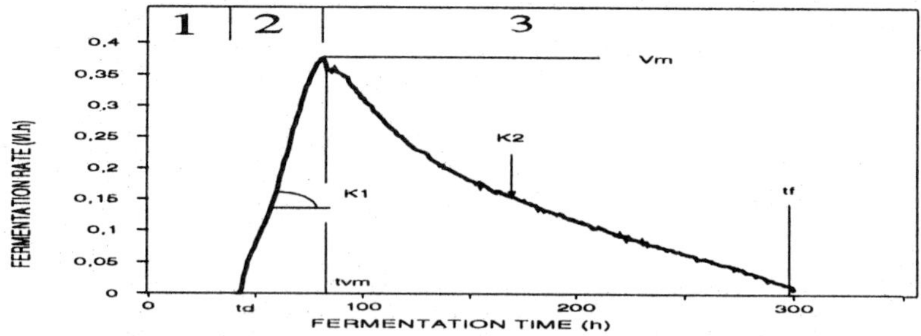

Figure 1 : typical fermentation rate evolution.

Phase 1 : any effective release of carbon dioxide

Phase 2 : linear increase of the rate of carbon dioxide production

Phase 3 : exponential decrease in the rate of carbon dioxide release.

Furthermore, the CO_2 emission measurement allow the determination of the produced ethanol and the residual sugar concentrations as functions of time (3). The last one, in regard to the initial sugar concentration S_0, leads to the final fermentation time (tf). For more convenience, we have used a normalized fermentation time : tnf = (tf-td)/S_0.

II - MODELLING THE FERMENTATION KINETIC

The real time predictive modelling of the fermentation kinetic consists in the K1 determination in the first hours following the onset of the alcoholic fermentation. For that purpose, a linear regression of the values of rate of CO_2 production is executed. The $K1_c$ computed parameter is then used to predict Vm_c, $K2_c$ and tf_c using relationships of table 1.

The adaptation of this method to 14 tests realized on a pilot fermentation plant (700 liters and 5000 liters tanks) allowed the calculation of $K1_c$ and the determination of Vm_c, $K2_c$ and tf_c. The mean errors were respectively 7.5%, 6%, 7% and 8% (Figure 2).

A significant improvement is obtained if the $K1_c$ determination is replaced by the experimental measurement of Vm. This maximum fermentation rate is reached early enough regarding the total process time : at about 25 to 30% of the fermentation progress. The models involving Vm to compute $K2_c$ and tf_c are linear (figure 3). The mean errors are 4% for $K2_c$ and 7% for tf_c. The corresponding relationships are shown in table 1.

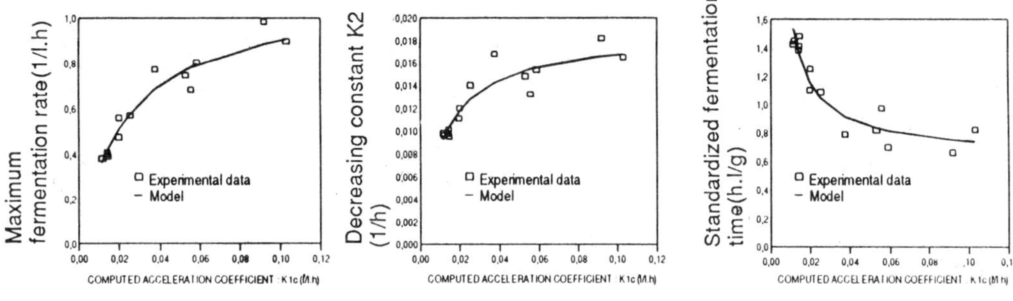

Figure 2 : determination of Vm,K2 and tnf from K1c

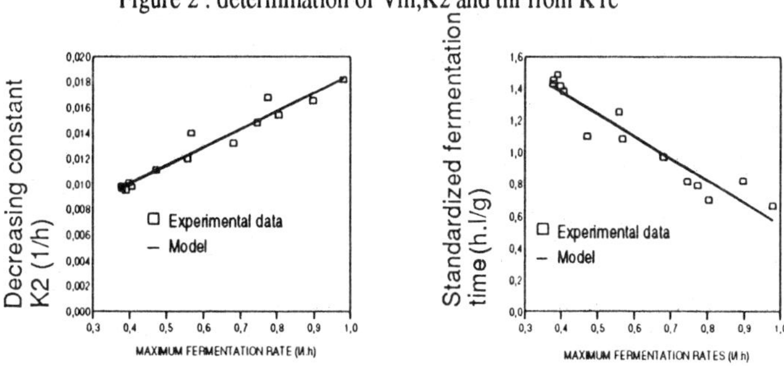

Figure 3 : determination of K2 and tn from Vm

Models involving K1c :	
Vm = K1c/(0.0316 + 0.894K1c)	$r^2 = 0.961$
K2 = K1c/(0.630 + 53.4K1c)	$r^2 = 0.896$
tnf = 0.636 + 0.103/K1c	$r^2 = 0.905$
Models involving Vm :	
K2 = 0.004 + 0.0142Vm	$r^2 = 0.948$
tnf = 1.93 - 1.39Vm	$r^2 = 0.910$

Table 1 : Empirical relationships used to predict the kinetic fermentation parameters

The successive working up of the two methods is of course possible. It leads to reduce the error on the further fermentation kinetic parameters when the fermentation progresses. Figure 4 shows two examples using the models from on-line $K1_c$ and Vm measurements. Those two examples give an account of two different situations with regard to the quality of experimental data and models fitting (4).

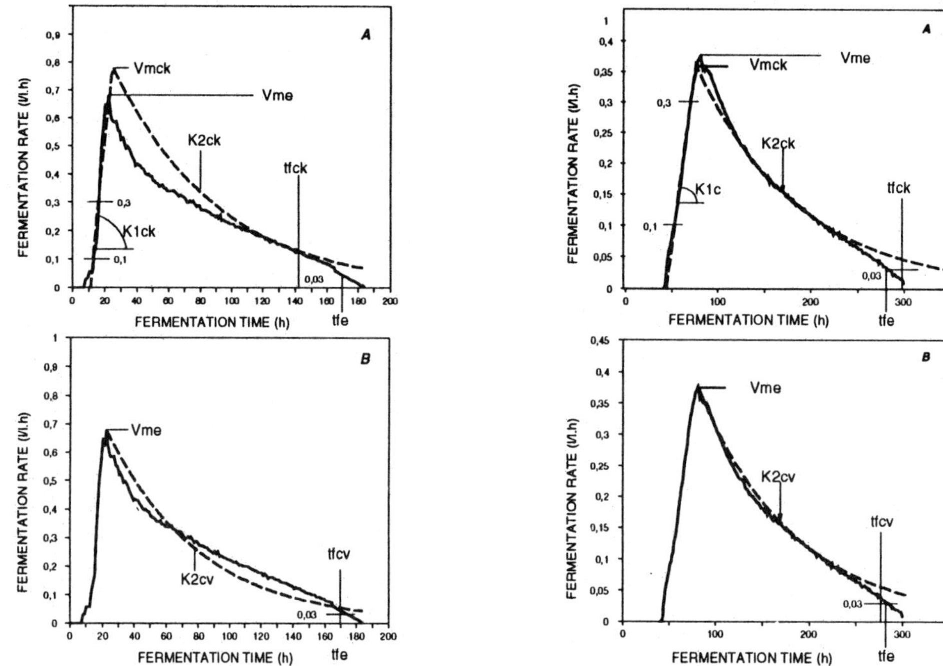

Figure 4 : comparison between experimental (——) and calculated kinetics (- - - -).
A : Models involving $K1_c$.
B : Models involving Vm.

The application of these methods to non isothermal fermentations (temperature and/or kinetic control) is now being considered.

REFERENCES

1. Barre P., Chabas J., Corrieu G., Davenel A., El Haloui N.E., Navarro J.M., Picque D., Sablayrolles J.M., Sevila F et Vannobel C. - Procédé de prévision et de contrôle en ligne des fermentations alcooliques et dispositif pour sa mise en oeuvre. Brevet INRA/CEMAGREF, n°86 157 79 et n°87 402 543 0.

2. El Haloui N.E., Corrieu G., Cleran Y. and Cheruy A. - Method for on-line prediction of kinetics of alcoholic fermentation in wine making. J. Ferm. and Bioeng., 1989, 68, **2**, 131-135.

3. El Haloui N.E., Picque D. and Corrieu G. - Alcoholic fermentation in wine making : on-line measurement of density and carbon dioxyde evolution. J. Food Eng., 1988, **8**, 2, 91 108.

4. Cleran Y. - Contribution au développement d'un procédé de suivi et de contrôle de la fermentation alcoolique, Thesis, Institut National Polytechnique de Grenoble, 1990.

5. Sablayrolles J.M. et Barre P. - Pilotage automatique de la température de fermentation en conditions oenologiques. Sci. Aliments, 1989, **9**, 239-251.

MEASUREMENT OF CELL DENSITY IN THE BROTH OF AGGREGATIVE ORGANISM BY CONTINUOUS-DILUTION-PHOTOMETRIC-ASSAY

TAKUO YANO, AGUS MASDUKI, YOSHINORI NISHIZAWA, AND HISAO OHTAKE
Department of Fermentation Technology, Faculty of Engineering,
Hiroshima University, 4-1, Kagamiyama 1 chome,
Higashihiroshima 724, Japan

ABSTRACT

A photometric measurement system of cell density was developed and applied to the culture broth of an aggregative organism, *Favolus arcularius*. The system was made up of a photometer with a flow cell, two tubing pumps and a circulating magnetic pump. The culture broth introduced from a culture flask was circulated to break up the aggregated cells into small pieces by the circulating pump for several minutes. The value of cell density measured by this method gave good agreement with that obtained from the drying method over the wide range of cell density from 5 to 40 g dry cells/*l*.

INTRODUCTION

To control and operate a fermentation successfully, the measurement of the cell density of culture broth is one of the most important operations. In the broth of an aggregative organism, the cell density was generally measured by the drying or the photometric method. The drying method takes at least 1-2 h to get the true value of cell density. The photometric method with homogenization of broth was troublesome. It is expected to develop a rapid, simple and precise on-line automatic system for the measurement of cell density in the culture broth.

In this study, the system developed previously (1) was applied for measuring the cell density of an aggregative organism, *Favolus arcularius*, a kind of mushrooms, and the effects of the operating conditions on the calculated value of cell density were studied.

MATERIALS AND METHODS

The system was made up of a photometer with a flow cell, two tubing pumps and a magnetic circulating pump (12 *l*/min of flow rate). At the first step of the measurement process, batch dilution was started, *i. e.*, the drain valve [No. 4 in Fig. 1] was closed and the pumps 1 and 2 were put in motion. After a certain time, t_b, pump 1 was stopped. If the dilution circuit was full up, pump 2 was stopped and pump 3 was put on to mix the culture broth diluted with water and to break up the aggregated cells into small pieces.

The mixing was continued until the optical density became approximately constant. This period was mixing time, t_m. When the optical density was higher than 0.8, pump 2 was put on again and continuous dilution was started. Surplus diluted broth was washed out from the top of the dilution circuit. The optical density of diluted broth was measured at 570 nm of wavelength by the photometer. All the procedures mentioned above were operated manually in this study.

The cell density of the culture, X_i was calculated as follows:
In batch dilution,

$$X_i = X_b \cdot (F_2 \cdot t_1 + V_b)/V_b \qquad (1)$$

In continuous dilution,

$$X_b = X_c \cdot \exp(D \cdot t_2) \qquad (2)$$

From Eqs. 1 and 2,

$$X_i = (V_t/V_b) X_c \cdot \exp(D \cdot t_2) \qquad (3)$$

Here, $V_t = F_2 \cdot t_1 + V_b$, $D = F_2/V_t$, $V_b = F_1 \cdot t_b$. F_1 and F_2 are flow rates (ml/min) of pump 1 and 2, respectively. t_1, t_2 and t_b are times (min) for batch dilution, continuous dilution and operation of pump 1, respectively. V_b and V_t are volumes (ml) of sample introduced from fermentor and dilution circuit, respectively. X_b and X_c are cell

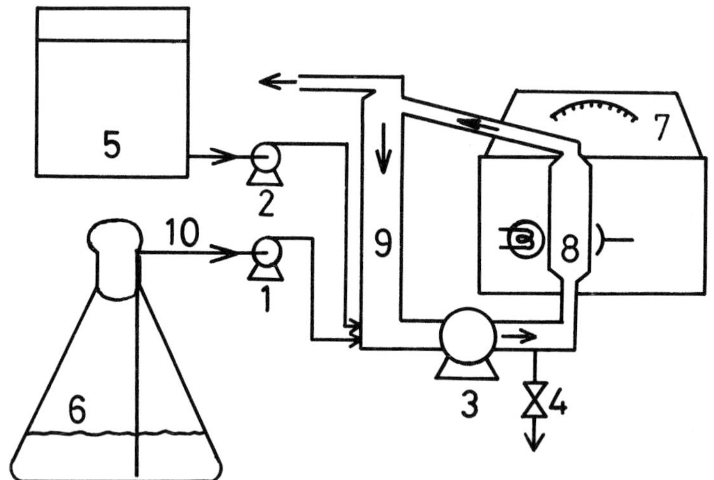

Figure 1. Outline of the system developed to measure the cell density in the culture broth. 1. Sampling pump; 2. Water feeding pump; 3. Circulation pump; 4. Drain valve; 5. Water reservoir; 6. Culture flask; 7. Photometer; 8. Flow cell; 9. Dilution circuit; 10. Sampling tube.

TABLE 1. Distribution of pellet (number of pellets/l) during mixing time.

mixing time (min)	pellet size (mm)							(– , not detected)	
	0.25–0.50	0.5–1.0	1.0–2.0	2.0–3.0	3.0–4.0	4.0–5.0	5.0–6.0	6.0–7.0	7.0–8.0
0	–	450	2350	450	200	200	1400	1500	100
2	5×10^7	1200	400	–	–	–	–	–	–
4	6×10^7	400	–	–	–	–	–	–	–
6	4×10^7	200	–	–	–	–	–	–	–
9	3×10^7	–	–	–	–	–	–	–	–
13	–	–	–	–	–	–	–	–	–

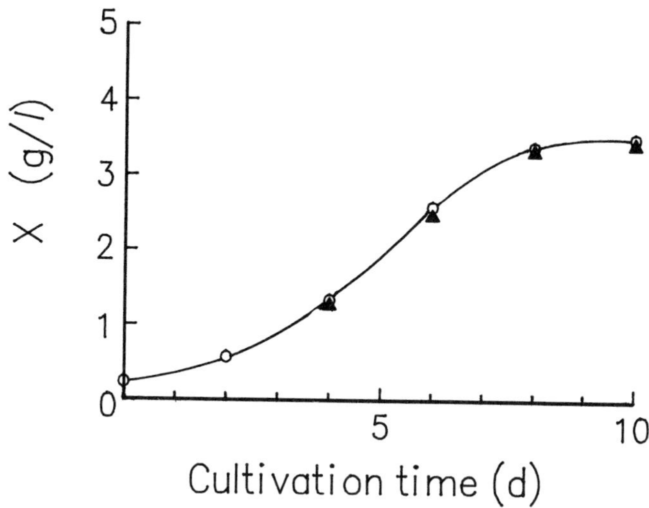

Figure 2. Cultivation of *F. arcularius*. Symbols: ○, by the drying method; ▲, by the system developed in this study under the conditions: F_2 = 50 ml/min, t_m = 13 min and V_b = 15 ml.

densities (g/*l*) after batch dilution and contiuous dilution, respectively. The cell density after dilution was obtained from the predetermined relationship between optical density and cell concentration.

The cultivation was performed on a rotary shaker (100 rpm) at 28℃ after transference of 100 ml seed culture into 900 ml of medium in an Erlenmeyer flask (3 *l*). To prepare the high cell density suspension, the culture broth was concentrated by a decantation.

RESULTS AND DISCUSSION

Distribution of pellet size during mixing time, t_m was studied (TABLE. 1). Eight-days cultured broth containing the pellet of large size was injected to the dilution circuit, and circulated by the circulating pump. The pellets were broken up to small pieces immediately and the value of optical density became stable after 4 min. The breakage of pellets seemed to be caused by impeller of the circulating pump rotated at high speed.

The effects of operation factors, end point of optical density, OD_f, volume of culture broth introdued from the fermentor, V_b and flow rate of water, F_2, on the performance of the system were studied. The value of OD_f in the range of 0.3 to 0.6, 15 ml of V_b and 50 ml/min of F_2 might be favored to calculate the cell density.

The developed system was applied to the batch culture of *F. arcularius*. Time course of the growth is shown in Fig. 2. The values measured by this method were in good agreement with those by the drying method throughout the cultivation.

REFERENCES

1. Yano, T., Agus Masuduki, and Nishizawa, Y.: Photometric measurement of high cell density by continuous dilution of broth with a circulating system. *J. Ferment. Bioeng.*, **74**, 100-103 (1992).

DYNAMIC MODELLING OF A LARGE SCALE AIR LIFT FERMENTER

G. TRYSTRAM, S. PIGACHE
ENSIA, Food Process Engineering, Massy, France

ABSTRACT

A model of a large scale air lift fermenter used for yeast production from whey is detailed. This model is a dynamic one, which predicts both biomass and ethanol production, as well as gas transfer. Parameter identification is performed from industrial experiments. Validation is characterized with a relative error: 6%.

INTRODUCTION

The Vendôme Bel Industries yeast plant produces lactic yeast by continuous aerobic fermentation of deproteinated wheys in airlift reactors. Considerable research work has been done for several years to improve the understanding of the process. A previous study reported elsewhere (1) allowed the determination of the oxygen profile inside the fermenter. In order to have a dynamic simulation tool able to build control strategies, a physiological model is developed in addition to the oxygen transfer model.

INDUSTRIAL FERMENTER AND CULTURE

Industrial fermenter

The fermenter used for the experiments is a 120 m^3 concentric-tube airlift reactor. Lactose concentration, substrate and ammonia flow rates are controlled. Inlet and exhaust gases are analysed. The dilution rate is chosen by adjusting the liquid volume in the fermenter through substrate flow rate. Temperature of the broth is maintained constant. pH is maintained close to the set point by adding sulfuric acid. Data logging is supported by a supervision system.

Culture

The population is composed of three complementary species (2): *Kluyveromyces fragilis* (90%), *Kluyveromyces lactis* (9%) and *Torulopsis bovina* (1%). The metabolism of these strains has been described at length and the authors have shown that the composition of the flora did not change for low variations of pH, temperature or dilution rate. The culture behaviour is governed by the predominant strain *K. fragilis* which is characterized by a marked Pasteur effect (3). Even if strict aerobic conditions are maintained, traces of ethanol are always produced (4) and the biomass production yield $Y_{x/s}$ is not at a maximum (5). In case of oxygen starvation, the phenomenon is accentuated and lactose accumulates in the medium and/or is converted to ethanol. If aerobic conditions are restored, ethanol will be consumed by *T. bovina* (5). *K. lactis* is assumed to grow on lactic acid always present in whey (5). Therefore, the combination of the three yeast strains allows the production of biomass whithout loss of yield.

MODELLING

Principle

The principle of the model is to describe the physiological behaviour of the cells in the fermenter and to take into account the different physiological states of the culture. Each of these states is reached as a result of various operating conditions or disturbances. This description coupled with mass balance is able to describe the time evolution of the culture parameters: biomass, lactose, ethanol. The principle of the model is presented in figure 1. Metabolism of lactic acid is neglected, because quantities are small (2%). Several approaches are previously presented but a multi species culture is difficult to model. It appears simpler more simple to consider the flora as a theoretical one with a specific metabolism which is chosen as similar to the predominant strain *K. fragilis* .

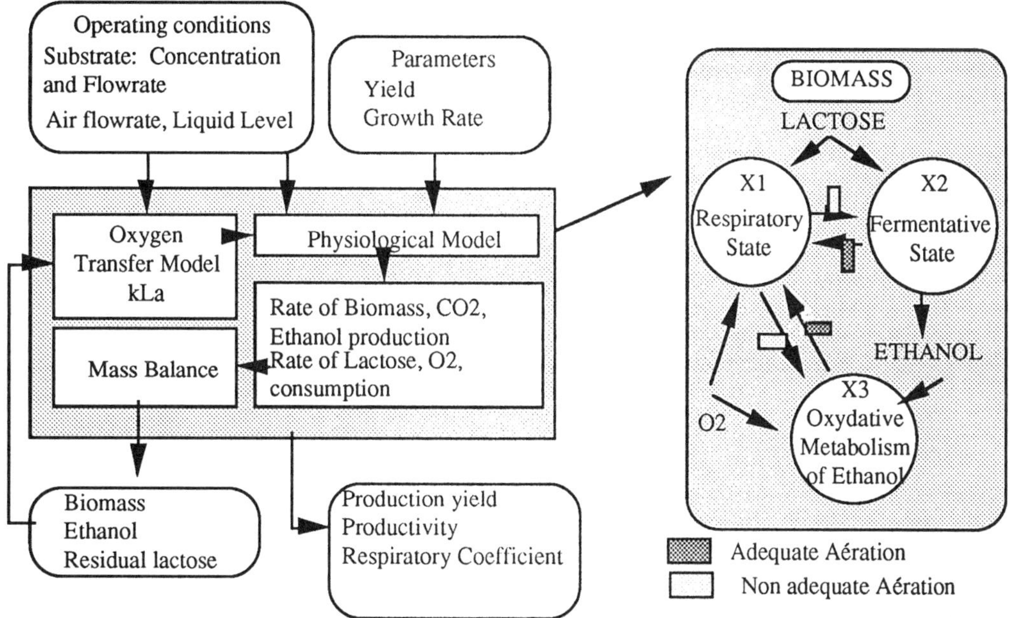

Figure 1 Principle of the dynamic model of the production of yeast from whey

As presented in figure 1, the yeast could reach three states:

state X1: in case of lactose respiration when oxygen is sufficient in regard to lactose concentration.

state X2: in case of fermentative evolution of the culture, when oxygen is not sufficient

state X3: which characterize the respiration of Ethanol. This is a transient state.

The total yeast biomass is the sum of X1, X2 and X3. It is necessary to describe how the biomass could progress from one state to another. Because of the choice of a theoretical strain, similar to *K. fragilis* , oxygen is the limiting factor. If Oxygen supply is sufficient, the culture reaches the state X1 and, if ethanol is present in the fermenter, state X3. If Oxygen supply is not sufficient, an evolution through fermentative state X2 is available. In practice three rates are determined at each time and comparisons are realized which permit determination of the evolution of the culture. These are :

-the rate of oxygen supply,

-the rate of oxygen consumption, necessary to oxydate all the lactose in the fermenter,

-the rate of oxygen consumption by yeast.

A mass balance is developed and 0 order kinetics are considered for the state changes. Figure 1 presents the relationship between the different parts of the model.

The oxygen transfer must be known. An experimental study allows the development of an empirical model for overall oxygen transfer versus operating conditions (1):

$$KLag = K1 + K2.Ln(Qgt) + K3.Ln(VL)$$

RESULTS

Most of the model parameters are identified from literature. Some adjustments are performed from experiments using simplex method (6). Figure 2 presents predictive simulations and comparisons with experiments. It can be seen that both large and small variations are well followed both for the liquid phase (yeast, ethanol and lactose) and the gas phase (O2 and CO2). The uncertainty is for yeast concentration 0.7 g/l, 0.2 g/l for lactose and 0.1g/l for ethanol. It gives a relative error of 5%. For a large scale fermenter, this is a good result which permits the study of control strategies or design of the air lift reactor.

CONCLUSION

Dynamic modelling of the fermenter is validated via industrial experiments and the model is used as a sensor to predict biomass concentration variations. The interpretation of experiments is facilitated by the model, and optimization is also performed (6% of increase for the yield). This work illustrates the ability to model large scale processes.

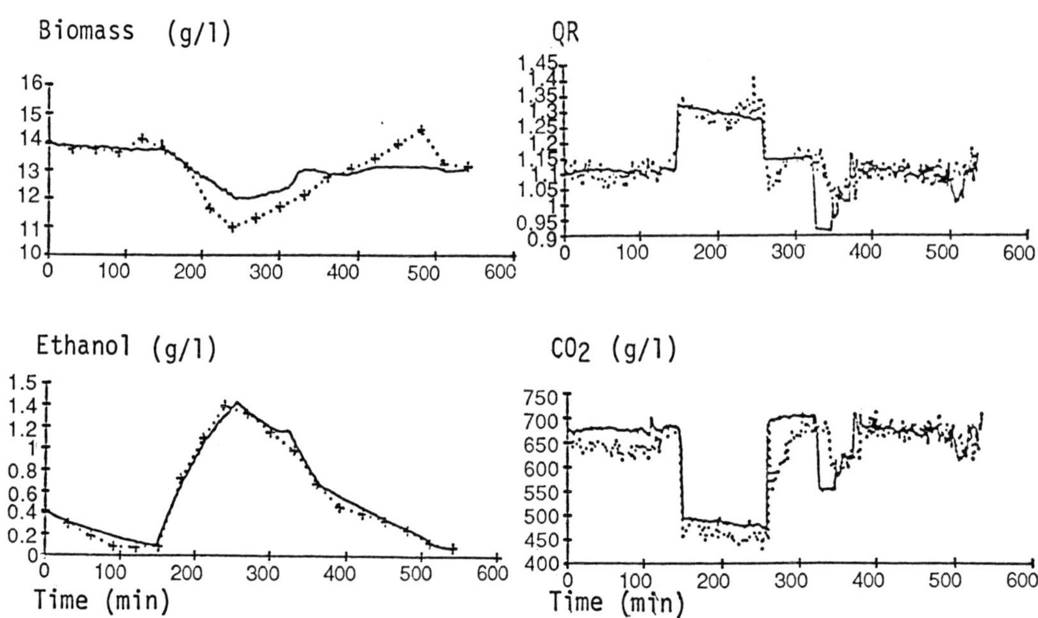

Figure 2: Comparisons between modelling and experimental results.

REFERENCES

1. Pigache, S., Trystram, G., Biotechnology and Bioengineering, 1992, 39,323-331.
2. Moulin, G., Malige, B., Galzy, P., Le lait, 1981, 66, 323-332.
3. Alexander, M. A., Jeffries, T.W., Enzime Microb. Technol., 1990, 12, 2-19.
5. Moulin, G., Malige, B., Galzy, P., J. of dairy science, 1983, 21-28.
6. Pigache, S., Doctoral thesis, ENSIA, Massy, France, 1993.

BIOCATALYTIC PRODUCTION OF CELLOBIOSE CONTAINING OLIGOSACCHARIDE MIXTURE

MARIANNE ROSSI, YU-YEN LINKO, PEKKA LINKO, TIMO VAARA*, MARJA TURUNEN*
Laboratory of Biotechnology and Food Engineering,
Department of Chemical Engineering
Helsinki University of Technology, SF-02150 Espoo, Finland
*Biotechnology Unit, Alko Ltd, SF-05200, Rajamäki, Finland

ABSTRACT

The production of cellobiose containing unique and novel oligosaccharide mixture from starch by cyclomaltodextrin-glucanotransferase (CGTase), using cellobiose as the acceptor sugar, is described. This is believed to be the first time such oligosaccharide production method has been published. Cellobiose containing oligosaccharides are linear molecules with one (OSG3) or several glucose units (OSG4-OSG7) adjoined to the acceptor sugar cellobiose. The effects of dry matter and enzyme concentration, the mass ratio of starch to acceptor sugar, and reaction time were studied to produce in controlled manner the oligosaccharide mixtures with a good yield. The best results were obtained in 48 h with starch of DE value 1, at 30 to 40 % total dry matter, starch/acceptor ratio of 1:1, 60 °C, and a CGTase concentration of 30 to 350 U per g of starch. Up to a 48 % yield of OSG3-5 was obtained.

INTRODUCTION

Oligosaccharides are important new raw materials applied in food, animal feed, pharmaceutical and chemical industries because of their low calorie content, mild sweetness, low cariogenecity, positive physiological effects on health, and good technical properties. A number of oligosaccharides are already commercially available, e.g. starch-based products such as maltose, maltotriose, maltotetraose, isomaltose, panose, saccharose-based products, such as "coupling sugar", fructooligosaccharides, palatinose, lactose-based oligosaccharides and sugar alcohols such as maltitol, and search for new products is still going on. In this paper, the production of cellobiose containing oligosaccharide mixture from starch by CGTase, using cellobiose as the acceptor sugar [1], is described.

METHODS

Production of Cellobiose Containing Oligosaccharides

Pretreatment of the starch (barley starch, Alko Ltd) was carried out by adding 30 U/g of CGTase (activity 7600 U/ml, Alko Ltd) to a slurry containing various quantities of starch in 50 mM imidazole buffer of pH 6.8, containing 1.5 mM $CaCl_2$. The CGTase was allowed to act at 85 °C for 30 minutes with agitation to obtain starch with DE=1.

The cellobiose containing oligosaccharides were produced by adding various amounts of cellobiose (Sigma) to the pretreated starch. The reaction was started by adding 50 U CGTase per gram of starch with agitation and carried out for 48 hours at 60 °C.

Characterization of the Oligosaccharide Mixture

The composition of the mixture was determined by Shimadzu HPLC using a μBondapak carbohydrate analysis column (Millipore) at room temperature and 65% acetonitrile as eluent with a flowrate of 0.9 ml/min.

RESULTS AND DISCUSSION

The Effect of Substrate Concentration

Table 1 shows the effect of the total substrate concentration on the composiotion of the oligosaccharides OSG3-OSG7 after 48 hours reaction time. The total quantity of the oligosaccharides OSG3-OSG5 in the mixture obtained at 10% dry matter content was 4.8 gram per 100 g of the reaction mixture (48.0% of the initial dry matter content). At 30% dry matter content the respective concentration was 14.6 g/100 g (48.7% of the initial dry matter).

TABLE 1

The effect of substrate concentration (starch + cellobiose) on composition of the oligosaccharide mixture produced in 48 hours at 60°C. The starch/acceptor mass ratio was 1:1

Substrate dry matter content in solution (%)	Glucose	Cellobiose	OSG3	OSG4	OSG5	OSG6	OSG7
	Product mixture (g/100 g of the reaction mixture)						
10	0.03	2.1	2.0	1.6	1.2	0.8	0.2
30	0.04	6.9	6.0	5.0	3.6	2.9	0.9

Time Course of Oligosaccharide Production

The yield of OSG3-OSG5 oligosaccharides and the cellobiose consumption (30% total substrate dry matter) during 48 hours incubation is shown in figure 1. Twelve hours was sufficient to obtain a maximum of about 50% yield.

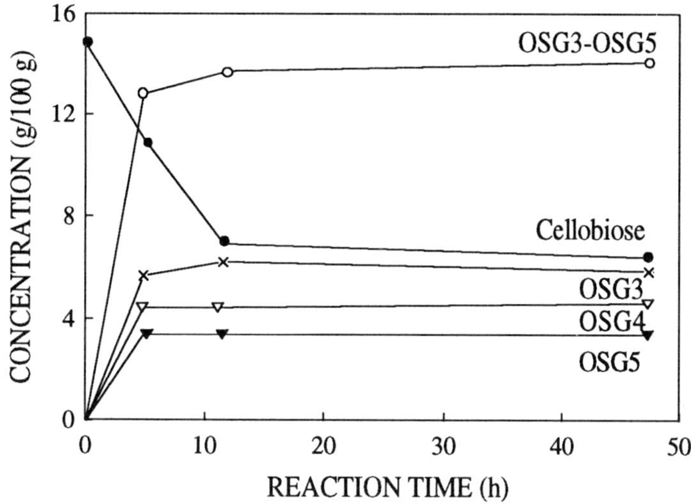

Figure 1. Time course of oligosaccharide production with 30% dry matter content

The Effect of Starch/Acceptor Mass Ratio

Starch/acceptor mass ratios of 1:4, 1:2 and 1:1 in oligosaccharide production with total substrate dry matter of 30% resulted in total quantities of oligosaccharides OSG3-OSG5 of 13.1 gram per 100 g of the reaction mixture, 14.4 g/100 g and 14.6 g/100 g, respectively. However, the starch/acceptor mass ratio of 1:2 was superior to the ratio of 1:1 during the first 24 hours of incubation.

Down-Stream Processing

It is possible by free or immobilized glucoamylase treatment in a controlled manner to increase the yield of OSG3-OSG5 oligosaccharides [2]. Furthermore, removal of free glucose from the oligosaccharide mixture with free or immobilized yeast reduces the cariogenicity and calorie content of the mixture. Ethanol (70%) may be used to precipitate the long chain oligosaccharides and polysaccharides out of the mixture. The product may be further treated with conventional methods by using active carbon, ion exchanger, spray-drying or vacuum concentration.

REFERENCES

1. Rossi, M., Linko, Y.-Y., Linko, P., Vaara, T. and Turunen, M., Oligosaccharide mixture, and procedure for its manufacturing. WO 92/03565, 5.3.1992.

2. Rossi, M., Linko, Y.-Y., Linko, P., Vaara, T. and Turunen, M., Oligosaccharide mixture, and procedure for its after-treatment. WO 92/07947, 14.5.1992.

ENZYMATIC PRODUCTION AND MEMBRANE CONCENTRATION OF OLIGOSACCHARIDES FROM MILK WHEY.

MIGUEL H. LOPEZ LEIVA and MONICA GUZMAN.
Department of Food Engineering, University of Lund,
P.O. Box 124, 221 00 Lund, Sweden

ABSTRACT

We have studied the enzymatic formation of oligosaccharides when whey UF-permeates are hydrolysed in an immobilised enzyme reactor.
The immobilised enzyme reactor used was of the "flow-through" type where the enzyme is immobilised in the pores of a microporous film and the substrate is forced to pass through the film. High mass transfer and short residence times can be obtained in such a reactor.
The concentrations of glucose + galactose, lactose and oligosaccharides in the hydrolysates were determined by HPLC.
It was found that oligosaccharide formation is mainly governed by the concentration of lactose and the residence time, and in a minor degree by the pH.
An attempt to fractionate the oligosaccharides from a whey permeate hydrolysate by nanofiltration was also made.

INTRODUCTION

Oligosaccharides are polymeric carbohydrates consisting of two to ten monomer residues joined through glycosidic bonds. During the enzymatic hydrolysis of the lactose present in whey and whey UF-permeate, galactosyl - oligosaccharides containing from two to seven units are formed. The ingestion of these galactosyl-oligosaccharides encourages the proliferation of *Lactobacillus bifidus* in the intestine.
These bacteria form the dominant flora of breast infants and are said to improve a number of body functions [4], among others growth inhibition of harmful entero pathogenic micro-organisms such as *Escherichia coli, Salmonella typhi* and *Staphylococcus aureus*.
Yoghurts with added strains of *Bifidobacteria* are commercialised world-wide. In 1989 there were already around 250 oligosaccharide-containing products on the Japanese market, including yoghurts and different kinds of drinks [2].
The aim of this work is to study the influence of the operating conditions in the immobilised enzyme system, during the enzymatic hydrolysis of whey UF-permeate, and to find the optimal parameters which favour oligosaccharide production.
Since the oligosaccharides are always obtained at rather low concentration, an attempt to separate or concentrate them by means of nanofiltration was also made.

MATERIALS AND METHODS

Immobilised Enzyme Reactor.

A porous film with immobilised lactase was purchased from Amerace Corporation (Butler U.S.A.). The immobilised enzyme is β-galactosidase from *A. oryzae* bound via glutaraldehyde covalent linkages to silica particles entrapped in the microporous of a PVC sheet. The film is 0.64 mm thick, with a porosity in the range of 70 to 80 %, and pore diameters between 5 and 24 μm (8 μm average).

The reactor is formed (fig. 1) by cutting this film into pieces of 2.5 or 4.7 cm diameter which are then mounted in Swinex disc filter holders (Millipore Corporation) with filtering areas of 3.4 and 13.8 cm^2 respectively. The substrate is forced to pass through the disc filter holder, where a good contact substrate - immobilised enzyme is obtained with minimal diffusive resistance.

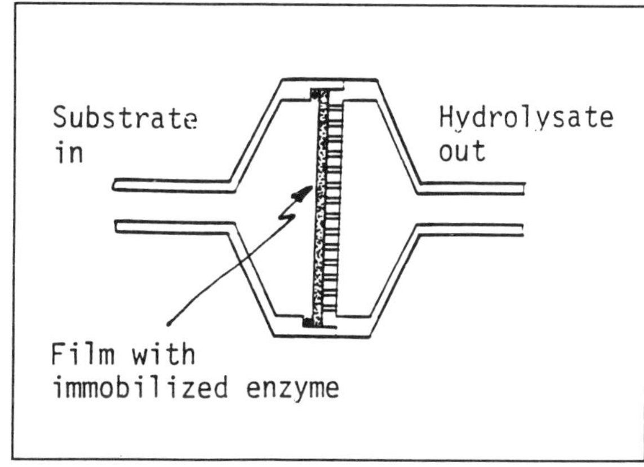

Figure 1.- Immobilized Enzyme Reactor

Whey.

The whey was obtained by rehydration of desalted whey powder. This was bought from the Swedish Dairies Association (SMR, Falkenberg), having the following composition: Lactose: 80%, Protein: 13%, Fat: 2%, Salts: 1% and Moisture: 4%.

Ultrafiltration

A batch UF circuit consisting of a Romicon PM 10 hollow fibre cartridge (cut off =10 000) of 5 sqft area was used for the ultrafiltration of the whey.

Whey powder re-hydrated at 20 % lactose was ultrafiltred at 55 °C and the permeate collected and stored at 4 °C for further utilisation. To avoid microbial growth, 0.1 % sodium azide was added to the solutions. New solutions were prepared every third week.

Whey hydrolysis.

The set-up depicted in figure 2 was used to study the lactose hydrolysis. Residence time, substrate concentration, and pH were the parameters studied. The temperature of operation was always 40 °C as suggested by the producers of the immobilised enzyme.

The whey UF-permeate was loaded into a 100 ml syringe (S) and then pumped, by means of a gear pump, through the immobilised enzyme reactor (IER), which is placed in a bath at constant temperature.

The residence times were varied either by changing the feed flow or by alternatively using two immobilised enzyme films. Oligosaccharide formation is especially important at the beginning of the reaction when the lactose is only partially hydrolysed [1,3]. Based on this we decided to work in the range of 2.5 to 70 seconds. Conversions were lower than 60 %.

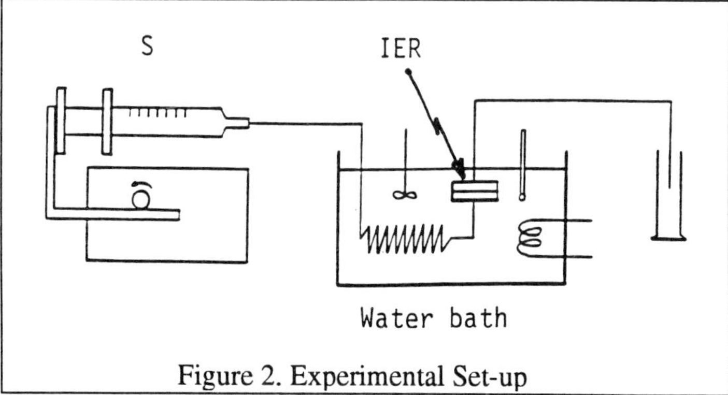

Figure 2. Experimental Set-up

RESULTS AND DISCUSSION

Effect of the Residence Time and Concentration.
In figure 3 we see the result obtained for the hydrolysis of a UF-permeate containing 20 % lactose. This figure typifies the results obtained also for other concentrations (6 and 15 %). The concentration of oligosaccharides reaches a maximum at around 12-seconds and then decreases, probably when they also begin to be hydrolysed.

The concentration of the lactose has the largest effect on oligosaccharide production; this is especially important for concentrations in the range of 6 to 15 %. The optimum residence time for maximum production of oligosaccharides depends on the initial lactose concentration, however, it is always around 10 - 15 sec. Since it is not possible to transform all the lactose in oligosaccharides, the use of purification and/or concentration steps after the enzymatic step become necessary.

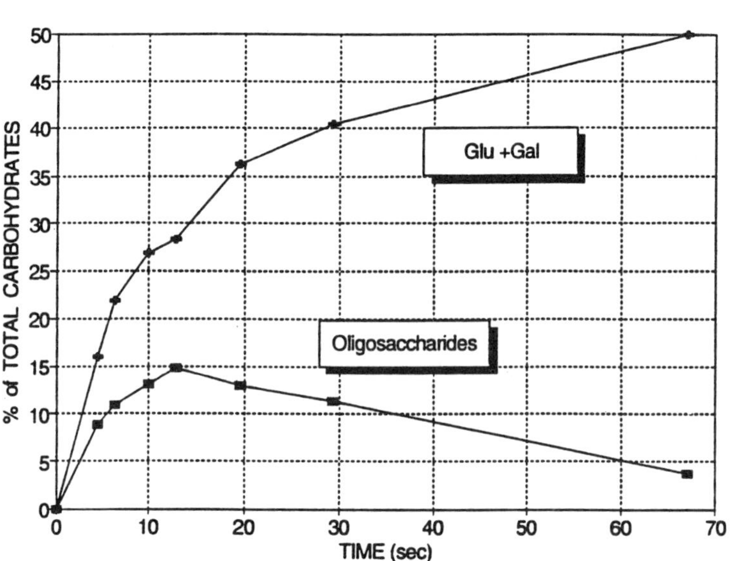

Figure 3.- Oligosaccharide formation for a whey UF-permeate containing 20 % lactose

Fractionation of UF-whey hydrolysate by membrane filtration.
An attempt to concentrate the oligosaccharides present in the whey permeate hydrolysates was done by filtrating the solution through a nanofiltration membrane (Nadir NF-PES-5 /PP 100, kindly supplied by Hoechst AB). A stainless steel module for reverse osmosis ("dead end" type) with 300 ml capacity was used. It was pressurised to up to approximately 20 bars with the help of a nitrogen cylinder.

Even though total separation of the oligosaccharides by nanofiltration was not possible they were concentrated to a certain extent. In one case the concentration was more than tripled. The present state of the art in nanofiltration does not allow 100 % rejection for the oligosaccharides. Therefore the recycling of the permeates from the nanofiltration step to the hydrolytic step and the extent to which this can be done should be further studied.

REFERENCES

1.- Burvall A, Asp N. G. and . A. 1977. Oligosaccharide formation during hydrolysis of lactose with *S. lactis* lactase I, II and III. *Lund University (Ph. D. thesis).*

2.- Comline news service,1989. Comline Biological & Medical. Sagetsu Building, 3-12-7 Kita-Ahoyama, Minato-Ku, Tokyo 107, Japan.

3.- López-Leiva M. H. and Zárate S. 1988. Mass transfer in an immobilized enzyme reactor. *Med. Fac. Landbouww. Riksuniv. Gent, 53 (4a):* 1639 - 1644

4.- Wijsman R. M. Hereijgers J. L. P. M. and Groote J. M. F. H. 1989. Selective enumeration of bifidobacteria in fermented dairy products. *Neth. Milk Dairy J. 43:* 395-405.

CONSTRUCTION OF INTEGRATED BIOREACTOR SYSTEM FOR BIOLOGICALLY ACTIVE PEPTIDES FROM ISOLATED SOYBEAN PROTEIN

KENJI SONOMOTO[1,2], YASUNORI OKAMOTO[2], SHINKICHI KOHNO[3], KANEO KANO[3]

[1]Department of Food Science and Technology, Faculty of Agriculture, Kyushu University, Hakozaki 6-10-1, Higashi-ku, Fukuoka 812, Japan

[2]Department of Biochemical Engineering and Science, Faculty of Computer Science and Systems Engineering, Kyushu Institute of Technology, Iizuka, Fukuoka 820, Japan

[3]Life Science Research Center, Nippon Steel Co., 1618 Ida, Nakahara-ku, Kawasaki 211, Japan

ABSTRACT

Among the commercially available endopeptidases from different sources tested, mixture of three kinds of proteases gave an efficient production of di-and/or tri-peptides mixture (average molecular weight, *ca.* 400) from a hardly water-soluble isolated soybean protein at 50 °C. The peptides mixture so obtained showed biologically active functions in mammalia, such as control of obesity and fatigue. Repeated batch system with ultrafiltration membrane (MWCO, 10,000) resulted in the peptides production with high potency during the operation for 20 days. Furthermore, the desired product was continuously obtained for one week by using a cross-flow membrane reactor over 90 % of the peptides yield.

INTRODUCTION

Functionalities of foods have been greatly attracted in food biotechnology. Especially, peptides activated by digestion and metabolism are known to have physiological functions such as control of obesity, adult diseases, immune system, cell differentiation and proliferation. Moreover, few studies were reported to construct new bioreactors having continuous production and separation abilities for small peptides. In this paper, we have attempted biologically active di-and/or tri-peptides production by enzymolysis of isolated soybean protein [1] and also construction of an integrated bioreactor with down-stream system for the continuous production by the use of ultrafiltration membrane.

MATERIALS AND METHODS

Isolated soybean protein (protein, 96 %) was obtained from Nishin Oil Co., Japan. Unless otherwise noted, enzymolysis batch reaction was performed at 50 °C in 100 ml of the reaction mixture containing 5.0 % w/w autoclaved substrate and 0.1 % w/v filtered enzyme mixture.

The resulting reaction mixture was centrifuged and used for evaluation of the enzymolysis, such as average molecular weight of the product, di-and/or tri-peptides yield and nitrogen yield. The average molecular weight was calculated as (protein concentration in the supernatant)/(molarity of amino group in the supernatant). Di-and/or tri-peptides yield means its content in the solubilized protein from the substrate and was determined by Sephadex G-10 gel chromatography. Minitan-S ultrafiltration system (Millipore Co., USA) was used as a cross-flow membrane module in a continuous production bioreactor.

RESULTS AND DISCUSSION

Selection of commercially available endopeptidases at 50 °C was done by considering the high viscosity of substrate suspension and its water-insolubility. As a result, three kinds of proteases, Biotamilase P-1000 (Nagase Biochemical Industry, Japan), Protease S (Amano Pharmaceutical Co., Ltd., Japan) and Protease N (Amano) were of greatly high potency. Furthermore, their mixture (23.1:14.7:70.9 w/w in the above-mentioned order) gave an efficient production of di-and/or tri-peptide mixture (average molecular weight, *ca.* 400), which showed biologically active functions in mammalia, such as control of obesity and fatigue. Amino acid composition of the peptides was almost same as of the substrate, containing essential amino acids well-balanced and abundantly . Amino acid mixture of the same composition as of the peptides as well as the substrate exhibited no biological activity. Optimum pH and temperature of the enzyme reaction were 7 and 60 °C, respectively. Figure 1 shows the typical enzymolysis of isolated soybean protein with the enzyme mixture. The reaction yield on nitrogen was about 90 % and the yield of the desired peptides was calculated over 85 % after 24 h of reaction. However, the residual proteolytic activity in the reactor gradually decreased and was about 40 % of the initial value after 24 h. The enzymolysis reaction was slightly inhibited by the peptides formed and succcessive addition of the substrate in the reactor stabilized the enzyme mixture. From this result, it is necessary for the construction of effectively continuous production system that the product of low molecular weight should be selectively removed from the reactor and that the enzymes and the substrate of high molecular weight are kept in a certain space.

Repeated batch production of di-and/or tri-peptides was done with a stirred concentrator with ultrafiltration membrane (MWCO, 10,000). Each reaction was carried out at 50 °C for 22 h in 10 ml of reaction mixture containing 0.1 % enzyme mixture and 2.5 % substrate, followed

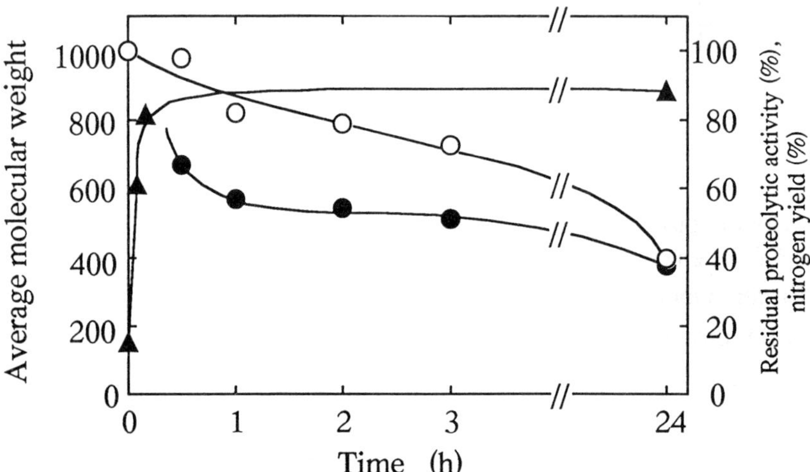

Figure 1. Time courses of enzymolysis of isolated soybean protein. (●) Average molecular weight, (O) residual proteolytic activity, (▲) nitrogen yield.

by filtration of the reaction mixture with nitrogen gas at 0 °C. Enzymes were proved to be kept in the concentrator by the membrane. Although the residual activity of the enzymes diminished slowly, this system showed a relatively good operational stability during 21 days. The average molecular weight of the resulting products after 10 and 18 days of reaction were 435 and 455, respectively. The peptides yield was more than 80 % even after 21 days.

We constructed a continuous production system with a cross-flow ultrafiltration membrane module to separate the product and to hold the enzymes and the substrate. The influence of several crucial factors such as enzyme concentration and dilution rate on the productivity and the stability of the system was investigated. As a result, the desired product was continuously obtained for one week with a high operational stability (Figure 2). The average molecular weight of the product was maintained about 400 and the peptide yield was over 85 % in the course of operation. The recovery of the peptide from the substrate was 67 % w/w. There was no bacterial contamination in the system at the end of reaction. Increase in the lifetime of the membrane reactor system is now being studied.

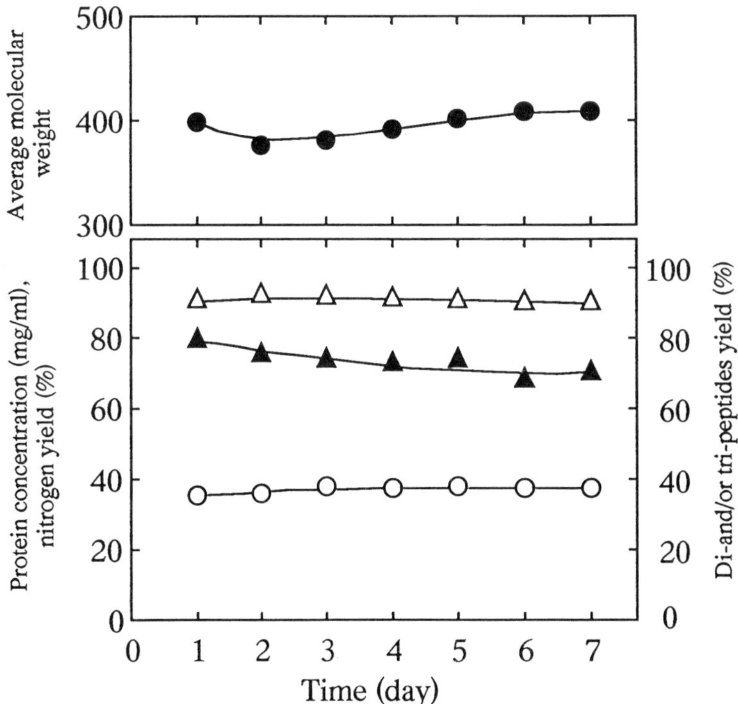

Figure 2. Continuous production of di-and/or tri-peptides with cross-flow membrane reactor. Substrate 5.0 %, enzyme concentration 0.1 %, dilution rate 0.5 day^{-1}, working volume 110 ml, ultrafiltration membrane molecular weight cut-off 10,000. (●) Average molecular weight of filtrate, (O) protein concentration in filtrate, (▲) nitrogen yield, (Δ) di-and/or tri-peptides yield in filtrate.

REFERENCE

1. Sonomoto, K., Okamoto, Y., Kohno, S. and Kano, K., Enzymatic production of biologically active peptides from isolated soybean protein. 9th International Biotechnology Symposium, Abstract 460, Crystal City, Virginia, USA, August 1992.

TRYPSIN-CATALYZED SYNTHESIS OF OLIGOPEPTIDES CONTAINING HYDROPHILIC AND ESSENTIAL AMINO ACID, L-LYSINE.

YUKITAKA KIMURA, MOTOHIRO SHIMA, TOMOYUKI NOTSU, SHUJI ADACHI, and RYUICHI MATSUNO

Department of Food Science and Technology, Faculty of Agriculture, Kyoto University, Sakyo-ku, Kyoto 606-01, Japan

ABSTRACT

The synthesis of peptides from *N*-(benzyloxycarbonyl)-L-lysine *n*-hexyl ester (Z-K-OHex) and L-lysine amide (K-NH_2) was done in 65% (v/v) acetonitrile. We mainly got two dipeptides (Z-KK and Z-KK-NH_2) and a tripeptide (Z-KKK-NH_2). We had Z-KK-NH_2 in the 90% yield but it rapidly decreased as the reaction proceeded. The decrease of the concentration of trypsin prolonged the period for which the yield was maintained at 90%. The yields of Z-KK and Z-KKK-NH_2 increased when the concentration of K-NH_2 increased. The above results were well simulated by a model of the reaction.

INTRODUCTION

Peptides have some properties superior to proteins and amino acids. Peptides are absorbed in the hydrolyzed form into the blood stream from intestine faster than amino acids[1). Solution of peptides has lower osmotic pressure than that of amino acids which consist of the same composition. So, peptides are good as a nutrient after a surgical operation.

Peptides are prepared by chemical, enzymatical, or genetical methods. A good method of them is enzymatic one because of its mild condition and low consumption of energy. Many researchers have studied enzymatic synthesis of peptides catalyzed by proteinases[2). However, it is difficult to synthesize peptides consisting of hydrophilic amino acids without protective groups of their side chains. In this study, we tried to synthesize the peptides containing a hydrophilic and basic amino acid, L-lysine, which is one of essential amino acids, using trypsin as a catalyst.

MATERIALS AND METHODS

Materials

Trypsin from porcine pancreas was purchased from Wako Pure Chemicals (Japan), and used without further purification. *N*–(benzyloxycarbonyl)–L–lysine *n*–hexyl ester (Z–K–OHex) was prepared by esterification of α–*N*–(benzyloxycarbonyl)– ε –*t*–(butyloxycarbonyl)–L–lysine (Kokusan Chemicals, Japan) with *n*–hexanol and by deblocking with formic acid. L–Lysine amide hydrochloride (K–NH_2·2HCl) was purchased from Kokusan Chemicals. Other Chemicals were of analytical grade.

Method of synthesis of lysine peptide

The synthesis of peptides from Z–K–OHex and L–lysine amide (K–NH_2) was done in 65% (v/v) acetonitrile. The high contents of acetonitrile gave the high yields of peptides but K–NH_2 could not dissolve in more than 65%. The reaction mixture included 50 mM Z–K–OHex, 100 mM K–NH_2·2HCl, 200 mM triethylamine (neutralizing HCl), and 1 mg/ml trypsin when the reaction started. Its pH was 9.4 and it was kept at 30°C. At appropriate times, a portion of the reaction mixture was sampled and analyzed by high performance liquid chromatography with a UV detector (258 nm). The yield shown in this study is based on the initial concentration of Z–K–OHex.

RESULTS AND DISCUSSION

Figure 1 shows the time course of synthesis. At the initial stage, a dipeptide, Z–KK–NH_2, was obtained with 90% yield but rapidly decreased. Then another dipeptide, Z–KK, and tripeptide, Z–KKK–NH_2 increased. The hydrolysate, Z–K, also increased constantly. Considering the order of appearances of the products and the substrate specificity of trypsin, we assumed a model for synthetic pathways as shown in Fig. 2. The model has ten first–order reactions. Their reverse reactions can be ignored because of their slow reaction rates. Factors in the concentration of enzyme was considered into the reaction rate constant, $\boldsymbol{k}$. The values of $\boldsymbol{k}$s were determined to get an agreement between experimental and calculated values. The calculated yield of products are shown by the solid lines in Fig. 1.

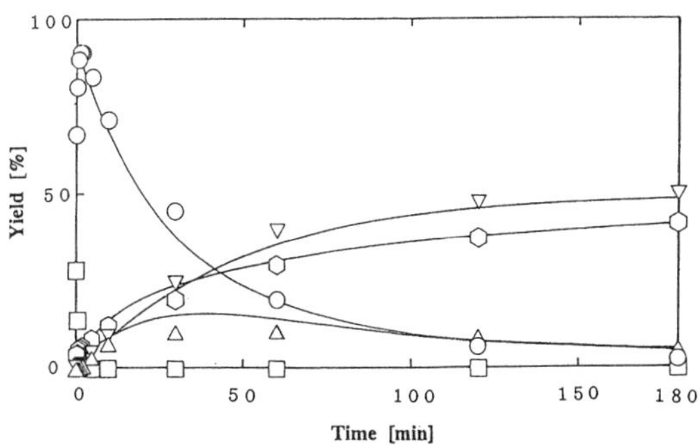

Figure 1. Time course of products in the synthesis catalyzed by free trypsin. The keys show the experimental results. ○, Z–KK–NH_2; ▽, Z–KK; △, Z–KKK–NH_2; ⬡, Z–K; □, Z–K–OHex. The solid lines show the calculated ones based on the model as shown in Fig.2.

The effects of the concentration of trypsin on the yields of the products were estimated by the model and the experiments. When the concentration decreased from 1 to 0.01 mg/ml, the period for which the yield of Z–KK–NH_2 was maintained at 90% prolonged from 1 min to more than 60 min. This phenomenon would result from that hydrolysis of the amide is slower than that of the ester.

Figure 3a shows the time courses of the products when the concentration of K–NH_2 increased to 200 mM. The yield of Z–KK–NH_2 increased to 95% and its decreasing rate became slower. The yield of Z–KKK–NH_2, one of the objective peptides, also increased to 23%. The undesired product, Z–K,was produced slowly. The calculated values in Fig. 3b show the same pattern of the experimental values.

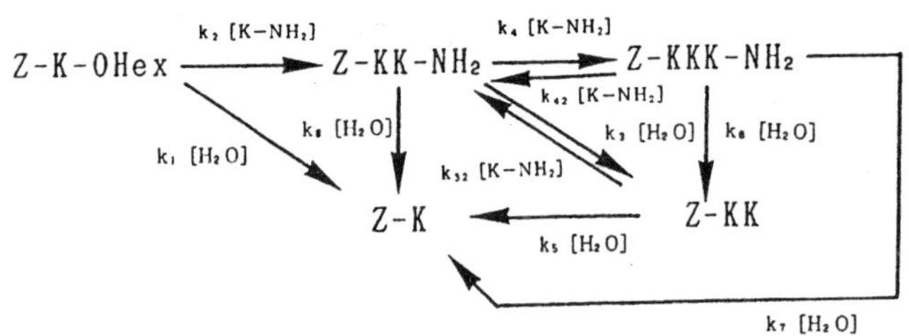

Figure 2. A model of pathways of the synthesis of lysine peptide. Z–, *N*–(benzyloxycarbonyl)–; K, L–lysine; –NH_2, amide; –OHex, *n*–hexyl ester; k, reaction rate constant.

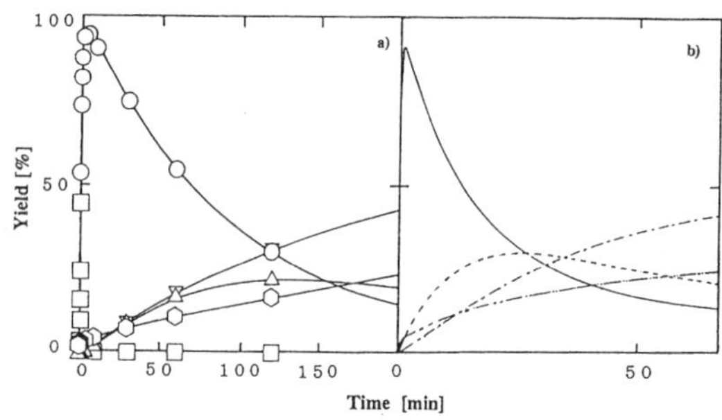

Figure 3. Time courses of synthesis when tge concentration of Z–K and K–NH_2 were 50 and 200 mM, respectively. The left (a) shows the experimental results and the right (b) shows the calculated ones. ○, —, Z–KK–NH_2; ▽, –·–,Z–KK; △, –··–, Z–KKK–NH_2; ⬡, –···–, Z–K; □, Z–K–OHex.

REFERENCES

1. Ganapathy, V., Burckhardt, G., Leibach, F.H., Characteristics of Glycylsarcosine Transport in Rabbit Intestinal Brush–Border Membrane Vesicles. J. Biol. Chem., 1984, **259**, 8954–8959

2. Morihara, K., Using proteases in peptide synthesis. TIBTECH, 1987, **5**, 164–170

INTEGRATION OF REACTION AND RECOVERY BY A CONTINUOUS EMULSION ENZYME REACTOR WITH IN-LINE REMOVAL OF THE OIL AND WATER PHASE BY MEMBRANE SEPARATION.

C.G.P.H. SCHROËN*, A. VAN DER PADT, K. VAN 'T RIET.
Wageningen Agricultural University,
Food and Bioprocess Engineering Group, Department of Food Science,
P.O. Box 8129, 6700 EV Wageningen, The Netherlands.

ABSTRACT

The emulsion/membrane bioreactor consists of an emulsion contained in a stirred tank and two membrane units. In the emulsion an enzyme catalyzed reaction is carried out. The oil phase and the water phase are separated by a hydrophobic and a hydrophilic membrane respectively.

The emulsion/membrane bioreactor can be used for both hydrolysis and esterification of acylglycerols. A cellulose membrane can be used as the hydrophilic membrane. The hydrophobic polypropylene membrane has to be modified with block copolymers in order to prevent protein adsorption and, therewith, to maintain selective separation.

INTRODUCTION

The enzyme, lipase, can catalyze both hydrolysis and esterification reactions. In case of esters of glycerol and fatty acids (acylglycerols) the following reactions are important:

Triglyceride + Water	⇄	Diglyceride + Fatty Acid
Diglyceride + Water	⇄	Monoglyceride + Fatty Acid
Monoglyceride + Water	⇄	Glycerol + Fatty Acid

* To whom correspondence should be addressed

Depending on the reaction conditions either mainly esters or mainly fatty acids are produced. Triglycerides are interesting for the edible oil industry. Monoglycerides can be used as e.g. emulsifiers. Fatty acids are used, in large quantities, in the soap industry.

The enzyme, lipase, can only be active at an oil/water interface where both substrates and the enzyme are available. It is important to create a large oil/water interface to obtain a high volumetric reactor activity. This can be done by using an emulsion in a stirred tank. The emulsion has to be separated in an oil phase and a water phase in order to obtain the products. In the emulsion/membrane bioreactor (figure 1) this separation is carried out by coupling the stirred tank to two membranes. The membranes are a hydrophilic cellulose membrane (cut-off value 6,000; manufactured by Enka) to separate the water phase and a hydrophobic polypropylene membrane (pore size 0.1μm, manufactured by Enka) to separate the oil phase. The hydrophilic membrane is not permeable for the enzyme due to its pore size. The enzyme does not permeate through the hydrophobic membrane also. Only the oil phase permeates through the hydrophobic membrane and the enzyme is not soluble in the oil. The retentate of both membranes contains the enzyme and is pumped back to the emulsion making a continuous process possible. This reactor concept can be used for both hydrolysis and esterification reactions.

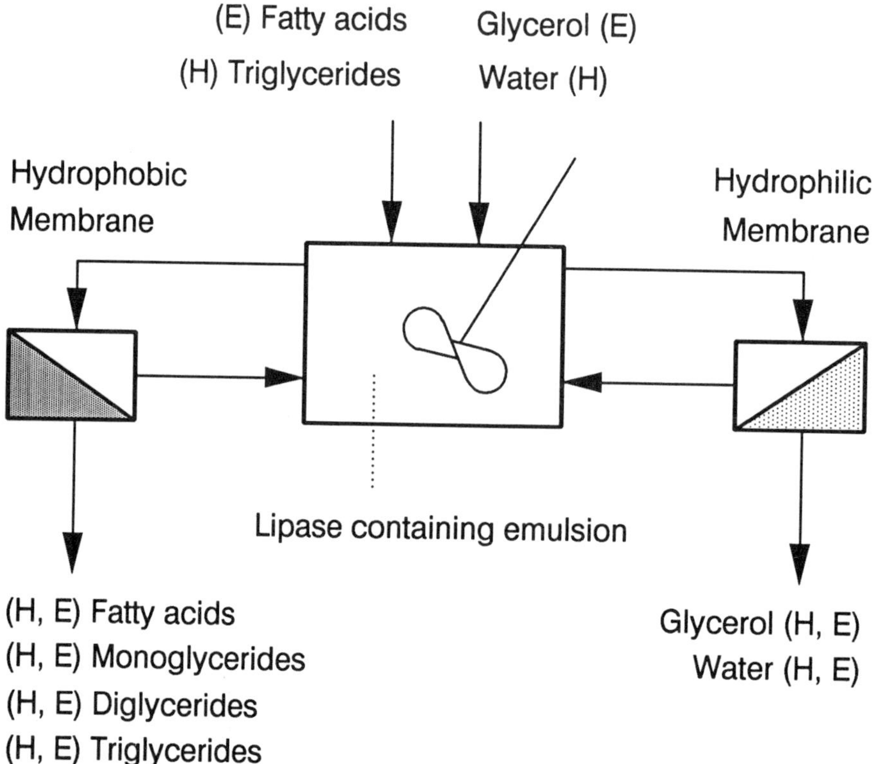

Figure 1. The emulsion/membrane bioreactor; H denotes substrates and products for the hydrolysis reactor, E denotes substrates and products for the esterification reactor.

RESULTS AND DISCUSSION

Both membranes were tested separately. The hydrophilic membrane was fouled by lipase but the flux decrease was limited. After an initial decrease in flux of approximately 10%, the flux value remained constant for seven consecutive days. Therefore, the hydrophilic cellulose membrane was judged to be appropriate for application in the emulsion/membrane bioreactor. The hydrophobic membrane, however, was fouled severely by enzyme adsorption, resulting in permeation of the water phase at low transmembrane pressure ($\approx 0.03 \times 10^5$ Pa). Enzyme adsorption has to be prevented. This was done by modifying the hydrophobic membrane with an appropriate block copolymer that gives a steric hindrance to enzyme molecules that come near the membrane. The block copolymer consists of a hydrophobic, polypropylene oxide, middle block and two hydrophilic, polyethylene oxide, blocks on both sides of the middle block. The hydrophobic middle block adsorbs onto the hydrophobic surface, both hydrophilic groups are extended into the water phase.

The block copolymer-coated membrane was not permeable for the water phase unless the Laplace pressure was exceeded. The maximum transmembrane pressure at which selective separation was possible (ΔP_{max}) was 0.5×10^5 Pa. After 10 days of continuous operation ΔP_{max} had not decreased. Therefore, it seems reasonable to assume that no enzyme adsorption took place at the modified membrane otherwise water could have permeated through the membrane at a lower transmembrane pressure. The modified membrane was used in continuous emulsion/membrane bioreactors for both hydrolysis and synthesis of acylglycerols.

CONCLUSIONS

Both hydrolysis and esterification of acylglycerols is carried out in an emulsion/membrane bioreactor. The flux through the hydrophilic membrane decreases slightly because of lipase adsorption but the flux level is high enough for continuous operation. The hydrophobic membrane is heavily fouled by lipase adsorption and has to be modified with block copolymers in order to prevent lipase adsorption. Through the modified membrane only the oil phase permeates, selective separation is possible up to a transmembrane pressure equal to the Laplace pressure.

Acknowledgement

The authors would like to thank the "Programme Committee on Membrane Research" (PCM) of the Netherlands and the "Dutch Foundation for Technical Sciences" (STW) for their financial support.

INTERESTERIFICATION OF TRIGLYCERIDES IN ORGANIC SOLVENT USING MODIFIED LIPASE

KENICHI MOGI**, SOBHI BASHEER*, AVA YAMAOKA*, FRED B PADLEY***, KAZUHIKO FUJIWARA**, MITSUTOSHI NAKAJIMA*
* National Food Research Institute, MAFF, Kannondai, Tsukuba, Ibaraki 305, Japan
** Nippon Lever B.V., 2–22–3, Shibuya, Tokyo 150, Japan
*** Unilever Research Colworth Laboratory, Colworth House, Bedford, MK44 1LQ, U.K.

ABSTRACT

Interesterification of tripalmitin and stearic acid in hexane was investigated using modified lipase. The modified lipase is a complex of lipase and surfactant, which has interesterification reactivity in organic solvent. Screening tests using various kinds of lipases and surfactants were done to optimize the activity of the modified lipase and its recovery. The modified lipase from *Rhizopus japonicus* using sorbitan monostearate as surfactant had the highest activity in hexane system. At low water concentration, interesterification occurred predominantly, and the production of diglycerides was very low. Hydrolysis of triglycerides was increased with the increase in water concentration. In a non–solvent system, interesterification occurred similarly as in hexane system under low water concentration.

INTRODUCTION

The physical properties of oil are related to the fatty acid composition of the triglycerides. By the interesterification reaction, especially using 1,3 specific lipase, the fatty acid composition of triglycerides can be changed. Several enzymatic interesterification methods are known; suspension of water in oil, W/O emulsion with surfactant, dispersion of powdered enzyme, immobilized enzyme, reversed micelles and PEG–modification systems have been investigated. In these systems, interesterification occurs under low water concentration. However, it has been reported that 10 to 20% of triglycerides were converted to diglycerides and monoglycerides by these conventional methods due to addition of water to the systems. Recently modified lipase, which is a complex of lipase and surfactant, was proposed by Okahata (1). Figure 1 shows the schematic model of modified lipase. By this modification, the lipase becomes soluble and/or well dispersed in the organic solvent. Then modified lipases can have much higher interesterification reactivity than the original lipases.

In this article, the modification process and evaluation of modified lipase in the interesterification of tripalmitin and stearic acid in n–hexane and non–solvent systems were investigated.

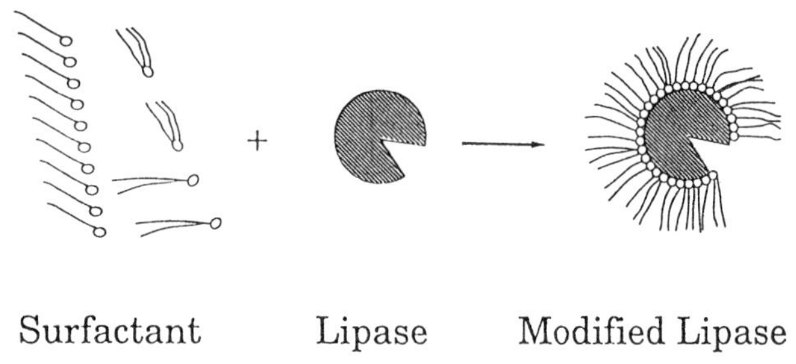

Figure 1. Schematic model of modified lipase.

MATERIALS AND METHODS

Modified lipases: The modification procedure is shown in Figure 2. Various lipases from different sources, such as *Rhizopus niveus, Rhizopus sp., Aspergillus niger, Mucor javanicus, Mucor miehei, Rhizopus delemar, Rhizopus japonicus, Aspergillus oryzae, Aspergillus sp., Candida rugosa, Candida cylindracea,* and *Alcaligenes sp.* were used. Sugar esters and sorbitan esters were used as surfactant.

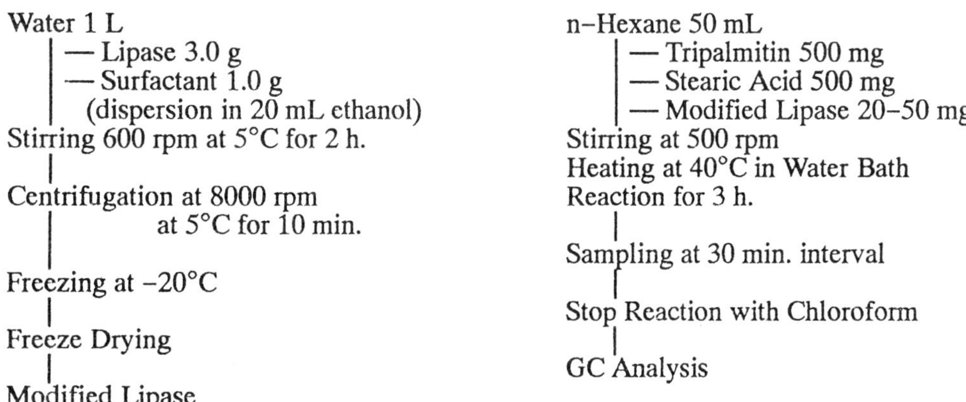

Figure 2. Lipase modification procedure.

Figure 3. Interesterification experiment.

Substrates: Tripalmitin, stearic acid and n–hexane supplied from Wako Pure Chemical, Japan, were used for interesterification experiments.

Interesterification reaction: Interesterification experiment is shown in Figure 3. 500 mg of tripalmitin, 500 mg of stearic acid and 20–50 mg of modified lipase were added in 50 mL n–hexane, in which the water content was decreased by molecular sieves (4A 1/16, Wako Pure Chemical, Japan). The reaction was carried out for 3 h at 40°C with 500rpm stirring.

Analysis: Interesterification reactivity was evaluated from the change of triglycerides composition, which was determined by GC (GC: R–14A with FID detector, Shimazu, Japan, Column: HR–TGC1 Capillary column, Shinwa Chemical, Japan). Water concentration in reaction system was determined by Karl Fischer Water Analyzer (684 KF coulometer, Metrohm, Switzerland). Protein contents of crude and modified lipases were analyzed by Automatic Nitrogen Analyzer (FP–428, Leco, USA).

RESULTS AND DISCUSSION

Figure 3 shows a typical time course of the interesterification reaction. The concentrations of tripalmitin and stearic acid decreased, and palmitic acid, PPS and SPS were produced with time. The reaction system nearly reached equilibrium after 2 h reaction. SSS was not produced, which means the modified lipase has 1,3 specific activity.

Monoglycerides were not detected at all. The amount of diglycerides produced was less than 10% of that of triglycerides.

In this system the water concentration was below 20 mg/L. The modified lipase had interesterification reactivity at such low water concentrations. The original crude lipase did not have interesterification reactivity under the same conditions.

From interesterification reaction and protein yield analysis, lipase from *Rhizopus japonicus* and sorbitan monostearate, HLB 4.7, as surfactant were selected for modified lipase formation.

Using the modified lipase, interesterification reaction in non-solvent system was also carried out. Figure 4 shows the comparison of interesterification reaction using hexane and non–solvent in the reaction systems. The modified lipase also had high reactivity in non-solvent system.

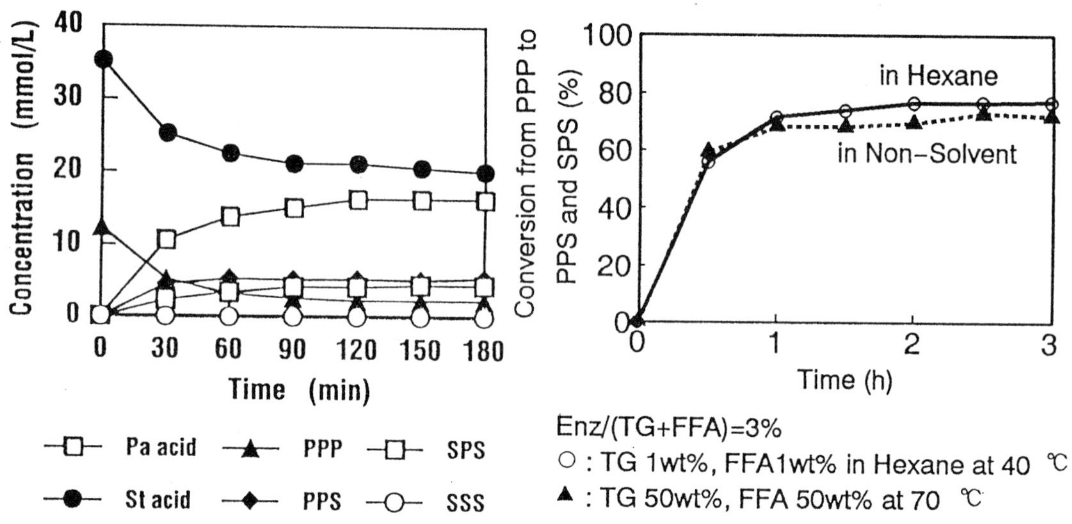

Figure 3. Interesterification reaction.

Figure 4. Comparison of hexane and non–solvent systems.

REFERENCE

1. Okahata, Y. and Ijiro, K., A lipid–coated lipase as a new catalyst for triglyceride synthesis in organic solvents. J. Chem. Soc., Chem. Commun., 1988, 1392–94.

THE EFFECT OF LIPIDS OXIDATION ON THE ACTIVITY OF INTER-ESTERIFICATION OF TRIGLYCERIDE BY IMMOBILIZED LIPASE

Toru Nezu,Satoru Kobori and Wataru Mastumoto
Food and Fat Laboratory, Asahi Denka Kogyo K.K.,
Higashi-ogu 8-4-1, Arakawaku, Tokyo 116, Japan

ABSTRACT

The lipase catalyzed interesterification of high oleic sunflower oil and stearic acid was carried out on a packed bed reactor, and the effects of lipid oxidation on the deactivation rate was examined. Oxidation of oils increase the deactivation rate; however the purification of oils stabilized lipase activity during continuous interesterification, deceased peroxide value and anisidine value of oils. There was a significant correlation between the deactivation rate and lipid oxidation parameters mentioned above. Furthermore, the incubation of lipase with 2-unsaturated aldehydes dissolved in hexane caused inactivation of lipase. It was concluded that the secondary lipid oxidation product strongly affect the lifetime of lipase activity.

INTRODUCTION

There has been great interest in lipase-catalyzed interesterification as a new method of industrial fat modification. Using this process selective exchange of acyl groups at 1,3-position of triglyceride is possible and there is also the potential for the economic synthesis of cocoa butter equivalents, which cannot be obtained by conventional chemical interesterification.

To minimize the reaction cost in this process, it is important to extend the lifetime of the lipase as much as possible. Recently, the poisoning effect of the lipids hydroperoxide[1),2),3),4)] and phospholipids[5)] on the lipase activity were reported.

In this study, continuous interesterification of high oleic sunflower oil and stearic acid was carried out, and the effects of lipid oxidation and purification were examined. The significant correlation between deactivation rate and lipids oxidation parameter was obtained.

MATERIALS AND METHODS

The lipase from Alcarigenes (LIPASE PL) absorbed on diatomaceous earth (RADIOLITE GK–02, Showa Kagaku Co.,Japan) was supplied from Meito Sangyo Co, Japan. High oleic sunflower oil (HOS) was oxidized by stirring at 140°C for 60 min in a vacuum of 300 mmHg. This oxidized HOS was purified by neutralization, bleaching and silica gel chromatography. The reaction mixtures were prepared by mixing HOS, stearic acid and hexane in a ratio of 1:1:4, and 800ppm of ethanol was added[6], This mixture was then pumped through a column containing 4.0g of immobilized lipase at 45°C. The glyceride composition of the reaction products was analyzed by reverse phase HPLC, and 2–oleoyl–distearin (SOS) content was measured.

The interesterification activity was estimated by following equation,

$$LN\,(1/1-X) = K / SV\,, \quad X = SOS / SOSeq \tag{1}$$

Where SV and SOSeq are space velocity(hr^{-1}) and SOS content at equilibrium, respectively.

The first–order kinetic model was applied to the decrease in the activity during continuous interesterification as follows;

$$LN\,(K / K0) = -KdT \tag{2}$$

Where K0 and T are the initial activity and reaction time, respectively.

RESULTS

The deactivation rates were increased by oxidation of oil. When oxidized and non–oxidized oils were used as a substrate, the first order constant of deactivation (Kd) were 0.0158 and 0.0285 respectively. The oxidized HOS (OX) was purified by neutralizing–bleaching (NB), bleaching (B) and silica–gel chromatography (SC), and it was then used as substrate. The purification of the substrate increased the stability of lipase in the order B<NB<SC, and reduced phosphorus content, ferrous content, peroxide value and anisidine value of oils (TABLE–1).

The deactivation rates were dependent on peroxide value and anisidine value, and the following correlation ($P<0.05$) was found to be significant:

$$Kd = 0.00156\,(POV) + 0.00076\,(AnV) - 0.00412 \tag{3}$$

TABLE–1. THE EFFECT OF OXIDATION AND PURIFICATION ON LIPASE DEACTIVATION.

Oil	POV (meq/Kg)	AnV (–)	Fe (ppm)	P (ppm)	Kd (1/day)
–	15.5	1.5	1.8	54.4	0.0158
OX	36.9	9.1	1.5	24.5	0.0569
B	0.4	16.2	1.3	11.3	0.0146
NB	ND	19.4	0.3	ND	0.0120
SC	ND	1.5	0.2	19.0	0.0059

POV : peroxide value AnV : anisidine value

The immobilized lipase was incubated in 100 mM of various aldehydes dissolved in hexane at 45°C for 3 days, and interesterification activity was measured (TABLE-2). In saturated aldehyde, only short chain aldehyde caused the deactivation of lipase, but even so the effects on lipase activity were relatively weak. In monounsaturated aldehyde, regardless of chain length, 2-unsaturated aldehydes were predominately responsible for the deactivation of lipase. The formation of 2-unsaturated aldehyde by lipid oxidation was reported.[7]

TABLE-2. THE EFFECT OF ALDEHYDE ON LIPASE ACTIVITY

	Relative activity			
	Saturated	Unsaturated		
Double-bond		t-2	c-4	t-12
Propionyl	62.6	-	-	-
Valer	54.9	-	-	-
Heptyl	82.5	3.0	60.0	-
Nonyl	95.0	3.1	-	-
Undecyl	124.8	6.8	-	-
Tridecyl	107.2	-	-	216.2
Control*	100.0	100.0	100.0	100.0

Control : The lipase was incubated with hexane alone.

DISCUSSION

In this study, it was suggested that the content of the lipids hydroperoxide and aldehyde strongly affected to the lifetime of lipase. While hydroperoxides are easily reduced by conventional purification, it is not economical to remove the secondary lipid oxidation product on an industrial scale. Therefore, it is important to use fresh nonoxidized oil as a substrate. Berger et al[8] reported that immobilization of Candida cylindracea lipase by covalent linkage involving the ε-amino residue of lysine lead to stabilization against the deactivation caused by acetaldehyde. The effects of aldehyde observed in this study may also contribute to the intereaction of ε-amino residue of lipase with aldehyde.

REFERENCE

1.Posorske,L.H., G.K.LeFebvre, C.A.Miller, T.T.Hansen and B.L.Grenvig, J. Am. Oil Chem. Soc.,1988,**65**,922
2.Wisdom,R.H., P.Dunnil and M.D.Lilly, Biotechnol. Bioengneer.,1987, **29**,1081
3.Ohta,Y., T.Yamane and S.Simizu, Agric. Biol. Chem.,1989, **53**,1855
4.Wang,Y. and M.H.Gordon, J. Agric. Food Chem.,1991,**39**,1693
5.Wang,Y. and M.H.Gordon, J. Am. Oil Chem. Soc.,1991,**68**,588
6.Mastumoto,W.,E.Nakai and T.Nezu,USP-5089404,1989
7.Frankel,E.N., Flavor Chemistry of Oils and Fats, ,Ed. D.B.Min and T.H.Smouse, American Oil Chemists' Society.,1985, pp.1
8.Berger,B. and K.Faberl, J. Chem. Soc., Chem. Commun., 1991,**1991**,1198

TRIACYLGLYCEROL SYNTHESIS FROM FATTY ACID AND GLYCEROL USING IMMOBILIZED LIPASE.

Yoshitsugu Kosugi*, Noboru Tomizuka* and Naoki Azuma**
*National Institute of Bioscience and Human-Technology, Agency of Industrial Science & Technology, 1-1 Higashi Tsukuba Ibaraki 305, Japan.
**Boso Oil & Fat Co. Ltd., 2-17-1 Hinode-chou Funabashi Chiba 273, Japan.

ABSTRACT

High free fatty acid (FFA) rice bran oil containing 30-50 % FFA was converted to an oil containing about 75-90 % triacylglycerol (TG) using immobilized lipase. Enzymatic refining of the FFA oil was performed continuously for more than one month using a reactor with two circulation loops, each connecting a fixed bed reactor and a dehydrator. The substrate was the stoichiometric amount of the FFA oil and glycerol required for synthesizing TG. The technology could be applied to synthesizing TG at more than 95 % yield from polyunsaturated fatty acids such as docosahexaenoic acid.

INTRODUCTION

Rice bran oil has attracted much attention since it was found to lower serum cholesterol(1). However, the rice bran oil available in most Asian countries contains a high proportion of free fatty acid (FFA) making the oil inedible. Successful enzymatic refining of high FFA rice bran oil can lead to the use of a potentially large and convenient domestic edible resource for south Asia (2). Icosapentaenoic acid (EPA) and docosahexaenoic acid (DHA) are characteristic biochemically active lipids. There is a demand for the conversion of these polyunsaturated fatty acids (PUFAs) into natural TG form.

This paper describes continuous enzymatic refining of high FFA rice bran oil and a high yield enzymatic preparation of TG from PUFA.

MATERIALS

High FFA rice bran oil used was prepared either as described in a previous paper (3) or mixing oleic acid with refined rice bran oil (1:1). The latter of a high FFA oil model was easily available in Japan and used for the continuous refining. Immobilized lipases from Rhizomucor miehei and Candida antarctica were donated by Novo Nordisk Bioindustry. Their commercial names are Lipozyme IM 60 and sp 382, respectively.

RESULTS

TG synthesis conditions using immobilized lipase

The optimum amount of substrate for synthesizing TG was the stoichiometric amount of fatty acid (FA) and glycerol (G) necessary for synthesizing TG. The reaction was carried out at 60 °C with continuous mixing and dehydrating of reactants. The mixing and the dehydration were performed by bubbling dry nitrogen. The dehydration rate was faster than the deacidification rate.

Continuous enzymatic refining of high FFA rice bran oil

During a batch reaction of enzymatic refining, production of monoacylglycerol (MG) was very low. Production of diacylglycerol (DG) increased rapidly during the first 3-4 hours. After that time production of DG

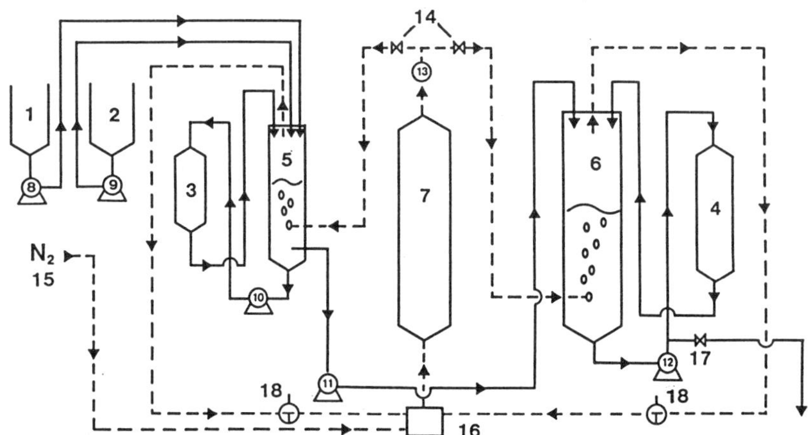

Figure 1. Schematic Diagram of Continuous Refining Reactor.

1) glycerol vessel; 2) high FFA rice bran oil vessel; 3) and 4) fixed bed reactor at 60 °C (Amounts of the Lipozyme IM 60 packed into 3) and 4) were 1.25 g and 5.0 g respectively.); 5) and 6), dehydrator (Working volumes for 5) and 6) were about 25 ml and 100 ml respectively.); 7) molecular sieve column; 8) liquid pump for glycerol(0.228 g/ hr); 9) liquid pump for high FFA oil (4.2g/hr); 10) and 12) liquid pump for circulation in the loop (6 ml/min); 11, liquid pump for pumping reactants from the first loop to the second loop (4.22 g/hr); 13) air pump; 14) needle valve; 15) nitrogen cylinder; 16) nitrogen damper; 17) outlet for product (4.22 g/ hr); 18) three way cock. The solid line represents a path of the reactants and the dotted line, a path of the nitrogen gas.

decreased. TG production increased continuously during the reaction for about 2 days. At the end of the reaction, the TG content reached 85-90 % and water content was less than 10 ppm. The consecutive reaction for the synthesis of TG from fatty acid (FA) and glycerol (G) are as follows:

$$2FA + G = DG + 2H_2O \text{ ------1}$$

$$DG + FA = TG + H_2O \text{ ------2}$$

The rate of reaction-1 was over 10 times faster than that of reaction-2.

Continuous enzymatic refining reactor is shown in Figure 1. In the first loop which combines the fixed bed reactor 3), pump 10) and dehydrator 5), about 30 % DG was continuously produced from FA and G. In the second loop, 74.2 % TG was continuously produced for more than one month.

Preparation of TG from G and PUFA or PUFA Ethyl Ester

The amount of DHA or EPA in the substrate was 5-10 % higher than the stoichiometric amount required to compensate for the PUFA loss during the reaction. In the case of DHA, the immobilized lipase from Candida antarctica was used because the immobilized lipase from Rhizomucor miehei did not synthesize TG from the DHA. Other conditions were the same as those used in the enzymatic refining of the high FFA oil. A TG yield of more than 95 % was achieved with EPA, DHA or arachidonic acid. Pure TG was obtained by passing the product through basic aluminum oxide column.

DISCUSSION

Pure TG obtained from PUFA is hopeful for such medical use as the intravenous infusion for preventing thrombotic disorder (4) as well as healthy food. The productivity for the continuous refining process is calculated as follows.

$$\text{Productivity} = (0.00422 \times 24) \times 0.742 \;/\; (0.005 + 0.00125)$$
$$= 12.0 \text{ (Kg of TG)/ day} \times \text{(kg of Immobilized lipase)}$$

The cost of the enzyme needed to produce 1 kg of TG can be calculated from the productivity, the life span of the immobilized lipase and the enzyme price. Enzymatic refining of high FFA rice bran oil can be industrialized if the enzyme cost is reduced less than 5-10 % the price of the rice bran oil.

REFERENCES

1. Haumann, B. F., Rice bran linked to lower cholesterol. J. Am. Oil Chem. Soc., 1989, 66, 615-618.
2. Gupta, H. P., Rice Bran Offers India an Oil Source, J. Am. Oil Chem. Soc., 1989, 66, 620-623.
3. Kosugi, Y., Igusa, H. and Tomizuka, N., Glyceride production of high free fatty acid rice bran oil using immobilized lipase. J. Jpn. Oil Chem. Soc. 1987, 36, 769-776.
4. Hamazaki, T., Fisher, S., Schweer, H., Meese,C.O., Urakase, M., Yokoyama, A. and Yano, S., The infusion of trieicosapentaenoylglycerol into humans and the in vivo formation of prostaglandin I_3 and thromboxane A_3. Biochem. Biophys. Res. Comm., 1988, 151, 1386-1394.

NAD(P)H REGENERATION IN YEAST CELLS USING ETHANOL AS ENERGY SOURCE

TADASHI KOMETANI, HIDEFUMI YOSHII, EITARO KITATSUJI, RYUICHI MATSUNO[*]
Department of Chemical and Biochemical Engineering, Toyama National College of Technology, Hongo 13, Toyama 939, Japan, [*]Department of Food Science and Technology, Faculty of Agriculture, Kyoto University, Sakyo-ku, Kyoto 606-01, Japan

ABSTRACT

Reduced nicotinamide cofactors, NAD(P)H, are regenerated from $NAD(P)^+$ through the pathway for ethanol oxidation in baker's yeast, *Saccharomyces cerevisiae*, that is coupled with bioreductions of carbonyl compounds. In this paper, the relationship between the ability using ethanol as energy source for the NAD(P)H regeneration and the culture condition on oxygen absorption rate was examined. The cells grown under the anaerobic condition could hardly use ethanol as energy source and those abilities increased with increasing the rate of oxygen absorption under the culture condition.

INTRODUCTION

Yeast-mediated bioreduction of prochiral ketones has been recognized as a useful technique for synthetic organic chemists, because many optically active compounds are conveniently prepared (1,2). In this microbial transformation, the regeneration of NAD(P)H from $NAD(P)^+$ in cells is necessary. The original procedure for the NAD(P)H regeneration in yeast cells was based on the hexose monophosphate pathway for glucose oxidation (3). Recently, we reported a new procedure using ethanol in an aerobic condition (4-6). In this procedure, NAD(P)H were regenerated from $NAD(P)^+$ through alcohol oxidation to carbon dioxide. The new procedure is more efficient and clean than the original one as discussed in the preceding papers (5,7).

This paper deals with the relationship between the ability using ethanol as energy source for NAD(P)H regeneration in the yeast cells, *Saccharomyces cerevisiae* HUT 7099, and the culture condition on oxygen absorption rate. The reduction reaction of ethyl acetoacetate (EA) to chiral product, (*S*)-(+)-ethyl 3-hydroxybutanoate (E 3-HB), an important starting material for many biologically active compounds, was examined to evaluate the ability of the cells grown under several conditions.

$$CH_3COCH_2COOEt \longrightarrow (S)\text{-}CH_3CH(OH)CH_2COOEt$$

Ethyl acetoacetate (EA) → (S)-(+)-Ethyl 3-hydroxybutanoate (E 3-HB)

MATERIALS AND METHODS

Culture medium A medium was composed of 1 %(w/v) glucose, 0.5% pepton, 0.3 % yeast extract, and 0.3 % malt extract. Its pH was adjusted to 5.5 by addition of hydrochloric acid.

Cultivation condition on oxygen absorption rate A 2–ℓ Sakaguchi culture flask containing 200 ml, 300 ml, 500 ml, or 1 ℓ medium was shaken at reciprocal shaking of 80 or 100 strokes/min. An aqueous solution of 0.20 M sodium sulfite in the presence of catalytic cuprous ion was used for a determination of oxygen absorption rate. Changes of concentration of sodium sulfite in those conditions were measured in the usual manner, and oxygen absorption rate was calculated.

Cultivation Yeast cells were precultured in a 50–ml Erlenemeyer flask with 10 ml medium by vigorous reciprocal shaking at 30 °C for 24 h. Into a 2–ℓ Sakaguchi flask containing a suitable amount of medium, the above culture, 2 %(v/v), was inoculated, and the culture was incubated at 30 °C with reciprocal shaking at a suitable speed for 48 h. The cells were harvested by suction filtration, washed with 0.9 % saline, and then used as described below. Weight of dry cells was determined after drying in an oven at 110 °C for 8 h

Bioreduction of ethyl acetoacetate A suspension of 1.00 g of wet yeast cells in a 20 ml of 77 mM EA and an energy source, which were dissolved in 0.1 M potassium phosphate buffer (pH 7.0), was placed in a 100–ml Erlenmeyer flask and shaken on a rotary shaker set at 30 °C and 80 rpm. The concentrations of energy sources were 200, 300, and 200 mM for ethanol, glucose, and acetate, respectively. As a control experiment, the reaction without addition of any energy source was also done. After 8 h, the reaction mixture was centrifuged and the supernatant was allowed to analysis.

Analysis The concentrations of EA, E 3–HB, and ethanol were measured by a gas chromatography as described in the preceding paper (5).

RESULTS AND DISCUSSION

Reduction rates by the yeast cells grown under various conditions on oxygen absorption rate were first obtained from concentrations of E 3–HB in reaction mixtures after 8 h per weights of dry cells. Then, reduction rates by additional energy sources were evaluated by deducting the rates of control experiments without addition of any energy source. Figure 1 shows the

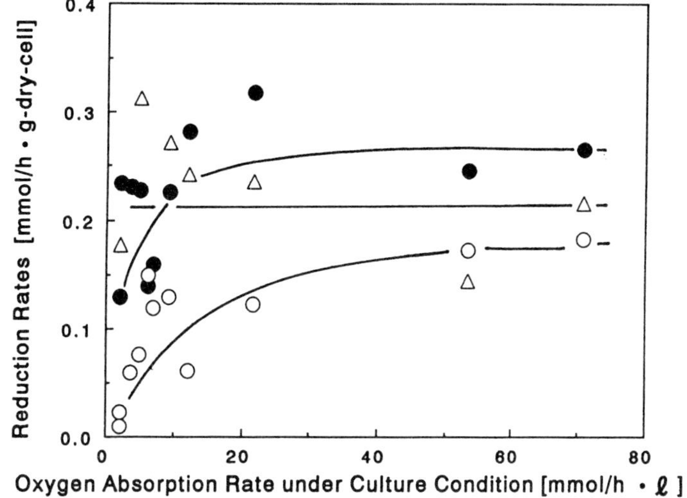

Fig. 1. Reduction rates of ethyl acetoacetate with various energy sources. Symbols: ○, ethanol; ●, glucose; △, acetate

effect of oxygen absorption rate under the culture condition on reduction rates of ethyl acetoacetate with various energy sources. The most important reslut was that the cells grown under almost anaerovic condition had hardly the ability using ethanol as energy source and the ability increased with increasing the rate of oxygen absorption under the culture condition. Then, the reduction rates in the presence of ethanol became nearly constant at the oxygen absorption rate above 20 mmol/h·ℓ. In the case of glucose as energy source, NAD(P)H would be regenerated through both glucose oxidation pathway and ethanol oxidation pathway. So, the reduction using glucose proceeds much faster than one using ethanol. In the case of acetate as energy source, the reduction rates with various yeast cells were almost constant. This result suggests that the ability of NAD(P)H regeneration using acetate is not influenced by the culture condition on oxygen absorption rate. Thus, we concluded that oxidation pathway from ethanol into acetate would be influenced by the culture condition on oxygen absorption rate. The further investigations are now in progress.

REFFERENCES

1. Servi, S., Baker's yeast as a reagent in organic synthesis. *Synthesis*, **1990**, 1–25.
2. Csuk, R. and Glänzer, B., Baker's yeast mediated transformations in organic chemistry. *Chem. Rev.,* 1991, **91**, 49–97.
3. Yamada, H. and Shimizu, S., Microbial and enzymatic processes for the production of biologically and chemically useful compounds. *Angew. Chem. Int. Ed. Engl.,* 1988, **27**, 622–642.
4. Kometani, T., Kitatsuji, E. and Matsuno, R., Baker's yeast mediated bioreduction. A new procedure using ethanol as an energy source. *Chem. Lett.,* **1989**, 1465–1466.
5. Kometani, T., Kitatsuji, E. and Matsuno, R., Baker's yeast mediated bioreduction: Practical procedure using EtOH as energy source. *J. Ferment. Bioeng.,* 1991, **71**, 197–199.
6. Kometani, T., Kitatsuji, E. and Matsuno, R., Bioreduction of ketones mediated by baker's yeast with acetate as ultimate reducing agent. *Agric. Biol. Chem.,* 1991, **55**, 867–868.
7. Kometani, T., Yoshii, H., Kitatsuji, E, Nishimura, H. and Matsuno, R., Large-scale preparation of (*S*)-ethyl 3-hydroxybutanoate with a high enantiomeric excess through baker's yeast-mediated bioreduction. *J. Ferment. Bioeng.,* in press.

LOOSE RO MEMBRANE REACTOR FOR L-AMINO ACID PRODUCTION WITH COENZYME RECYCLING

Takuya Harada and Osato Miyawaki
Dept. of Agricultural Chemistry, The Univ. of Tokyo
1-1-1 Yayoi, Bunkyo-ku, Tokyo 113, JAPAN

ABSTRACT

A microscale membrane bioreactor (vol. 2ml) was built up by using loose RO (LRO) membranes and applied to the production of L-amino acids (L-alanine form pyruvate and L-serine from hydroxypyruvate) with NADH recycling in a conjugated system of glucose dehydrogenase (GDH) and alanine dehydrogenase (ALDH). By using UTC-20 membrane, the reactor with immobilized GDH (695U/ml) and ALDH (206U/ml), operated at the retention time of 80min, produced 0.15M L-alanine at the productivity of 240g/l/day with NAD cycling number of 150,000 under continuous feed of the substrate solution containing 0.2M pyruvate, ammonia, glucose and 1μM NAD. This LRO membrane reactor is expected to have a wide applicability to coenzyme regeneration system for its general rejection characteristics for dissociable coenzymes.

INTRODUCTION

Immobilization of the dissociable coenzymes, such as NAD, NADP, ATP, are one of the key technologies for the development of the bioreactor system with multiple enzymes. To this end, there have been many attempts in which ultrafiltration (UF) membrane reactor was used combined with the macromolecule-bound coenzyme techniques or affinity chromatographic reactor [1] . With ultrafiltration membrane, however, the direct immobilization of coenzymes is difficult because molecular weight of the coenzymes are much smaller than the cut-off molecular weight of UF membranes. In this study, we employ LRO membranes for direct immobilization of coenzymes. Three different membranes (negatively charged membrane, NTR-7410, and two different type LRO membranes, UTC-20 and SSA) are used and compared with one another for their molecular rejection characteristics. A microscale membrane bioreactor is built up and applied to the production of L-amino acids with NADH recycling.

MATERIALS AND METHODS

GDH (*Bacillus sp.* 69.5U/mg) was gifted by Amano Pharmaceutical. ALDH (*Bacillus subtilis*

32U/mg) was purchased from Sigma Chemical. NAD, NADP, ATP were purchased from Böhringer Mannheim. NTR-7410 was gifted by Nitto Electric Industrial. UTC-20 and SSA were gifted by Toray Industries. A microscale membrane bioreactor, $15cm^2$ in membrane area and $2cm^3$ in volume was built and used for the L-amino acid production. HPLC pump was used to feed substrates.

RESULTS AND DISCUSSIONS

TABLE 1 shows the rejection ratio of the membranes for the coenzymes measured at different concentration of Tris buffer. The negative charged membrane, NTR-7410 rejects native coenzymes through electrostatic repulsion when no buffer coexists. However, the rejection ratio decreases drastically with increase in the ionic strength by increase in buffer concentration. SSA shows a higher rejection ratio than NTR-7410, but the rejection ratio decreases to 50% with increase in buffer concentration. On the other hand, the rejection ratio of UTC-20 not dependent on ionic strength was almost 90% to all the coenzymes tested.

TABLE 1
Rejection ratio of membranes for coenzymes at various ionic strength.

		conc. of Tris buffer		
		0.0M	0.1M	0.5M
NTR-7410	NAD	0.850	0.440	0.178
	NADP	0.885	0.242	0.131
	ATP	0.874	0.338	0.194
SSA	NAD	0.925	0.712	0.499
UTC-20	NAD	0.878	0.923	0.963
	NADP	0.906	0.904	-
	ATP	0.824	0.890	-

L-amino acid production with a microscale membrane bioreactor was carried out in the scheme shown in Fig.1. The system was composed of ALDH as a main reaction enzyme, and GDH as a regeneration enzyme for NADH. ALDH produces L-alanine from pyruvate and ammonia with the reducing power of NADH. NAD is oxidized back to NADH by GDH consuming glucose. In this reactor, the enzymes were completely immobilized in the system. The coenzyme was fed continuously and partially rejected or immobilized through the

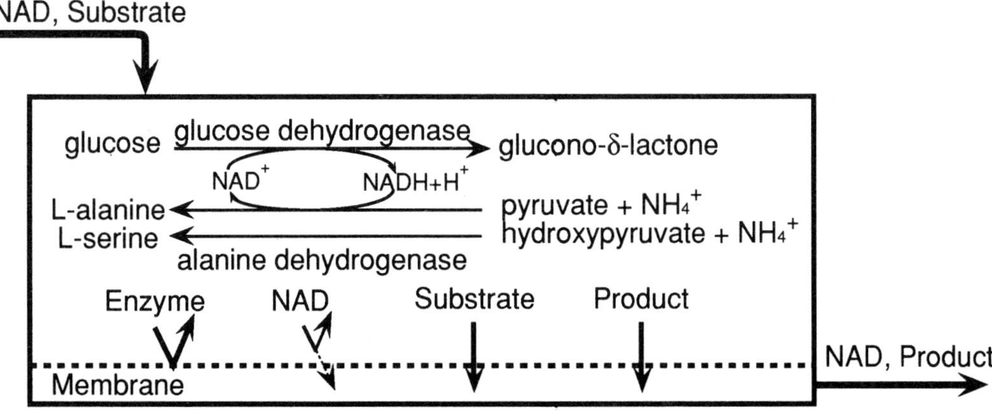

Fig.1 Membrane bioreactor with coenzyme recycling

rejection by the membrane used. Thus, the coenzyme was concentrated in the system depending on the rejection ratio of the membrane.

TABLE 2 shows the results of continuous L-alanine production at the retention time of 160min with the microscale reactor provided with different membranes. The NAD concentration with NTR-7410 was 10μM, tenfold higher than those with UTC-20 and SSA, due to its low rejection ratio for the coenzyme at the high buffer concentration. The system with UTC-20 showed the best result both in NAD cycling number and productivity so that it is used for further tests.

TABLE 2
Continuous L-alanine production in membrane bioreactors with different membranes.

	NTR-7410	SSA	UTC-20
substrates [M]	0.1	0.2	0.2
NAD [μM]	10	1	1
$\tau_{1/2}$ [hr]	192	-	> 800
conversion [%]	60	45	50
productivity [g/l/day]	48	72	80
turn over [s^{-1}]	0.86	1.29	1.43
cycling number [1]	6000	90000	100000

The effect of the retention time for UTC-20 system was tested. NAD cycling number was the highest to be 150,000 at the retention time of 80min, that means that the cost of NAD is reduced to 1/150,000. The reactor productivity showed the highest value of 260 g/l/day at the retention time of 53 min. This productivity is high enough compared with the productivity in fermentation reactors.
Hydroxypyruvate is substrate analog of pyruvate. Therefore, the same system is applicable to L-serine production when pyruvate is replaced with hydroxypyruvate. With ALDH at 206U/ml and GDH at 695U/ml, the conversion ratio of 20%, NAD cycling number of 6,700, and the productivity of 57g/l/day was achieved under the continuous feed of 0.1M substrates and 3μM NAD.

CONCLUSION

A microscale membrane bioreactor with LRO membrane was constructed and applied to L-alanine production and L-serine production with NAD regeneration. A high productivity was observed with a high cycling number of NAD. Because of the general rejection characteristics of the membrane used for dissociable coenzymes, the present system is expected to be widely applicable to the bioreactor system with coenzyme recycling.

REFERENCE

1. Nakamura, K., Aizawa, M. and Miyawaki, O., Electoenzymology/Coenzyme Regeneration, Biotechnology Monographs, vol.4, Springer-Verlag, Heiderberg, 1988

STUDIES ON THE BIOLOGICAL REDUCTION OF THE NITRATE CONTENT OF CARROT JUICE AND OF A MODEL SOLUTION USING *PARACOCCUS DENITRIFICANS* DSM 65

MIN, S. H. and SPIEß, W.E.L.
Institute of Process Engineering, Federal Research Centre for Nutrition, Engsesserstr. 20, 76131 Karlsruhe, Germany

ABSTRACT

The influence of pH, temperature and various electron-donators on the nitrate reductase activity of *Paracoccus denitrificans* DSM 65 was investigated. Nitrate reduction rate was studied in a model solution and in carrot juice. The denitrifying efficiency of *P. denitrificans* DSM 65 depended on pH, temperature and mainly on the H donator. Enzyme activity was highest at 34°C, pH 7.0 and with fructose as H donator. Fructose has been found to have the greatest influence on nitrate; with fructose as H donator, the specific nitrate degradation rate was twice to three times as high as with glucose or sucrose. At the higher temperatures and pH values the nitrate degradation rate increased as well. This applied to both model solution and carrot juice.

INTRODUCTION

Because of its nutritional quality and its high vitamin content, vegetable food is greatly appreciated. Some vegetables contain high nitrate concentrations, however. Although nitrate itself is of little primary toxicity for humans [1], uptake of nitrate-containing food may be harmful as the nitrate may be reduced to reactive nitrite by the human oral and intestinal flora [2]. In babies, the nitrite may lead to cyanosis [3]. N-nitrosamine formation represents another problem. N-nitrosamines form in acid media from nitrous acid and secondary amines [4]. Therefore elimination of nitrates and nitrite in food is desirable for reasons of health. Because of previous studies into its metabolism and its applicability as biocatalyst, *Paracoccus denitrificans DSM* 65 was selected as the most suitable microorganism for nitrate degradation in vegetable food [5].

The present work was concerned with factors influencing nitrate reductase activity of *Paracoccus denitrificans* DSM 65. We also tried to improve the growth conditions of the bacterium in order to obtain more biomass of increased nitrate and nitrite reductase activity.

MATERIALS AND METHODS

Nutrient medium (modified according to Burnell, 1975)

The medium consists of the following 3 solutions (Table 1) and trace-element solution SL-6. The three solutions were autoclaved separately and mixed together shortly before use. The trace element solution was filtrated sterilely and added at a concentration of 2ml/l.

Table 1
Contents of 3 solutions

Solution A		Solution B		Solution C	
$KH_2PO_4 \times H_2O$	2 g	EDTA	0.002 g	glucose	2.0 g
$K_2HPO_4 \times 2H_2O$	5 g	$MgSO_4 \times 7H_2O$	0.2 g	fructose	8.0 g
Aqua dest	700 ml	$FeSO_4 \times 7H_2O$	0.02 g	sucrose	2.0 g
		Na_2SO_4	0.054 g	Aqua dest	150 ml
		$CaCl_2 \times 2H_2O$	0.025 g		
		NH_4Cl	1.6 g		
		KNO_3	6.0 g		
		Aqua dest	150 ml		

Conditions of growth and cell harvest

The microorganism was grown in this medium at 30°C under compressed air (6 l/min) for the first 6 hours. After the supply of compressed air was stopped, the culture was further incubated until nitrate in the medium was completely degraded. To increase the biomass and nitrate reductase activity, 6 g/l KNO_3, 8 g/l fructose, and 2 ml/l trace element solution were once more added. Growth in the liquid medium was monitored by determination of the optical density at 578 nm. After complete degradation of the nitrate added, cells were separated from the medium by centrifugation at 6,000 rpm for 20 min at 4°C. The cells were washed using 0.1 M potassium phosphate buffer of pH 7.0, again centrifuged and frozen at -18°C.

Determination of specific nitrate reduction rate

All experiments to determine the nitrate reduction rate were conducted in a model solution and in carrot juice at different pHs. The influence of various H donators on the nitrate reduction rate was also studied in a model solution by enzymatic determination.

Nitrate degradation test:

Model solution: 0.1 g potassium nitrate, resp., and 1 g glucose, 1 g sucrose, 1 g fructose, or 2 g each of glucose and fructose plus 2 g sucrose as H donators were added to 100 ml 0.1 M potassium phosphate. The pH of the model solution was adjusted to 5.5, 6.0, 6.5 and 7.0.

Carrot juice: Commercial carrot juice was centrifuged to remove solid components (15,000 rpm, 15 min). pH of aliquots was adjusted to 5.5, 6.0, 6.5 and 7.0, resp. 0.1 g potassium nitrate were added to 100 ml carrot juice.

RESULTS AND DISCUSSION

Formation and activity of nitrate reductase are influenced by various factors such as composition of the medium, incubation conditions, kind of H donator, pH and temperature. The optimal pH and temperature for nitrate reduction using *P. denitrificans* DSM 65 was reported to be pH 7.5-8.0 and 30-40°C [5]. As denitrification in this range of pH and temperature is feasible only to limited extent for hygienic and sensory reasons, growth conditions and contents of the medium were optimized, so that enzymatic activity was improved at pH 6.0, 6.5 and 7.0 and at temperatures of 25 and 30°C.
At the temperatures of 25, 30 and 34°C used for the present experiments *P. denitrificans* DSM 65 has been found to show relatively good growth and high enzyme activity. The pH of freshly pressed carrot juice is between 6.0 and 6.5; the juice contains 2.33 g invert sugar and 2.42 g sucrose per 100 ml [7]. Therefore glucose, fructose, sucrose, or a combination of these were used as H donators.

The results of these studies have shown that there are virtually no differences in the nitrate reduction rate between the results obtained from carrot juice and model solution containing 1 g glucose, 1 g fructose and 2 g sucrose. This sugar combination corresponds to that of carrot juice. When glucose or sucrose alone wes used as H donator in the model solution, a significantly lower nitrate reduction rate was observed. Fructose had the great influence on nitrate reduction in the nitrate-degradation test, although the activity determinations had not shown any major difference among the kinds of sugar. In the presence of fructose, *P. denitrificans* degrades nitrate faster than with glucose or sucrose. This will be subject of further investigations. Because of increased enzyme activity and greater biomass yield under the improved growth conditions, the applicabilities of *P. denitrificans* DSM 65 to denitrification of vegetable foods have clearly improved.

REFERENCES

[1] Corre, W.J.; Breimer, T., 1979; Nitrate and nitrite in vegetables. Wageningen: Centre for Agricultural Publishing and Documentation
[2] Swann, P.F., 1975: The toxicology of nitrate, nitrite and N-nitroso compounds. J. Sci. Fd. Agric. 26, 1761-1770
[3] Rimpel, R.; Bruchmann, H.; Neinass, R., 1981: Zur Problematik der Nitrat- und Nitritkontamination bei enzymatisch produziertem Möhrensaft. Lebensmittelindustrie 28 Nr. 6, 263-268
[4] Preußmann, R., 1982; Nitrosaminbedingte Cancerogenese. VCH, Weinheim, 143-148
[5] Mayer-Miebach, E. und Schubert, H., 1991; Biokatalytische Verringerung der Nitratkonzentration in pflanzlichen Lebensmittelrohstoffen. Bio Engineering 2, 34-43
[6] Burnell, J.N., 1975; The reversibilty of active sulphate transport in membrane vesicles of *Paracoccus denitrificans*. Biochem. J., 150, 527-536
[7] Souci, S.W., Fachmann, W., Kraut, H., 1989/1990: Food composition and nutrition tables. 568, Wissenschaftliches Verlagsgesellschaft mbH, Stuttgart

DEVELOPMENT OF A BIOREACTOR FOR SEMI-SOLID FERMENTATION PURPOSES: BACTERIAL INSECTICIDE FERMENTATION

DEISE M.F.CAPALBO
Centro Nacional de Pesquisa de Defesa da Agricultura
CP 69 CEP 13820 000 Jaguariuna SP Brazil

IRACEMA O. MORAES
Universidade Estadual Paulista - UNESP / IBILCE-DETA
CP 136 CEP 15054 000 S.J.Rio Preto SP Brazil

REGINA O. MORAES
Universidade Estadual de Campinas - UNICAMP / FEAGRI
CP 6011 CEP 13081-970 Campinas SP Brazil

ABSTRACT

The development of *Bacillus thuringiensis* to obtain bacterial insecticide was well studied in Brazil. In the submerged process it was studied in a 250-liter fermentor, that is semi-pilot size production, and offers the opportunity scale up to a commercial scale.

When comparing with semi-solid processes, it can be concluded that in developing countries this is a lower cost process that uses cheaper sources of carbohydrates and proteins available in those regions. Therefore, it was decided to study parameters and variables to solve the difficulties encountered in this process. Heat and mass transfer,and sterility level are important engineering aspects to be considered. Yield and productivity are fundamental to cost studies and make the semi-solid process more feasible in developing countries, using a properly designed bioreactor.

INTRODUCTION

Chemical insecticides proved to be toxic to non-target species, particularly to man, in addition to the loss of effectiveness with constant use. The environmental damage observed gave impetus to the search for new means of insect control.

Biological control is one alternative that has been widely studied and used in recent years.

The most studied and commercially available biocontrol agent is *Bacillus thuringiensis* (Bt). Its toxic activity takes place primarily in the parasporal crystal formed within the mother cell during sporulation, either *in vivo* or *in vitro*. A mixture of Bt spores and crystals are commercially produced by fermentation means throughout the world; however, in Brazil, economic reasons have restricted its wider use.

The costs for Bt production depend to a large extent on the fermentation conditions (temperature, aeration, etc.), and to the cost of the medium. These topics were studied and are described elsewhere [1,2,3]. They generated two patents on Bt production in Brazil [4,5].

Based on the promising results obtained in those laboratory and pilot scale production by submerged fermentation, the CNPDA, with the cooperation of UNICAMP and UNESP initiated studies to employ new methods of fermentation to avoid the need for large aerated fermentors, by using the semi-solid fermentation (SSF).

The preliminary studies were developed using small static flasks, and have been described already [6,7]. The selection of low cost substrate, the variety of Bt, engineering parameters, temperature, humidification, and analytical methods were considered.

Those studies showed the adequacy of the SSF for Bt growth, sporulation, and highly active endotoxin production [7,8]. Some advantages and problems were also mentioned.

To scale up the process, it was necessary to stablish all the parameters involved: the enginnering parameters like aeration and agitation rate, diameter of the bioreactor, loading limit, and the physical characteristics of the solid substrate.

MATERIAL AND METHODS

Two types of column bioreactors have been studied: an aerated fixed bed, and a fluidized bed fermentor.

Fluidized Bed Column Reactor

A fluidized bed column reactor was tested for its ability to overcome the limitation of aeration that occured in static flasks. This kind of fermentation increased the mixing and the heat transfer rate, and permitted effective aeration of the mash. The reactor consisted of a jacket glass column (inside diameter = 7 cm; length = 25 cm) with a special internal support at the bottom. The apparatus was connected to an air-compressed line with controlled humidity and temperature [FIG.1].

FIGURE 1. Schematic diagram of the fluidized bed column reactor used for SSF of *Bacillus thuringiensis* var. *kurstaki*.

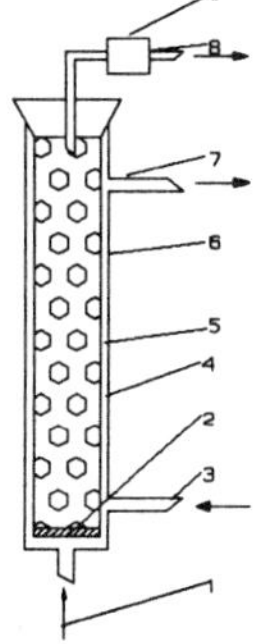

1. Humidified air
2. Internal support
3. Water from a water-bath
4. Glass column
5. Glass jacket
6. Fluidized mash
7. Water
8. Air
9. Filter

Fixed Bed Column Reactor

An aerated, loosely packed fixed bed column will be used. The mash will be composed of the preliminary selected substrates [7,8] , plus the inoculum of

Bt from a liquid fermentation, and tap water [3].

This reactor will be gently packed, aerated with humidified saturated air, as proposed by Raimbault and Alazard [9]. Previous researchers succeeded in determining some of the parameters (such as pH and growth kinetics) for fungi using a similar apparatus [10]. This bioreactor is currently being assembled to be used with the bacterium Bt.

RESULTS

Experimental data of the preliminary tests developed are being processed to change or ameliorate the conditions of the columns tested.

BIBLIOGRAPHY

1. Moraes, I.O.; Santana, M.H.A. and Hokka, C.O., The influence of oxygen concentration on microbial insecticide production. Advances in Biotechnology, 1981, **1**, 75-79.

2. Martucci, E.T. and Moraes, I.O., Substrate sterilization for Bacillus thuringiensis. Process Biochem., 1982, **17(6)**, 35-36.

3. Moraes, I.O. and Capalbo, D.M.F., The use of agricultural by-products as culture media for bioinsecticide production. In Food Engineering and Process Applications: unit operations, ed. M. LeMauguer and P.Jelen, Elsevier Publishers, London, 1986, v. 2, pp. 377-81.

4. Moraes, I.O., Patent BR PI 7608688. Processo de fermentacao submersa para producao de inseticida bacteriano. 10p. 1984.

5. Moraes, I.O., Patent BR PI 8500663. Processo de producao de toxina termoestavel de Bacillus thuringiensis. 8p. 1985.

6. Capalbo, D.M.F. and Moraes, I.O., Production of proteic protoxin by Bacillus thuringiensis by semi-solid fermentation. 12th Simposio Anual da Academia de Ciencias do Estado de Sao Paulo, Campinas, 1988. v. II, p. 46-55.

7. Capalbo, D.M.F., Desenvolvimento de um processo de fermentacao semi-solida para obtencao de Bacillus thuringiensis Berliner. Campinas, UNICAMP/FEA. 1989, 159p. (PhD thesis).

8. Capalbo, D.M.F.; Moraes, I.O. and Moraes, R.O., Recent observations on bacterial insecticide production by semi-solid fermentation technique. Annales del I Reunion Latinoamericana y del Caribe em Biotecnologia, Industria y Politicas Publcias para el Control Biologico de Plagas, Barquisimeto, Venezuela, 1991.

9. Raimbault, M. and Alazard, D., Culture method to study fungal growth in solid fermentation. European J. Appl. Microbiol. Biotecn., 1980, **9**, 199-209.

10. Crooke, P.S.; Hong, K.; Malaney, G.W. and Tanner, R.D. Solid and semi-solid state bioreactors: static, rotating and fluidized bed fermentors. J. Biomass Energy Soc. China, 1991, **10(1-2)**, 1-17.

COTINUOUS ETHANOL PRODUCTION BY *ZYMOMONAS MOBILIS* IN A BIOREACTOR WITH FLAME MAKING CERAMIC CARRIERS

AKIHIKO MORI and TETSUYA AOKI
Department of Chemical Engineering, Niigata University
Ikarashi 2-nocho, Niigata-shi, Niigata 950-21, JAPAN

ABSTRACT

Continuous ethanol production in a fixed bed bioreactor with flame making ceramic carriers, APHROCELL, adhesively immobilized by *Zymomonas mobilis* was studied. In the single bioreactor, the growth activity and specific conversion rate showed nearly same level at a higher dilution rate more than 0.6 h^{-1}. In the multistage bioreactors, final ethanol concentrations was nearly same level of 7.9g·l^{-1} (76% yield) at a higher dilution rate more than 0.6 h^{-1}, while ethanol productivity per total working volume increased as a dilution rate increased. It is suggested that high productivity and high yield can be obtained simultaneously at a high dilution rate with fixed bed multi-stage reactors.

INTRODUCTION

Many researchers have studied on ethanol production by *Zymomonas mobilis* in anaerobic or aerobic submerged culture, and also in a bioreactor since 1965 [1-3].

While a flame making porous caramic carrier newly developed, APHROCELL, has superior characteristics for an immobilizing carrier because it has large specific area, many spaces in the configuration, then a large capacity for gas holding and cell keeping. Many microorganisms have been used for immobilization to APHROCELL carriers, but genus Zymomonas have not yet.

In this paper, we studied ethanol production using a fixed bed bioreactor with APHROCELL carriers adhesively immobilized *Zymomonas mobilis*. Usually, in continuous culture, rising productivity and abundant yield of products are antinomic. But we will like to propose both high productivity and high yield by the bioreactor, and we found widely availability of APHROCELL in industrial use.

MATERIALS AND METHODS

Microorganism and medium

The microorganisms used in this study was *Zymomonas mobilis* ATCC 10988. RM medium used in this study consisted of 20 g glucose, 10 g yeast extract, 2 g KH_2PO_4, 0.5 g L(+)ascorbic acid as reductant and water in 1 *l*.

Bioreactor system

A bioreactor used in this study was a glass column of total volume of 300 ml packed with a cylindrical cartridge of APHROCELL No.13 of working volume of 200 ml. It was continuity of flame making porous ceramics. It consisted of the hyperbolic paraboloids bodies with 208 pores in 25 mm, and developed by Kirin Brewery Co., Ltd. and Bridgestone Corp. The outside of cartridge was coated by coating silicon (KE45, Shin-Etsu Chemical Co., Ltd.) then fixed to the inside of the reactor with the silicon.

Medium was supplied to the lower side face of the column and flowed out from the upper side face by a peristaltic pump with 2 channels. Nitrogen gas was supplied to the bottom of the column and exhausted from the top. Sampling was carried out from the outflow tube at the point after leaving the peristaltic pump. Multi-stage continuous culture was carried out by connection of a new column with nitrogen gas supply to the outflow of the working one.

Cultivation

The seed culture with a test tube was inoculated by a loop of stock culture, and incubated statically till foam generated, the preliminary culture with a test tube was inoculated by 3 loops of seed culture, and incubated statically for 18 h. Batch culture of 150 ml with a Erlenmyer flask was inoculated by a tube of the preliminary culture and incubated for 18 h under nitrogen gas supply of 0.2 vvm. Then, the flask was connected with APHROCELL bioreactor and the culture broth was flowed in circulation for 30 h under the nitrogen gas supply of 0.2 vvm, so that the bacteria might adhere sufficiently. Thereafter, the bioreactor was separated from the Erlenmyer flask and washed out with fresh medium for 10 h at a dilution rate of 1.5 h^{-1} so that almost free cells in the bioreactor were flowed out. After that, the bioreactor was operated at various dilution rates under nitrogen supply of 0.2 vvm till the steady state was obtained. All cultivations were performed at 30°C.

Analytical methods

Concentration of free cells was measured by optical density at 610 nm and calculated dry cell weight by calibration curve previously obtained. Cell mass kept in the APHROCELL carriers was washed out by 1000 ml of 0.1% Tween 80 solution for 4 h in circulation after finishing culture, then harvested, centrifuged, dried for 2 h at 105°C and weighed.

Specific growth rate (growth activity) was estimated by measurement of optical density of a sample which was harvested from a sample point of each reactor in steady state, inoculated into a shake flask with 100 ml RM medium, incubated by shaking at 140 reciprocation per min and sampled at 0.5 h intervals for 10 h.

Ethanol concentration was estimated by potassium dichromate oxidation method after distillation. Glucose was estimated by Somogyi method.

Every dilution rate was represented by flow rate per volume of a single reactor.

RESULTS AND DISCSSION

Kinetic characteristics of *Z. mobilis*

In RM medium, the growth rate of *Z. mobilis* was limited by yeast extract concentration, the maximum specific growth rate, μ_m, was 0.625 h^{-1} and K_s was 1.86 g yeast extract$\cdot l^{-1}$ broth. Growth yield was rather limited by glucose concentration than by yeast extract, and $Y_{X/S}$ was 0.026 g cell$\cdot g^{-1}$ glucose.

Continuous culture

In a single stage bioreactor, ethanol and free cell concentrations of outflow had peaks at a dilution rate of 0.6 h^{-1}. Then ethanol gradually and free cell steeply decreased as dilution increased, but cells did not be washed out at higher dilution rate. After 1800 h from start of the culture, total cell mass kept in carriers was 0.5 g, which was calculated to be 2.6 g$\cdot l^{-1}$ working volume. This was about 10-fold higher than free cells in the outflow at a high dilution rate.

Measurement of growth activity

In the single stage bioreactor, growth activity measurement was carried out. The bacteria showed synchronized growth at a high dilution rate (D = 1.5 or 5.0 h^{-1}), but did not show at a low dilution rate (D = 0.15 or 0.5 h^{-1}). The plots of specific growth rate of free cells in the outflow against dilution rate had a peak value of 0.48 h^{-1} at D = 1.0 h^{-1} and showed slightly down to 0.42 h^{-1} at D = 5.0 h^{-1}. It was suggested that free cells had nearly same growth activity and a lot of the cells were newly released from the carriers.

Multistage continuous culture

Multistage continuous culture was carried out with 3 bioreactors in series. The concentration of final outflow was nearly same level of about 7.9 $g \cdot l^{-1}$ (76% yield) at more than 0.6 h^{-1} of dilution rate (Fig. 1). While, ethanol productivity per total working volume of reactors increased linearly as dilution rate increased, and obtained 7.77 $g \cdot l^{-1} \cdot h^{-1}$ at D = 3.0 h^{-1} (Fig. 2). These results suggest that operation at a high dilution rate with multistage bioreactors can obtain high productivity and high yield simultaneously in the continuous culture with fixed bed bioreactors.

From these results APHROCELL bioreactor is expected to be available in the industrial continuous production by *Z. mobilis*.

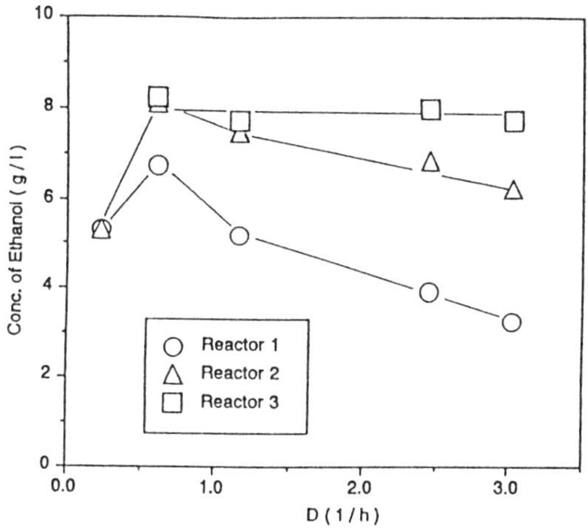

Figure 1. Relationship between dilution rate (D) and ethanol concentration in the multistage production.

Figure 2. Relationship between dilution rate (D) and ethanol productivity in the multistage production.

ACKNOWLEDGEMENT

We are very grateful to Kirin Brewery Co., Ltd. for their support, and to Mr. A. Suzuki of Kirin Brewery Co., Ltd. for supplying the carriers and reactors.

REFERENCES

1.Swings, J. and DeLey, J., The biology of *Zymomonas. Bacteriol. Rev.*, 1977, **41**, 1-46

2.Sahm, H., Ethanol production by bacteria. In *Bioprocess Engineering, The First Generation,* ed. T. Ghose, Ellis Horwood Ltd., Chichester, 1989, pp. 301-315.

3.Amin, G. and Doelle, H.W., Production of high ethanol concentrations from glucose using *Zymomonas mobilis* entrapped in a vertical rotating immobilized cell reactor. *Enzyme Microb. Technil.*, 1990, **12**, 443-446.

DESIGN AND SCALE-UP OF BIOREACTORS FOR MICROBIAL POLYSACCHARIDE FERMENTATION

YOSHINORI KAWASE AND MASANARI TSUJIMURA
Biochemical Engineering Research Center
Department of Applied Chemistry
Toyo University, Kawagoe, Saitama 350,Japan.

ABSTRACT

We have discussed the effects of rheological properties of microbial polysaccharides on the design and scale-up of bioreactors. Oxygen transfer rates and gas holdups are measured using a bubble column and two external-loop airlift bioreactors. The volumetric mass transfer coefficients significantly decrease with increasing viscous non-Newtonian flow behavior of fermentation media.

INTRODUCTION

Xanthan is an anionic exopolysaccharide produced by bacteria of the genus *Xanthomonas*. Xanthan gum has many applications as a thickener in the food, cosmetics, and pharmaceutical industries. Pneumatically-agitated fermentors increasingly are being considered for use in the fermentation industries in place of the traditional mechanically-agitated bioreactor design because of the potential savings in both capital and operating cost. However, the heat and mass transfer rates may not be sufficient.

In this study, we have made an extensive investigation of the frictional gas holdup in the riser section of external-circulation-loop airlifts, and of the volumetric liquid-phase mass transfer coefficient using liquid phases having non-Newtonian characteristics.

EXPERIMENTAL

The two external-circulation-loop airlifts used in this work were composed of a riser having a diameter of 0.15 m and a downcomer having inside diameters of 0.105 m (Airlift I) or 0.07 m (Airlift II). The values of the downcomer-to-riser cross-sectional area ratio (Ad/Ar) are 0.459 and 0.204, respectively. The airlifts were equipped with a perforated plate sparger located at the base of the riser section. The k_La coefficients were determined by the dynamic method.

Tap water, carboxymethylcellulose (SIGMA Chemical Co.), and Gum Xanthan (SIGMA Chemical Co.) were used as the liquid. Aqueous solutions of CMC and Gum Xanthan represented non-Newtonian flow behavior.

RESULTS and DISCUSSION

Figure 1 shows the k_La data obtained in Airlift I with 0.1 wt% Xanthan gum aqueous solution (n=0.402) and 0.15wt% CMC aqueous solution (n=0.796). It can be seen that the k_La coefficient significantly decreases with increasing viscous non-Newtonian flow behavior or decreasing n. As shown in Fig. 2, the k_La coefficients for highly viscous non-Newtonian fermentation media in Airlift II are less than those for water, as well as the results for Airlift I shown in Fig. 1. In order to improve oxygen transfer performance, four short draft tubes covered with perforated plates were inserted in the riser. It is seen in Fig. 2 that, as expected, the performance of oxygen transfer was enhanced. In high viscous non-Newtonian media, large bubbles were formed by coalescence, and a large number of very tiny bubbles resulted from break-up. For comparison, the correlations of Bello et al. (1) and Popovic and Robinson (2) are also given in Figs. 1 and 2. The predictions of correlation for water proposed by Bello et al.(1) agree very well with the experimental results. On the other hand, the correlation for non-Newtonian fermentation media (2) predicts considerably higher k_La coefficients compared with the present data for Xanthan gum and CMC. A reliable correlation for k_La coefficients is required.

Nomenclature

k_La	volumetric mass transfer coefficient
n	flow index in power law model
U_{sg}	superficial gas velocity

REFERENCES

1. Bello, R.A., Robinson, C.W. and Moo-Young, M. Prediction of the volumetric mass transfer coefficient in pneumatic contactors. Chem. Eng. Sci., 1985,40, 53-58.

2. Popovic, M. and Robinson, C.W. Mass transfer studies of external-loop airlifts and a bubble column. AIChE J, 1989, 35, 393-405.

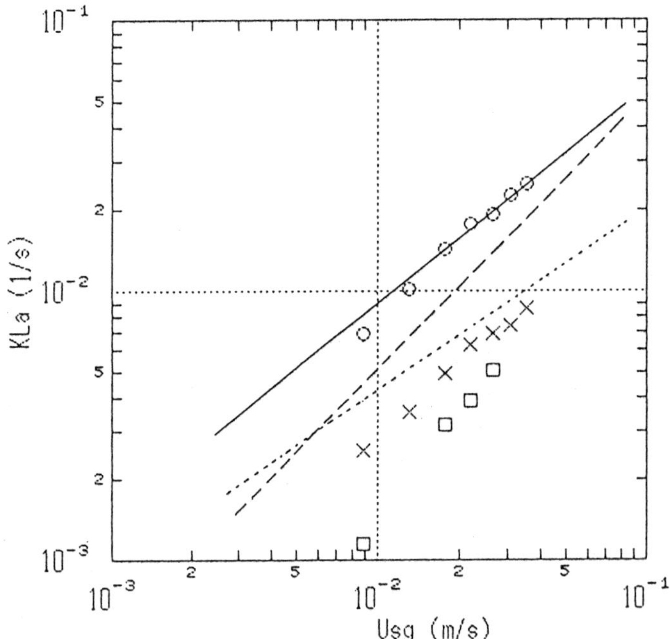

Figure 1. Volumetric mass transfer coefficients in AirliftII
O water ☐ Xanthan gum × CMC
——— Bello et al.(1)
---Xanthan gum, ······CMC Popovic and Robinson(2)

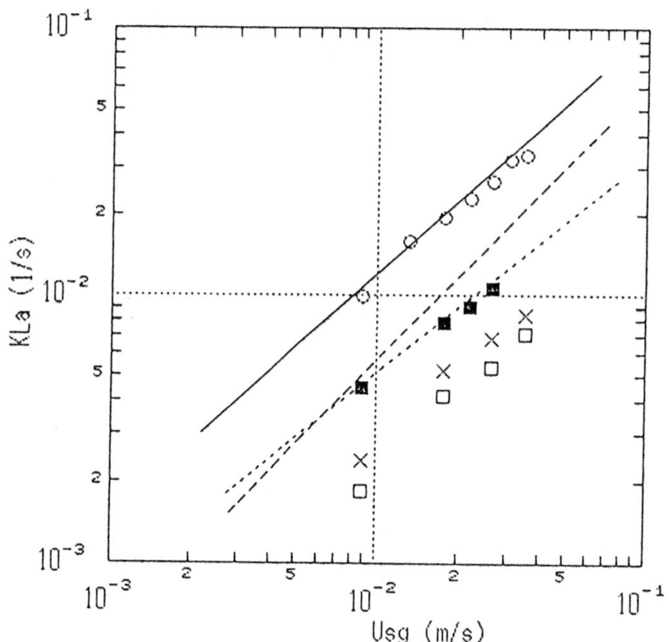

Figure 2. Volumetric mass transfer coefficients in AirliftII
(symbols as in Fig. 1.)
■ draft tubes with perforated plates

APPLICATION OF A BIOREACTOR SYSTEM TO SOY SAUCE PRODUCTION

TAKASHI HAMADA, YAICHI FUKUSHIMA AND HIROSHI MOTAI
Research Division, Kikkoman Corporation, 399 Noda,
Noda-shi, Chiba, 278, Japan

ABSTRACT

Continuous fermentation of soy sauce with immobilized glutaminase and immobilized cells of *Pediococcus halophilus, Zygosaccharomyces rouxii* and *Candida versatilis* was investigated. Glutamic acid production and both lactic acid and alcohol fermentation continued for over 100 days without any problems. The total time required for the production of soy sauce by the bioreactor system was much less than that of the conventional method (about 1/10). The soy sauce made by this system was similar to a conventional one in regard to chemical components and the quality of flavor.

INTRODUCTION

In the conventional method of brewing soy sauce, cooked soybean and roasted wheat are mixed with seed spores of *Aspergillus sojae* and/or *Aspergillus oryzae* to make *koji* in 2 days of solid culture. Then *koji* is mixed with brine to make *moromi*. At the first stage of *moromi* mash, *Pediococcus halophilus* grows and produces lactic acid which lowers the pH. Accompanying the decrease in the *moromi* pH, vigorous alcohol fermentation by *Zygosaccharomyces rouxii* occurs. As the result, 2-3% ethanol and many kinds of aroma components are produced by this yeast. At the same time phenolic compounds such as 4-ethylguaiacol (4-EG) and 4-ethylphenol, which add characteristic aroma to soy sauce, are also produced by other types of yeasts such as *Candida versatilis* and *Candida etchellsii*.

It takes over 6 months for the entire fermentation and aging of the *moromi* mash. Therefore, shortening this period is significantly important. In this study, we investigated a bioreactor system for soy sauce production, which consisted of reactors containing immobilized glutaminase and immobilized cells of *P. halophilus, Z. rouxii* and *C. versatilis*.

MATERIALS AND METHODS

Immobilization and reactors

Glutaminase from *Candida famata* was immobilized with Chitopearl followed by cross-linking with glutaraldehyde. *P. halophilus* cells were immobilized in AS gel [1] consisted of alginate and colloidal silica, and *Z. rouxii* and *C. versatilis* cells were immobilized in alginate gel [2]. Plug flow reactors (total vol.: 1.8 l; H/D=10.8 and total vol.: 7.5 l; H/D=5.4) were used for immobilized glutaminase and immobilized *P. halophilus* cells, respectively. Airlift reactors with a draft tube (total vol.: 27 l; H/D=6.1 and total vol.: 1 l; H/D=3.6) were used for immobilized cells of *Z. rouxii* and *C. versatilis*, respectively.

Preparation of raw liquid

Cooked soy bean and roasted wheat were mixed with culture broth of *A. oryzae* [3] and were enzymatically hydrolyzed in the presence of less than 10% NaCl and at a temperature 45°C for 3 days. This liquid fraction was used as raw liquid for the bioreactor.

RESULTS AND DISCUSSION

Manufacturing processes

The processes of soy sauce production using the bioreactor and the conventional methods are quite different as to the periods of lactic acid and alcohol fermentation. The entire fermentation with the conventional method takes several months. On the other hand, only about 2 days are needed for the bioreactor method using the immobilized whole cells.

In the bioreactor method, the first step is preparation of raw liquid for bioreactors. Raw liquid was successively passed through, firstly, a glutaminase reactor to increase glutamic acid, secondly, a *P. halophilus* reactor to carry out lactic acid fermentation and, thirdly, a *Z. rouxii* reactor to carry out alcohol fermentation, and a *C. versatilis* reactor to produce phenolic compounds such as 4-ethylguaiacol. Two reactors containing immobilized yeast cells were set in parallel and the flow rate of the feed solution to *Z. rouxii* and *C. versatilis* reactors was set in the ratio of ten to one.

Continuous fermentation

The fermentation by immobilized cells of *P. halophilus*, *Z. rouxii* and *C. versatilis* continued for over 100 days without any troubles such as microbial contamination. A consistently increased level of glutamic acid (the increase being in the range of 0.3–0.4%) was found in the effluent from glutaminase reactor, with a residence time of 0.7 h. Lactic acid was produced by immobilized cells of *P. halophilus* in quantities of 0.7–1.0% at a residence time of about 6 h and consequently the pH declined to 4.9–5.0, similar to that of the conventionally brewed soy sauce. Ethanol, often used as an index of alcohol fermentation, was produced constantly at a residence time of about 26 h by immobilized cells of *Z. rouxii* in quantities of 2.5–2.7%. This is the standard content in conventional soy sauce. About 10 ppm of 4-EG was produced by immobilized cells of *C. versatilis* at a residence time of about 10 h after passing through the reactor, and the final 4-EG content resulting after mixing the two fermented liquid issuing from the reactors of *Z. rouxii* and *C. versatilis* was about 1 ppm, which is the optimum concentration in conventional soy sauce. The total residence time in lactic acid and alcohol fermentation was about 30 h in this system, compared with the conventional fermentation periods of 2–3 months required to produce the same amounts of lactic acid and ethanol.

High numbers of viable cells were present in the gel and in the liquid for each reactor. The number was 10–100 fold higher than that in *moromi* mash. The shortening of the fermentation period in the bioreactor method is possibly due to the high density of immobilized cells in the gel and free cells in the liquid. Furthermore, from our previous results [4], free cells of *Z. rouxii* and immobilized cells of *C. versatilis* appear to contribute mainly to the production of ethanol and 4-EG in each reactor.

Properties

The main chemical components of the fermented liquid after passing through bioreactors were examined. The content of lactic acid, glucose, ethanol and formol nitrogen which were important components of soy sauce were kept in proper proportion. Aroma components present in both the bioreactor-produced soy sauce and the conventional sauce were not different qualitatively. However, the former sauce had more isoamyl alcohol and acetoin, and less isobutyl alcohol, ethyl lactate, 4-hydroxy-2(or5)-ethyl-5(or2)-methyl-3(2H)-furanone and 4-hydroxy-5-methyl-3(2H)-furanone than the latter. In order to evaluate the quality of soy sauce, a sensory test was carried out. For example, the intensity of the alcoholic, fresh, sweet, acid, and sharp odors were compared between the bioreactor and conventional soy sauces. The odors are important for the judgment of the quality of soy sauce. Although bioreactor soy sauce was a little weaker in aroma and fresh odor than conventional soy sauce, the quality of the former was generally judged to be similar to that of the latter [5].

CONCLUSIONS

Soy sauce production by a bioreactor system was investigated. Both lactic acid and alcohol fermentation and 4-EG production continued for over 100 days at a satisfactory level. The total time required for the production of soy sauce by the bioreactor system including enzymatic hydrolysis of raw materials, fermentation with immobilized whole cells and refining process is only about 2 weeks. This is considerably shorter than about 6 months in the conventional method of soy sauce brewing consisted of *koji* making, fermentation and aging in *moromi* mash, and refining process. The bioreactor soy sauce was considered to be similar to the conventional sauce in regard to chemical components and the quality of flavor. Therefore, it seems that this bioreactor system is practical for the production of soy sauce.

REFERENCES

1. Fukushima, Y., Okamura, K., Imai, K. and Motai, H., A new immobilization technique of whole cells and enzymes with colloidal silica and alginate. Biotechnol. Bioeng., 1988, **32**, 584-594.
2. Hamada, T., Sugishita, M. and Motai, H., Continuous production of 4-ethylguaiacol by immobilized cells of salt-tolerant *Candida versatilis* in an airlift reactor. J. Ferment. Bioeng., 1990, **69**, 166-169.
3. Fukushima, Y., Itoh, H., Fukase, T. and Motai, H., Continuous protease production in a carbon-limited chemostat culture by salt tolerant *Aspergillus oryzae*. Appl. Microbiol. Biotechnol., 1989, **30**, 604-608.
4. Hamada, T., Sugishita, M. and Motai, H., Contribution of immobilized and free cells of salt-tolerant *Zygosaccharomyces rouxii* and *Candida versatilis* to the production of ethanol and 4-ethylguaiacol. Appl. Microbiol. Biotechnol., 1990, **33**, 624-628.
5. Hamada, T., Sugishita, M., Fukushima, Y. and Motai, H., Continuous production of soy sauce by a bioreactor system. Process Biochem., 1991, **26**, 39-45.

SUBSTRATE REMOVAL CHARACTERISTICS IN AN IMMOBILISED CELL FLUIDIZED BED REACTOR

S M Rao Bhamidimarri
Process and Environmental Technology Department
Massey University, Palmerston North, New Zealand

ABSTRACT

Substrate bioxidation characteristics were studied in an immobilized cell fluidized bed bioreactor. The reaction-diffusion mechanisms in conjunction with the reactor flow characteristics were modelled to predict the substrate transport and bioreaction in the reactor. The significant feature of the model is that it permits evaluation of active biomass in the reactor. A combination of orthogonal collocation and Runge-Kutta-Gill techniques was used to solve the models. Experimental evidence is presented to demonstrate the validity of the models.

INTRODUCTION

The application of immobilized cell fluidized bed reactors for substrate conversion, in particular in wastewater treatment has been attempted by many researchers in recent years. The processes studied include denitrification (1), nitrification and organic carbon removal (2). In fluidized beds, in which small particles offer a large surface area. The immobilized cells utilize the substrate in the bulk liquid and grow into a biofilm resulting in reactor biomass concentrations of an order of magnitude greater compared to the conventional dispersed growth reactors. However, the volumetric reaction rates in such reactors are not proportional to the increase in biomass concentration. In order to predict the substrate removal in these reactors, it is necessary to identify various process mechanisms and to quantify them. Although, models based on reaction - diffusion mechanisms were developed (3), the fluidized-bed bioreactors have been continued to be investigated largely on an input-output basis. Reliable experimental techniques for the determination of dry and wet densities of biofilm and the effective substrate diffusivity into biofilm have been reported (4).

Predictive models for substrate removal in a fluidized bed reactor with experimental verification for substrate removal are discussed in this paper.

MODEL EQUATIONS

The biofilms supported on granular media assume spherical geometry in liquid fluidized beds. The methodology used for chemical reactors containing porous spherical catalyst particles can be adapted for analysing the substrate conversion in an immobilized cell fluidized-bed reactor.

Diffusion - Reaction

For a biofilm immobilized on support particle and steady-state diffusion of substrate and its conversion for a pure growth process is described by

$$\frac{d^2S}{d^2r} + \frac{2}{r}\frac{ds}{dr} = \frac{\varrho_f \mu_m S}{Y_t (K_s + S) D_s} \tag{1}$$

where ϱ is the biofilm dry density. Monod type growth results in the non-linear term on the right hand side of the above equation, which in the absence of external mass transfer resistance, is subject to the following boundary conditions:

$$at \quad r = r_{bp} \qquad S = S_b \quad and$$
$$at\ r = r_i \qquad \frac{ds}{dr} = 0 \tag{2}$$
$$S = S_{r_i}$$

r_i the inactive radius represents the radius of the core of the particle in which substrate biooxidation does not occur. The inactive core includes the inert support and the inner layers of biofilm which may not receive sufficient oxygen to catalyse substrate oxidation.

The substrate concentration at the inactive boundary is calculated based on the relative diffusivities of oxygen and substrate and the stoichiometry of substrate biodegradation. In a thin section of the fluidized-bed column reactor, if the bulk substrate concentration is assumed to be uniform S_{ri} and therefore r_i can be evaluated from equations (1) to (4) via numerical integration.

Reactor Model
For a uniform fluidized bed with no significant radial gradients and axial dispersion.
where η, the effectiveness factor, is defined as the ratio of reaction rate when diffusional resistances are significant to the rate of reaction in the absence of diffusion effects. Therefore equation (5) becomes

$$U_{sl}\frac{dS_b}{dH} + \frac{r_{bp}^3 \times \frac{ds}{dr}\Big|_{r_{bp}}}{(r_{bp}^3 - r_c^3)\,\varrho_f} = 0 \tag{3}$$

With the initial condition

$$H = 0 \qquad S_b = S_{b_i} \tag{4}$$

The particle model (equation 1) and the reactor model (equation 3) are solved simultaneously to obtain the concentration profile through the reactor.

Experimental
The substrate removal was studied in a fluidized bed reactor system as described by Bhamidimarri (4). The substrate used was phenol and the oxygen was supplied by dissolving pure oxygen in the feed solution to the fluidized bed column reactor.

RESULTS AND DISCUSSION

Model Analysis
The active biomass concentration varies with the height of the reactor because of varying oxygen and phenol concentrations and this gives rise to changing inner boundary condition for the particle model (equation 1). A continuous solution to such a moving boundary problem with non-linear kinetics is not easy to find. Therefore, a step-wise solution was developed for the analysis of the model equations. The column reactor was considered to be a series of thin sections. Assuming the substrate concentration in each thin section to be uniform, the reactor models were analysed for the first section to evaluate the concentration, which was then used as the inlet concentration for the next section. The

diffusion-reaction equation was solved by an orthogonal collocation technique and the Runge-Kutta-Gill routine was used to solve the reactor flow model. This procedure was repeated to generate concentration profile in the reactor.

Substrate Removal in the Reactor

The stoichiometry of phenol bio-oxidation is well established and it is reported that 1.45kg of oxygen are required per kg of phenol (5). Therefore, oxygen concentration is readily evaluated.

The measured oxygen concentration profiles along the height of the reactor were compared with the predicted values. The parameters used in the model analysis are those of Bhamidimarri (4). Typical results shown in Figures 1 prove the validity of the methodology adopted for the development of the models.

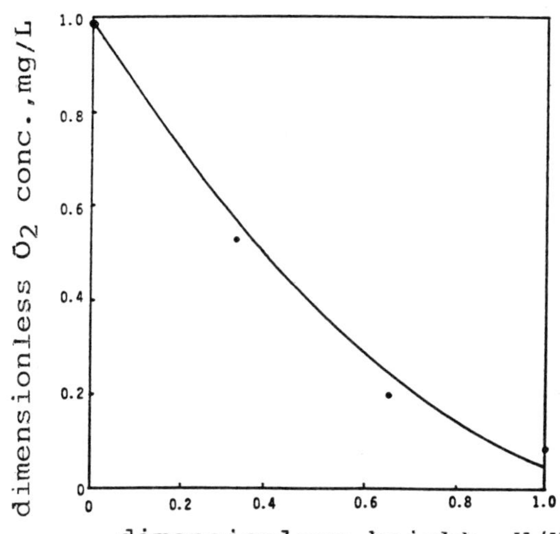

Figure1. Oxygen concentration profile in the reactor for r_c=0.14mm, biofilm thickness=0.15mm, O_{bi}=28mg/L, o=observed, line=predicted.

NOMENCLATURE

K_s	=	saturation constant, mg L^{-1}
r	=	particle radius, cm
r_{bp}	=	radius of particle with biofilm, cm
r_c	=	support particle radius, cm
S_b	=	substrate concentration, g cm^{-3}
S_{bi}	=	influent bulk substrate concentration, g cm^{-3}
U_{si}	=	superficial liquid velocity, cm s^{-1}
Y_t	=	cell yield

REFERENCES

1. Jeris, J.S., Beer, C and Mueller, J.A. (1974). High rate biological denitrification using a granular fluidized-bed. *J. Water Poll. Control Fed.*, **46**, 2118-2128.
2. Jeris, J.S., Owens, R.W., Hickey, R. and Flood, F. (1977). Biological fluidized-bed treatment for BOD and nitrogen removal. *J. Water Poll. Control Fed.*, **49**, 816-831.
3. Mulcahy, L.T. (1978). Mathematical model of the fluidized-bed biofilm reactor. *PhD Thesis*, University of Massachussetts, Amherst, MA.
4. Rao Bhamidimarri, S.M. (1985). Biofilm characteristics and their role in a fluidized-bed bioreactor. *PhD Thesis*, University of Queensland, St Lucia, Qld.
5. Luthy, R.G. (1981). Treatment of coal coking and coal gasification wastewaters. *J. Water Poll. Control Fed.* **53**, 325-339.

SOLVENT EXTRACTION OF FLAXSEED

F. Shahidi, P.K.J.P.D. Wanasundara and R. Amarowicz
Department of Biochemistry, Memorial University of Newfoundland
St. John's, NF, Canada, A1B 3X9

ABSTRACT

Extraction of ground flaxseed with hexanes or a two-phase solvent system consisting of hexanes and a polar solvent mixture was carried out. The polar solvent was alkanol, 95% (v/v) alkanol or 95% (v/v) alkanol containing 10% (w/v) dissolved ammonia. The latter solvent extraction system using methanol, produced a meal with high content of proteins and also retained its amino acid compositional integrity. It also lowered cyanogenic glycosides and phenolic compounds of the seed. Changing the conditions of parameters of extraction with methanol-ammonia-water/hexane resulted in over 90% removal of the major antinutrient of flaxseed.

INTRODUCTION

Flax or linseed (*Linum usitatissimum* L.) is the third major oilseed crop in Canada. The oil of flax (linseed oil) is rich in polyunsaturated fatty acids especially α-linolenic acid and is used mainly as an industrial oil [1]. Flaxseed meal is currently used as an animal feed. Presence of antinutrients such as cyanogenic glycosides, phenolic compounds and anti-vitamin B_6 [2] has hindered its utilisation in food formulations [3]. The main objective of this study was to study the effect of alkanol-ammonia-water/hexane extraction on the removal of antinutrients of flaxseed meal.

MATERIALS AND METHODS

Flaxseed obtained from Omega Nutrition Company (Vancouver, BC) was extracted by a two-phase solvent extraction system consisting of alkanol-ammonia-water/hexanes [4]. The alcohols used were methanol, ethanol and isopropanol with or without ammonia (10%, w/w) and water (5%, v/v). The contents of crude proteins [5], cyanogenic glycosides [4], condensed tannins [6] and total phenolic acids [7] of the extracted meals were determined. The efficiency of methanol-ammonia-water/hexane extraction system was studied for the removal of cyanogenic glycosides by changing the extraction parameters; water 5, 10 and 15% (v/v) in methanol phase, 15 and 30 min quiescent period and 6.7 and 13.3 polar phase-to-seed ratio (R). Effect of a multistage extraction (2 and 3 times) was also investigated.

RESULTS AND DISCUSSION

The yields of oil, meal and solids following extraction with different solvent systems are provided in Table 1. Methanol-ammonia-water/hexane removed more polar components (solids) and enhanced protein content of the meals (15%). Amino acid profile indicated that flaxseed meal is high in glutamic acid but low in sulphur-containing amino acids. However, solvents used for extraction did not change the amino acid composition of the recovered meals.

TABLE 1
Mass balance of flaxseed as affected by extraction with different solvent systems

Solvent	Yield, on a dry weight basis (%)			Loss (%)
	Meal	Oil	Solids	
(A) Hexanes only	48.9±1.0	49.2±0.6	0.0	1.9
(B) Methanol/hexanes	46.7±1.0	45.9±1.9	5.2±0.1	2.2
(C) Methanol-water/hexanes	46.7±2.0	48.0±0.5	4.8±0.1	0.5
(D) Methanol-ammonia/hexanes	47.6±2.0	46.6±0.8	4.8±0.7	1.0
(E) Methanol-ammonia-water/hexanes	46.4±2.0	47.1±0.8	5.7±0.1	0.8
(F) Ethanol-ammonia-water/hexanes	48.1±1.0	46.8±1.8	4.2±0.1	0.9
(G) Isopropanol-ammonia-water/hexanes	50.0±3.0	48.8±0.5	no phase separation	1.2

Figures 1 and 2 show the percentage removal of anti-nutrients of flaxseed meal. The content of condensed tannins of flaxseed was 136±13 mg/100g, as (+)-catechin equivalents. The solvent extraction systems employed were able to remove 24-74% (Figure 1) of condensed tannins present in the meal. The total content of phenolic acids of flaxseed meal was about 220 ± 13 mg/100g as ferulic acid equivalents, and 12-48% (Figure 1) of these were removed depending on the solvent system employed.

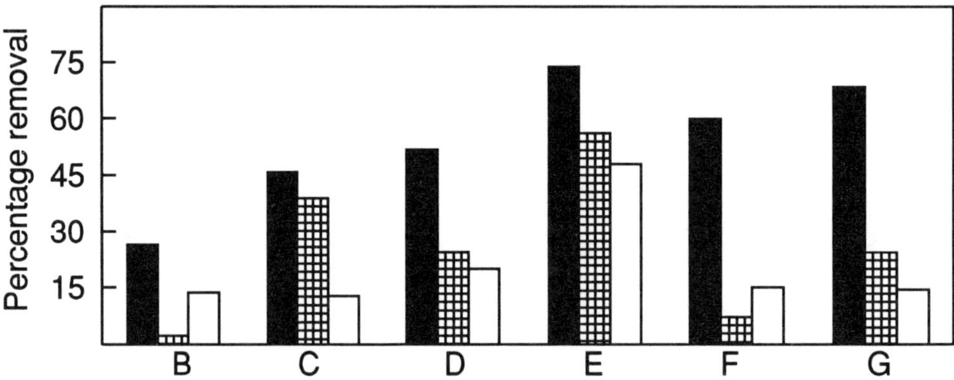

Figure 1. Percentage removal of condensed tannins, ■ ; total cyanogenic glycosides, ▦; and total phenolic acids, □ ; of flaxseed meal by different solvent extractions (B-G refers to combinations in Table 1).

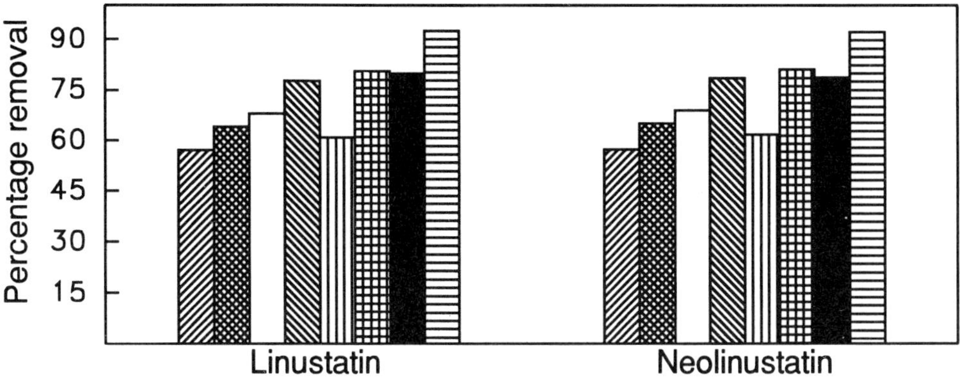

Figure 2. Percentage removal of linustatin and neolinustatin under different extraction conditions for 15 min at R=6.7 with 10% (w/w) ammonia in methanol containing 5% water, ; 10% water, ; 15% water, ; 5% water and R=13.3, ; 5% water and 30 min, ; 5% water, 30 min, R=13.3, ; 5% water, twice, ; and 5% water, three times, .

The predominant cyanogenic glycosides in flaxseed meal were linustatin 4.42 ± 0.08 mg/g) and neolinustatin (1.90 ± 0.03 mg/g). The content of both these glycosides was reduced by 57% after one extraction with methanol-ammonia-water/hexane. Increasing the water content in the polar phase removed more cyanogenic glycosides (Figure 2) but gave a sticky meal. Doubling the volume of the extraction solvent and duration of the extraction period resulted in the removal of over 80% of cyanogenic glycosides. A three stage extraction gave a meal with 90% of its cyanogenic glycosides removed. The efficiency of the removal of condensed tannins, cyanogenic glycosides and total phenolic acids by the solvent mixtures was in the order of methanol-ammonia-water >> isopropanol-ammonia-water > ethanol-ammonia-water. Removal of the antinutrients of flaxseed meal provides a mean for better utilisation of meals in feed and possibly for human food formulation.

REFERENCES

1. Anderson, L., Composition of linseed oil and its relation to usage present and future. Flax Inst. US Proc., 1971, **41**, 16-20.

2. Klosterman, H.J., Lamoureux, G.L. and Parsons, J.I., Isolation, characterisation and synthesis of linatine, a vitamin B_6 antagonist from flaxseed. Biochem., 1967, **6**, 170-77.

3. Singh, N., Linseed as a protein source. Indian Food Packer, 1979, **33**, 54-7.

4. Wanasundara, P.K.J.P.D., Amarowicz, R., Kara, M.T. and Shahidi, F., Removal of cyanogenic glycosides of flaxseed meal. Food Chem., 1993, In press.

5. AOAC, Official Methods of Analysis. 15th ed. Association of Official Analytical Chemists, Arlington, Virginia, 1990.

6. Naczk, M. and Shahidi, F., The effect of methanol-ammonia-water treatment on content of phenolic acids of canola. Food Chem., 1989, **31**, 159-64.

7. Krygier, K., Sosulski, F.W. and Hogge, L., Free, esterified and insoluble bound phenolic acids 2. Composition of phenolic acids in rapeseed flour and hulls. J. Agric. Food Chem. 1982, **30**, 334-36.

LARGE SCALE PREPARATION OF HIGHLY-PURIFIED EGG YOLK PHOSPHATIDYLCHOLINE BY HPLC

HIDEAKI KOBAYASHI[1], HITOSHI NARABE[1], MINEO HASEGAWA[1], FUMIYA HARADA[2], KAHOKO NOMURA[2], KEISHI KITAGAWA[2]
[1]Research Institute of Q.P.Corp., 5-13-1,Sumiyoshi-cho, Fuchu-shi, Tokyo 183, Japan, [2] Kumiyama Laboratory of YMC Co.,LTD., 249, Morimurahigashi, Kumiyama-cho, Kuse-gun, Kyoto 613, Japan

ABSTRACT

We studied to develop the system which refines egg phosphatidylcholine (PC) for pharmaceutical and food use. HPLC was applied with silicagel particles(YMC-GEL S-50) as the column packing and ethanol/water(92:8) as the safety mobile phase. Six hundred grams of egg lecithin PL-100H (Q.P. Corp.) were separated into PC and phosphatidylethanolamine(PE) at a time. The purity of obtained PC was about 100%. And we could easily recover the mobile phase by distillation and use it repeatedly.

Then we examined oxidative stability. After 35 days in the presence of the air at 20 ℃, the POV of purified PC was less than 2meq/kg.

INTRODUCTION

Phospholipids such as phosphatidylcholine(PC) have the tendency to form molecular bilayers called liposome. And phospholipids also have some nutritional functions such as the serum cholesterol reducing effect and so on. In recent years, purified PC was needed for the drug delivery system in the pharmaceutical field and for a functional food use. So we tried to develop the system which refines PC by HPLC with safety organic solvents.

MATERIALS AND METHODS

Materials

Egg yolk lecithin was obtained from Q.P.Corp.(PL-100H). It contains about 80% PC, 15% phosphatidylethanolamine(PE) and a slight amount of cholesterol(CH), sphingomyelin(SPM) and lyso-phosphatidylcholine(LPC). A commercial purified egg PC for pharmaceutical use was purchased from NOF Corp.(NC-10S). This purity was more than 98%.

Large scale preparation of egg PC
Six hundred grams of egg yolk lecithin PL-100H were dissolved in ethanol and purified by preparative HPLC (PLC-200P made by YMC Co.,LTD.)[YMC-GEL S-50(1000×100mm I.D.) as the column packing, ethanol/water(92:8) as the mobile phase, flow rate= 600ml/min, monitoring by UV215nm : This system is explosion-proof and the maximum flux is about 3l/min.]

Oxidative stability of egg PC
The oxidative stability of PC was evaluated as follows; The samples were kept in glass bottles in the presence of the air with no light at 4 ℃ and 20 ℃. After 0,7,14,and 35 days, the peroxide-value(POV) was measured.

RESULTS AND DISCUSSION

Large scale preparation of egg PC
Figure 1 shows the HPLC chromatogram of PL-100H in preparative-scale. By using preparative column(1000 ×100mm I.D.), 600g of egg yolk lecithin PL-100H were separated into PC and PE at a time. The fraction of the second half of PC peak(2.2h-5.4h) was evaporated to dryness. Then 375g of purified egg PC was obtained(yield: 62.5%) and the purity was 99.9% by Iatroscan method. More than 98% of the mobile phase were recovered by evaporation. After the ratio of ethanol/water were adjusted, it was able to be used as the mobile phase repeatedly.

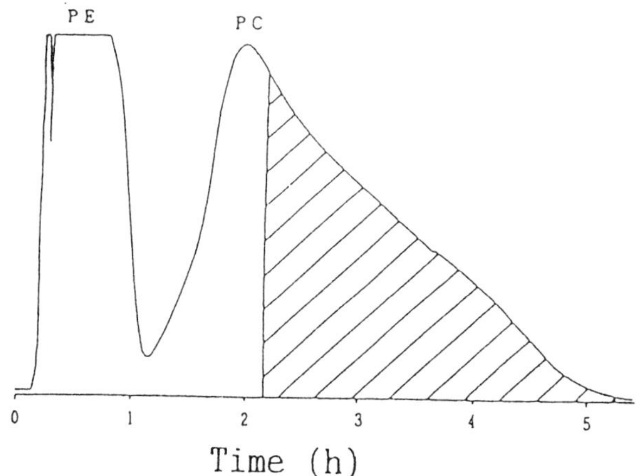

Figure 1. HPLC chromatogram of PL-100H in preparative-scale

Molecular species of egg PC
There was not much difference between fatty acid composition of HPLC-purified egg PC and that of a control (data not shown, a control egg PC; separated by TLC [chloroform/methanol/water(65:25:4)] from PL-100H, and the PC was extracted by chloroform/methanol(2:1) from TLC plate and evaporated to dryness).

Table 1 shows the molecular species of HPLC-purified egg PC. The molecular species of PC was analyzed by HPLC using the method of Smith,M. et al. [1] There were some differences between HPLC-purified egg PC and a control.

Table 1
Molecular species of HPLC-purified egg PC

Molecular species C1-position	C2-position	HPLC-purified egg PC (%)	Control egg PC (%)
C16-0 /	C22-6	3.9	13.6
C16-0 /	C20-4	2.2	5.8
C16-0 /	C18-2	22.5	24.3
C16-0 /	C18-1	44.9	35.9
C18-0 /	C20-4	2.2	3.9
C18-0 /	C18-2	11.2	6.8
C18-0 /	C18-1	11.2	5.8
others		1.9	3.9

Oxidative stability of egg PC

Table 2 shows changes of POV of HPLC-purified egg PC in the presence of the air. The content of unsaturated fatty acids of egg PC is about 50% (data not shown). So it is easy for egg PC to be oxidized.

But the POV of HPLC-purified egg PC did not rise after 35 days at 20℃. And oxidative stability of HPLC-purified egg PC was significantly higher than that of a control(a commercial egg PC). This result could not be explained on the fatty acid composition of PC. So we consider that the purity and molecular species of PC may have some effects on the oxidative stability.

Table 2
Changes of POV of HPLC-purified egg PC in the presence of the air

days	HPLC-purified egg PC (meq/kg)		Control[a] (meq/kg)
	at 4℃	at 20℃	at 20 ℃
0	0	0	0
7	0	0	0
14	0	0	2
35	0	0	76

a = a commercial egg PC

REFERENCES

1. Smith,M. and Jungalwala,F.B., J.lipid Res.,1981, 22, 697-704

APPENDIX

This work has been done in part as one of the R&D subjects of 'The Japanese Research and Development Association for the High Separation System in Food Industry'.(1988-1992)

EXTRACTION AND CONCENTRATION OF OMEGA-3 FATTY ACIDS OF SEAL BLUBBER

F. Shahidi,[1] R.A. Amarowicz,[1] J. Synowiecki[1] and M. Naczk[1,2]
[1]Department of Biochemistry, Memorial University of Newfoundland
St. John's, Newfoundland, Canada, A1B 3X9
and
[2]Department of Nutrition and Consumer Studies, St. Francis Xavier University
Antigonish, Nova Scotia, Canada, B2G 1C0

ABSTRACT

Seal blubber lipids consist of approximately 99% neutral triacylglycerols and 1% polar components. Blubber lipids contained 24% long-chain omega-3 fatty acids consisting of approximately 9% eicosapentaenoic acid (EPA), 5% decosapentaenoic acid (DPA), and 10% docosahexaenoic acid (DHA). The content of DPA was higher in seal blubber lipids as compared with other sources of marine oil, except mammalian species. Intramuscular lipids of seal meat were also rich in their content of DPA and other omega-3 fatty acids. Omega-3 concentrated blubber lipids containing 76% long-chain omega-3 fatty acids were prepared by extraction/precipitation of saturated/monounsaturated lipid fatty acids. Stability of seal blubber lipids was better than that of fish oils.

INTRODUCTION

Marine products play an important role in nutrition and health status of humans. Polyunsaturated fatty acids have long been recognized as desirable dietary components. Presence of long-chain omega-3 fatty acids namely eicosapentaenoic acid (EPA), decosapentaenoic acid (DPA) and docosahexaenoic acid (DHA) in seafoods is responsible for their health-pro properties. These fatty acids have been shown to be responsible for biochemical and physiological changes in the body. While DHA is essential for proper functioning of the eye and may have a structural role in the brain, EPA serves as a precursor to prostaglandins and thromboxanes; DPA in an intermediary species between EPA and DHA. Therefore, omega-3 fatty acids have beneficial effects in the prevention or possible treatment of coronary heart diseases, cancer, diabetes, high blood pressure, and autoimmune diseases [1]. Seal blubber lipid may have the added advantage of preferential absorption by the body since its long-chain omega-3 fatty acids are present mainly in 1- and 3-positions of the triacylglycerol molecule. The omega-3 fatty acids of fish oils are known to be randomly distributed and being present more in the 2-position of the triacylglycerols.

The present study was undertaken to examine the fatty acid composition of blubber of harp seal (*Phoca groenlandica*). Effects of urea complexation on enrichment of omega-3 fatty acids in seal blubber lipids were also examined.

MATERIALS AND METHODS

Seal oil was prepared from seal blubber by denaturation and precipitation of protein residues, their subsequent filtration and repeated washing of the oil with water, possibly in the presence of antioxidant/sequestrant combinations. Enrichment of omega-3 fatty acids was achieved by a urea complexation procedure which involved hydrolysis, complexation, filtration and esterification steps [2]. Fatty acid methyl esters in preparations were quantified by transmethylation of lipids and were separated on a 30 mm x 25 mm i.d. fused silica capillary column using a Perkin Elmer 8500 gas chromatograph as described elsewhere [3].

RESULTS AND DISCUSSION

Blubber from six animals were used in these studies. Proximate composition of blubber from these animals is summarized in Table 1. The content of neutral lipids accounted for 98.89 ± 0.16% of the total lipids. The proportion of long-chain omega-3 fatty acids in blubber was in the decreasing order of DHA > EPA > DPA. Presence of DPA, an intermediary species between DHA and EPA, in relatively large amounts may have added beneficial effects in human health; studies for better understanding of its role in human diet is on-going in our laboratory. It is of interest to note that DPA is also found in relative abundance in human milk [4].

The omega-3 fatty acid esters prepared by urea complexation of hydrolysed blubber lipids had dominant quantities of EPA, DPA and DHA (see Table 2). However, depending on process conditions, the enrichment of specific fatty acids varied significantly

TABLE 1

Proximate Composition of Seal Blubber.

Component	Content, %
Moisture	2.3 ± 0.1
Protein (N x 6.25)	1.3 ± 0.1
Lipid	95.5 ± 0.7
Minerals	0.8 ± 0.1

TABLE 2
Major Fatty Acids of Seal Blubber and its Concentrates.

Fatty Acid	Raw Blubber	Omega-3 Concentrates	
		Preparation A	Preparation B
14:0	5.0	0.7	0.8
16:0	9.5	—	—
16:1	13.1	1.4	1.3
18:1	19.1	—	—
20:1	11.9	—	—
20:5	9.1	27.7	24.6
22:1	6.5	—	—
22:5	5.0	9.3	4.8
22:6	10.0	38.9	46.1

(Table 2). Meanwhile the omega-3 enriched fraction retained most of the cholesterol present in the original oil. Furthermore, oxidative stability of the original seal oil, as determined by TOTOX values, was better than fish oils, perhaps due to the presence of natural antioxidants in the former.

ACKNOWLEDGEMENT

We are grateful to the Newfoundland Department of Fisheries, Atlantic Canada Opportunities Agency, Canadian Sealers Association and Canadian Centre for Fisheries Innovation for facilitating or providing financial support.

REFERENCES

1. Gordon, D.T. and Ratiff, V., The implications of omega-3 fatty acids in human health. In Advances in Seafood Biochemistry: Composition and Quality, eds. G.J. Flick, Jr. and R.E. Martin, Technomic Publishing Company, Inc., Lancaster and Basel, 1992.

2. Haagsma, N., van Gent, C.M., Luten, J.B., de Jong, R.W. and van Doorn, E., Preparation of an ω3 fatty acid concentrate from cod liver oil. J. Amer. Oil Chem. Soc., 1982, **59**, 117-118.

3. Shahidi, F. and Synowiecki, J., Cholesterol and lipid fatty acid composition of processed seal meat. Can. Inst. Food Sci. Technol. J., 1991, **24**, 269-272.

4. Koletzko, B., Thiel, I. and Springer, S., Lipids in human milk: a model for infant formulae? European J. Clin. Nutr., 1992, **46**, 545-555.

SEPARATION OF EPA AND DHA FROM FISH OIL USING SUPERCRITICAL EXTRACTION WITH Ag COMPLEX PRETREATMENT

KUNIO NAGAHAMA, TATSURU SUZUKI, SATOSHI KIKUCHI, YOSHIHIRA TANAKA, KAORU NAKANO, HIDETAKA NORITOMI and SATORU KATO
Dept. of Ind. Chem., Tokyo Metropolitan University,
1-1, Minamiohsawa, Hachioji, Tokyo 192-03, JAPAN

ABSTRACT

Supercritical fluid extraction (SFE) of polyunsaturated fatty acid (PUFA) ethyl esters coupled with pretreatment of Ag^+ – π–complex formation was studied. PUFA esters such as eicosapentaenoic acid (EPA) and docosahexaenoic acid (DHA) in fish oil were selectively dissolved into aqueous $AgNO_3$ solution. After that semi batch SFE was carried out to recover PUFA esters from the pretreated aqueous solution using supercritical (SC) CO_2, ethane and ethylene. All the PUFA esters could be recovered by SFE, and extracted products were consist of PUFA esters and Ag free water only. The results of SFE showed that the complex stability affected strongly on the SFE behavior. Selective separation for EPA–Et and DHA–Et was obtained when SC–CO_2 and ethane were used, and 90 wt.% of DHA–Et could be produced at the final SFE stage. SC–ethylene has little selectivity for the PUFA esters, however, the extraction rates of PUFA esters were the fastest. This is because ethylene makes π–complex with Ag^+ ion competitively.

INTRODUCTION

Fish oils have attracted wide commercial and academic interests as a rich source of PUFAs, particularly EPA and DHA, which are reported to possess beneficial physiological activities. However, the isolation of PUFAs from fish oil is very difficult because it is a complicated mixture of saturated and unsaturated fatty acids.

To develop a efficient method for the isolation of PUFAs, SFE coupled with pretreatment of Ag^+ – complex was studied. PUFA esters originating from sardine oil were preconcentrated in an aqueous $AgNO_3$ solution and it was extracted using supercritical CO_2 [1], ethane and ethylene. The feasibility of SFE to isolate EPA–Et and DHA–Et form aqueous Ag^+ solution was discussed.

EXPERIMENTAL SECTION

Experimental apparatus and procedures have been described in detail elsewhere [1]. Briefly sardine oil ethyl ester (10 mL) was contacted with 5 M of aqueous $AgNO_3$ solution (20 mL) for 12 hours

TABLE 1 Composition of fatty acid ethyl esters [wt.%]

	Saturated fatty acids	C18:4	EPA (C20:5)	DHA (C22:6)	Unsaturated fatty acids
Sardine oil (feed)	31.2	2.4	19.5	8.4	34.2
Aq. $AgNO_3$ phase* (5M)	0.0	4.7	55.8	25.8	11.1
Residue in oil phase	46.6	1.4	3.0	0.3	39.6

* water and $AgNO_3$ free basis

at 298 K. Then the aqueous phase (5 mL) was separated and used for feed solution as SFE feed stock. Only PUFA esters having more than 4 double bonds in the oil are soluble in aqueous $AgNO_3$ solution and almost all PUFA esters dissolve in 5 M the solution as listed in Table 1. Semi batch SFE was carried out with a flow-type system. Extracted samples were fractionated every 5 - 120 minute intervals and analyzed both composition and weight. We investigated the presence of Ag in the extracted samples by an atomic flame emission technique, and Ag was not detected (less than 0.5 mg/L). This means that PUFA ester in the aqueous $AgNO_3$ solution was extracted not as Ag complex form but as free PUFA ester.

RESULTS AND DISCUSSION

A typical extraction curve is presented in Fig. 1. Efficient extraction of the PUFA esters occurs at the early stage of the extraction, thereafter the rate of extraction decreases markedly, while water was extracted at a constant rate throughout the extraction. Similar curves were obtained for all experiments. A wide range of operating temperatures and pressures were examined for the CO_2 SFE, and we obtained as the results that extraction pressure did not have significant effect on the extraction rate of the PUFA esters, whereas higher temperature resulted in higher extraction rate [1]. The results showed that the complex stability affected strongly in the SFE behavior.

The extraction yield curves of the PUFA esters for three supercritical fluid, CO_2, ethane and ethylene under the same pressure (24.5 MPa) and temperature (313 K) were plotted in Fig. 2,

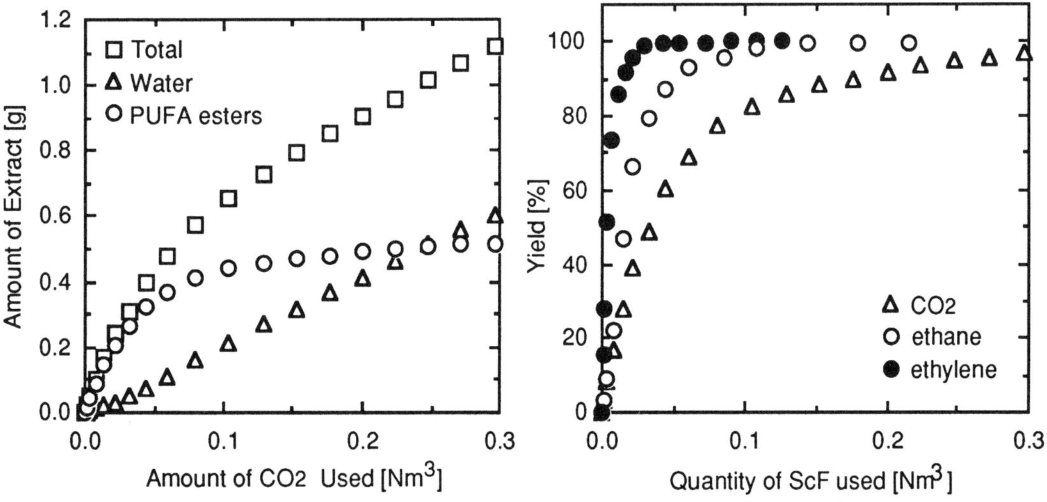

Fig. 1 CO2 extraction curve at 313K, 24.5MPa

Fig. 2 Extraction yield curve of PUFA esters at 313 K and 24.5 MPa

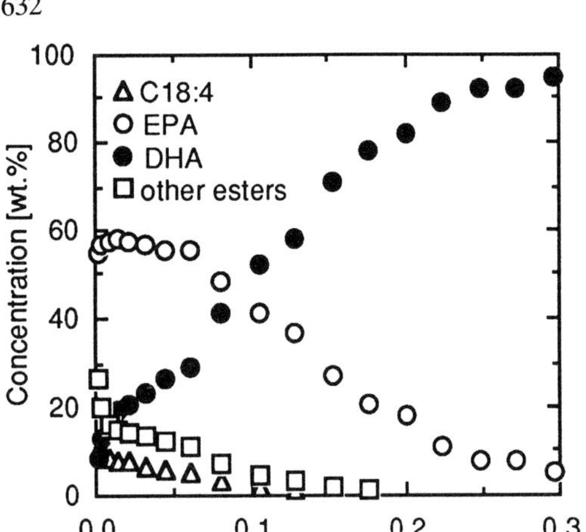

Fig. 3 Concentration changes of PUFA esters for CO2 extraction at 313 K and 24.5 MPa

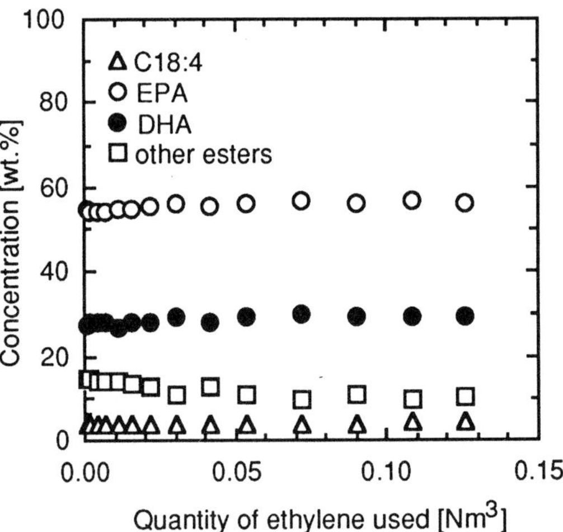

Fig. 4 Concentration changes for ethylene extraction at 313 K and 24.5 MPa

where yield was defined as the weight of total extracted PUFA esters divided by that of the esters in feed aqueous solution. The extraction rate of PUFA esters was the order of CO_2 < ethane < ethylene.

The concentration changes of extracted PUFA esters for a SC–CO_2 extraction are shown in Fig. 3, where the concentration was calculated on water and $AgNO_3$ free basis. Selective separation for EPA and DHA esters was obtained when SC–CO_2 and SC–ethane were used, and more than 90 wt.% of DHA ester could be produced at the final SFE stage. On the other hand, SC–ethylene has little selectivity for the PUFA esters as shown in Fig. 4. It is suggested that free PUFA esters in the aqueous phase are increased because ethylene makes π–complex with Ag^+ ion competitively.

CONCLUSIONS

SFE of PUFA ethyl esters coupled with pretreatment of aqueous $AgNO_3$ solution was studied. PUFA esters in fish oil were selectively dissolved into Ag^+ aqueous solution. After that semi batch SFE was carried out to recover PUFA esters from pretreated aqueous solution using SC–CO_2, ethane and ethylene. All the PUFA esters in the aqueous solution could be recovered by SFE, and extracted products were consist of PUFA esters and Ag free water only. The results of SFE showed that the Ag^+–complex stability affected strongly on the SFE behavior. Selective separation for EPA and DHA esters was obtained when SC–CO_2 and ethane were used, and more than 90 wt.% of DHA ester could be produced at the final SFE stage. SC–ethylene has little selectivity for PUFAs, however, the extraction rates of PUFA esters were the fastest. This is because ethylene makes π–complex with Ag^+ ion competitively.

REFERENCE

1. Suzuki T., Kikuchi S., Nakano K., Kato S., Nagahama K., Supercritical fluid extraction of polyunsaturated fatty acid ethyl esters from aqueous silver nitrate solution, Bioseparation, 1993, in press

SELECTIVE EXTRACTION OF PROTEINS FROM COMPLEX SOLUTIONS BY REVERSE MICELLES

Kazumitsu NAOE[1] , Masanao IMAI, Masaru SHIMIZU
Department of Chemical Engineering,
Division of Chemical & Biological Science,
Tokyo University of Agriculture & Technology,
24-16 2-Chome Nakamachi Koganei-C. Tokyo, 184 JAPAN

ABSTRACT

Desirable operating conditions for selective extraction of lysozyme from protein mixture were found to be around minimal AOT concentration for dilute feed system ($<10^{-4}$M for lysozyme). Efficient selective extractions of lysozyme from protein mixtures were successfully carried out by reverse micelles.

INTRODUCTION

Reverse micellar extraction process is expected as a novel protein separation technique for its ability of large-scale, continuous processing. In practical protein separation, the target protein is separated from multicomponent system. The labile nature of protein must be considered in process design and development of protein separation technique. Therefore, recovery conditions to get high percent extraction and activity in complex system are important for its industrial applications. In this study, the efficient operating conditions for selective extraction of protein from protein mixture were investigated.

MATERIALS AND METHODS

In this study, AOT(di-2-ethylhexyl sodium sulfosuccinate)/iso-octane reverse micellar system was used. Lysozyme, cytochrome *c*, and ribonuclease A were used as a model protein. The procedures of extraction experiment and enzymatic activity measurement were described in previous paper[1]. The protein concentration was determined by UV absorbance. The water content in the organic phase was measured by Karl-Fischer titration. Analytical separation and quantification of protein mixture in the aqueous solution was made by HPLC. All experiments

---

[1] Present address: Department of Chemical Engineering, Nara National College of Technology, Yata, Yamato-Koriyama, Nara, 639-11 JAPAN

were carried out at 298K(25°C).

RESULTS AND DISCUSSION

Single Protein Component in Reverse Micellar Phase

Suitable AOT concentration in the organic phase to achieve high recovery amount and activity of protein: The forward and backward extraction of lysozyme were examined at the various AOT concentrations. Minimal AOT concentration[2] of lysozyme was found for each lysozyme feed aqueous concentration.

The relative specific activity (RSA) of lysozyme was measured at $5x10^{-4}$M lysozyme in the initial aqueous phase. The RSA is the ratio of specific activity of recovered lysozyme to that of feed lysozyme[1]. In the higher AOT concentration range (>ca.100mM), the RSA of lysozyme decreased. The extraction at lower AOT concentrations is desirable to obtain higher RSA. The recovered lysozyme activity was preserved at the minimal AOT concentration.

The recovery of activity(= the product of RSA and percent total extraction) took a maximum value in this experimental range of AOT concentration. Authors propose to call this AOT concentration "the optimal AOT concentration". The optimal AOT concentration in this system was around the minimal AOT concentration.

Suitable feed protein concentration to get high recovery of activity: The protein concentration of feed solution is a significant factor to get a high recovery of activity in practical extraction, as the concentration of target protein determines the minimal AOT concentration accordingly. The backward extractions of lysozyme were carried out under the minimal AOT concentrations. In less than $1.0x10^{-4}$M of lysozyme concentration in the reverse micellar organic phase, the percent backward extraction increased and consequently the recovery of activity apparently increased up to almost 100%. The recovery of activity was dominated mainly by the percent backward extraction in this system. As a result, the reverse micellar extraction processes can be effectively applied to a dilute feed system.

Binary Protein Component in Reverse Micellar Phase

Minimal AOT concentration in binary protein component system: The effect of AOT concentration on the simultaneous forward extraction of cytochrome *c* and lysozyme into the reverse micellar phase was investigated. There was a minimum of initial AOT concentration for complete forward extraction of the protein mixture. As to the solubilization into reverse micellar phase, the mixture system behaved like single component system. Hence, this AOT concentration is recognized as a minimal AOT concentration for the protein mixture of lysozyme and cytochrome *c*.

Effect of protein composition in micellar phase: The recovery of lysozyme from the reverse micellar phase containing binary protein mixture (lysozyme & cytochrome *c*) was performed under the minimal AOT concentrations varying the protein composition of mixture. The percent total extraction of lysozyme was dependent only on the lysozyme concentration in the micellar phase, not on the total protein concentration.

Selective Extraction of Lysozyme from Protein Mixture
On the basis of findings in above sections, the extraction of lysozyme from protein mixtures (mixture 1: lysozyme & ribonuclease A, and mixture 2: lysozyme & cytochrome *c*) was performed under the condition of minimal AOT concentrations. In the both systems, the complete recovery of lysozyme was obtained(Figure 1) and the activity of the recovered lysozyme was well preserved.

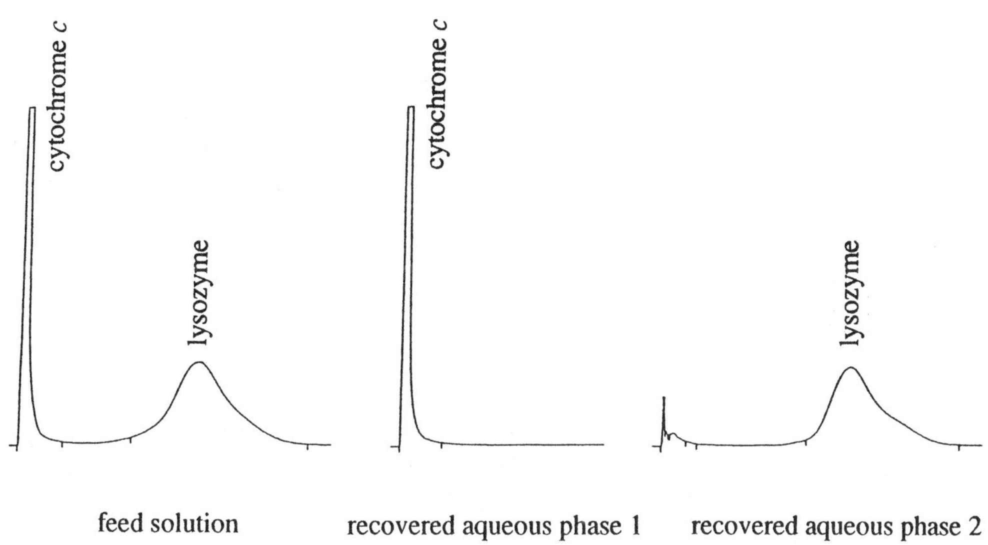

Figure 1. HPLC analysis of aqueous solutions obtained in the separation of a binary protein mixture (lysozyme & cytochrome *c*) under minimal AOT condition.

CONCLUSIONS

The desirable operating conditions for selective extraction of proteins using reverse micelles were investigated. Around minimal AOT concentration and dilute feed system (target protein conc. $<10^{-4}$M for lysozyme) were efficient conditions for high recovery of activity. On the basis of these findings, the efficient selective extractions of lysozyme from protein mixtures were successfully carried out and the activity of lysozyme was well preserved.

REFERENCES

1. Naoe, K., M. Imai, and M. Shimizu, In: "Biochemical Engineering for 2001", S. Furusaki, I.Endo, R.Matsuno(Eds.), Pp.584-586, Springer-Verlag(1992)
2. Ichikawa, S., M. Imai, and M. Shimizu, Biotechnol. & Bioeng., 39, 20-26(1992)

FRACTIONATION OF β-LACTOGLOBULIN IN BOVINE WHEY BY INTERMITTENT UP FLOW SYSTEM WITH MULTI-STAGE ION EXCHANGE COLUMN

HIDEO OHTOMO AND TAMOTSU KUWATA
Meiji Milk Products Co., Ltd., Central Research Institute,
1-21-3, Sakae-cho, Higashimurayama-shi, Tokyo 189, Japan
ICHIRO KURIHARA, TAKASHI KIKUCHI, AND SABURO FURUSHO
Nippon Rensui Co., Ltd., Laboratory, 1000, Kamoshida, Midori-ku,
Yokohama-shi, Kanagawa 227, Japan

ABSTRACT

We have screened some separating gels for fractionation of bovine β-lactoglobulin (β-LG) in whey, and found that the most suitable ones are cellulose cation exchangers. Although they are easily compacted in the conventional single column (CSC) by scaling up, some problems led from gel compaction could be resolved with "Intermittent up flow system with multi-stage column (IUFS)".

INTRODUCTION

β-LG can be fractionated by cellulose cation exchanger (1). However, compaction of the gel bed is often caused when the CSC packed with such a soft gel is scaled up. Consequently, decrease in the flow rate, large pressure drop, and channeling are observed. To resolve these problems, we have developed IUFS.

MATERIALS AND METHODS

Electrodialyzed (ED) Gouda cheese whey or reconstituted ED whey was used as starting materials. In the laboratory trials, 5- or 10-stage acrylic column (50mm $\phi \times$ 200mm) was used. For the pilot scale trials, 5-stage polyvinyl chloride column (400mm $\phi \times$ 200mm) packed with 50 ℓ of separating gel was installed. Sample was fed into the multi-stage column with down flow, and the aliquot of non-adsorbed fraction, i.e. whey passed through the column was stored in a tank. To loose the pressing gel bed, stored whey was fed into the column with up flow at intervals just before the gel bed was compacted. β-LG adsorbed on the gel was eluted with NaCl and/or NaOH. The eluent was fed into the column with circulative down flow.

RESULTS

Separating gel

Among several separating gels tested, Indion S3 (Life Technology Inc.) was one of the most suitable gels. Pressure drop of the CSC packed with Indion S3 was lower than that with CM-Cellulofine CH (Chisso Co., Ltd.) which has larger adsorption capacity of β-LG than Indion S3.

Laboratory scale IUFS

Several testing conditions, ending points of chromatographies, and the pressure drop of the column were shown in Table 1. Inner pressure of IUFS was lower and more stable than that of CSC.

Pilot scale IUFS

Although preferential β-LG adsorption on Indion S3 was not observed by using the pilot scale CSC, about 90% of β-LG was trapped by IUFS without the compaction of the gel bed. The amount of adsorbed β-LG was equal to that of laboratory trials (78g/ℓ·gel). The eluate was concentrated and demineralized by ultrafiltration and diafiltration. Freeze dried β-LG isolate showed excellent solubility and gelation properties. β-LG accounted for 84% of total protein of the isolate (Table 2).

TABLE 1
Testing conditions, ending points of chromatographies, and the pressure drop of laboratory scale IUFS

	Test 1	Test 2	Test 3	Test 4	CSC*4
Number of stage	10	5	5	5	1
Flow rate(m/h)					
Down	0.55	0.55	0.61	0.49	0.49
Up	0.55	0.55	0.61	-	-
Feed time(min)					
Down	56	56	54	cont*2	cont
Up	4	4	6	-	-
Average space velocity(1/h)	1.19	1.19	1.22	1.22	1.22
Ending point(BV)*1	56.7	54.0	58.3	25.0*3	50.0
Pressure drop (kgf/cm^2)	0.03 ~0.07	0.06 ~0.09	0.09 ~0.13	0.1 ~2.0<	0.10 ~0.30

*1:Sample feeding was stopped when β-LG concentration in non-adsorbed fraction reached to 0.4g/ℓ (β-LG concentration in starting materials was 2.7g/ℓ). (Volume of non-adsorbed fraction)/(gel volume) was shown as "BV" unit.
*2:Continuous feed with down flow.
*3:Pressure drop exceeded $2kgf/cm^2$ at 25BV unit.
*4:Conventional single column.

DISCUSSION

To elute β-LG in short time, it was effective to use higher concentration of NaCl and/or NaOH as the eluent. However, resultant β-LG was partially denaturated, therefore the functional properties were deteriorated. On the other hand, the lower concentrations of eluent elongated the elution time. Thus, to prevent protein denaturation and to shorten elution time, we used 2 step elution. To give buffering ability to the eluate, small amount of NaCl was fed into the column at first to elute β-LG to some degree. At next step, NaOH was added gradually to raise the pH of circulating eluate to 9 and to keep it constant. It is very important to consider the interval of up flow and the concentration of the eluent at the first elution step to perform β-LG separation more efficiently.

CONCLUSIONS

When the experiments with pilot scale CSC packed with a soft gel were carried out, shrinkage and/or cracking of gel bed were observed. On the other hand, in the case of IUFS, these phenomena did not occur. Although such chromatographies as a size exclusion mode do not perform by disorder of gel bed in general, intermittent up flow did not affect β-LG separation by ion exchange mode.

TABLE 2

Protein compositions of β-LG isolates obtained by pilot scale IUFS

	Ig[*3]	BSA[*4]	α-LA[*5]	β-LG[*6]
Test 1[*1]	5.9	6.2	3.7	84.3
Test 2[*2]	5.8	5.6	4.2	84.5

*1:down flow;SV=1.48, 108min, up flow;SV=1.66, 6min

*2:down flow;SV=1.98, up flow;SV=1.66, 6min(Intermittent up flow was induced when the inner pressure of the column had reached to 0.6kgf/cm^2.)

*3:Immunoglobulins, *4:Bovine Serum Albumin

*5:α-Lactalbumin, *6:β-Lactoglobulin

ACKNOWLEDGEMENT

This study was made by the great help of The Japanese Research and Development Association for The High Separation System in Food Industry.

REFERENCE

1. Ohtomo, H., Hamamatsu, K., Hori, E., and Kuwata, T., Studies on the Elimination of β-lactoglobulin from Whey Using Carboxymethyl Cellulose Cation Exchanger. J. Jap. Soc. Food Sci. Technol., 1988, 35(11), 755-762.

RECOVERY OF LYSOZYME AND AVIDIN FROM EGG WHITE BY ION-EXCHANGE CHROMATOGRAPHY

SHUICHI YAMAMOTO, TOMOYUKI SUEHISA AND YUJI SANO
Department of Chemical Engineering,
Yamaguchi University, Tokiwadai, Ube 755, JAPAN

ABSTRACT

A method for determining the mobile phase composition in stepwise-elution chromatography is presented and tested for cation-exchange chromatographic separation of lysozyme and avidin from egg white. he distribution coefficient (K) as a function of the salt concentration (I) was determined from linear salt gradient elution experiments at a fixed pH. Based on the $K - I$ relationships, two purification schemes were designed and successfully carried out.

INTRODUCTION

Lysozyme and avidin are two basic proteins contained in egg white. They are important for therapeutic and diagnostic clinical applications. The process for recovering lysozyme and avidin from egg white must be such that the properties of the recovered egg white are comparable to the egg white untreated[1,2].

In this paper, a method for determining the mobile phase composition in stepwise-elution chromatography is presented and tested for cation-exchange chromatographic separation of the above-mentioned separation system.

RESULTS AND DISCUSSION

We have shown[3-7] that a large amount of important information can be extracted from linear salt gradient elution experiments. The distribution coefficient (K) as a function of the salt concentration (I) can be obtained from the peak salt concentration (I_R) - the normalized gradient slope (GH) relationships (**Figs.1** and **2**). Stepwise elution chromatography can be easily designed on the basis of the $K - I$ data [3,4,6,7]. Based on the $K - I$ data for ovalbumin, lysozyme and avidin at various pH values, two purification schemes were designed and carried out.

In the first scheme, the I of the elution buffer (I_E) of the fixed pH was changed (**Table 1**). In the second scheme, both the I_E and the pH were varied(**Table 2**). The productivity(recovered amount per column volume per process time) of the second scheme is much higher than that of the first scheme.

The performance of liquid chromatographic separation of proteins is usually governed by stationary (gel) phase diffusion, and the operating conditions such as flow-rate or residence time are adjusted so that a desired resolution can be achieved. However, the present study has shown that the productivity of chromatographic separations can be drastically increased by choosing suitable elution buffer conditions even when such parameters as flow-rate, column dimension and sample loading are not optimized.

Note: The details of the experimental procedure, the apparatus, the assay methods and the materials employed in this study are given in [3-6]. The purification factor(PF) is defined as the activity of the recovered fraction per protein content divided by that of the sample. The recovery ratio (Q_R) is the ratio of the total activity recovered to that applied to the column.

TABLE 1 Purification results at a fixed pH=8 [a]

Operation	V [mL]	time [min]	V/V_t [-]
sample application[b]	5.0	38	4.0
washing[c]	5.0	38	4.0
elution for avidin[d]	1.4	11	1.1
elution for lysozyme[e]	6.1	47	4.8
total	18	134	13.9

a)V=elution volume, V_t=column volume Column(0.9 cm diameter, 2 cm height), flow-rate=0.13 mL/min
b)Egg white was diluted to one-second of the original concentration with buffer A (10mM phosphate pH8) and used as a sample.
c)buffer A containing 0.03M NaCl
d)buffer A containing 0.22M NaCl Q_R=79%, PF=26
e)buffer A containing 0.50M NaCl Q_R=100%, PF=23

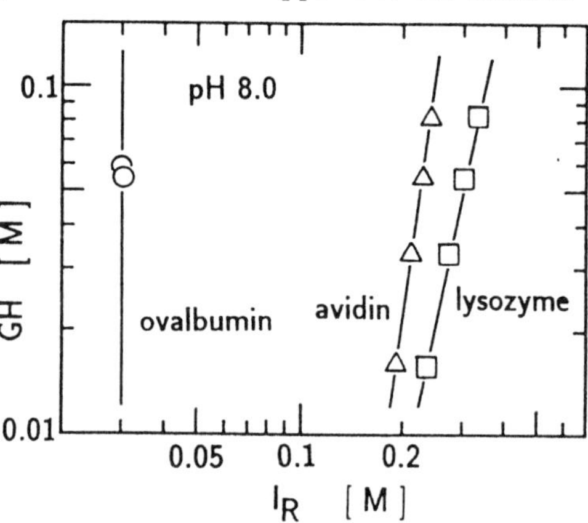

Fig.1 Relationships between GH and I_R. GH=slope of the salt gradient(M/mL) divided by the column gel volume, I_R= salt concentration at the peak retention time. These curves are not dependent on the flow-rate, the column size and the particle diameter[6,7].

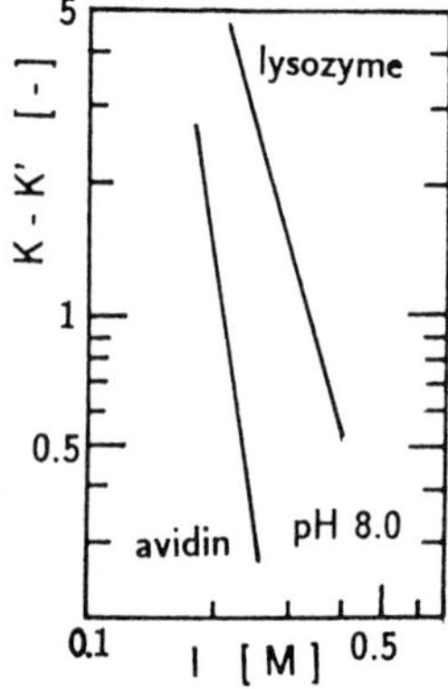

Fig.2 Relationships between $K - K'$ and I. The curves were determined from the $GH - I_R$ curves in Fig.1 according to the relation $dGH/dI = 1/[K(I) - K']$ where K' is the distribution coefficient of the salt (gradient substance or mobile phase modulator)[3,4,6,7].

TABLE 2 Purification results[a]

Operation	V [mL]	time [min]	V/V_t [-]
sample application[b]	3.8	14	3.0
washing[c]	3.2	12	2.5
elution for avidin[d]	3.2	11	2.5
elution for lysozyme[e]	5.9	22	4.6
total	16	60	12.6

a)Column(0.9 cm diameter, 2 cm height), flow-rate=0.27 mL/min
b)Egg white was used as a sample.
c)buffer A containing 0.03M NaCl
d)buffer B(50 mM carbonate buffer,pH 10) containing 0.05M NaCl Q_R=89%,PF=87
e)buffer B containing 0.50M NaCl Q_R=92%,PF=21

REFERENCES

1. Chan,E.Li, Nakai,S., Sim,J., Bragg, D.B. and Lo, K.V., *J.Food Sci.*, 1986, **51**,1032.
2. Ahvenainen,R., Heikonen, M., Kreula, M., Linko, M. and Linko, P., *Food Process Engng.*, 1980, **2**,301.
3. Yamamoto,S., Nomura,M. and Sano,Y.,*J.Chromatogr.*, 1990, **512**, 89.
4. Yamamoto,S., Nomura,M. and Sano,Y., *Chem.Eng.Sci.*, 1992, **47**,185.
5. Yamamoto,S., Suehisa,T. and Sano,Y., *Chem.Eng.Commun.*, 1993, **119**,221.
6. Yamamoto,S., Nomura,M. and Sano,Y., *AIChE J*, 1987, **33**,1426.
7. Yamamoto, S., Nakanishi, K. and Matsuno,R., *Ion Exchange Chromatography of Proteins*, Marcel Dekker, New York, 1988.

DEVELOPMENT AND APPLICATION OF CYCLODEXTRIN POLYMER

HISASHI OKEMOTO, HITOSHI HASHIMOTO
Ensuiko Sugar Refining Co., Ltd.
13-46, Daikoku-cho, Tsurumi-ku, Yokohama, Japan

ABSTRACT

The objective of this study is to establish a technology for purifying substances which are difficult to separate in food processing by utilizing the inclusion and/or catalytic function of cyclodextrin (CD). We examined methods to synthesize CD polymers by fixing CD to four kinds of organic matrixes ; acrylic polymer, agarose gel, vinyl polymer and chitosan beads. We applied these CD polymers to separating or purifing various substances from foods. Chitosan CD polymer was the most effective in our polymers.

INTRODUCTION

CDs are cyclic oligosaccharides composed of six, seven or eight glucopyranose units and called α-, β- and γ-CD, respectively. The internal cavity of the structure of CD is hydrophobic whereas the external surface is hydrophilic. The most characteristic or unique function of CDs is the formation of inclusion complexes by taking various substances into its hydrophobic cavity. This function can be reffered as one of lock and key recognition ability. This CD's ability can be used for separation of optical and structural isomers or substances which have resembling molecular structure or similar molecular weight.

Up to now, a considerable number of reports have been published on the manufacturing methods and properties of CD polymers, but due to high costs or to problems in physical strength they have not been actually utilized for industrial purposes. Accordingly, investigations have been done on industrial methods of manufacturing novel CD polymers to be used as separating and isolating materials in food industry.
The present studies were conducted in collaboration with Japan Organo Co., Ltd.

RESULTS AND DISCUSSION

Firstly, it was found that the fixation of CDs to acrylic macro-porous polymer, highly hydrophilic agarose gel and vinyl polymer by using our technique could not increase the amount of CDs to be fixed and the physical structures of products were weak.

Then chitosan beads were chosen as a new carrier for the fixation to be tested. Chitosan is a natural polysaccharide like agarose, has a relatively high physical strength and has been utilized as a carrier for immobilized enzymes. Suitable conditions of synthesizing CD polymer were studied by using chitosan beads.

The fixation of CD to chitosan beads was performed by using a spacer. Polyethylene glycol-di-glycidyl ether was employed as the spacer. After investigating for adequate reaction time for the introduction of spacer and for adequate spacer length, a maximum amount of fixed CD, about 110 μmol/g dried carrier, was obtained when a spacer of short chain length was used in a reaction for 5-6 hours.

An attempt to increase the amount of introduced spacer revealed that increase of introduced spacer did not necessarily increase the amount of fixed CD, and about 110 μmol/g dried carrier of fixed CD was maximum amount that could be obtained. This is thought to be due to that the structure of the carrier inhibits fixation of CD or that plural spacers combine with a molecule of CD, with the result of low efficiency of spacer utilization.

Then, for the purpose of examining the standard performance of chitosan CD polymer, inclusion tests were conducted by employing nitrophenol as a standard substance. The results showed that estimation of the amount of adsorbed nitrophenol in 0.01 M borate buffer (pH 9.18) at 25°C for 30 minutes of reaction time made evaluation of the standard inclusion capacity of the CD polymer possible. For the adsorption characteristics of o-, m- and p-nitrophenol to chitosan β-CD polymer, there was found a considerable difference between o- and m-nitrophenol and p-nitrophenol, and as the standard substance p-nitrophenol was found to be most suitable. Existence of the large difference in the adsorbed amount among isomers suggests that it is possible to isolate p-isomer from a solution of mixed isomers by means of adsorption and elution or chromatographic separation.

Then separation of nitrophenol isomers by using a column of chitosan β-CD polymer was conducted. Application of a mixed solution of o-, m- and p-nitrophenol to a column packed with chitosan β-CD polymer allowed o- and m-nitrophenol to be eluted in the earlier half of eluate while p-nitrophenol to be eluted in the later half; in this way p-nitrophenol could be isolated by being separated from other isomers. In comparison, a similar experiment was conducted by using a column packed with the chitosan beads, the starting material of making the CD-polymer, under the same conditions to find that all isomers were not separated at all, being eluted at an identical position.

In this way utilization of CD polymer in chromatographic separation was shown possible, and so its application to the field of food industry was studied.

CD polymer has adsorbing ability of naringin. This is thought to be due to that benzenediol that forms the terminal part of the naringin molecule is included by fixed CD. Then separation of naringin and hesperidin, which have benzenediol-like structures in common and have close molecular weights, by column chromatography was studied. Hesperidin is a compound contained in fruit juices and is responsible to juice turbidity like naringin. Chitosan β-CD polymer was packed into a column and a mixed solution of naringin and hesperidin was applied onto it to obtain a distinct separation pattern. For the sake of comparison , similar chromatographic experiments were done with a column of chitosan beads and a column of spacer-introduced chitosan beads to find no separation of naringin and hesperidin from each other at all; both were eluted at near void volume. From the molecular structures, naringin and hesperidin are thought only to be lightly included by chitosan β-CD polymer, but the slight difference in the structures is recognized by the polymer and subsequently they were separated. The fact should provide a preferable condition to chromatographic separation of substances.

CONCLUSIONS

Chitosan beads is considered to be excellent material for making CD polymer in reference to handling easiness and cost. Chitosan beads is more profitable since the material does not show hydrophobic adsorption. However, the chitosan CD polymer available at the present time allows undesirably acids to be adsorbed by the base material of chitosan and shows somewhat unsatisfactorily weak physical and chemical strength, and so more improvements are required in future.

At the start of this development the target was isolation and purification by adsorption and elution, and so the main theme of research was to find how to increase the amount of fixed CD. Unfortunately marked increase of the fixed amount was not achieved but it was made clear that CD polymer could be used as a separating material in chromatography. When used as a separating material in chromatography, its application to continuous processing in the simulated moving bed method may be possible, and extensive utilization of CD polymer may be expected.

More studies for improving strength, durability and manufacturing cost are needed, but the basic technique of isolation and fractionation by using CD polymer as a separating material in chromatography has now been established. More and more contribution of CD polymer not only in the field of food industry but also in other fields is expected.

TURBULENT HIGH-SCHMIDT NUMBER MASS TRANSFER IN UF

CHRISTER ROSÉN, CHRISTIAN TRÄGÅRDH and PETR DEJMEK
Department of Food Engineering,Lund University
P.O.Box 124, S-221 00 Lund, Sweden

ABSTRACT

A computer simulation model have been developed predicting the mass transfer in terms of turbulent diffusion for high molecular weight substances. The method involves a numerical solution of the time averaged equations of motion with a spatial resolution down in the very thin concentration boundary layer appearing in turbulent wall boundary layers.

For the prediction of required Reynolds stresses a so called low Reynolds number k-ε turbulence model has been used. The turbulent diffusion of mass is obtained from the Reynolds' stresses through a turbulent Schmidt number relationship.

INTRODUCTION

The most common method of predicting mass transfer across the concentration polarization boundary layer is to use the film theory equation. A basic assumption in the derivation of the film theory equation is that physical properties are constant in the film. The macromolecules filtered in UF have physical properties that are highly concentration dependent, and thus the assumption of constant physical properties is not valid. In the method presented here, the general conservation equations for species concentration and momentum are solved throughout the boundary layer with the aid of a computer program based on a finite volume formulation. This formulation has the important quality that concentration-dependent physical properties can be calculated throughout the concentration boundary layer at spatially resolved points.

THEORIES AND METHODS

Governing equations

A statistical approach is taken for the instantaneous velocity,U , and pressure P, separated into a mean and a fluctuating part; U = U + u' and P = P + p'. In analogy with this is the instantaneous concentration described as; C = C + c'. The following set of equations, in the Cartesian coordinate system, for a steady, two-dimensional boundary layer flows;

$$\overline{U}\frac{\partial U}{\partial x}+\overline{V}\frac{\partial U}{\partial y}=-\frac{1}{\rho}\frac{\partial P}{\partial x}+\frac{\partial}{\partial y}(\nu\frac{\partial U}{\partial y})-\frac{\partial}{\partial y}(\overline{u'v'})$$

$$\overline{U}\frac{\partial \overline{C}}{\partial x}+\overline{V}\frac{\partial \overline{C}}{\partial y}=\frac{\partial}{\partial y}(\frac{\nu}{\sigma}\frac{\partial \overline{C}}{\partial y})-\frac{\partial}{\partial y}(\overline{c'v'})$$

The averaging process introduces the unknown correlations u'v' and v'c'.

Turbulence models

In Rosén [1] different low Reynolds number k-ε turbulence models were tested in respect to their ability to predict turbulent momentum transport in the viscous sublayer.As a result the model by Chien [2] was chosen. For predicting u'v', the eddy viscosity concept is used. The k-ε model employs two transport equations, one for the turbulent kinetic energy k and one for the turbulent kinetic energy dissipation rate ε which relate to the eddy viscosity as;

$$\nu_t=C_\mu f_\mu\frac{k^2}{\epsilon}$$

which in turn relates to the u'v'and c'v' correlation as;

$$-\overline{u'v'}=\nu_t\frac{\partial \overline{U}}{\partial y};-\overline{c'v'}=\epsilon_t\frac{\partial \overline{C}}{\partial y}$$

The turbulent momentum transport is related to the turbulent transport of species concentration by a turbulent Schmidt number expression, derived by Rosén [1];

$$\sigma_t\equiv\frac{\nu_t}{\epsilon_t}=\frac{1.153(1-\exp(-\sqrt{(Re-3000)/2000}))}{\sigma^{-0.112}}$$

RESULTS

The simulated results of the membrane concentration, are compared with results calculated from the film theory equation for two Sherwood correlations in Figure 1; the Deissler [3] analogy $Sh = 0.122\ (f/2)^{0.5}\ Re\ \sigma^{0.25}$ and the correlation of Shaw and Hanratty [4] $Sh = 0.0889\ (f/2)^{0.5}\ Re\ \sigma^{0.296}$. The latter model was chosen as the derived turbulent Schmidt number expression is obtained from their experimental results. A good agreement is found for low permeate fluxes while going to higher a continuously larger deviation appears. In Figure 2 the spatial resolved near-membrane gradients are shown and in Figure 3 wall concentration variations over a membrane having a Normal distributed varying permeability. The poly-saccharide Dextran T70 at a concentration of 0.142% was the working fluid for which physical data was taken. Simulated rectangular flow channel has a height of 5.9 mm a width of 60 mm and a membrane length of 100 mm. Applied pressure was 0.6 MPa. 100 % retention was assumed.

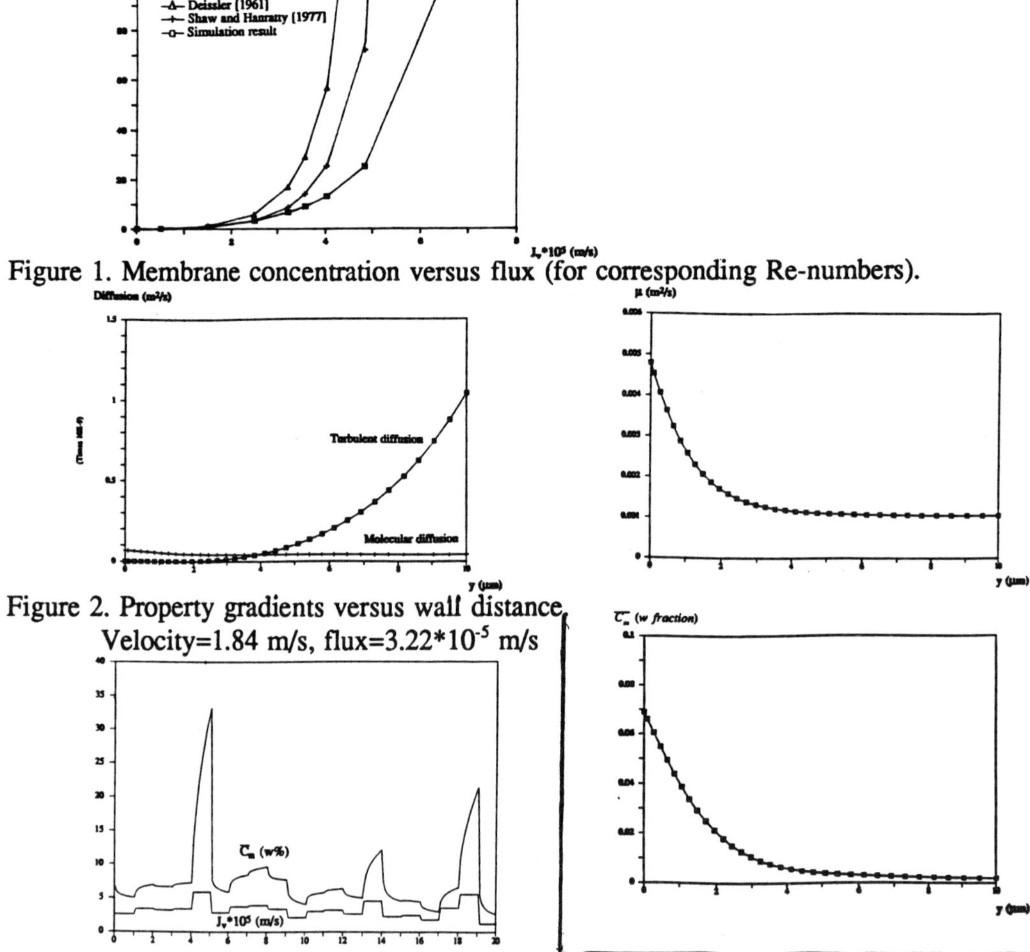

Figure 1. Membrane concentration versus flux (for corresponding Re-numbers).

Figure 2. Property gradients versus wall distance,
Velocity=1.84 m/s, flux=$3.22*10^{-5}$ m/s

Figure 3. The variation of membrane concentration with a membrane flux distribution (J_v=3.22±1.12)$*10^{-5}$m/s.

REFERENCES

1. Rosén, C., Turbulent High-Schmidt-Number Mass Transfer in the Concentration Polarization Boundary Layer over Ultrafiltration Membranes, Ph.D. thesis, Lund University, 1992

2. Chien, K.-Y., Prediction of channel and boundary layer flows with a low-Reynolds-number two-equation model of turbulence, AIAA Paper No 80-0134, 1980

3. Deissler, R., Analysis of turbulent heat transfer, mass transfer and friction in smooth tubes at high Prandtl and Schmidt numbers. In Advances in Heat and Mass Transfer, McGraw-Hill, New York, 1961

4. Shaw, D.A., Hanratty, T.J., Turbulent mass transfer rates to a wall for large Schmidt numbers, AIChE Journal, 1977, **23**, pp.28-37

The pore size of ultrafiltration membranes - a novel approach

KENNETH M PERSSON AND GUN TRÄGÅRDH
Dept. of Food Engineering, Chemical Centre, Lund University
PO Box 124, S-221 00 LUND, SWEDEN

ABSTRACT

The pore size of flat sheet membranes of different compositions with nominal molecular weight cut-offs between 20 kDa and 100 kDa were evaluated in a cross-flow mode, using monodisperse incompressible silica sols with mean size from 11 to 45 nm. The observed silica retention, from 97.6 to 98.7 % gave an indication of the largest pore size of the membranes. The silica content was low in the permeate at any condition, yet always above the monomeric equilibrium concentration of about 130 ppm, even for colloidal sizes much larger than the nominal pore size of the membranes.

INTRODUCTION

Ultrafiltration (UF) is used for the separation and concentration of macromolecules in *e.g.* food industry. The characterization of the UF membranes involves pore size distribution measurements, which can be measured directly as pores, or indirectly as the retention of well characterized macromolecules of known sizes (dextran, BSA etc.) [1]. Since the tertiary structure of these molecules can change, especially when undergoing shear [2], the macromolecular permeation in practice is a function of size and actual shape. By using incompressible silica particles of various sizes, this problem should be limited.

The nominal pore size of a UF membrane is either expressed as a molecular weight cut-off, nmwco (in kDa) or as a pore diameter (in nm). All the direct porosimetric methods assume some physical structure of the membrane, mostly some specific geometric shape of the pores. An advantage with indirect methods is that they do not preassume any given pore shape.

Silica has been used frequently as a model fouling agent, yet seldom for retention measurements. Iler [3] used UF-membranes with pore sizes from 0.9 nm to 2.0 nm for the fractionation of concentrated silica sols with mean size 1-2 nm.

The colloids are in equilibrium with silica monomers in water solution. About 130 ppm silica is solved as monomer in water, the rest are particles [4].

With C_p as the permeate concentration and C_b the bulk concentration of silica, the observed retention R_{obs} is defined:

$$R_{obs}=1-\frac{C_p}{C_b} \tag{1}$$

EXPERIMENTAL

Three polymeric flat sheet [(R)]NADIR ultrafiltration membranes from Hoechst AG, Germany were tested: polyamid UF-PA-20 (nmwco 20 kDa with PVP K 30 retention measurements); hydrophilic polysulfone UF-PS-100M (nmwco 100 kDa with dextran T 2000); cellulose acetate UF-CA-100 (nmwco 100 kDa with dextran T 110).

Colloidal silica sols Bindzil® from EKA-Nobel (Sweden) in different sizes were used as particle source, see table 1. The particle size distribution was analyzed with a Malvern photon correlation spectrometer (PCS 100), giving the hydrodynamic radius. Particles were also studied by SEM and compared with a flow field fractionation measurement that had been performed previously [5]. The different methods gave consistent sizes. The silica content in permeate and retentate was analysed according to Iler [3]. The monomeric silica was also analysed separately for some cases, without previous decomposition of the particles.

An automatic UF laboratory plant constructed at the department was used for the experiments [6]. With controlled temperature, flux, and pressure, and with tangential velocity measurement devices and computerized permeate levelling, the total accuracy in the flux measurement was about $5*10^{-7}$ m/s.

Each experiment followed the same procedure at a circulation velocity of 11 m/s, a mean transmembrane pressure of 0.14 MPa and a temperature of 18.5°C. A fresh membrane was mounted and conditionized with water for 30 minutes before permeation. The silica sol was added to a final concentration of 1 %, and permeate was collected for analyses after 1 minute, 10 minutes, 60 minutes and 240 minutes. The plant was thereafter rinsed 3 x 2 minutes with water. Finally, the pure water flux was measured again for at least one hour. All water used was MilliQ-filtered.

RESULTS AND DISCUSSION

The silica content in the permeate reached steady state for the particles used within 30 minutes. Values after 240 minutes are found in table 1. At the high cross-flow velocity used, the membranes were hardly fouled at all, indicating no polarization; the membrane fluxes were stable even after hours, except for a compaction effect with time. The analyse of the monomers gave a theoretical retention of 98.7 % if maximum monomer concentration was assumed.

A fouled membrane would result in a higher retention, not lower. The lower actual retention could be explained by the prescence of particles in the permeates. The particles might pass through the small number of large pores, which have been shown for UF-membranes [7]. The highest silica retention was found for the polyamid membrane, which could be explained by its lower nmwco. It is not possible to distinguish between the CA100 and the PS100 with the particles used.

TABLE 1.
Particle size and observed retention after 240 minutes

Size data (in nm)		Observed retention (in %)		
Particle name	Size±stand.dev.	CA100	PS100	PA20
15/500	11.4±4.5	97.6±0.1	97.7±0.2	97.8±0.1
30/360	18.9±5.6	98.1±0.2	98.0±0.1	98.1±0.1
30/220	21.3±6.0	98.4±0.1	98.3±0.1	98.4±0.1
40/130	26.1±4.5	98.3±0.1	98.4±0.1	98.6±0.2
50/80	45±>10	98.3±0.1	98.3±0.1	98.7±0.1

CONCLUSIONS

Silica particles can be used to investigate the presence of large pores in UF membranes. Silica is retained to a high extent, yet traces of particles can be detected in the permeate even for low-nmwco membranes. The results are best explained by a postulation of a pore size distribution for the membranes. The method would be improved by the use of smaller particles, about 4 nm in size or less.

REFERENCES

1. G. Trägårdh (1988). Survey of characterization methods for ultrafiltration membranes, in Characterization of ultrafiltration membranes, Studentlitteratur, Lund 9-38.
2. K.M. Persson and V. Gekas (1993). Factors influencing aggregation of macromolecules in solution, accepted in Process Biochemistry.
3. Ralph K. Iler (1982). Colloidal Components in Solutions of Sodium Silicate, ACS Symposium Series 194 96-113.
4. Ralph K. Iler (1979). The chemistry of silica, chapter 1, John Wiley & Sons, New York.
5. B. Olsson, K.-G. Wahlund and A. Litzén (1992) Fast size characterization of silica sols by asymmetrical flow field fractionation (FFF), poster presented at 3:rd int. symp. on FFF, Salt Lake City, Utah, 5-7 October 1992.
6. G. Trägårdh and K. Ölund (1986). A method for characterization of ultrafiltration membranes, Desalination 58 187-198.
7. J. Nilsson (1989). A study of ultrafiltration membrane fouling, *thesis*, Lund University, Sweden.

REVERSE OSMOSIS SYSTEM FOR HIGHLY CONCENTRATED JUICE

HIROSHI NABETANI*, HARUYUKI IGAMI**, MITSUTOSHI NAKAJIMA*
KIROU HAYAKAWA**, YASUNORI YAMADA**, YUKIO ISHIGURO**
* National Food Research Institute, Ministry of Agriculture, Forestry and Fisheries, Tsukuba, Ibaraki, 305 Japan
** Research Institute, Kagome Co.,Ltd., Nasu, Tochigi, 329-27 Japan

ABSTRACT

To develop the multistage reverse osmosis (RO) system, which is regarded as useful for highly concentrated fruit juice, the permeate flux value and the permeate concentration of the system were analyzed in this study according to the membrane transport equation and the concentration polarization equation. The pressure drop along the flow channel in the membrane module was also quantified. Based on these results, a two-stage RO system for clarified apple juice has been proposed. This continuous system is composed of a tight RO membrane at the first stage and a loose RO membrane at the second stage, and the membrane modules were arranged in a tapered fashion. The energy efficiency of the system will be compared with those of conventional methods.

INTRODUCTION

RO has advantages over other methods of removing water from liquid foods. Its application in food industries has been developed successfully. However, RO requires operating pressure higher than the osmotic pressure of the solution and the resisting pressure of the module is 6-7 MPa. Therefore, RO use is limited to low concentrations. However, if a loose RO membrane is combined with a conventional tight RO membrane, higher concentrations of juice can be obtained. The performance of both tight and loose RO membranes was analyzed in this study according to the membrane transport equation and the concentration polarization equation. Based on the results, a two-stage RO system of high energy efficiency has been proposed.

THEORETICAL

Permeate flux, J_V, and observed rejection, R_{obs}, are predicted by the membrane transport equation based on nonequlibrium thermodynamics and the concentration polarization equation. The mass transfer coefficient, k, of the turbulent flow in the tubular module is calculated using Deissler's correlation.

The membrane transport equation

$$J_V = L_p \{2\mu_w / (\mu_p + \mu_m)\} \{\Delta P - \Sigma (\sigma_i \Delta \Pi_i)\} \quad (1)$$

$$R = \sigma_i (1 - F_i) / (1 - \sigma_i F_i) \quad (2)$$

$$F_i = \exp \{-(1 - \sigma_i) J_V / P_i\} \quad (3)$$

The concentration polarization equation

$$(C_{m,i} - C_{p,i}) / (C_{b,i} - C_{p,i}) = \exp (J_V / k) \quad (4)$$

Deissler's correlation

$$N_{Sh} = 0.023\, N_{Re}^{0.875}\, N_{Sc}^{0.25} \quad (5)$$

Real rejection

$$R_i = 1 - C_{p,i} / C_{m,i} \quad (6)$$

Observed rejection

$$R_{obs,i} = 1 - C_{p,i} / C_{b,i} \quad (7)$$

MATERIALS AND METHODS

An RO unit equipped with a tubular module and a high pressure pump with variable speed motor was used. The tight RO membrane (Type AFC-99) and the loose RO membrane (Type AFC-30) were supplied by PCI Ltd. The inner diameter, length and area of the membrane were 1.25 cm, 55.5 cm and 0.0218 m^2, respectively. Clarified apple juice (weight ratio of monosaccharides(ms) to disaccharide(ds) = 5 : 1) was used as the sample solution.

Total circulation experiments were performed by returning the permeate to the feed tank to prevent change in concentration of clarified apple juice. The conditions of the tight RO membrane treatment and the loose RO membrane treatment were as follows: feed concentration, C_b: 12-26 wt.%; applied pressure, ΔP: 2-7 MPa; flow velocity, u: 1.0 and 0.5 m s^{-1}; feed temperature: 25 ℃. The permeate fluxes and the observed rejections at steady state were measured under each condition.

RESULTS AND DISCUSSION

1. RO membrane treatment of clarified apple juice

The pure water permeability was measured after each treatment. The reflection coefficient and the solute permeability of loose RO membrane were determined by the use of the curve fitting method and were as follows: $\sigma_{ms}=0.999$, $P_{ms}=8.1\times10^{-7}$ m s^{-1}, $\sigma_{ds}=0.999$, $P_{ds}=1.7\times10^{-7}$ m s^{-1}.

The relationship between applied pressure and permeate flux for the tight RO treatment of clarified apple juice was calculated by solving Eqs. (1) and (4)-(7) (i = ms, ds). The experimental relationships between applied pressure, permeate flux and observed rejection for the loose RO treatment of clarified apple juice were calculated by solving Eqs. (1)-(7) (i = ms, ds). The experimental values of the tight RO treatment and the loose RO treatment were in good agreement with the calculated values.

2. Development of Two-Stage RO System for Highly Concentrated Juice

1) Two-stage RO system

In order to concentrate clarified apple juice from 10 wt.% (ms 8.33 wt.%, ds 1.67 wt.%) to 40 wt.%, a two-stage RO system which is composed of a tight RO membrane (Type AFC-99) at the first stage and a loose RO membrane (Type AFC-30) at the second stage has been proposed. The first stage of this system has more membrane units than the second stage, that is the tree type system. In addition, the permeate of the second stage is returned to the feed tank of the first stage.

The conditions of this system are as follows: applied pressure of the first and second stage: 7.85 MPa; mass feed rate of clarified apple juice: 1000 kg h^{-1} per one membrane unit of the second stage.

2) Mass balance, concentration of bulk solution and pressure drop

Mass balance, concentration of bulk solution and pressure drop in the module can be expressed by Eqs. (8)-(10).

Mass balance

$$du/dA = -J_V/S \qquad \text{B.C.} \quad u = u_0 \quad \text{at } A = 0 \qquad (8)$$

Concentration of bulk solution

$$C_b = (u_0\rho_0 C_0 - \int C_p J_V \rho_p \, dA) / (u\rho) \qquad (9)$$

Pressure drop

$$dP_d/dA = \lambda u^2 \rho / (2\pi d_h^2) \qquad \text{B.C.} \quad P_d = 0 \quad \text{at } A = 0 \qquad (10)$$

3) Optimization of the two-stage RO system for highly concentrated juice

The first stage area was calculated by solving Eqs. (1) and (4)-(10), and the second stage area was calculated by solving Eqs. (1)- (10), using a personal computer. The relationship between the membrane area and the concentration of clarified apple juice at the inlet of the second stage is shown in Fig. 1.

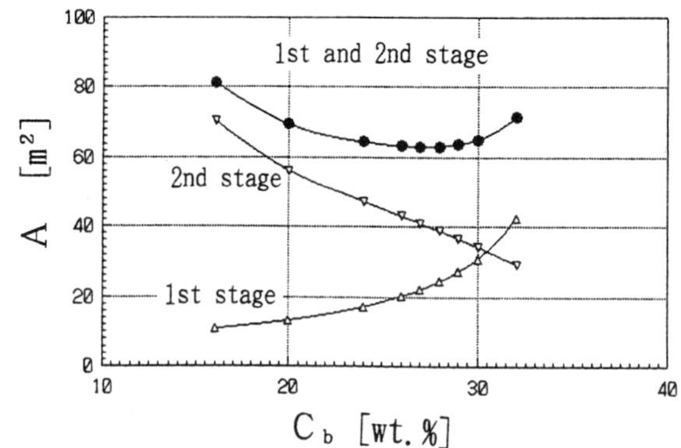

Fig. 1 Relationship between membrane area and concentration of clarified apple juice at the inlet of the 2nd stage

4) Energy consumption

The energy required to remove 1 kg of water from clarified apple juice (For the case of clarified apple juice concentrated from 10 wt.% to 40 wt.%) is shown Table 1. The energy requirements of this system and freeze concentration were calculated assuming that the efficiency of electric power generation and pump (or compressor) are 0.4 and 0.7, respectively.

Table 1 Energy required to remove 1 kg of water from clarified apple juice

This system	63 kJ
Evaporation	
Single effect	2627 kJ
Triple effect	997 kJ
Freeze concentration	272 kJ

CONCLUSION

The permeate flux and the observed rejection in the tight RO membrane treatment and the loose RO membrane treatment of clarified apple juice can be predicted by the membrane transport equation based on nonequlibrium thermodynamics and the concentration polarization equation. Based on these results, a two-stage RO system of high energy efficiency for the concentration of clarified apple juice has been proposed.

REFERENCE

1) Nabetani, H., M. Nakajima, A. Watanabe, S. Nakao and S. Kimura: *J. Chem. Eng. Japan*, 25, 575(1992)

FACTORS AFFECTING THE PERFORMANCE OF CROSSFLOW FILTRATION OF YEAST CELL SUSPENSION

TAKAAKI TANAKA AND KAZUHIRO NAKANISHI
Department of Biotechnology, Faculty of Engineering, Okayama University, Tsushima-naka, Okayama 700, Japan

ABSTRACT

Factors affecting the performance of crossflow filtration were investigated in a module with a thin channel using *Saccharomyces cerevisiae* cells cultivated in two media. When a suspension of *S. cerevisiae* cells cultivated in YPD medium was cross-filtered, the steady-state flux agreed well with the value predicted by conventional filtration theory using the specific resistance of the cake of cells and the amount of the cake deposited on the microfiltration membrane. The flux decreased considerably due to the small particles (<1 μm) in molasses in the case of crossflow filtration of a broth of cells cultivated in molasses medium.

INTRODUCTION

In fermentation industry crossflow filtration has been applying to the separation of cells from broth. However, optimization of crossflow filtration has not been achieved as the mechanism of the crossflow filtration has not been clarified. In some cases the permeation flux is not sufficiently high because particles derived from medium such as molasses cause additional increase in the resistance to permeation. Here, we showed the factors affecting the performance of crossflow filtration of yeast cell suspension. We also intended to increase the permeation flux in crossflow filtration of broth of yeast cells cultivated in molasses by backwashing.

MATERIALS AND METHODS

Yeast Cells

A strain of *S. cerevisiae* donated by Kanegafuchi Chemical Industry Co., Ltd. was cultivated with YPD (yeast extract, polypeptone, dextrose) medium or molasses medium containing 2% of sugar in a 5-dm^3 jar fermentor at 30 °C for 24 h. The aeration rate was 0.5 vvm. The YPD medium consisted of 1% yeast extract, 2% polypeptone, and 2% dextrose. Antifoam was added at 50 ppm.

Measurement of Specific Resistance

Cells were washed three times with four volumes of saline. Then, the mean specific resistance was measured at 20 °C with a dead-end filtration module by the steady-state method reported previously [1].

Filtration Module and Operations

The module used for crossflow filtration was of thin channel type (24 mm in width, 305 mm in length, and 2.5 mm in depth) [2]. The membranes used were synthetic membranes from Fuji Photo Film Co., Ltd: FM45, FM80, FM120, FM300, and FM500 made of cellulose acetate and with a nominal pore size of 0.45, 0.80, 1.2, 3, and 5 μm, respectively, and an SE45 membrane made of polysulfone and with a nominal pore size of 0.45 μm. Yeast suspension or broth was circulated by a rotary pump at 20 °C. The permeate was usually recycled to the reservoir to maintain the cell concentration. Yeast cell suspensions were prepared by washing and suspending cells cultivated in YPD medium into saline.

The backwashing was usually done as follows. After 10 min of permeation, the circulation flow was introduced to a bypass to stop the flow in the filtration module; 7.3 ml of permeate was allowed to flow from the permeate exit into the module by a hydrostatic pressure difference. Then, the valve for permeate was closed, and the circulation flow was again introduced into the module. Permeation was started 4 min after the start of bypassing.

RESULTS AND DISCUSSION

Crossflow Filtration of Suspension of Yeast Cells Cultivated in YPD medium

The specific resistance of *S. cerevisiae* cells cultivated in YPD medium was 2.8×10^{11} m/kg at 49 kPa. When a suspension of *S. cerevisiae* cells cultivated in YPD medium cross-filtered, the flux was decreased rapidly after the start of filtration and usually reached a steady state within a few minutes. The steady-state flux was 1.1×10^{-4} m/s at the circulation flow rate of 30 cm^3/s, the cell concentration of 80 kg/m^3 (wet weight), and the transmembrane pressure of 49 kPa. It agreed well with the value predicted by conventional filtration theory using the specific resistance of cake and the amount of the cake deposited on the microfiltration membrane.

As the circulation flow rate increased, the permeation flux increased due to decrease of cell deposition. The flux considerably decreased at a circulation flow rate of 15 cm^3/s or lower. A part of the module inside was filled up with cells, which caused channeling of a circulation flow. The channeling was also observed at a high cell concentration. When the transmembrane pressure difference increased the permeation flux increased but not in proportion to the pressure difference because of increases of cell deposition and specific resistance.

Crossflow Filtration of Broth of Yeast Cells Cultivated in YPD Medium

When the yeast cells were cultivated in YPD medium, the cell concentration of the broth was 30 kg/m^3 and the permeation flux showed a flux similar to that observed with yeast cell suspension (Figure 1).

Crossflow Filtration of Broth of Yeast Cells Cultivated in Molasses

The molasses used for cultivation contained 1 kg/m^3 of small particles (<1 μm). The cell concentration after 24 h of cultivation was 12 kg/m^3. Figure 1 shows the change in flux during crossflow filtration of broth of cells cultivated in molasses at the circulation flow rate of 30 cm^3/s and at the transmembrane pressure of 49 kPa. The flux decreased gradually with time. After 15 min of filtration a steady state was reached. The steady-state flux was nearly 10^{-5} m/s for all the membranes tested. This value was much lower than that observed in crossflow filtration of broth of cells cultivated in YPD medium.

Effect of Membrane Pore Size on Recovery of Permeation Flux by Backwashing

To increase the permeation flux, we tried backwashing. The recovery of flux strongly depended on pore size of membranes. The flux decreased as in the filtration without backwashing when the nominal pore size of membrane was 0.8 μm or smaller. The flux was, however, almost recovered when the pore size was 3 to 5 μm.

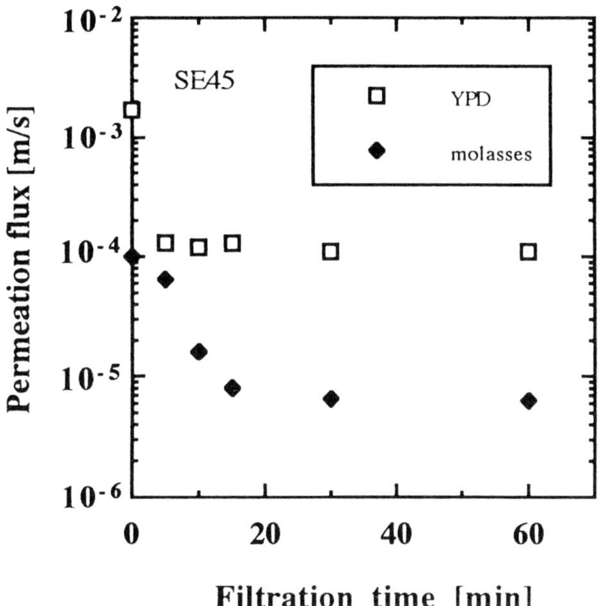

Figure 1. Crossflow filtration of broths of yeast cells cultivated in YPD medium and molasses medium. Broths were cross-filtered at the circulation flow rate of 30 cm^3/s and the transmembrane pressure of 49 kPa.

CONCLUSION

We cross-filtered suspensions of yeast cells, and broths of cells cultivated in YPD medium and molasses medium. The steady-state flux agreed well with the value predicted by conventional filtration theory in crossflow filtration of cell suspensions and broth of cells cultivated in YPD medium. The flux considerably decreased due to the small particles in molasses in the case of crossflow filtration of a broth of cells cultivated in molasses medium.

REFERENCES

1. Nakanishi, K., Tadokoro, T., Matsuno., R., On the specific resistance of cakes of microorganisms. *Chem. Eng. Commun.*, 1987, **62**: 187-201.

2. Tanaka, T., Kamimura, R., Itoh, K., Nakanishi, K., Matsuno, R., Factors affecting the performance of crossflow filtration of yeast cell suspension. *Biotechnol. Bioeng.*, 1993, **41**: 617-624.

CROSS-FLOW MEMBRANE FILTRATION OF SOY SAUCE LEES

TOSHIO FURUKAWA, KAORU KATOU, KEIZOU KUMAKURA, AND KATSUMICHI OSAKI
Technology Center, Production Engineering Division,
Kikkoman Corporation
2470 Imagami, Noda-shi, Chiba 278, Japan

ABSTRACT

The amount of soy sauce lees coagulated and precipitated at about 50℃ during 3-8 days reaches around 5-15% of the volume of raw soy sauce. Four different types of polymeric membrane module for cross-flow filtration were applied for the recovery process of soy sauce from lees. Cross-flow filtration for lees is discussed from two aspects; energy consumption on permeate and product properties. Permeate per energy input defined by P/E (L/kW•h) is compared for tubular, hollow fiber, rotary, and plate & frame-types of modules. The plate & frame module, mounted with ultra and micro-filtration membranes, was found to reduce membrane fouling effectively. This particular type of module provides the least changes in color and taste in soy sauce and the highest P/E at volume reduction ratio of 5 among the four types of module.

INTRODUCTION

Japanese soy sauce has been used for over 300 years as a food flavoring and is currently served throughout the world as an all purpose seasoning. The four major characteristics are aroma, flavor, color, and stability, and these properties should be considered first when introducing a new process.

Membrane filtration for soy sauce has been researched and applied in the final refining process since 1978[1] in Japan. The cross-flow ultrafiltration was first substituted for diatomaceous earth filtration of raw soy sauce and soy sauce lees about 15 years ago. In view of the developing membrane technology, the membrane filtration process for lees had to be restructured.

EXPERIMENTAL METHODS

Experimental device and operating conditions: Four different types of module were tested to measure permeate flux, the possibility of concentration of soy sauce lees and pumping energy. Each module was operated as a batch process at 45±2℃, under conditions described in Table 1.

Analytical procedures: Permeate was analyzed for color change, molecular weight distribution and sensory evaluation. Ingredients of the permeate were analyzed by standard methods (SOY SAUCE STANDARD ANALYSIS). Molecular weight distribution was measured by HPLC (TOSO:PFK-gel G3000SW). Color was measured by Color Number Meter(OMRON:3WCOL-10). Sensory evaluation was performed by 10 skilled panels.

TABLE 1
Specifications of membrane modules and operating conditions of each module

Module configuration / Trade name	Membrane materials /Nominal pore size	Cross-flow velocity(m/s)	Treated volume/ Area:V/A(L/m²)
	Membrane area (m²)	Pumping power(kW)	Inlet / Outlet pressure(kP)
Hollow fiber MEMCOR	polyethylene 0.2 micron 3.0 m²	0.5m/sec 1.0 kW	350(L/m²) 300/200(kP)
Tubular MITSUBISHI RAYON ENGINEERING	polyolefin 150 kDa 19.2 m²	0.5-1.5m/sec 7.0 kW	200(L/m²) 350/100(kP)
Plate & frame NITTO DENKO	a. PTFE b. PTFE & polysulfone a. 0.2micron b. a/50 kDa a. 3.0 m² b. 12.0m²(6m²+6m²)	1.0-2.5m/sec a. 2.0kW b. 7-7.5kW	170/350(L/m²) 450/50 (kP)
Rotary disk HITACHI PLANT ENG. NITTO DENKO	polysulfone 750 kDa 2.2 m²	4.0-6.0m/sec 3.0 kW	450 (L/m²) 100-200 (kP)

RESULTS AND DISCUSSION

Selection of a suitable module for soy sauce lees: Permeate fluxes were compared with each other against the volume reduction ratios. The rotary disc type module indicated a stable flux up to 20 volume reduction ratio(VRR). The plate and frame module was successful in maintaining a high flux up to 5 VRR. The tubular and the hollow fiber modules had no significant flux for any VRR. However, the level of energy consumption for each module were different from the gradient of flux vs. VRR.

Permeate per energy input (P/E) defined by

$$\frac{\text{permeate flux}(L/m^2 \cdot h)}{\text{pumping energy per unit membrane area } (kW/m^2)} \qquad (1)$$

is given in Table 2.

TABLE 2
Mean flux, energy input per unit area and permeate per energy input P/E of each module at 5 VRR at 45℃

Membrane Module	Mean flux(LMH)	E/A(kW/m²)	P/E(L/kW•h)
Hollow fiber	8-10	0.31	25.8-32.3
Tubular	10-15	0.28-0.36	35.7-41.7
Plate & frame	18-30	0.35-0.63	34.5-51.4
Rotary disk	25-35	1.55	16.1-22.6

Differences between the UF/MF and the MF/UF- module: The plate and frame module of 12m² with 73 plates is divided into UF membranes (6m²/36p) and MF membranes(6m²/36p). The liquid inlet of the UF/MF module is on the side of the UF membranes and the P/E value of the UF/MF module is higher between 30-50℃ up to 5 VRR than that of the MF/UF module.
Evaluation of permeate properties: The permeability of high molecular weight materials around 300kDa varied between the UF/MF and the MF/UF-modules. As high molecular weight materials were extruded into the permeate by force at a higher operational pressure, the permeability of those materials was accelerated in the MF/UF module. Although the UF permeate color in both the UF/MF and the MF/UF-modules underwent a similar change, the MF permeate in the MF/UF module changed to a lighter color than that of the UF/MF module. These results are considered as a kind of dynamic membrane phenomena.

CONCLUSION

The index of P/E, defined by Eq.(1) is useful for the evaluation of various membrane modules. The membrane module with a high P/E value provides low energy consumption and the cost reduction of membrane modules and the membrane replacement. As the P/E value is particularly influenced by the change of viscosity of the lees which depends mainly on the temperature and VRR, the P&F membrane filtration should be operated at around 40℃ up to 10 VRR, where the change of the permeate properties is relatively small. Soy sauce lees have been successfully filtrated to 5-10 VRR in the commercial size of P&F module with UF/MF membranes since 1992.

REFERENCES

1. Tadanobu Nakadai, and Hironaga Hashiba, JAPAN PATENT NO.53-1360(1978)

APPLICATIONS OF THE CERAMIC MEMBRANES TO SOY SAUCE

NOBUHIKO KANEKUNI, HISASHI NOGAKI, KENJI TABATA and ATSUO WATANABE,
Research and Development Div., TOTO Ltd.
2-8-1 Honson, Chigasaki-shi, Kanagawa 253 Japan

ABSTRACT

We developed two types of cross-flow filtration modules with ceramic membranes. Each module was applied to recover soy sauce from sediment. It was confirmed that the permeation flux of the proposed modules were higher than that of the conventional modules, and soy sauce could be recovered from sediment at the yield of more than 90%. It seemed that the proposed modules could work more effectively than the conventional type.

INTRODUCTION

Recently ceramic membranes have attracted special interest because of their superior characteristics. However, few cross-flow filtration modules have been used for recovering or retaining any substance with high concentrated liquid or viscous fluid, due to a difficulty of circulating viscous fluid.

In this paper, we propose two novel cross-flow filtration modules to apply for low fluidity liquid. And the application of the modules for the recovery of soy sauce from the sediment, which is known to be highly viscous liquid, is discussed.

EXPERIMENTAL

Fig.1 shows the schematic diagrams of two types of cross-flow filtra-

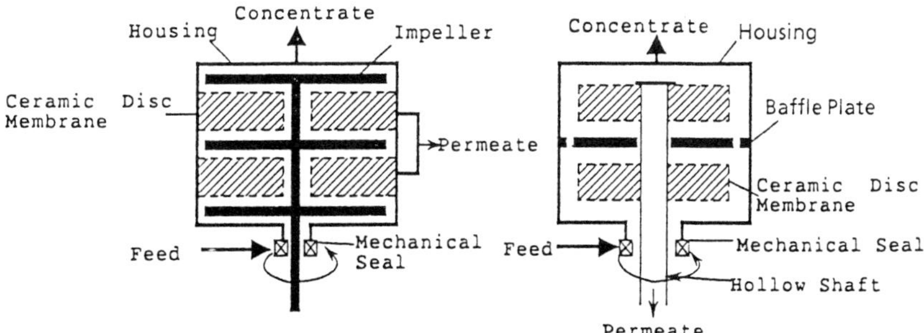

Fig.1 Schematic diagrams of two types of cross-flow filtration modules with ceramic membranes

tion modules with ceramic membranes[1]):

1)Stationary Disc Membrane Type (SDM-type) ; The stationary membranes are enclosed in a housing with rotating impellers configurated in the vicinity of membrane surface.

2)Rotary Disc Membrane Type (RDM-type) ; The membrane is rotating in the stationary shell. The filtrate is drawn off through a hollow shaft

The modules were examined by means of two methods : total circulating filtration and batch filtration. The specifications of membrane modules and operating conditions are shown in Tables 1 and 2, respectively.

The sediment of soy sauce employed in this study was a blackish fluid with viscosity of 5 - 15cP. The particles suspended in the sediment have a size distributed from 0.3 to 45μm.

Table 1 Characteristics of membrane modules

Material	Aluminium-oxide
Pore Diameter	0.1μm
Sharp	Disc(External Pressure Type)
Membrane Area	2.0m^2(SDM-type),0.3m^2(RDM-type)

Table 2 Operating conditions for cross-flow filtrations

Total Circulation Filtration	Trans-Membrane Pressure Revolution	50 - 200kPa 50 - 500rpm
Batch Filtration	Trans-membrane Pressure Revolution	200kPa 500rpm

RESULTS AND DISCUSSION

Fig.2 shows the effects of trans-membrane pressure and revolution on permeation flux for the SDM-type module. The permeation flux increased with increasing trans-membrane pressure and with increasing revolution. The pressure dependence of permeation flux was similar between the two modules, although the permeation flux for the RDM-type module was 30% smaller than that for the SDM-type module.

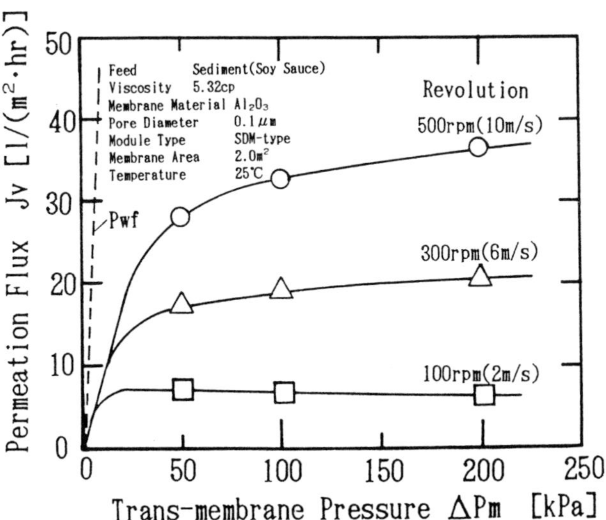

Fig.2 Effects of trans-membrane pressure and revolution on permeation flux for the SDM-type module

Fig.3 shows the relationships between permeation flux and volume yield of permeate. The symbol "■" shows the filtration characteristics of an organic membrane with hollow fiber[2]. The permeation flux of all modules decreased as the yield increased. The permeation flux of the organic membrane was about as half as that of the SDM-type module. Furthermore, the upper limit of the yield was 75% in the organic membrane, while the proposed modules could filtrate to attain the volume yield of 90%.

Fig.4 shows the influence of volume yield of permeate on the viscosity of sediment. A steep increase in the viscosity was observed at the yield of 75%. And at the yield of 90%, the viscosity of the concentrated sediment became more than 1000 times as much as that of the initial feed sediment. The conventional hollow fiber module concentrated the sediment the viscosity of which was less than several poise, whereas the proposed modules, SDM- and RDM-type modules, could filtrate the sediment with the viscosity of more than 100 poise.

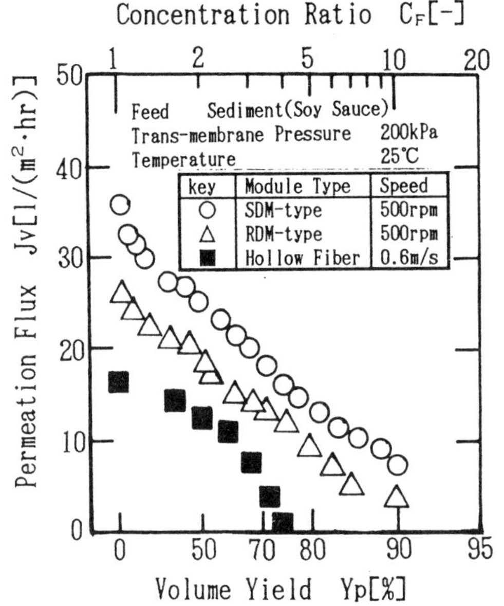

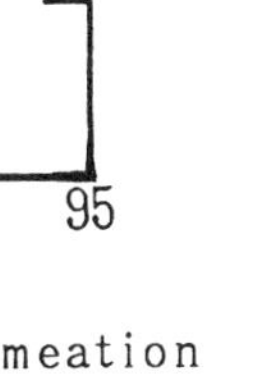

Fig.3 Relationships between permeation flux and volume yield

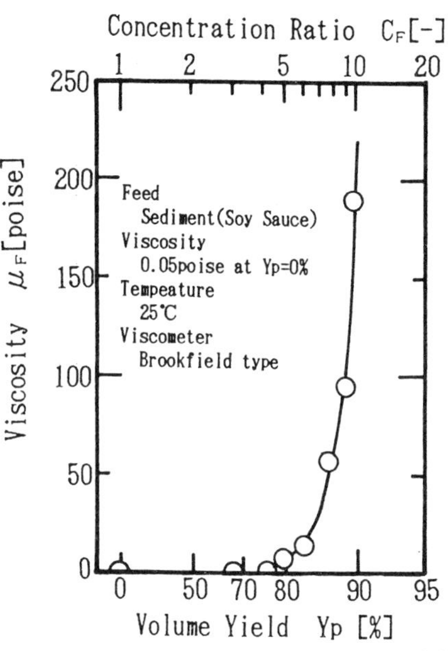

Fig.4 Influence of volume yield of permeate on viscosity of sediment

CONCLUSION

In order to recover soy sauce from sediment, we developed two novel cross-flow filtration modules with ceramic membranes. Compared with the conventional membrane module, the newly developed modules had the higher yield as well as the higher permeation flux. The proposed modules seem to be suitable for the concentration processes in food and medical field.

ACKNOWLEDGMENTS

We would like to express our appreciation to Mr.T.Takatoh of Shizuoka-kensan Soy Sauce Industrial Co. for his help in the experiments.

REFERENCES

1)Matsushita,K.,et al., Preprints of the 58th Annual Meeting of the Society of Chemical Engineers,Japan,B-206,Kagoshima(1993)
2)Takatoh,T.,et al., Shoken,**19**,No.1(1993)

CERAMIC FILTERS FOR FOOD AND BEVERAGE APPLICATIONS

M.FUSHIJIMA
Millipore Corporation 80 Ashby Rd. Bedford, MA 01730 USA

Abstract

Ceramic membrane filtration devices address environmental issues and process improvements in the food and beverage industry. Features and operating modes of ceramic microfiltration systems are described. Installed production process designs in Europe, USA and Japan are detailed for juice, soup, beer bottom, wine and vinegar processes to show the benefits of systems using ceramic filters.

1. Background

Diatomaceous earth (DE) filter media has traditionally been used to clarify beverage products such as beer and wine. However, environmental concerns are forcing users to seek alternative methods because disposal of media is becoming more restricted every year. Membrane microfiltration in a cross flow mode has been noticed as a potential replacement because it gives the same or "more sharp" separation consistently along with other benefits. Two negative factors related to conventional polymeric membranes have prevented a wide usage of cross flow membrane devices in beverage applications. They are 1) short membrane life time due to limited temperature and chemical resistance and 2) flavor changes caused by the extraction of polymers, especially adhesive.

Ceramic filters address all of these issues. Environmental concerns are minimized by eliminating the use and disposal of filter media, while offering the additional benefits of improved yield, consistent quality and cost savings in both labor and energy. Ceramic filters can have significantly longer lives without releasing polymers to product streams. Thus microfiltration using new ceramic materials has become not only an alternative, but a primary technology for these applications.

Ceramic filters features

Thermal resistant
Chemical/biological resistant
Sanitary design, open channel
Mechanical strength, backpulsable
Steamable, sanitizable
No flavor change
Energy efficient, 2-6 mm lumen
Minimum pretreatment
Long life 3-5 years

2. Features of ceramic filters

The most significant advantage of ceramic filter is their extraordinary thermal resistance, enabling SIP (steam in place) sterilization and high temperature operation. The filter module configuration is an open channel, the best sanitary design. Choices of lumen diameter of 2-6 mm allow optimizing energy efficiency and prefilter requirements. Larger diameters, 4-6 mm are used for viscous feeds like yeast, while 2-3 mm are selected for watery streams like wine. Membranes are available in a variety of materials; harder ones like α–alumina have superior durability compared to γ-alumina. Membrane pore sizes of 0.2μm, 0.45μm and 1μm are used depending on the application.

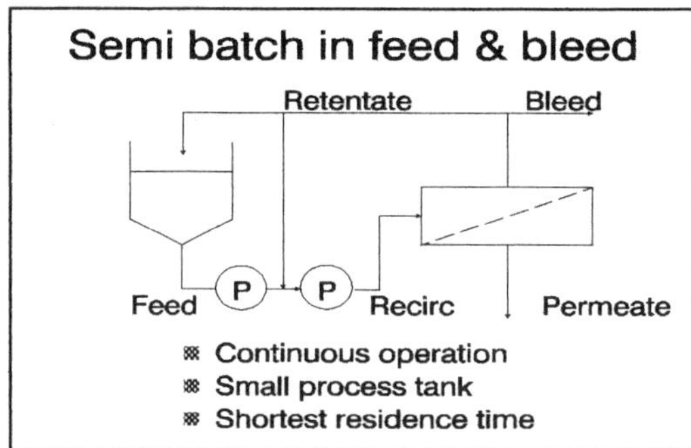

Many ceramic filtration systems are operated in a semi batch mode. Others include complete batch, feed & bleed continuous and multi-stage continuous operating modes. Feed is introduced to a process tank which is also used as a CIP tank to minimize tankage requirements. Retentate is returned to the process tank rather than the feed tank to avoid property changes. If necessary, diafiltration water is added after concentration to achieve higher yields. A system can be sanitized by steam or chemicals such as chlorine. Then it can be drained and stored in a dry mode eliminating storage steps. A significant advantage comes from the ceramic filter configuration is the "backpulse". This is possible because the membrane is sintered onto the substrate not adhered. By applying backpulse, higher and more stable fluxes can be achieved, while chemical cleaning is reduced.

3. Application

Cranberry juice The first application of ceramic microfiltration was to clarify cranberry juice. Initially, replacement of a DE based pressure filter was intended. However maintaining consistent clarity became more important, and the ceramic system is now used to polish output of the pressure filter. This ceramic microfiltration system in the USA processes 6 m^3/hr of cranberry juice bringing 10-20 NTU turbidity down to less than 1 NTU. Other benefits are reduced usage of DE and extended life of the down stream evaporator.

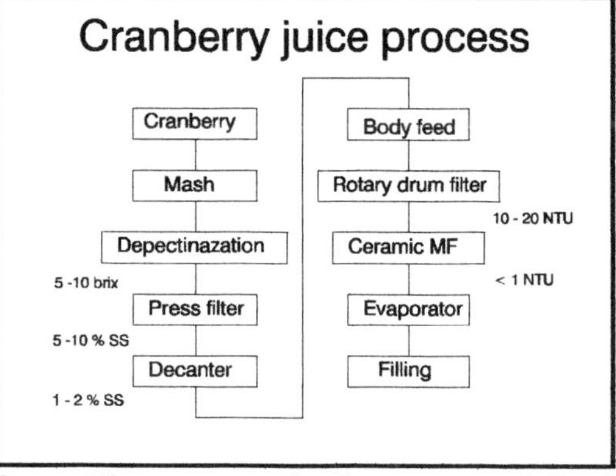

Soup The next application commercialized was soup clarification in Canada. A traditional method to clarify soup is to absorb fat and other cloudy material by adding high quality lean ground beef. This process required a full time operator and had to rely on his "sense by experience". Replacing this step with a ceramic filter that can withstand 50-60 °C operating temperature, significant savings were realized by eliminating raw material while maintaining consistent quality product.

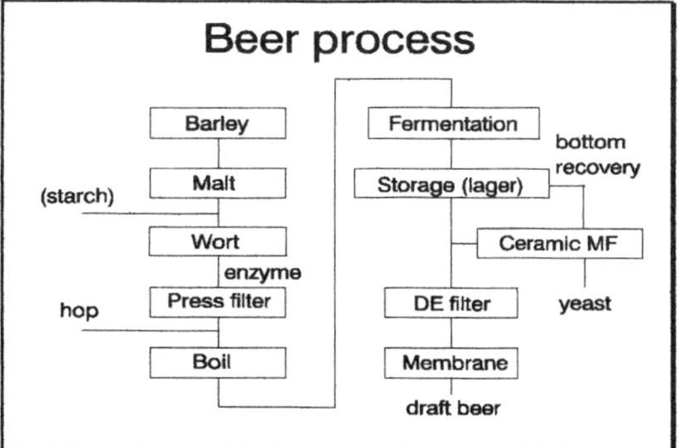

Beer bottoms Many attempts have been made to apply cross flow filtration to clarify beer or to concentrate beer bottoms as many reports can be cited in the literature. Beer bottoms concentration is first being commercialized in Europe and Japan because it requires relatively small capital investment, and gains from recovering beer often makes the pay back period less than one year. Beer bottoms or yeast are used as additives to medicine and natural foods. By concentrating two fold, a saving in transportation and fuel and recovery of beer are both achieved at the same time.

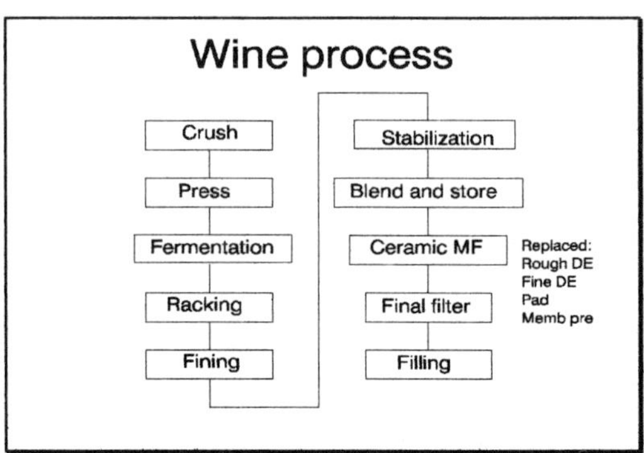

Wine A great deal of effort has been made to apply membrane technology in the wine making process to improve existing operations. The applications that have been investigated are juice and must clarification, clarification of wine and sterilization of wine before filling. Also with reverse osmosis, concentration of juice prior to fermentation and dealcoholization of wine have been attempted. Among many possibilities, two applications in the wine industry are being established. Clarification of wine after fermentation employs a traditional DE filter (coarse and fine) followed by a pad filter.

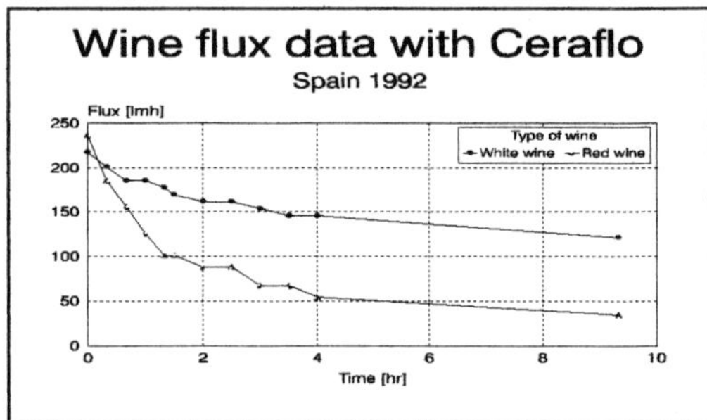

In Spain and Italy, ceramic microfiltration has been introduced to replace these steps. The benefits are elimination of DE disposal, reduced labor cost to one fifth, elimination of wine loss in the DE media or pad filter, improved wine clarity (fining index 10-20 to 1-5) and extended life time of the final membrane filter. Another hidden but important point is that wine quality became more consistent (separation is done by absolute pore size rather than depth filter mechanism). A typical performance and economic comparison are shown.

Wine economics
US cents / kl

	DE+Pad	Ceramic
Capital	36 (DE)	102
	8 (Pad)	
Energy	260	145
DE media	132	
Pad	50	
Ceramic		88
Labor	100	20
Total	586	355

Vinegar The last application mentioned here is vinegar clarification. After fermentation to acetic acid, broth can be clarified with either a DE filter or ceramic microfiltration as the process is shown. Another application in the vinegar process is stabilization by removing protein via ultrafiltration. In Italy and Japan, actual systems have been installed.

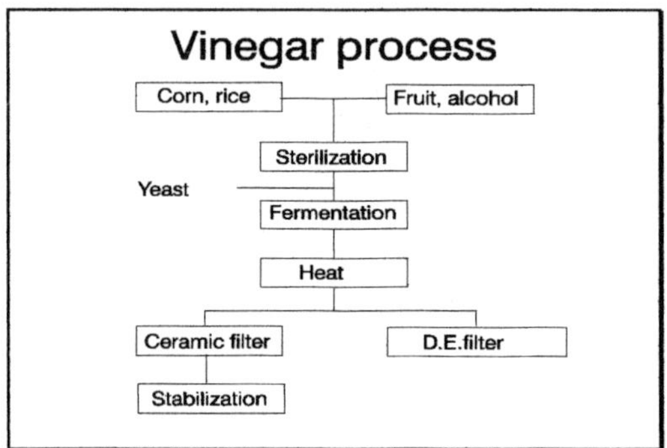

4. Conclusion

Ceramic cross flow filtration has found several commercial applications in food and beverage processes. The advantages of both material and configuration fit many needs by food equipment. Therefore, more and more new applications will be developed over the next several years.

Recovery and Functional Properties of Protein from the Wastewater of Mungbean Starch Processing by Ultrafiltration

W.C. KO, W.J. CHEN, and T.H. LAI
Department of Food Science,
National Chung Hsing University,
250 Kuokuang Road, Taichung, Taiwan, ROC.

ABSTRACT

Wet process, a conventional method for mungbean starch preparation, may cause about 80% protein loss in wastewater. In this study, ultrafiltration(UF) was used to recover the protein from the simulated wastewater of mungbean starch processing. 40℃ and 4 kg/cm^2 were found to be the appropriate condition for operation. Recovery of protein was 87.8% and 50.3% for molecular weight cut-off (MWCO) 30,000 and 500,000 membranes respectively. Protein remained in the wastewater could almost be separated by UF with the MWCO 30,000 membrane. It seems feasible to recover the protein by UF. This would not only prevent contamination but also improve the utilization of mungbean.

INTRODUCTION

Mungbean starch noodle is one kind of traditional Chinese food products. Wet process, the most popular method being used to prepare mungbean starch in Taiwan, may cause water pollution and loss of a nice protein source(1). It is necessary to recover the protein to prevent contamination and make the complete utilization of mungbean possible.

There had been some reports about protein recovery from the wastewater(2). In spite of over 80% protein being recovered, those methods were not adapted in actual practice because protein denaturation might be resulted and large amounts of coagulants were needed.

Ultrafiltration could be operated at considerably low temperatures and could also prevent the protein from denaturation. The objectives were to study the performance and process engineering characteristics of the spiral-wound modules used for isolation of protein from the wastewater of mungbean starch processing. Functional properties of recovered protein were also determined.

RESULTS AND DISCUSSION

The effect of inlet pressure on permeate flux is shown in Fig.1. Permeate flux increased with the increase of operating pressure under pressures below 4 and 3kg/cm², and then stabilized at values of 7.0 and 4.3L/m²/hr for the membranes of MWCO 30,000 and 500,000 respectively. It indicates that flux initially increased with pressure and then became independent. The major resistance to permeate flow in the pressure-independent regions was due to the concentration polarization-gellayer.

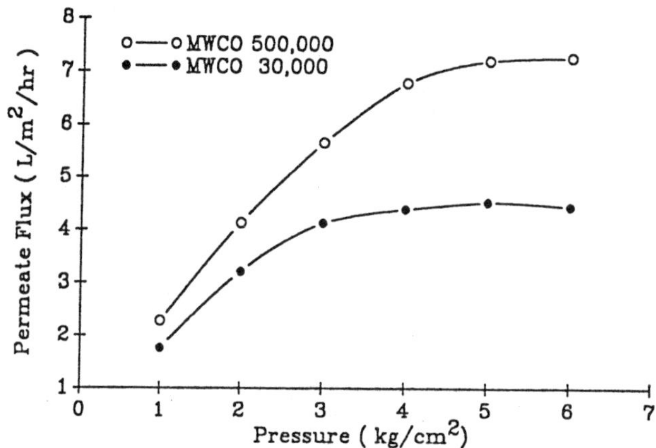

Figure 1. Relationship of operating pressure and permeate flux.

The effect of inlet temperature on permeate flux is shown in Fig. 2. Inlet temperature enchanced the permeate flux with a ratio of 3%/℃ for both membranes.

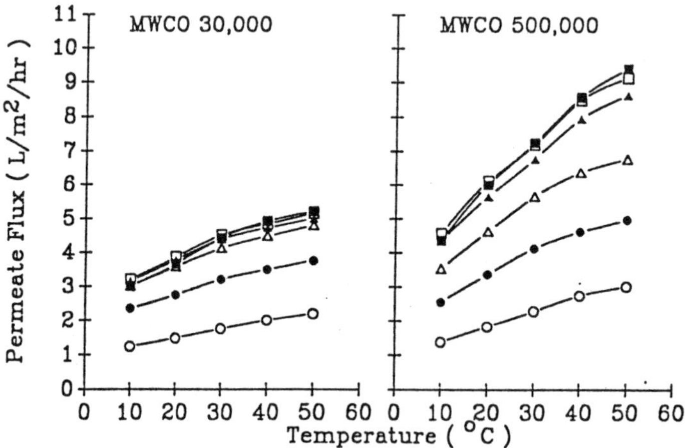

Figure 2. Relationship of inlet temperature and permeate flux.
○—○ 1 kg/cm²; ●—● 2 kg/cm² ; △—△ 3 kg/cm²
▲—▲ 4 kg/cm²; □—□ 5 kg/cm² ; ■—■ 6 kg/cm²

Rejection coefficient for total solids, protein and ash for the two membranes are illustrated in Table 1. MWCO 30,000 membrane could recover 90% protein and most total solids. On the other hand, MWCO 500,000 membrane could recover only 60% protein because 70~80% of protein was agglomerated into protein bodies(3).

TABLE 1

Recovery of the compositions during UF processing at VCR of 12.

MWCO	retentate 30,000	permeate 30,000	retentate 500,000	permeate 500,000
Total solids,%	62.5	37.3	55.8	42.3
Ash,%	34.1	59.3	34.9	62.1
Protein,%	92.1	5.2	55.9	43.7

VCR : volume concentration ratio

Functional properties, including solubility, emulsify activity, water-binding capacity and foam expansion, of spray- and freeze-dried products were determined. Because of the protein recovered by MWCO 500,000 membrane was mostly existed in the protein recovered by MWCO 30,000 membrane, functional properties of the products obtained from two membranes were not significantly different. Solubilities of the four products were over 85% and higher than the bubble separation method (2).

Table 2

Functional properties of drided retentates from UF treatment

MWCO drying method	30,000 spray	30,000 freeze	500,000 spray	500,000 freeze
solubility,%	88.71	85.24	86.33	85.15
water-binding capacity	7.23	7.09	8.46	7.89
foam expansion,%	76.5	80.3	78.9	79.4
emulsify activity O.D. (500 nm)	0.640	0.602	0.651	0.615

REFERENCES

1. Hill, J. and Vananuvat, P. Preliminary investigation of mungbean protein preparations. Report No.1, Applied Scientific Research Corporation of Thailand, Bangkok, 1968.
2. Peng, W.Y. and Chiang, W.C. Recovery of protein from the wastewater of mungbean starch noodle factory by absorption bubble separation. Food Sci.(Taiwan), R.O.C. 1982,8,93-102.
3. Harris, N. and Chrispeels, M.J. Histochemical and biochemical observations on storage protein metabolism and protein body autolysis in cotyledons of germinating mungbeans. Plant Physiol. 1975,56,292-7.

IMPROVED EXTRACTION AND ULTRAFILTRATION PROCESS FOR PROTEIN RECOVERY FROM SOY FLOUR

JOHN L. HARRIS[1], S.K. SAYED RAZAVI[1] & F. SHERKAT[2]
1. Department of Chemical & Metallurgical Engineering 2. Food Technology Unit
RMIT, 124 Latrobe Street, Melbourne 3000, AUSTRALIA.

ABSTRACT

By using a feedstock of dry milled enzyme-active full-fat soy flour, median particle size of 50 microns, an extraction process was developed which gave high solids and protein recovery. Aqueous extraction for 10 min at 80°C, and homogenization at 10 MPa followed by filtration gave 95.6% solids recovery and 86.9% protein recovery with a nitrogen solubility index of 93%. The native lipoxygenase was inactivated and the activity of trypsin inhibitors was lowered by 50%. The extract was concentrated to 20 wt% solids by ultrafiltration. The concentrate changed from dilatant to pseudoplastic behaviour at 14-15 wt% solids.

INTRODUCTION

Soybeans are highly valued for their protein content (about 40%) which can be extracted with water. The traditional method of extraction by wet-milling the soaked soybeans, heating the mash for 15-20 min near its boiling point, followed by filtration gives a low solids recovery of 26% and protein recovery of 33% (1). Improved yields of solids (58%) and proteins (44%) were reported for a hot grind process (1). Another method where the soaked soybeans were blanched at 90°C for 3 mins, ground hot for 5 mins and filtered gave a solids recovery of 73%, protein recovery of 81.6%, inactivation of lipoxygenase and reduction in trypsin inhibitor activity of 62% (2). In a recent method, 80% protein solubility was obtained by extraction at 80-85°C for 2 min (3). A continuous steam infusion cooking system, known as rapid-hydration hydrothermal cooking (RHHC), made a hot water slurry with full-fat soy flour at ≥80°C followed by cooking at 154°C for 18 s, and resulted in lipoxygenase inactivation with 100% solids recovery as no separation steps were involved (4).

The heating (temperature-time relationship) required for inactivation of trypsin inhibitor can have a deleterious effect on product quality in terms of protein solubility. The desired degree of solubility depends on the end use of the product (5).

Ultrafiltration has been used successfully to remove undesirable factors and produce specific products from soy extract. 96% removal of oligosaccharides, 90% removal of phytic acid and 17% reduction of trypsin inhibitor activity have been achieved (6).

MATERIALS AND METHODS

The enzyme active full-fat soy flour was milled from cleaned, dehulled soybeans and had a proximate composition of 40% protein, 20% crude lipid, 27% carbohydrate, 5% ash and 8% moisture (supplied by Soy Products of Australia Pty. Ltd.). The particle size specification of the flour was 95% pasing through 100 mesh sieve. The extraction method involved preheating the water to the required temperature (30-90°C), mixing the soy flour with water (1:12 w/w), stirring at 1200 rpm with a propeller stirrer for 10 min (excessive frothing occurred for longer extraction times), cooling in a cold water bath to 40°C, homogenizing with a Manton Gaulin V90 homogeniser and filtering through a 150 micron sieve. The soy extract contained approximately 6.2% total solids. Ultrafiltration was performed at 50°C and 400 kPa in a DDS flat plate module with 2.25 m^2 area of nominal 50,000 M.W. cut-off polysulphone membrane.

Analyses performed included total solids [AOAC gravimetric method 925.23 (1990)], trypsin inhibitor activity [AOCS method Ba 12-75 (1983)], total nitrogen by Kjeldahl method, total protein as *6.25 x Total Nitrogen* and nitrogen solubility index (NSI) [AOCS Method Ba 11-65 (1985)]. Physical properties measured included particle size using Malvern Instruments Master Sizer X Ver. 1.0 and viscosity using Contraves Rotational Viscometer Rheomat 115.

RESULTS AND DISCUSSION

There is a significant difference between the particle size distribution curves of unfiltered suspensions of full-fat and defatted flour prepared at 20°C as shown in Fig.1. Both suspensions exhibited peaks at 50 μm, but there is an absence of a 10 μm peak for the defatted flour. The soy extracts prepared at temperatures up to 90°C also exhibited peaks at 50 μm as shown in Fig.2, but the 90°C extract contained more larger particles due to protein denaturation. The effect of homogenization at 10 MPa was to slightly reduce the median particle size, but homogenization at 17.5 MPa lead to an increase in frequency of large particles (>200μm).

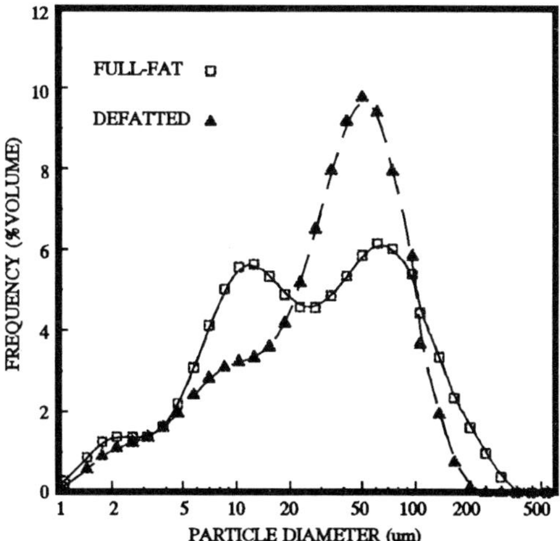

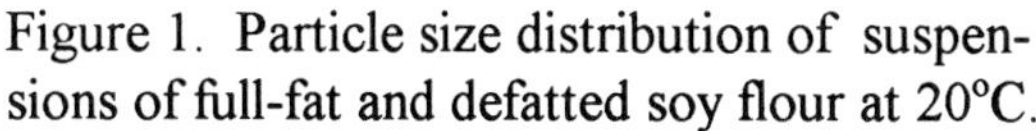
Figure 1. Particle size distribution of suspensions of full-fat and defatted soy flour at 20°C.

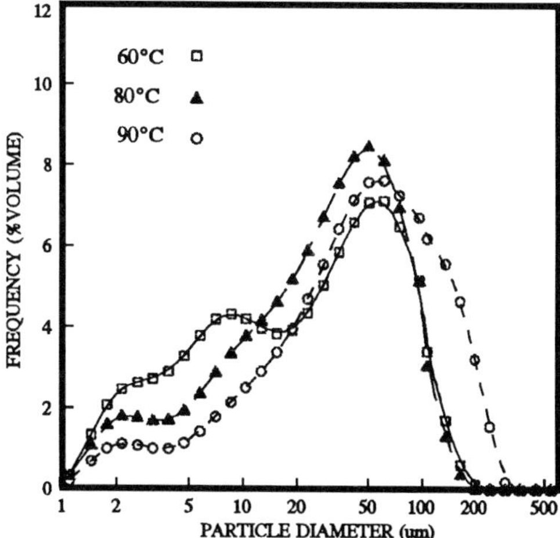

Figure 2. Effect of extraction temperature on the particle size distribution.

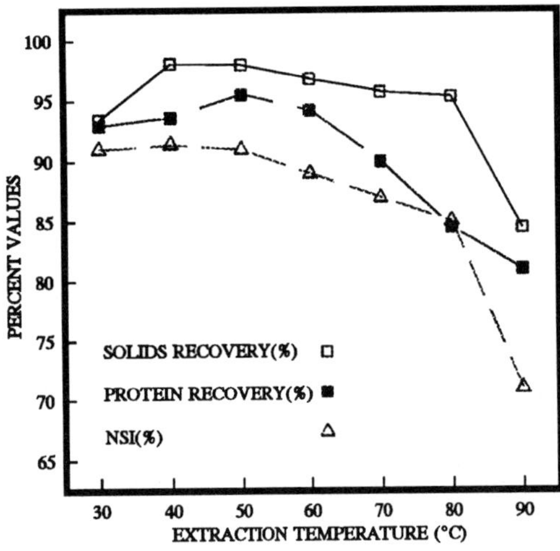

Figure 3. Effect of extraction temperature on solids recovery, protein recovery and NSI.

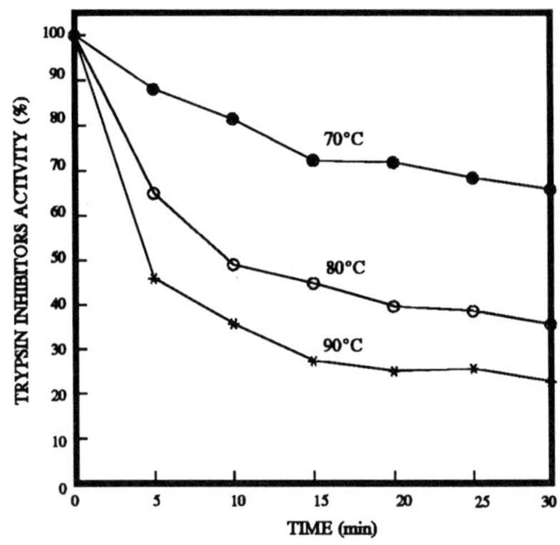

Figure 4. Heat inactivation of trypsin inhibitor.

Solids recovery exceeded 95% for extraction temperatures up to 80°C, and declined thereafter (Fig.3). Protein recovery declined from 50°C indicating heat denaturation of protein molecules. This is supported by the NSI decreasing from 50°C, with sharp decrease above 80°C. Homogenization at 10 MPa gave a 9% increase in NSI at 80°C. Trypsin inhibitor activity was lowered by 50% for heat treatment of 10 min at 80°C (Fig.4).

The extract was concentrated from 6 to 20 wt% solids by ultrafiltration. A substantial reduction in the permeate flux rate was observed at 14 wt% solids. Interestingly, the rheological behaviour of the concentrate measured at 25°C changed from dilatant (<14 wt% solids), to Newtonian for 14-15 wt%, to pseudoplastic (14 - 20 wt% solids).

CONCLUSIONS

Water extraction at 80°C for 10 min with homogenization at 10 MPa resulted in a solids recovery of 95%, protein recovery of 86.9% and nitrogen solubility index of 93%. The residual trypsin inhibitor activity was 50%. The next step of process improvement is to investigate extraction temperatures above 80°C for extraction times less than 10 min.

REFERENCES

1. Johnson, K.W. and Snyder, H.E., Soymilk: A comparison of processing methods on yields and composition. J. Food Sci., 1978, **43**, 349.
2. Al-Kishtaini, S.F., Methods of preparation and properties of water extracts of soybeans. Ph.D. thesis, Univ. of Illinois, Urbana, IL.,1971.
3. Xie, Z.L. and Frelzdorff, B., Optimizing soybean blanching in relation to lipoxygenase inactivation and protein solubility for soymilk production. Zeitschrift fur Lebensmittel-Untersuchung und-Forschung.1992, **194(1)**, 43-46.
4. Kim, C. Physico-chemical properties of soybean extracts processed by rapid-hydration hydrothermal cooking. Thesis, Iowa State Univ. of Sci & Tech., Ames, IA., 1989.
5. Snyder, H.E. and Kwon, T.W. Soybean Utilization. AVI Publ. Co. New York, 1987.
6. Omosaiye, O., and Cheryan, M. Ultrafiltration of soybean water extracts: Processing characteristics and yields. J. Food Sci., 1979, **44**, 1027.

THE PRODUCTION OF PROTEIN ISOLATES FROM CHINESE RAPESEED

Levente L. Diosady,
Department of Chemical Engineering, University of Toronto,
Toronto, Ontario, Canada, M5S 1A4

A process tailored to the chemical characteristics of rapeseed and canola proteins was developed, consisting of extraction of oil-free meal at pH 10.5-12.5, isoelectric precipitation to recover the isoelectric proteins and ultrafiltration followed by diafiltration to concentrate and purify the acid-soluble proteins remaining in solution. These steps complement one another to give three products with excellent protein recovery. Using 5 types of canola meal, "isoelectric" and "soluble" protein isolates containing 87-104% protein (Nx6.25) and a meal residue were obtained. All fractions were free of glucosinolates. The two isolates were low in phytate, light in color, and mostly bland in taste. The isolate yield was dependent on the starting meal. The "soluble" protein isolate has a high nitrogen solubility index throughout the pH range of 2.5 to 11, which makes it suitable for use in carbonated soft-drinks. This makes it possible to produce safe and nutritious soft drinks for regions of the world that are chronically short of protein. The process is especially suited to Chinese rapeseed meal, which is now primarily wasted as a fertilizer, due to its large glucosinolate content.

INTRODUCTION

Canola is the source of 60% of the vegetable oil consumed in Canada, and it is the third largest oilseed crop world-wide. The meal produced by extracting the oil is an excellent source of proteins, but unfortunately it contains glucosinolates which are toxic; phytin, which is an anti-nutritional substance; polyphenols, which produce a dark colour and a bitter flavour; and some 30% hull, which is high in fibre. These restrict its use in feed, and prevent its use in food products.

We, in the Food Engineering Group at the University of Toronto have developed a novel approach to the processing of canola. Using a solvent system consisting of methanol containing ammonia, and hexane a high-quality degummed edible oil and a meal, that is essentially free of glucosinolates and low in polyphenols, are produced. The meal contains about 50% protein and is light in colour and bland in taste. It is much superior to conventional canola meal, and its protein content is nutritionally equivalent to soybean meal, [1].

China is the largest producer of rapeseed, producing mainly varieties that contain very high levels of glucosinolates. The use of the meal in feed is severely restricted, and it has been primarily used as fertilizer - a waste of some 1.5 million tons of excellent protein annually. The methanol-ammonia/hexane extraction process has been successfully applied to Chinese rapeseed, to produce a meal superior to canola meal, which may be used freely in animal feed. Unfortunately, due to the high phytate and fibre content of the meal, as well as the residual glucosinolates, the meal is unsuitable for human consumption.

Canola and rapeseed protein has a somewhat better amino acid distribution than soy protein. Unfortunately, unlike the protein in soybeans, it is very difficult to extract, and to recover by the conventional protein isolation techniques. The losses in the process has made its production uneconomical.

By combining membrane concentration and purification techniques with conventional protein isolation methods we have produced two high-quality protein products:

1. an *isoelectric protein isolate*, (PPI) which contains ~95% protein (Nx6.25, dry basis), which has similar functional properties to commercially available soy protein isolates; and

2. an *acid soluble protein isolate*, (SPI) which contains ~95% protein (Nx6.25, dry basis), which is soluble under mildly acidic conditions, down to pH 3.

Both isolates are free of glucosinolates or their breakdown products, and have low phytate levels (1.0-1.5%). They are white in colour and bland in taste. The goal of our project was to develop a process for making food-grade protein isolates from canola and conventional (Chinese) rapeseed varieties.

MATERIALS AND METHODS

Chinese rapeseed (Nin-U 7 from Jiangsu province) was extracted with methanol containing 10% ammonia (w/v) and 5% water (v/v) as described by Rubin et al., [2]. The resulting meal was extracted with 30 volumes of aqueous NaOH at pH values between 10.0 and 12.5. The extract was filtered, and 0 to 15% $CaCl_2$ was added, and the pH decreased to a predetermined value between 3.5 and 7.0, thus precipitating a significant fraction of the dissolved protein. The remaining solution was ultrafiltered to a concentration factor of 10, and diafiltered with a diavolume of 5 to produce a clear solution. The acid-soluble isolate was recovered by freeze drying of this solution. The procedure closely followed that published earlier, [3].

The composition and the functional properties of the resulting isolates were determined according to the techniques described earlier [4].

RESULTS AND DISCUSSION

The composition of canola and Chinese rapeseed and the meals produces are shown in Table I.

Sample	Rapeseed variety	Oil (%)	Protein (%)	Glucosinolate (μmol/g)	Phytic acid (%)
Seed	Chinese	38.8±0.2	25.2±0.2	-	-
	Canola	44.5±0.3	21.6±0.2	-	-
Hexane-defatted meal	Chinese	-	48.1±0.4	113.2±5.1	3.12±0.14
	Canola	-	44.7±0.1	13.0±0.5	3.86±0.05
$CH_3/NH_3/H_2O$-hexane-extracted meal	Chinese	-	54.8±0.3	4.0±0.3	3.86±0.05
	Canola	-	48.4–0.2	0.0±0.2	4.39±0.04

At pH 12, the protein extractability of methanol/ammonia/water-hexane-extracted Chinese rapeseed meal was found to be over 70%, which was some 15% higher than that of canola meal treated by two-solvent-phase-extraction. Although Chinese rapeseed meal also showed a slightly greater phytic acid extractability than canola meal, since its initial phytate content was some 30% lower, the amount of phytic acid dissolved in the alkaline extract was actually somewhat lower than in the case of canola.

The extracted Chinese rapeseed protein exhibited distinctly different solubility and precipitation yields than typical canola protein. While ~55% of the dissolved canola protein was precipitated at pH 3.5, more of the Chinese rapeseed protein remained in solution at this pH, but was insoluble at neutral pH values. We found that the phytate content of the isolates was significantly lower than that of canola isolates, and thus $CaCl_2$ treatment to dissociate phytate-protein complexes was unnecessary, thus simplifying the process.

The pH of a 10% dispersion of $CH_3OH/NH_3/H_2O$-hexane-extracted Chinese rapeseed meal had a pH value in the alkaline range, which was more than one unit higher than its canola counterpart. The slurries of the isolates were in the neutral pH range, as expected. While soybean protein had a minimum nitrogen solubility index (NSI) at pH 5, the NSI of Chinese rapeseed PPI reached its lowest point at pH 6. Over the almost entire pH range, the NSI of Chinese rapeseed SPI remained consistently high. This will make it possible to incorporate this product in carbonated beverages as a true solution.

$CH_3OH/NH_3/H_2O$-hexane-extracted Chinese rapeseed meal was slightly inferior in most functional properties to similarly prepared canola meals The water absorption of Chinese rapeseed PPI was 220%, some 80% lower than that of the canola PPI. All rapeseed protein products investigated were superior to the soybean protein isolate in fat absorption with Chinese rapeseed SPI giving the highest value. Both Chinese rapeseed protein isolates showed a excellent whippability. The foaming properties of the soybean protein isolate were intermediate between those of the rapeseed meals and isolates. All samples produced foams that were stable for more than 2 hours.

CONCLUSIONS

A process for producing acid-soluble, food grade protein isolates from high-glucosinolate Chinese rapeseed has been developed. China, and India, the largest rapeseed producers are chronically short of protein. A soft drink containing *acid soluble protein isolate* could provide the same protein nutrition as milk, in a safe source of water. It has the additional advantage to being digestible by many people that cannot tolerate milk or soy proteins.

REFERENCES

[1] L.J.Rubin, L.L.Diosady and C.R.Phillips - "Solvent Extraction of Oil Bearing Seeds" U.S. **4,460,504** (1984)

[2] L.J.Rubin, L.L.Diosady, M.Naczk and M.Halfani - "The Alkanol-Ammonia-Water/Hexane Treatment of Canola", Can.Inst.Food Sci.Technol.J., **19**:1, 57-61, (1986)

[3] Y-M. Tzeng, L.L. Diosady and L.J. Rubin - Production of canola protein materials by alkaline extraction, precipitation, and membrane processing, J. Food Science, **55:4** 1147-56, 1990.

[4] M.Naczk, L.L.Diosady, and L.J.Rubin.-"The Functional Properties of Canola Meals Produced by a Two-Phase Extraction System", J.Food Sci. **50**, 1685-1689 (1985)

SEPARATION AND CONCENTRATION OF POLYUNSATURATED FATTY ACIDS BY A COMBINED SYSTEM OF LIQUID-LIQUID EXTRACTION AND MEMBRANE SEPARATION

YUKO SAHASHI, HIROTOSHI ISHIZUKA, SEIJI KOIKE*, and KAZUAKI SUZUKI*

Medical Reseach Lab., Nitto Denko Corp., Shimohozumi, Ibaraki, Osaka 567, Japan

*Fat & Food Lab., Asahi Denka Kogyo K.K., Higashi-ogu, Arakawa-ku, Tokyo 116, Japan

ABSTRACT

PUFAs-rich glycerides were separated from fish oil hydrolysate by membrane separation system. This system is a combination of extraction using ethanol solution of 75 (v/v)%, and separation using hydrophilic UF membrane. The results of PUFAs recovery in the system agreed well with the expected results from analytical data of hydrolysate. It was also found that our technique was more effective than fractional crystallization in PUFAs recovery from fish oil. The quality of oil was not damaged throughout these processes.

INTRODUCTION

n-3 Polyunsaturated fatty acids (PUFAs) have recently attracted special interest because of their physiological activities. It is reported that PUFAs in the form of glycerides are absorbed to the small intestine better than those in the form of ethylester.

The aim of our work is to develop an economical separation technique, concentration and purification of PUFAs in glycerides without damaging their qualities. The diagram of our research, composed of lipase hydrolysis and membrane separation system, is shown in Fig. 1. Fish oil was first selectively hydrolysed with lipase. The hydrolysate was mainly composed of PUFAs-rich glycerides and other free fatty acids (FFAs). The glycerides and FFAs were separated by membrane separation technique, so that the PUFAs-rich glycerides were separated. The utilized membrane separation system will further be discussed in this report.

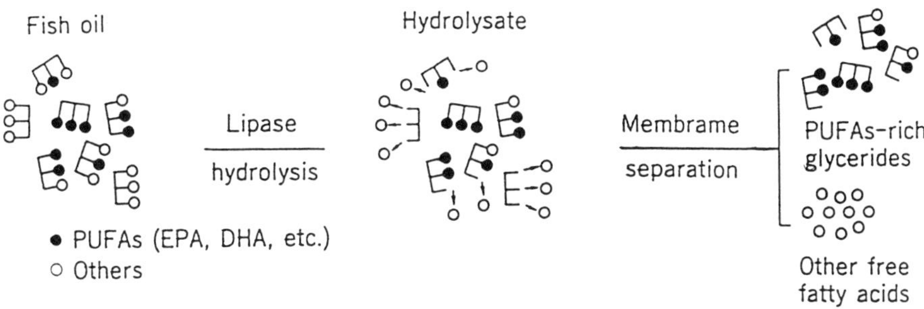

Figure 1. A combined system of selective hydrolysis with lipase and membrane separation of PUFAs-rich glycerides.

MATERIALS AND METHODS

A schematic flowsheet of the membrane separation combined with solvent extraction is shown in Fig. 2. Hydrolysate of fish oil was mixed with extraction solvent to extract only FFAs. The mixture in form of O/W emulsion was forced by applying pressure to permeate through a membrane. FFAs permeated through the membrane with solvent. Whearas glycerides in oil phase were rejected by the membrane. The PUFAs-glycerides were concentrated in concentrate. The solvent was recycled after passing through an ion-exchange column to remove FFAs. Cross-flow filtration and rotating disk membrane systems were tested in a large scale process.

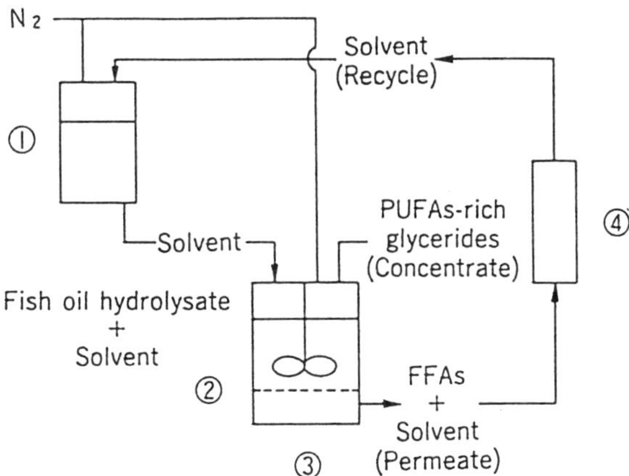

Figure 2. Schematic flowsheet of a combined system liquid-liquid extraction and membrane separation.

RESULTS AND DISCUSSION

Among various solvents, ethanol solution of 75 (v/v)% was most suitable for separation of glycerides and FFAs. Glycerides content in raffinate increased to 98% when 20 ml solvent per g-oil was used. Solvent extraction was combined with membrane separation technique. It was found that hydrophilic UF membrane (M.W. cut-off=20, 000, Material: Polyimide), was most suitable for separation of FFAs and glycerides, where almost all FFAs permeated through the membrane, and glycerides composed of mainly triglycerides retained in concentrate.

In hydrolysis ratio of 54%, PUFAs content in concentrate increased from 34% to 52%, and PUFAs recovery yield from fish oil was 71%. Relation between PUFAs recovery yield from fish oil and PUFAs content in glycerides in our membrane separation was compared with that in fractional crystallization method as shown in Fig.3. The composition of hydrolysate was analyzed. PUFAs content in glycerides and PUFAs recovery yield were determined. The closed circles and reverse triangles show the expected results from analytical data of hydrolysate. After membrane separation was completed, the result was reasonable with a little losses (shown as open reverse triangles). When PUFAs content was nearly 50%, PUFAs recovery in our system increased above 2 times as large as that in fractional crystallization.

It was also found that utilizing a rotating disk membrane system in a large scale process was effective for separation. The quality of the separated oil was good.

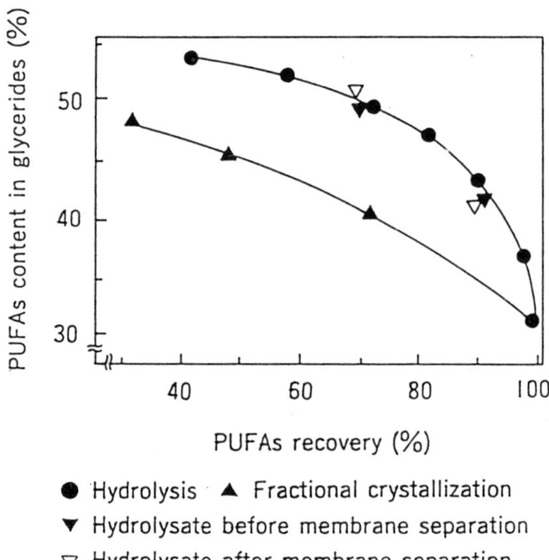

Figure 3. Relation between PUFAs recovery yield and PUFAs content in glycerides.

CONCLUSION

It was found that utilizing a combined system of liquid-liquid extraction and membrane separation was effective for separation for PUFAs-rich glycerides without damage.

REFINING VEGETABLE OILS BY MEMBRANE TECHNOLOGY

M.CHERYAN, L.P.RAMAN and N.RAJAGOPALAN
University of Illinois, Agricultural Bioprocess Laboratory
1302 W. Pennsylvania Avenue, Urbana IL 61801, USA

ABSTRACT

Membranes can solve some of the major drawbacks of current vegetable oil processing methods. They can greatly enhance the efficiency of hexane recovery, combine degumming and bleaching into one step, solve the pollution problems of alkali deacidification, and also be used to treat gas and vapor streams.

INTRODUCTION

Vegetable oil processing is usually conducted in three stages: Pre-processing; Extraction (where the oil is recovered from the oilseeds); and Refining (where the extracted oil is purified to remove the undesirable components before being packed for consumer use). The purification steps include degumming, deacidification, bleaching and deodorization which are presently accomplished by physical and chemical means.

There are several drawbacks to current oil processing technology: (a) The operations are energy intensive. The soybean oil industry in the U.S. alone uses 2.2 trillion kilojoules of energy annually for the recovery of solvent and oil from the miscella. (b) Significant amounts of neutral oil are lost during the purification steps, due to saponification of tri-glycerides and occlusion of oil in the soapstock. Higher acidity results in higher refining losses. Hence, it is often uneconomical to refine certain oils like rice bran oil, which can have up to 40% free fatty acids (FFA), with current technology. (c) Large amounts of water and chemicals are utilized, which increases production costs considerably, and (d) Several processing steps result in heavily contaminated discharges; e,g., the soapstock splitting process results in a heavily polluting effluent.

MEMBRANES IN VEGETABLE OIL PROCESSING

Membrane technology can solve or reduce these problems, by exploiting the differences in physical and chemical properties at a molecular level between various components of the feed streams in an oil refinery. The earliest publication in this field is probably the U.S. patent by Sen Gupta in 1978 [1] for a degumming process. Conceptually, membranes can be used in almost all the stages of oil production and purification, such as the recovery of the solvent used in extraction, removal of phospholipids (degumming), refining (deacidification), bleaching/removal of color pigments, treatment of effluents and production of nitrogen for packaging (Figure 1). Except for nitrogen production by gas separations (GS), we are unaware of any commercial membrane installations in the vegetable oil industry.

Degumming

In the conventional method, oil is first separated from the miscella and then the crude oil is degummed. In contrast, membrane degumming can be done with the miscella itself, taking advantage of the amphiphilic nature of the phosphatides, which causes them to form micelles in the miscella. The phospholipids-micelles act like macromolecules with molecular weights exceeding 20,000 daltons. Thus, ultrafiltration membranes could be employed to separate them from oil and hexane [1,2].

Bleaching

"Bleaching" refers primarily to the process of reducing the color of oil by adsorbing the coloring components such as carotenoids and chlorophyll on to activated clay or carbon. Some of the color compounds are removed in the membrane degumming process. An economic analysis of the membrane bleaching process for a plant processing 250 tons of oil per day indicates potential savings of \$(US)730,000 annually [3]. The analysis is for the bleaching process alone and possible additional savings from the membrane degumming is not included.

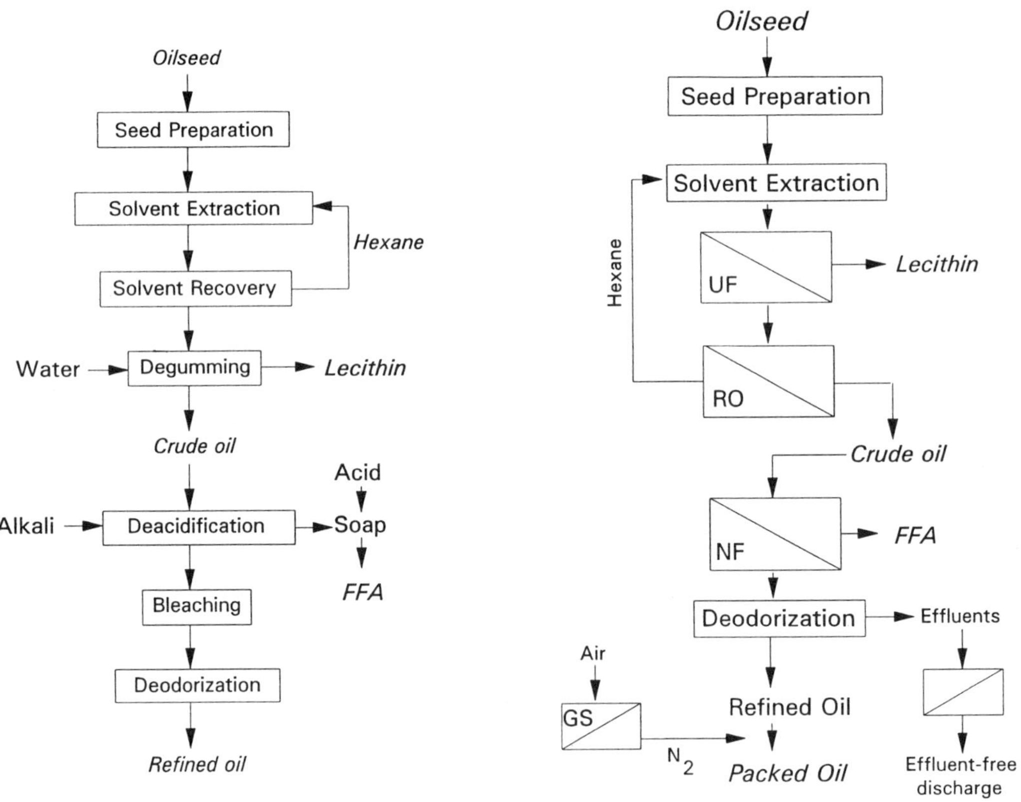

Figure 1. Left: Conventional vegetable oil processing. Right: Membrane applications in vegetable oil processing

Solvent recovery
If membranes could be used to pre-concentrate the miscella before being sent to evaporators, it would reduce the energy needed for this process by 43% [4]. In addition, since the volume that is being processed is reduced, the size of the evaporators can be reduced. This reduction in energy and capital costs is likely to make the alcohol process in cottonseed extraction economically viable. However, recovery of hexane from oil-hexane miscella is difficult [5] because very few membranes with the desirable separation characteristics are stable in hexane [4].

Deacidification
Deacidification has the maximum economic impact on oil production and any inefficiency in this process has a great bearing on any subsequent oil purifying or modification processes. "Single" membrane systems [6,7], "two-membrane" and "three-membrane" systems [8] have been proposed in recent years; each has its own advantages and limitations and may require fundamental changes from current methods.

Other membrane applications
These include dewaxing [9], removal of metals [10], catalyst recovery from hydrogenated oil [11], production of nitrogen for packaging [12], membrane recovery system for scrubbing solvent vapors [12] and treatment of frying oils for re-use [13].

REFERENCES

1. Sen Gupta, A.K., Purification process. U.S. Patent, 1978, 4,093,540.
2. Iwama, A. and Y. Kazuse., Process for the purification of crude glyceride oil compositions. U.S. Patent, 1983, 4,414,157.
3. Raman, L.P., M.S. Thesis, University of Illinois, 1993.
4. Cheryan,M., Raman,L.P. and Rajagopalan,N., Solvent recovery in edible oil manufacture. Patent disclosure, University of Illinois, Urbana, 1992.
5. Koseoglu, S.S., Lawhon,J.T. and Lusas,E.W., Membrane processing of crude vegetable oils: Pilot plant scale removal of solvent from oil miscellas. Journal of American Oil Chemist's Society, 1990, **67**, 315-322.
6. Sen Gupta, A.K., Refining. U.S. Patent, 1985, 4,533,501.
7. Cheryan,M., Raman,L.P. and Rajagopalan,N., Refined vegetable oil. Patent disclosure, University of Illinois, Urbana, 1992.
8. Keurentjes, J.T.F., Ph.D. Thesis, Agricultural University of Wageningen, 1991.
9. Mutoh, Y., Matsuda,K., Ohshima,M. and H. Ohuchi,H., Method of dewaxing a vegetable oil. U.S. Patent, 1985, 4,545,940.
10. Keurentjes,J.T.F., Bosklopper,T.G.J., van Drop, L.J. and van't Riet,K., The removal of metals from edible oils by a membrane extraction procedure. Journal of American Oil Chemist's Society, 1990, **67**, 28-32.
11. Koseoglu, S.S. and Varva,C.J., Catalyst removal from hydrogenated oil using membrane processing. INFORM, 1992, **3**, 536.
12. Mohr,C., Leeper,S.A.,Engelgau,D.E. and Charboneau,B.L., Membrane applications and research in food processing: An assessment,DOE Report DOE/ID 10210, 1988.
13. Koseoglu,S.S., Membrane processing of used frying oils. INFORM, 1991, **2**, 334.

This research was supported by the Illinois Soybean Program Operating Board and the Illinois Agricultural Experiment Station at Urbana-Champaign.

MEMBRANE SEPARATION OF SUNFLOWER OIL HYDROLYSATES IN ORGANIC SOLVENTS

Seiji Koike*, Masakatsu Yokoo, Hiroshi Nabetani, Mitsutoshi Nakajima.
National Food Res. Inst., MAFF; Tsukuba, Ibaraki 305, Japan,
*Asahi Denka Kogyo K.K.; Arakawa-ku, Tokyo 116, Japan

ABSTRACT

Various types of commercial membranes (18) were used to separate the constituents of oils including triglycerides (TG), diglycerides (DG), monoglycerides (MG) and free fatty acids (FFA) in organic solvents. Cellulose acetate (CA) membrane in ethanol and gas separation membrane in hexane gave higher permeate fluxes, and large rejection differences between solutes. When membrane separation of the oil–ethanol mixture using CA membrane (NTR-1698) was repeated 16 times in batch mode (diafiltration mode, initial retentate weight 100g, and total permeate weight 504g), the composition (TG:DG:MG:FFA) of retentate changed from 31:28:9:32 to 65:29:2:4, respectively.

INTRODUCTION

Solvent recovery in extraction process and fractionation of fats and oils are industrially energy consuming steps, and more economical processes such as membrane technology are required. Recently, some researchers have reported that removal of phospholipids (degumming) was possible by using ultrafiltration (UF) membranes.[1), 2)] However, few reports are available on the use of reverse osmosis (RO) membranes for these processes. In this study, RO membranes were used to separate the constituents of oils, such as triglycerides (TG), diglycerides (DG), monoglycerides (MG), and free fatty acids (FFA) in organic solvents.

MATERIALS AND METHODS

1. Materials

Lipase hydrolysate of high oleic sunflower oil, in which more than 80% of the fatty acid composition was oleic acid, was used as a lipid mixture for the membrane separation test. The hydrolysate contained TG, DG, MG and FFA in the weight fraction of 31, 28, 9 and 32, respectively.

2. Membrane Separation Test

Membranes (ϕ75 mm, 32 cm^2 area) were evaluated using a flat membrane test cell (Nitto Denko Corporation, C70–B). The testing unit was operated in batch mode by charging the cell with 100 cm^3 of oil–solvent mixture and applying pressure as required by nitrogen gas. The permeated oil–solvent mixture was collected through a port beneath the membrane support, and the permeate flux was calculated. The compositions of FFA, MG, DG and TG of initial and final retentates, and permeates were determined by TLC–FID method.[3] The solvent in the permeate was evaporated, and then solutes weight was measured. From these determinations the mass–balance was calculated. Each solute weight was corrected by partitioning proportionally. Observed rejections were determined as follows, assuming that observed rejections are constant during the experiment (Eq.1):

$$R_{S,\,obs} = \ln(C_S / C_{S,\,0}) / \ln(W_0 / W) \qquad (1)$$

where, $C_{S,\,0}$ and C_S are the initial and final concentrations (wt./wt. solvent %) of each solute in the retentates, and W_0 and W are the initial and final solvent weights in the retentate, respectively.

RESULTS AND DISCUSSION

1. Membrane screening

Each membrane was tested with ethanol or hexane as a solvent containing 6.67% (wt./wt. ethanol) or 7.52% (wt./wt. hexane) oil. Results of the flux and the rejection using different types of membranes are shown in Figures 1 and 2. In the case of oil–ethanol mixture, the permeate fluxes were in the range of 1–13 g m^{-2} s^{-1}. Cellulose acetate (CA) membranes gave higher flux, and large difference of rejection among FFA, MG, DG and TG. Polyvinyl alcohol, polyamide and polyether membranes yielded a relatively high rejection, but there was a little difference between solutes.

Lower fluxes and rejections were usually observed in the hexane system compared to the

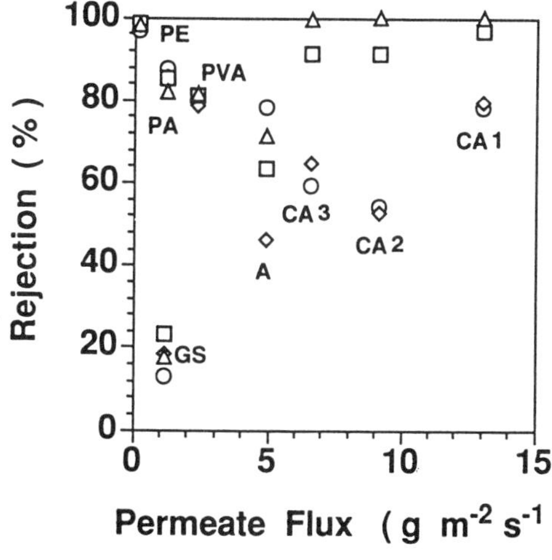

△ : TG, □ : DG, ◇ : MG, ○ : FFA,
CA1 : SC-3000, CA2 : NTR-1698, CA3 : DRC-97,
A : UTC-70T, PVA : NTR-729HF, PE : PEC-1000,
PA : NTR-759HR, GS : NTGS-2100

Figure 1. Permeate Flux and Rejection in Ethanol System.

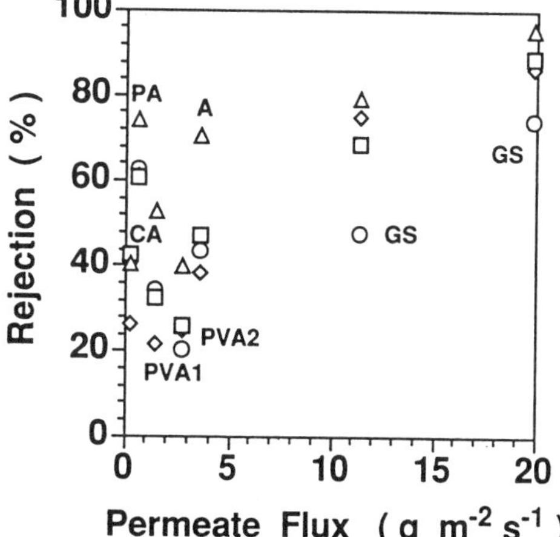

△ : TG, □ : DG, ◇ : MG, ○ : FFA,
CA : NTR-1698, PVA1 : NTR-729HF,
PVA2 : NTR-7250, A : UTC-70T
PA : NTR-759HR, GS : NTGS-2100

Figure 2. Permeate Flux and Rejection in Hexane System.

ethanol system. The gas separation membrane gave the highest flux and high rejection (70.2% for FFA, 94.4% for TG), the differences between substances, however, was not so remarkable compared to the CA membrane. The CA membrane in ethanol system was therefore used in the following experiments.

2. Effect of oil concentration on rejection and flux

Figure 3 shows the effect of oil concentration on rejection and permeate flux for the CA membrane. The oil concentration per weight of ethanol was changed from 6.7% to 84.4%. Permeate flux decreased according to oil concentration, perhaps due to the effect of osmotic pressure. TG shows the highest rejection of the 4 substances at all concentrations. Rejection values of TG were higher than 98%. The rejection of DG was 90% at low concentration, which was noticeably lower than that of TG. At concentrations greater than 30%, rejection of DG increased and rejection values were nearly equal to that of TG. Rejections of MG and FFA were lower than DG and TG, and varied in the range of 50% to 70%.

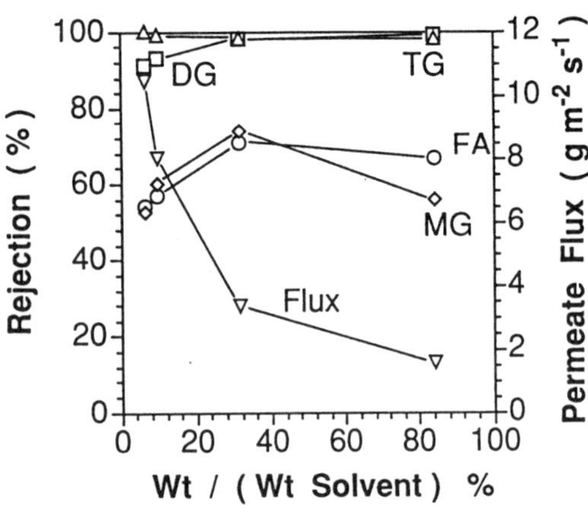

Figure 3. The Effect of Oil Concentration on Permeate Flux and Rejection for CA Membrane.

3. Discontinuous diafiltration

Diafiltration was operated in batch mode by charging the cell with 100g of 10% (wt./wt. ethanol) oil–solvent mixture. In one batch, about 35g of oil–solvent mixture permeated through the membrane, and an equal weight of solvent was fed to the cell. This operation was repeated 16 times (total permeate weight:504g). Permeate flux per batch decreased with increasing oil concentration of retentate. Initial permeate flux in each batch almost recovered by feeding solvent to the retentate. However, the flux gradually decreased with the number of operation, perhaps due to membrane fouling. The fractions of MG and FFA in the retentate decreased gradually, whereas the fraction of TG increased. Finally, the composition (TG:DG:MG:FFA) of the retentate changed from 31:28:9:32 to 65:29:2:4, respectively.

CONCLUSIONS

In this study, 18 types of commercial membranes were tested for their ability to separate the constituents of oils. A cellulose acetate membrane with ethanol as solvent was found to be the most suitable system for oil separation. From the results of discontinuous diafiltration, the fractions of FFA and MG in the retentate decreased, whereas that of TG increased.

REFERENCES

1) A.K.S. Gupta, Fett Seifen Anstrichm., 1986, **88**, 79–86
2) A. Iwama, J. Jpn. Oil Chem. Soc. (Yukagaku), 1985, **10**, 852–58
3) M. Tanaka, T. Itoh, and H. Kaneko, J. Jpn. Oil Chem. Soc. (Yukagaku), 1976, **25**, 263–65

CATALYST REMOVAL FROM HYDROGENATED OILS USING MEMBRANE TECHNOLOGY

CARL VAVRA and S. SEFA KOSEOGLU
Food Protein Research and Development Center
Texas A&M University System
College Station, Texas 77845

ABSTRACT

Nickel catalyst currently is recovered from hydrogenated oil-catalyst mixtures by high capacity filter presses, but the process is highly labor-intensive. The use of membranes potentially could reduce labor expenses and the amount of oil lost during catalyst recovery, and can extend the useful life of catalysts. In the current study, catalyst ranging from 1 to 100 microns in particle size, did not pass through the membrane. All the nickel catalyst was recovered in the retentate fraction, resulting in a clear permeate stream. Experiments were performed over various pressure and temperature ranges, with flow rates ranging from 3.5 to 119 lmh.

INTRODUCTION

Application of membrane technology to edible oil processing is expanding and is expected to make a considerable impact on profitability and efficiency of the edible oil refineries (Koseoglu 1991; Iwama, 1988: Gupta, 1986). A summary of the membranes evaluated in edible oil refining during the last ten years is summarized in Table 1.

TABLE 1. Examples of membranes evaluated in edible oil refining and their producers

Producers	Membrane Name	Membrane Materials	Membrane Form
Rhone-Poulenc (F)	IRIS 3042	PAN	Flat Sheet
PCI (GB)	T6B/BX3	PAN/Polysulfone	Tubular
SFEC (F)	Carbosep	Inorganic	Tubular
Romicon (USA)	PM 1, 10, SM10	Polysulfone	Hollow Fiber
Bergof (D)	BM 50, 100, 1000	Polyamide	Hollow Fiber
Koch (USA)	HFM 180	Polyvinylfloride	Tubular
Nitto Denko (J)	NTU 4220	Polyimide	Tubular
Wafilin (NL)	WFA 3010	Polyimide	Tubular

Oil hydrogenation is usually performed in batch reactors and when hydrogenation is complete, the catalyst is recovered by recirculating the oil through a high capacity filter until the oil is free of nickel. The filter cake containing the catalyst, is recovered manually and can be reused a number of times. After final use, the catalyst usually is trucked to another site for disposal. Exhausted catalyst can be processed to recover its nickel content. However, due to concerns about leaching of heavy metals into the soil and water, increased disposal costs, and tighter federal and state regulations, alternative methods are being evaluated. For example, nickel catalyst is being recovered from wastes by digestion with either organic or inorganic acids. The rate of recovery and economics of the process are dependent on nickel concentration and oil content of the filter aid-catalyst mixture. Therefore, its value is directly related to its nickel content.

EXPERIMENTAL

Several membranes, including: 1) ceramic; 2) carbon; and 3) polymeric materials, membranes were evaluated in this project. The ceramic, PEI and Carbon coated Zirconia membranes were obtained from Ceramem Corporation, Walton, MA, ROCHEM Separations, Torrence, CA and Rhone-Poulenc Inc., Cranbury, NJ respectively. The ultrasonicly sealed ROCHEM membranes were packed in plate and frame MF/UF module. The ceramic membranes were available in honeycomb monolithic elements with 1.9 mm passage way. The Catalyst evaluation experiments were performed at pressure range of 25-125 psi and temperature range of 55-100°C. Permeate flow rates were measured, and samples were analyzed by Atomic Absorption Spectroscopy to determine the nickel removal selectivity of the each membrane system. Membranes were cleaned according to manufacturer's recommendations.

RESULTS AND DISCUSSION

All of the membranes, tested removed the catalyst from the hydrogenated soybean oil. Analyses of the permeate, retentate and feed samples are summarized in Table 1. All of the membranes were found to be effective for separating catalyst from the oil with reasonable flow rates. Membrane CM2 produced permeate with 0.9 ppm nickel catalysts. However membrane RC also performed well and resulted in a residual catalyst concentration of 9.4 ppm, which is negligible since it can be easily removed during citric acid treatment and polishing filtration. The remaining few ppm of nickel is usually in the form of nickel soaps that can not be filtered due to their solubility in fats and oils. The citric acid or any chelating agents can react with nickel and make them insoluble that could be removed with conventional dead end filters.

Flow rates were largely dependent on membrane pore size, pump speeds, temperature, and pressure. One of the most important criteria in these experiments was maintaining a high temperature during processing. Membranes CM2, NCM2, RP8, RP14 gave stable flow rates ranging from 8.5 lmh to 42.5 lmh. The membrane RC showed a very large flux of 113 lmh.

TABLE 1. Analysis of permeates and retentates obtained from various commercial membrane systems

Membrane Company	Membrane Type	Feed	Nickel Content, ppm Permeate	Retentate
ROCHEM Separations (RC)	PEI 0.01 micron	297	9.4	219
Ceramem Corporation (CM5)	Ceramic 0.05 micron	178	2.2	210
Ceramem Corporation (CM2)	Ceramic 0.2 micron	275	0.9	883
Ceramem Corporation (NCM2)	New Ceramic 0.2 micron	656	8.0	1068
Rhone-Poulenc Inc. (RP8)	Carbosep 0.08 micron	178	3.4	229
Rhone-Poulenc Inc. (RP14)	Carbosep 0.14 micron	330	2.7	304

Figure 1 shows typical graph of flux and percent rejection rate versus concentration factor. The volume of the catalyst containing oil was reduced from 32 liters to 2.7 liters at the end of the evaluations. The initial rejection rate of 82% was increased to 100 percent as the catalyst concentration increased in the feed.

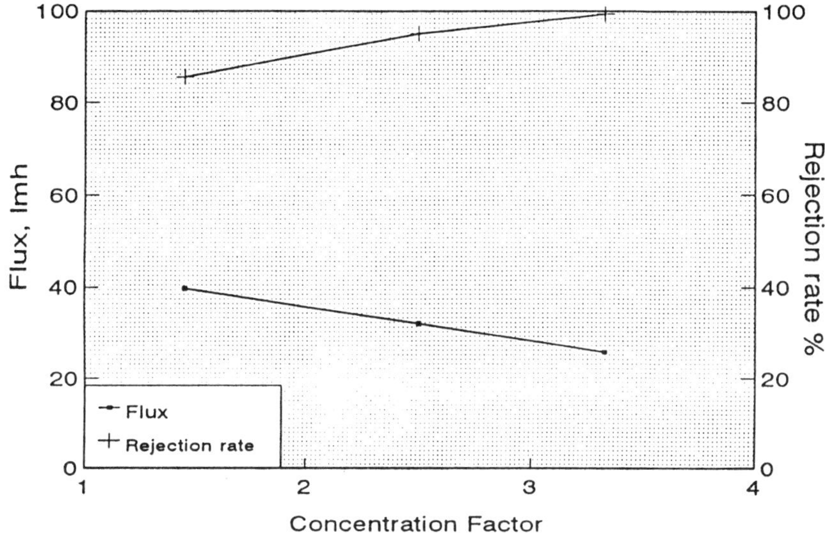

Figure 1. Flux and percent rejection rate versus concentration factor.

The present study indicated that the hydrogenated oil containing catalyst could be reduced to 1/100 of the original volume. This reduction allows the processor to reuse the undiluted catalyst for a longer period of time without selectivity changes that usually come from catalyst poisons adsorbed on the surface of the catalyst and the filter aid. This process also eliminates or considerably reduces oil losses, and could potentially eliminate use of the labor intensive filter presses. The envisioned process could combine a cross flow membrane system with a very small filter press, and can be run continuously with minimal labor cost. Further, the used catalyst mixture would be more valuable to both refiners and to reclaimers of nickel. This work indicates that membranes are capable of separating catalysts from hydrogenated soybean oils. Preliminary experiments show that membranes are easy to clean and the original flux can be maintained. Membrane life, durability, efficiency, and selectivity, and feedback received from the membrane manufacturer, are being evaluated presently.

REFERENCES

1. Koseoglu. S.S., Membrane Technology in Edible Oil Refining. Oils & Fats International, 1991, **5**, 16-21.

2. Iwama, A., New Process for Purifying Soybean Oil by Membrane Separation and an Economical Evaluation of the Process, In Proceedings of World Conference on Biotechnology for the Fats and Oils Industry, Ed. T. H. Applewhite, Am. Oil Chem. Soc., Champaign, IL, 1988, pp. 244-250.

3. Gupta, A.K.S. Neuere Entwicklungen auf dem Gebiet der Raffination der Speiseole. Fette Seifen Anstrichmittel, 1986, **3**, 79-86.

CONTINUOUS SYNTHESIS OF ASPARTAME PRECURSOR WITH MEMBRANE ENZYME REACTOR - MEMBRANE EXTRACTOR SYSTEM

YASUYUKI ISONO*, HIROSHI NABETANI and MITSUTOSHI NAKAJIMA
National Food Research Institute, Ministry of Agriculture, Forestry and Fisheries
2-1-2 Kannondai Tsukuba Ibaraki 305 Japan
*Dainichiseika Color & Chemicals Mfg. Co., Ltd.
1-9-4 Horinouchi Adachi-ku Tokyo 123 Japan

ABSTRACT

An enzyme membrane reactor and a membrane extractor were used simultaneously for the synthesis of N-(benzyloxycarbonyl)-L-aspartyl-L-phenylalanine methyl ester (ZAPM), the precursor of the artificial sweetener, aspartame. The synthesis of ZAPM in the biphasic membrane reactor proceeded by an enzymatic reaction between N-(benzyloxycarbonyl)-L-aspartic acid (ZA) and L-phenylalanine methyl ester (PM) in the water phase at pH 5-6. The organic phase, where ZAPM and PM are extracted, comes into contact with the water phase at pH 7 in the membrane extractor. Because of this pH control, purification of the ZAPM and recycling of substrates can be achieved. A simulation study was performed to confirm the suitability of the system.

INTRODUCTION

Aspartame is a synthetic sweetener which is about 200 times as sweet as sucrose. In recent years, several investigations on enzymatic synthesis of the aspartame precursor (ZAPM) using by the reverse reaction of hydrolysis have been reported [1-3]. However, few studies for a purification of ZAPM or reusing of PM combined with synthesis have been reported. In order to reduce production costs, reuse of PM is necessary. In our present study, continuous synthesis of ZAPM with product purification and substrates recycle using biphasic membrane enzyme reactor - biphasic membrane extractor system is proposed. The performance of this system was modeled and simulated.

SYSTEM

A schematic flow chart of the membrane enzyme reactor - membrane extractor system is shown in Figure 1. The reactor consists of a water phase (0.05M acetic buffer saturated with butyl acetate, pH 5 - 6) containing an enzyme (thermolysin) and an organic phase (butyl acetate saturated with acetic buffer). The substrates (ZA and PM) are supplied in the water phase and the major part of the product of enzymatic reaction (ZAPM) and some part of PM are extracted into the organic phase. The organic phase is in contact with a water phase (0.05M acetic buffer saturated with butyl acetate, pH 7). About 75% of ZAPM is extracted to the water phase, however almost PM remains in the organic phase. Therefore, this system enables us to purify ZAPM and to reuse PM.

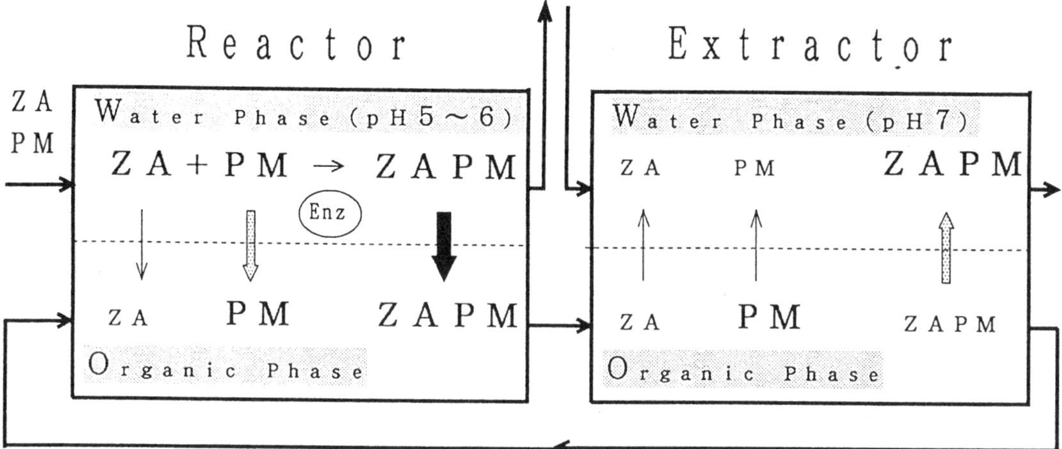

Figure 1. A schematic flow chart of the system

SIMULATION

A simulation of this system was performed on the assumption as follows: 1. Each phase is completely mixed, 2. Extraction equilibrium states exist between the water phase and the organic phase. The following equations represent the system:

Enzyme reaction rate in water phase

$$= \frac{k_2[E_0][ZA][PM]-k_{-2}[W][E_0][ZPAM]Km_{ZA}\cdot Km_{PM}/Km_{ZAPM}}{(Km_{ZA}+[ZA])(Km_{PM}+[PM]+[ZA]Km_{PM}/Ki_{ZA})+[ZAPM]Km_{ZA}\cdot Km_{PM}/Km_{ZAPM}} \quad (1)$$

where k_2 and k_{-2} are the reaction rate constants of the forward and reverse reactions, Km is the Michaelis constant, Ki is the inhibition constant of ZA, E_0 is enzyme concentration, and [ZA], [PM], [ZAPM] are the concentrations of ZA, PM, and ZAPM in the water phase, respectively.

$$\text{Partition coefficient of ZA} = \frac{1+10^{pH-pK1}+10^{2pH-pK1-pK2}}{K_{non}} \quad (2)$$

Where the K_{non} is the partition coefficient of the non-ionized form of ZA. Partition coefficients of PM and ZAPM are similar.

From the mass balance equations and partitional balance, the process was simulated.

RESULTS AND DISCUSSION

Figure 2 shows the calculated equilibrium concentrations of each material where the pH of the reactor is five and the pH of the extractor is seven. High conversion and similar productivity was obtained compared to the literature, because of the PM recycling with extractor. The extractor combined with reactor is effective for the reaction. Further research on the enzyme activity and long term stability is to be investigated.

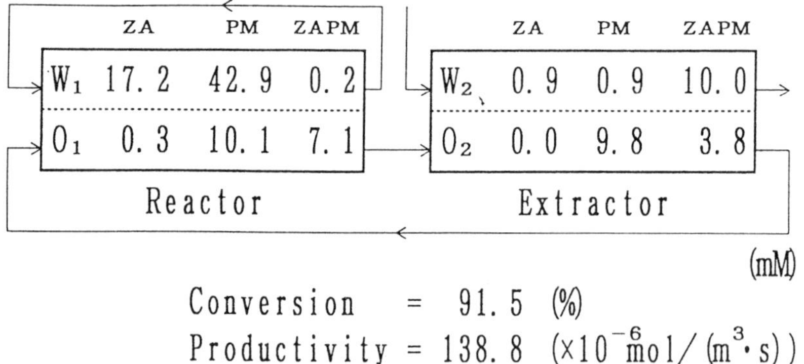

Figure 2. Calculated equilibrium concentrations

REFERENCES

1. Nakanishi, K., Kimura, Y. and Matsuno, R., Kinetics and equilibrium of enzymatic synthesis of peptides inaqueous/organic biphasic system. Eur. J. Biochem., 1986, **161**, 541-549

2. Oyama, K., Irino, S. and Hagi, N., Production of aspartame by immobilized thermoase. Methods in Enzymology, 1987, **136**, 503-516

3. Hirata, A., Hirata, M. and Furuzawa, H., Production of Aspartame Precursor by Semi-Continuous Pulsed Extraction Column Bioreactor Retaining Free Enzyme in an Aqueous Phase. Kagaku Kogaku Ronbunshu, 1991, **17**(3), 586-588

ELECTRO-ULTRAFILTRATION BIOREACTOR FOR ENZYMATIC REACTION IN REVERSE-MICELLES

MASARU HAKODA, YOSHIHIRO OGAWA, TATSUYA AKASHI
AND KOZO NAKAMURA
Department of Biological and Chemical Engineering
Gunma University, Kiryu, Gunma 376, Japan

ABSTRACT

Electro-ultrafiltration (EUF) was applied to separate the Aerosol OT (AOT) reverse micelles containing lipase from isooctane, and the effect of electric field on the rejection of reverse micelles and the permeation flux was examined experimentally. It could be said from the experimental results that EUF was effective for separation of the reverse micelles containing the enzyme and that the motive forces for the negatively charged reverse micelles were electrophoretic and dielectrophoretic.

INTRODUCTION

EUF is effective in decreasing the gel layer formation and in increasing the filtration flux, owing to the electrokinetic phenomena such as electrophoresis and electroosmosis[1,2]. We applied EUF to the enzymatic reaction of starch hydrolysis and reported the improvement of the permeation flux with some damage of glucoamylase in the electric field[3]. We reported also the experimental data of UF and EUF separation of the reverse micelles containing enzyme[4]. In this work EUF was further applied to separation of the Aerosol OT (AOT) reverse micelles containing lipase in isooctane, and the effect of electric field on the rejection of reverse micelles and the permeation flux was examined in detail experimentally.

MATERIALS AND METHODS

The experimental apparatus consisted of a EUF module, a feed tank, a feed pump and a D.C.electric power supply unit etc. The Carbosep M5 membrane (cutoff M.W. 20,000) are made of carbon with the skin layer of zirconium oxide. A stainless steel rod of 3mm diameter was installed as an electrode in the central axis of the tubular module (inner diameter 6 mm, thickness 2 mm, Sumitomo Heavy Industries Ltd., Japan). The porous substrate of the membrane was made of sintered carbon and used as a counter electrode. The surfactant AOT used was the reagent of Nacalai Tesque, INC., Japan (purity 98.8%) and it was used without further purification. The lipase used (600U/mg, MW 44,000) was purchased from Seikagaku Kogyo Co., Ltd.. All experiments were carried out by the total circulation method under the condition of temperature 310K and trans-membrane pressure 50kPa . The rejection, R, is defined as follows;

$$R=1-(Cp / Cb) \qquad (1)$$

where Cp and Cb are the concentrations of solute in the permeate solution and in the feed solution, respectively. It is noted here that the majority of reverse micelles does not contain the enzyme under the condition of this experiment and their size and separation property could be different from those of the reverse micelles containing the enzyme.

RESULTS AND DISCUSSION

The EUF of the reversed micelles containing lipase was experimentally investigated with three different modes of voltage application.

Effect of AOT concentration

Fig.1 shows the effect of AOT concentration on the permeation flux for three modes of voltage application. In either mode the flux increased with increase in AOT concentration, probably due to decrease in the size of reverse micelles. The application of voltage improved the flux, and this result can be well explained if the approach of the reverse micelles containing enzyme to the membrane surface is electrically hindered. The flux improvement was obtained irrespective of the direction of applied voltage, probably because the dielectrophoresis could occur in the nonuniform electric field.

Effect of Water concentration

Fig.2(a) shows the effect of water concentration on the permeation flux in three modes of voltage application. The flux decreased in any mode with increase in the water concentration,

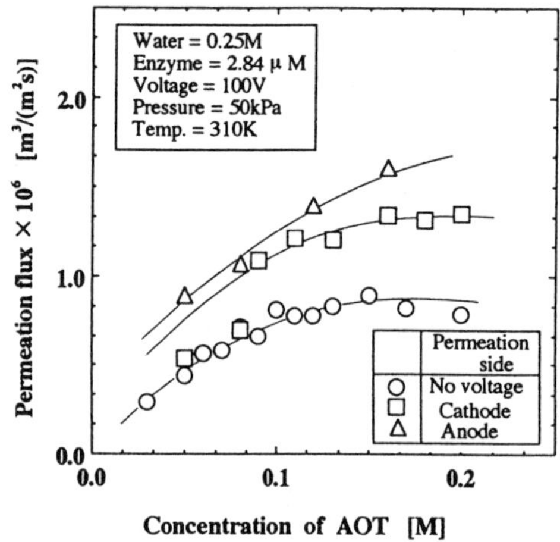

Figure 1. Effect of AOT concentration on permeation flux.

and the application of electricity improved the flux. When the anode was set in the permeation side, the flux was larger than that obtained by the other modes. Fig.2(b) shows the effect of the water concentration on the rejection of water. The rejection of water increased with increase in the water concentration in the case of no voltage application, probably due to the increasing size of reverse micelles. When the anode was set in the permeate side, the rejection of water decreased with increase in the water concentration. The effect of the water concentration were, however, reversed by setting the cathode in the permeate side. These results could be well explained with the assumptions; (1) The increase in the water concentration contributes to increase the electrophoretic force and to decrease the dielectrophoretic force. (2)The reverse micelles as well as the free AOT molecules possess the negative charges.

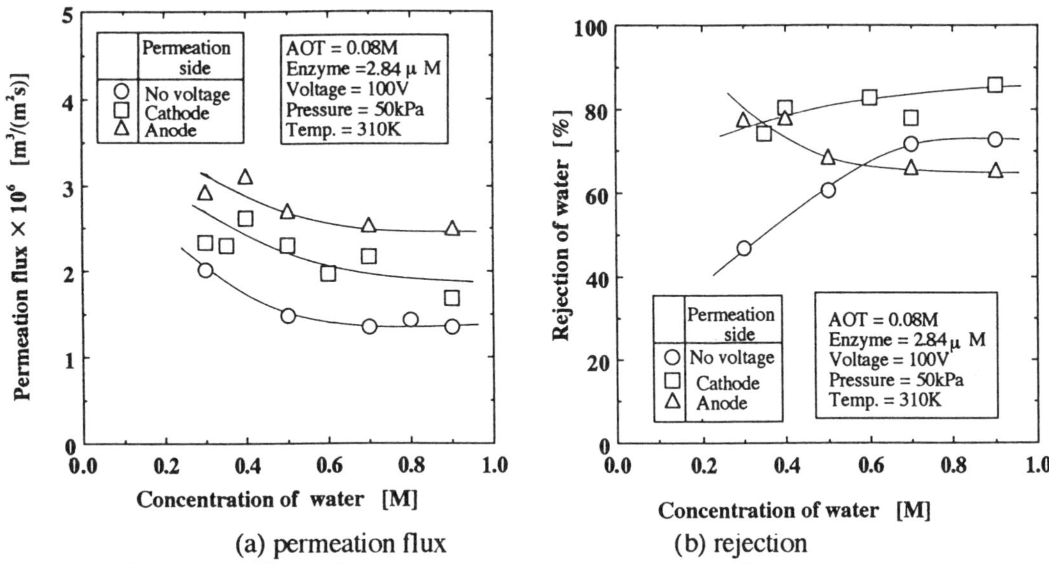

(a) permeation flux (b) rejection

Figure 2. Effects of water concentration on permeation flux and rejection

CONCLUSIONS

1) The application of voltage improved the permeation flux and also the rejection of reversed micelles in the membrane separation .
2) The voltage gradient set against the direction of permeation improved the flux more than the reversed gradient.
3) The reverse micelles were supposed to possess the negative charges and to move by the dielectrophoretic force as well as the electrophoretic force.

REFERENCES

1. Yukawa,H., Obuchi,H., Kobayashi,K., *Kagaku Kougaku Ronbunshu*, 1980,**6**, 288-293.
2. Kimura,S. and Nomura,T., *Membrane*, 1982, **7**, 245-250.
3. Hakoda,M., Chiba,T., Nakamura,K., *Kagaku Kougaku Ronbunshu*,1991, **17**, 470-476.
4. Nakamura,K. and Hakoda, M., (eds., Furusaki,S., Endo,I., Matsuno,R.), *Biochemical Engineering for 2001*, Springer-Verlag., 1992, 433-436.

EVALUATION OF THE INTEGRATED-TIME-TEMPERATURE EFFECT OF THERMAL PROCESSES ON FOODS: STATE OF THE ART

HENDRICKX, M., MAESMANS, G., DE CORDT, S. AND TOBBACK, P.
Katholieke Universiteit Leuven, Faculty of Agricultural Sciences, Center for Food Science and Technology, Unit Food Preservation
Kardinaal Mercierlaan 92, B-3001 Leuven, Belgium

ABSTRACT

The need for wireless Time-Temperature Integrators as thermal process evaluation tools to overcome inherent disadvantages of *in situ* and physical-mathematical process impact quantification techniques is indicated. A critical overview of existing TTI's is given. Possibilities of an enzyme-based TTI in heat penetration studies and in monitoring the impact of processing conditions are illustrated by a case-study.

INTRODUCTION

The widespread use of thermal treatments to eliminate enzymes, (pathogenic) microorganisms and induce palatability makes proper quality and safety quantification techniques for this physical food preservation variant of paramount importance for the nutritional well-being of consumers. The prevalent concept to quantify the integrated impact of a heat process on a food quality attribute is the processing value F. Approaches to obtain this processing value can be classified in (i) *in situ* evaluation, (ii) physical-mathematical techniques and (iii) by using Time-Temperature-Integrators.

The *in situ* thermal process evaluation technique relies on direct determination of the change in a food quality attribute concentration. Disadvantages associated with this methodology are the large sample size required, isolation and analysis time and detection limits. Only when information on temperature profile shape and maximum centre temperature are available, the effect of a previous heat treatment on food safety can be monitored by the changing concentration profile of innate chemical and biochemical markers.

The physical-mathematical method is based on determining the temperature program imposed on a food from direct registration or constructive computation. This approach is limited by difficulties in measurement of the temperature profile, especially for heterogeneous foods (rotating retorts, aseptic) and/or lack of accurate input parameters and models for the constructive computation option.

Time-Temperature-Integrators (TTI's) have been suggested as alternative evaluation technique. A TTI can be defined as '*a small device that shows a time-*

temperature dependent, easily and correctly measurable, irreversible change that mimics the changes of a target quality parameter undergoing the same variable temperature exposure' [7,2]. Theoretical analysis [8] indicates that TTI and target shall exhibit the same temperature dependency coefficient (activation energy Ea or z-value).

TTI'S : STATE OF THE ART

Based on the nature of the temperature dependent phenomenon, TTI's can be subdivided in biological, chemical and physical sensing systems. Examples of each category are listed in Table 1.

The use of *microbiological systems* to monitor efficacy of a thermal treatment is the most endeavoured application of TTI's in food and pharmaceutical industry. Different tools have been designed to control location and (contamination-free) recovery of microorganisms or their spores at a specified site in the food product instead of dispersing it in the food or drug (inoculated pack). With the advent of aseptic processing technologies for heterogeneous foods, carrier systems for microbiological TTI's have been miniaturized. To avoid complicated quantitative microbiology, also enzymic TTI's were proposed.

Temperature dependent diffusion is the basis of *physical TTI's* described thus far in literature.

Detection of the concentration change of a *chemical compound* as a measure for the impact of a thermal process was advocated two decades ago [5]. Chemical compounds have been used as single or multi-component TTI before, but often fail in application because of abuse: no generally valid technique is yet available to correlate the change in status of one or more TTI-component to a change in target quality attribute when activation energy (or z-value) of TTI and target are different [4].

TABLE 1. Examples of TTI's

Time Temperature Integrator	Temperature coefficient	Reference
Micro-sized Microbiological TTI's		
B. stearothermophilus	10°C (z)	[3]
C. sporogenes	12.5-12.7°C (z)	[1]
Immobilized peroxidase in dodecane	10°C	[8]
Chemical TTI's		
Thiamine	48°C (z)	[5]
Blue#2 degradation	58.2-74.5 kJ/mol (Ea)	[6]
Physical TTI's		
Coloured compound diffusion	291.2 kJ/mol	[9]

THERMAL PROCESS EVALUATION TECHNIQUES BASED ON ENZYMIC TTI'S

Recent work on peroxidase [8] and α-amylase [2] indicates that control of state (immobilization) and micro-environment (solvent engineering) of the enzyme is a promising basis for TTI development. Figure 1 illustrates application of an enzyme-based TTI in determining the coldest particle without restricting its free rotation/translation and studying the influence of end-over-end rotation for a heterogeneous food model system. α-amylase ($z=7°C$, $D_{100°C}$ = 13 min) was encapsulated in the central cavity (2.5 mm diameter) of a 25 mm diameter teflon sphere. Ten of these TTI-carrier systems were spatially distributed among 22 others in a 600 ml glass jar. The coldest point shifts from the lower part of the can (particle 2) in the zero rotation mode to the centre of the container (particle 8). The TTI response clearly reflects the homogenized and enhanced heat transfer rates induced by rotation.

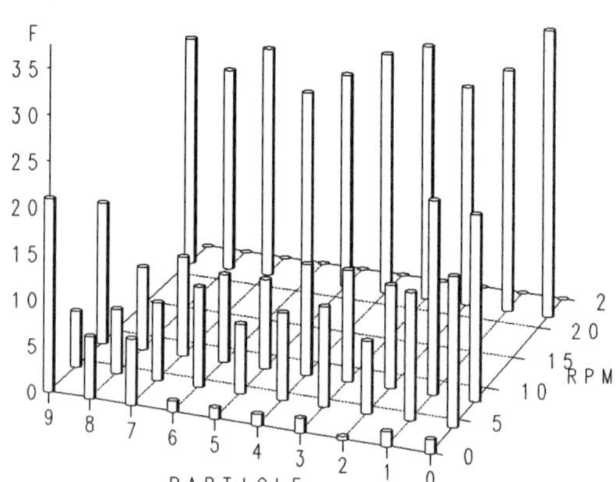

CONCLUSION

Time Temperature Integrators can contribute substantially to our understanding of thermal treatment efficacy, especially in processing conditions where *in situ* or physical-mathematical evaluation techniques are restricted. The basic requirement of proper TTI-functioning, *i.e.* matching activation energy of TTI and target, is often overlooked. The search for a non-microbiological TTI to monitor sterilization processes continues.

Figure 1 : Spatial processing value distribution ($^{7°C}F_{100°C}$) in a 600 ml glass jar as a function of the End-over-End rotational speed, measured with α-amylase as TTI. The cascading water heat treatment consisted of a C.U.T. of 8'06", 30'05" holding time at 105°C and 20' cooling time to 15°C. Silicone oil (350 CST) was used as brine and the solid/liquid ratio was 1.7 .

REFERENCES

1 Brown, K.L., Ayres, C.A., Gaze, J.E., Newman, M.E. (1984). Thermal destruction of bacterial spores immobilized in food/Alginate particles. Food Microbiology, **1**, pp 187-198.

2 De Cordt, S. ,Hendrickx., M., Maesmans, G., Tobback, P. (1992). Immobilized alfa-amylase for *Bacillus licheniformis*: a potential enzymic time-temperature integrator for thermal processing. International Journal of Food Science and Technology, **27**, pp 661-673.

3 Hersom, A.C., Shore, D.T. (1981). Aseptic processing of foods comprising sauce and solids. Food Technology, **35**(4), pp 53-62.

4 Maesmans, G., Hendrickx, M., De Cordt, S. and Tobback, P. (1993) Multi-component Time Temperature Integrators in the evaluation of thermal processes. Accepted for publication in Journal of Food Processing and Preservation.

5 Mulley, A., Stumbo, C., Hunting, W. (1975). Thiamine: a chemical index of the sterilization efficacy of thermal processing. Journal of Food Science, **40**(5), pp 993-996.

6 Sadeghi, F., Swartzel, K.R. (1990). Time-temperature equivalence of discrete particles during thermal processing. Journal of Food Science, **55**(6), pp 1696-1698, 1739.

7 Taoukis, P.S., Labuza, T.P. (1989). Applicability of time-temperature indicators as shelf life monitors of food products. Journal of Food Science, **54**(4), pp 783-788.

8 Tobback, P., Hendrickx, M.E., Weng, Z., Maesmans, G.J., De Cordt, S.V. (1992). The use of immobilized enzymes as Time-Temperature Indicator system in thermal processing. In 'Advances in Food Engineering', Eds. Singh, R.P. and Wirakartakusumah, M.A., CRC Press, Boca Raton, Florida, USA, pp 561-574.

9 Wittonsky, R.J. (1977). A new tool for the validation of the sterilization of parenterals. Bulletin of the Parenteral Drug Association, **11**(6), pp 274-281.

INACTIVATION KINETICS OF BACTERIAL SPORES IN SKIM MILK, CONCENTRATES, IN DEPENDENCE ON WATER ACTIVITIES AND UNDER SEALINGS

H.G. KESSLER, R. BEHRINGER[1] and J. PFEIFER[2]
Institute for Dairy Science and Food Process Engineering
Technical University Munich, D-85350 Freising-Weihenstephan, Germany
[1]Nestec, Lausanne; [2]Kraft General Foods, Munich

ABSTRACT

The heat resistance of mesophilic spores increased as the concentration of skim milk was raised up to 40 % total solids. The heat resistance of thermophilic spores, however, decreased considerably in concentrated products. This suggests that the product concentration stabilized the mesophilic mixed flora and destabilized the thermophilic flora. Decimal reduction times of Bacillus spores were determined in a wide range of relative humidities and air temperatures. A maximum of heat resistance was found at relative humidities from 20 % to 40 %. Very interesting results could be found with microorganisms under sealings, where the heat resistance was clearly higher.

1 Inactivation of Bacterial Spores in Skim milk (SM) and Skim milk Concentrates (SMC)

Our first objective was to find out which heat treatments are required to destroy the mesophilic and the thermophilic spores of the innate flora in SM and SMC (40 % TS). The mesophilic flora was by a factor of 1.5 more resistant in SMC than in SM. Surprisingly, the thermophilic flora was approx. threefold less resistant in the concentrate. Four species of the Genus Bacillus were isolated and SM and SMC were inoculated with spores of pure cultures to find out which of these species exhibit a behaviour similar to the innate flora. The ARRHENIUS plot of the (first order) rate constants (Fig. 1) reveals that *B. stearothermophilus* is the only species which is less resistant in SMC than in SM, whereas all other spores were more resistant in the concentrate. Further investigations were done to establish which factors are responsible for the particular behaviour. The value of each factor like pH-value, lactose, calcium, phosphate and proteins was separately changed from the SM level to the SMC level. It was concluded that in the case of *B. stearothermophilus* the destabilizing effect of the reduced pH-value and the increased calcium-phosphate concentration predominate, whereas in the case of all other sporeformers the reduced water activity is the dominant factor.

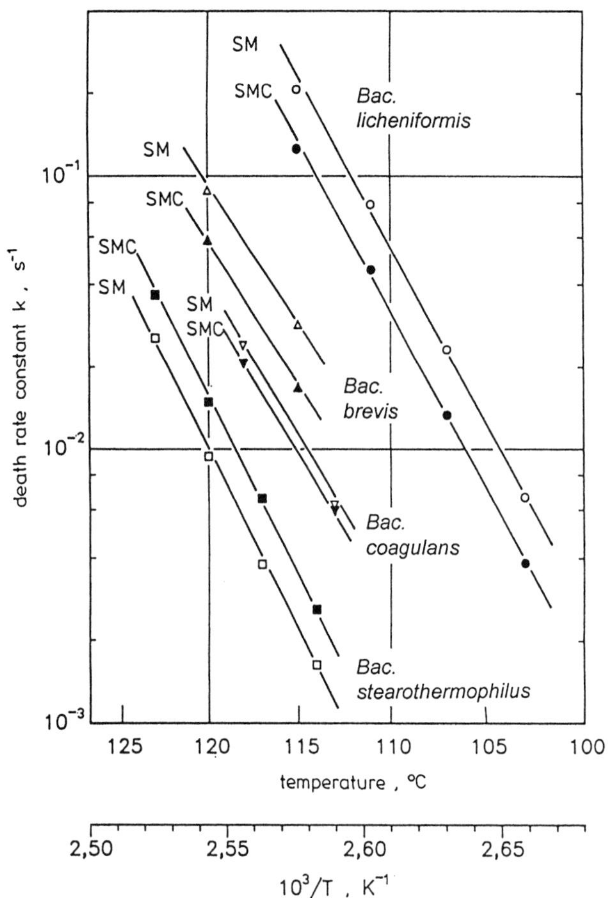

Figure 1. ARRHENIUS plot of death rate constants of the destruction of Bacillus spores in skim milk and skim milk concentrate.

2 Inactivation of Spores as a Function of the Relative Humidity of Hot Air

Investigations with *Bacillus cereus* spores were done covering the whole range from 1 % to 100 % r.h.. A special heating device was built up. Hot air of well-defined temperature and relative humidity was applied as the heating medium. Subsequently the relative humidity of the hot air was step by step reduced down to 1 % r.h. (Fig. 2). The relative humidity influenced the frequency factor of the ARRHENIUS equation as well as the activation energy. As a result minimal rate constants and concurrently a maximal resistance occurs in the range from 20 % to 40 % r.h.. This behaviour is ascribed to the occurence of two separate inactivation mechanisms.

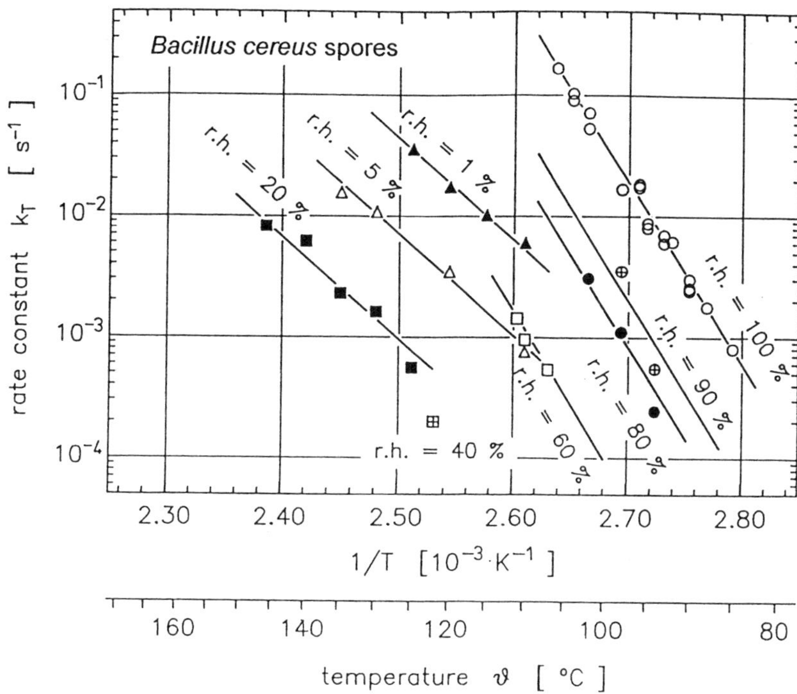

Figure 2. ARRHENIUS plot of the rate constants of the inactivation of *Bacillus cereus* spores at various relative humidities

3 Heat Resistance of Spores Under Sealings

Components with sealings are often regarded as critical points. Therefore, we investigated the heat resistance of bacterial spores, which are enclosed between a sealing surface of stainless steel and sealings of different materials. The experiments revealed that the resistance of spores under sealings strongly depends on properties of the sealing material, and that spores under sealings may withstand a heat treatment much longer than in an aqueous environment. The rate constants of experiments at various temperatures were compared with the rate constants depending on the relative humidity. No deviations from the 100 % r.h.-line was observed with the natural rubber sealings. The experimental points of the silicone rubber sealings, however, moved towards lower relative humidities at higher temperatures resulting in a 20fold increase in heat resistance. After these findings the environment between the sealing surface and the sealing material was characterized by direct measurement of the water activity. The water activity of the sealing materials, which were placed in a hermetically sealed chamber, was recorded as a function of the temperature. The water content of the sealings remained almost constant. It was found that the water activity of the natural rubber was practically independent of the temperature. By contrast the water activity of the silicone rubber continuously decreased as the temperature was raised. Consequently the high resistance under the silicone rubber sealings is due to the reduced water activity.

DETERMINATION OF THE THERMAL DEATH KINETICS OF BACTERIAL SPORES BY INDIRECT HEATING METHODS

JÜRGEN HAAS[+], RAINER BÜLTERMANN[*], HELMAR SCHUBERT[*]
[+]Department of Biotechnological Production, Dr. Karl Thomae GmbH, Birkendorfer Str. 65, D - 88400 Biberach, Germany
[*]Institute of Food Process Engineering, University of Karlsruhe, Kaiserstr. 12, D - 76128 Karlsruhe, Germany

ABSTRACT

The thermal death kinetics of bacterial spores were investigated by two indirect heating methods. In the lower temperature range a batchwise capillary tube method was used. Theoretical and experimental results have shown a negligible lethal effect during heating up and cooling down (quasi-isothermal conditions) for temperatures of up to 135 °C. For investigations at higher temperatures a continuous capillary tube method (turbulent flow conditions in a tube of an inner diameter of 0.2 mm) was developed. This device allows to realize quasi-isothermal conditions up to 170 °C. In the temperature range of 135 - 155 °C experiments show, that for short holding times *Bacillus stearothermophilus* spores are mainly activated rather than inactivated. A reaction model to describe this effect is discussed and compared to the classical First-Order Reaction kinetics.

INTRODUCTION

The quality of food preserved by ultra-high-temperature- (UHT-) treatment might be increased by applying even higher temperatures for shorter times, so–called extreme-high-temperature- (EHT-) treatment. Therefore, in a first step, the thermal death kinetics of bacterial spores must also be investigated in the EHT-range.

For an accurate examination of the thermal death kinetics so called quasi-isothermal conditions are necessary. This means that the survival-rate $\log_{10} (N/N_0)$ during heating up and cooling down is greater than - 0.1 (N = number of spores, N_0 = number of spores at time t = 0) and can therefore be neglected.

MATERIALS AND METHODS

In the sub-UHT-range inactivation experiments in a capillary tube of 1.4 mm inner diameter were carried out batchwise. The tube is immersed into the heating, the holding and the cooling bath by a computer controlled pneumatic mechanism [1].

To realize quasi-isothermal inactivation conditions in the UHT- and EHT-range too the spore suspension was heated up and cooled down in tubes of an inner diameter of 0.2 mm under turbulent flow conditions [1].

RESULTS AND DISCUSSION

Experimental survival curves of *Bacillus stearothermophilus* spores are shown in figure 1. In the batch capillary tube the spores were heated up to 135 °C. Experiments up to 155 °C were carried out in the flow-through capillary tube. To show all results in one figure a logarithmic time scale is used.

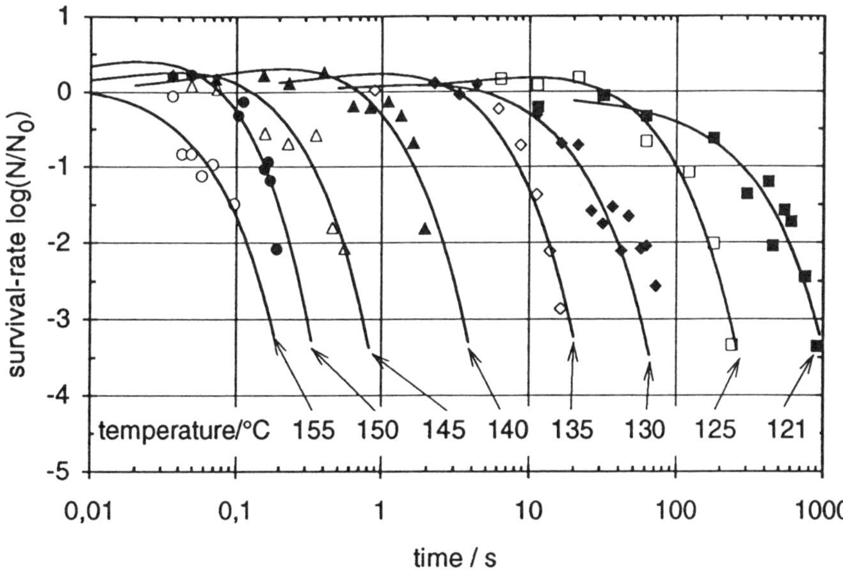

Figure 1. Survival curves of *Bacillus stearothermophilus* spores (Merck, No. 11499) in physiological NaCl-solution. Symbols are experimental results, lines are calculated after evaluation with the Activation-Inactivation model.

For short holding times positive inactivation-rates were obtained. This phenomenon cannot be described with the classical First-Order-Inactivation model but with the Activation-Inactivation-model [2]. There can coexist dormant spores (S_D) and activated spores (S_A). The dormant spores have to be activated first (by heat treatment) before they underlie inactivation (inactivated spores S_I):

$$S_D \dashrightarrow S_A \dashrightarrow S_I \qquad (1)$$

The results of both models are compared in figure 2 for survival-rates < -4. In addition curves for the undesired degradation of thiamine are shown [3].

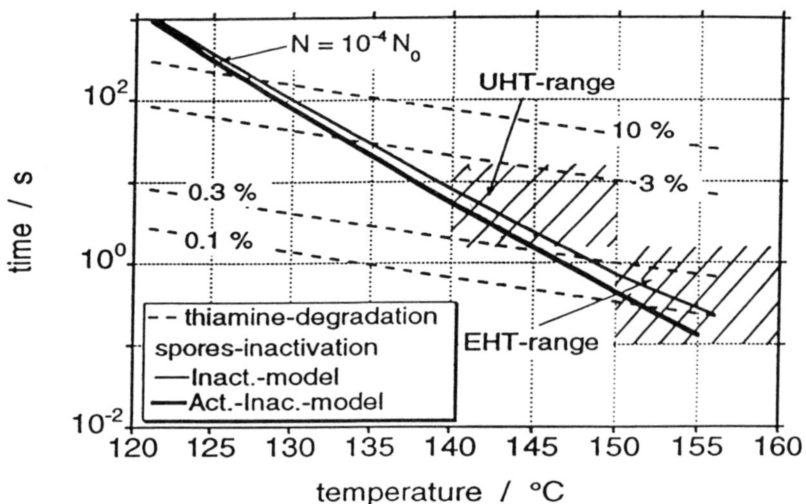

Figure 2. Comparison of the Inactivation and the Activation-Inactivation model.

For survival rates < -1 the Activation-Inact. model predicts shorter times than the First-Order-Inactivation model to obtain the same inactivation of *B. stearothermophilus* spores [1]. With both models destruction of the spores is much more dependant on temperature than the chemical degradation reactions are. So heat treatment in the EHT-range might provide an even higher food quality than UHT-treatment.

Theoretical considerations have shown that quasi-isothermal conditions are obtained up to 170 °C using the flow-through capillary tube plant [1].

Acknowledgement: The authors thank the Deutsche Forschungsgemeinschaft for their financial support.

REFERENCES

1. Haas, J., Untersuchungen zur thermischen Abtötung hitzeresistenter Bakteriensporen. Dissertation, University of Karlsruhe, 1992.
2. Abraham, G., Debray, E., Candau, Y., Piar, G., Mathematic model of thermal destruction of *Bacillus stearothermophilus* spores. Applied and Environmental Microbiology, 6 (1990), 3073 - 3080.
3. Kessler, H.G., Lebensmittel- und Bioverfahrenstechnik - Molkereitechnologie. A. Kessler, Freising, 1988.

A First Order Probabilistic Perturbation Analysis of the Growth and Thermal Inactivation of Lactobacillus Cells during Cold Storage and Reheating of Lasagna

BART M. NICOLAI, JAN F. VAN IMPE[1] & JOSSE DE BAERDEMAEKER
Agricultural Engineering Department, K.U.Leuven, Kardinaal Mercierlaan 92, B-3001 Heverlee, Belgium

ABSTRACT

A first order probabilistic perturbation method was developed for the analysis of microbial growth and inactivation as a consequence of conductive heat transfer under random conditions. The algorithm was used to investigate the effect of random variable refrigerator and oven temperatures on the growth and inactivation of *Lactobacillus* cells during cold storage and reheating of lasagna.

INTRODUCTION

Numerical simulation techniques are believed to provide a powerful tool for the design and analysis of food processes involving several consecutive process steps of heating, cooling, and cold storage. In the case of solid-like foods such as lasagna, an Italian dish composed of alternating pasta and tomato sauce layers, this involves the solution of the heat conduction equation together with an appropriate model for microbial growth and inactivation. The simulations are usually carried out under the assumption of well-known values of the model parameters. Because of the very nonlinear temperature dependency of the microbial growth as well as the inactivation, it can be expected that unpredictable disturbances of these parameters may cause large variabilities on the shelf-life of the food. In this paper, a probabilistic perturbation method is developed to investigate the effect of random variable model parameter uncertainties on the microbial load in the product during the consecutive process steps.

METHOD

Heat conduction in solid foods is governed by the Fourier equation with appropriate initial and boundary conditions and is often computed using the finite element method [1]. In the finite element method, the continuum is subdivided in (finite) elements of variable size and shape which are interconnected in a finite number m of nodal points. After spatial discretization and numerical solution of the corresponding differential system, an approximation for the temperature at every node and time step is found.

Recently, a dynamic model was described for the prediction of both microbial growth and inactivation in varying temperature conditions [2]. The model consists of two differential equations in two dependent variables. One variable is the microbial load $y = \log(N)$, with N the number of organisms per unit volume; the other variable y_r can be considered as a reference level from which growth restarts after thermal inactivation. The model equations corresponding to the temperature history T at a specific place in the food can be written concisely as follows:

$$\dot{\mathbf{y}} = \mathbf{g}(\mathbf{y}, T) \tag{1}$$

with $\mathbf{y} = [\ y\ y_r\ ]^T$ and $\mathbf{g}$ a 2×1 matrix function.

The propagation of random variable parameter uncertainties through the heat transfer process and the associated microbial kinetics can be investigated using Monte Carlo simulation techniques [3]. The method relies on the generation of a considerable number (e.g. 1000) of pseudo–random samples of the

[1] Senior research assistant with the N.F.W.O. (Belgian National Fund for Scientific Research)

random parameter vector by means of a computer. The model equations are solved for each sample parameter vector. Finally, the statistical characteristics (mean value, variance,...) of the temperature and microbial load at an arbitrary place in the food can be estimated for each time step by means of the common statistical procedures. The Monte Carlo method has a very high computational burden.

Probabilistic perturbation methods for heat conduction problems with random variable thermophysical parameters [4] and other process parameters [5] have been shown to be a fast alternative to the Monte Carlo method and are very well suited for incorporation in computer aided food process design software [6]. The perturbation method can be applied to the microbial kinetics as follows. It is assumed that T is random due to random variable parameters of the heat transfer model and that its mean $\overline{T}$ and variance $\mathbf{V}_{T,T}$ can be computed from the mean values and covariances of the random parameters by means of the algorithms outlined in ref. [4-5]. Further, the parameter vector $\mathbf{p}_M$ of the microbial model and the initial microbial condition $\mathbf{y}_0$ are assumed to be subjected to random variable disturbances. It is assumed that the mean vectors $\overline{\mathbf{p}}_M$ and $\overline{\mathbf{y}}_0$ and the covariance matrices $\mathbf{V}_{\mathbf{p}_M,\mathbf{p}_M}$ and $\mathbf{V}_{\mathbf{y}_0,\mathbf{y}_0}$ of $\mathbf{p}_M$ and $\mathbf{y}_0$, respectively, are known.

First, the dependency of $\mathbf{y}$ on T, $\mathbf{p}_M$, and $\mathbf{y}_0$ can be formally explicited. A first order Taylor expansion of $\mathbf{y}(T,\mathbf{p}_M,\mathbf{y}_0)$ around the solution $\overline{\mathbf{y}}(\overline{T},\overline{\mathbf{p}}_M,\overline{\mathbf{y}}_0)$ yields

$$\mathbf{y} \cong \overline{\mathbf{y}} + \frac{\partial \mathbf{y}}{\partial T}\Delta T + \frac{\partial \mathbf{y}}{\partial \mathbf{p}_M}\Delta \mathbf{p}_M + \frac{\partial \mathbf{y}}{\partial \mathbf{y}_0}\Delta \mathbf{y}_0 \tag{2}$$

in which the partial derivatives must be evaluated using $\overline{T}, \overline{\mathbf{p}}_M$, and $\overline{\mathbf{y}}_0$. Similarly, $\mathbf{g}(\mathbf{y}(T,\mathbf{p}_M,\mathbf{y}_0),T,\mathbf{p}_M)$ is expanded around $\overline{\mathbf{g}} = \mathbf{g}(\overline{\mathbf{y}},\overline{T},\overline{\mathbf{p}}_M)$

$$\mathbf{g} \cong \overline{\mathbf{g}} + \frac{\partial \mathbf{g}}{\partial \mathbf{y}}\left(\frac{\partial \mathbf{y}}{\partial T}\Delta T + \frac{\partial \mathbf{y}}{\partial \mathbf{p}_M}\Delta \mathbf{p}_M + \frac{\partial \mathbf{y}}{\partial \mathbf{y}_0}\Delta \mathbf{y}_0\right) + \frac{\partial \mathbf{g}}{\partial T}\Delta T + \frac{\partial \mathbf{g}}{\partial \mathbf{p}_M}\Delta \mathbf{p}_M. \tag{3}$$

Note that the partial derivative vectors and matrices are time–dependent in general. Substitution of equation (2) and (3) in equation (1) and combining appropriate terms in ΔT, $\Delta \mathbf{p}_M$, and $\Delta \mathbf{y}$ yields the following algorithm

$$\dot{\overline{\mathbf{y}}} = \overline{\mathbf{g}}(\overline{\mathbf{y}},\overline{T}) \tag{4}$$

$$\frac{d}{dt}\left(\frac{\partial \mathbf{y}}{\partial T}\right) = \frac{\partial \mathbf{g}}{\partial \mathbf{y}}\frac{\partial \mathbf{y}}{\partial T} + \frac{\partial \mathbf{g}}{\partial T} \tag{5}$$

$$\frac{d}{dt}\left(\frac{\partial \mathbf{y}}{\partial \mathbf{p}_M}\right) = \frac{\partial \mathbf{g}}{\partial \mathbf{y}}\frac{\partial \mathbf{y}}{\partial \mathbf{p}_M} + \frac{\partial \mathbf{g}}{\partial \mathbf{p}_M} \tag{6}$$

$$\frac{d}{dt}\left(\frac{\partial \mathbf{y}}{\partial \mathbf{y}_0}\right) = \frac{\partial \mathbf{g}}{\partial \mathbf{y}}\frac{\partial \mathbf{y}}{\partial \mathbf{y}_0} \tag{7}$$

with $\partial \mathbf{y}/\partial T$ and $\partial \mathbf{y}/\partial \mathbf{p}_M$ equal to zero vectors of appropriate dimension at $t = 0$. $\partial \mathbf{y}/\partial \mathbf{y}_0$ is equal to a 2×2 unity matrix at $t = 0$.

An expression for the covariance matrix $\mathbf{V}_{\mathbf{y},\mathbf{y}}$ can then be derived from equation (2)

$$\mathbf{V}_{\mathbf{y},\mathbf{y}} = \frac{\partial \mathbf{y}}{\partial T}\mathbf{V}_{T,T}\frac{\partial \mathbf{y}}{\partial T}^T + \frac{\partial \mathbf{y}}{\partial \mathbf{y}_0}\mathbf{V}_{\mathbf{y}_0,\mathbf{y}_0}\frac{\partial \mathbf{y}}{\partial \mathbf{y}_0}^T + \frac{\partial \mathbf{y}}{\partial \mathbf{p}_M}\mathbf{V}_{\mathbf{p}_M,\mathbf{p}_M}\frac{\partial \mathbf{y}}{\partial \mathbf{p}_M}^T. \tag{8}$$

NUMERICAL RESULTS AND DISCUSSION

Simulations have been carried out to compare the results for the mean values and variances obtained by the perturbation method presented above with those obtained by the Monte Carlo method.

The test problem consisted of a lasagna in a ceramic recipient which was stored for 4 days in a refrigerator and subsequently heated for 60 min in an oven. All parameters except the refrigerator and oven temperature were assumed to be deterministic. The mean and standard deviation of the refrigerator temperature were respectively $\overline{T}_{fridge}$ = 8°C and σ_{fridge} = 1°C ; the surface heat transfer coefficient was equal to 10 W/m^2°C . For the oven, $\overline{T}_{oven}$ = 100°C , σ_{oven} = 1°C and the surface heat transfer coefficient was equal to 25 W/m^2°C . Both the refrigerator and oven temperature were considered to have a Gaussian distribution. The shape, dimensions, thermophysical parameters and finite element grid were described elsewhere [7]. The equations were solved using a 4th order Runge-Kutta-Gill solver.

In the first Monte Carlo experiment, 100 runs were carried out; in a second experiment, 1000 runs. From Fig. 1 it is clear that the microbial load variances computed by both Monte Carlo experiments

(a)

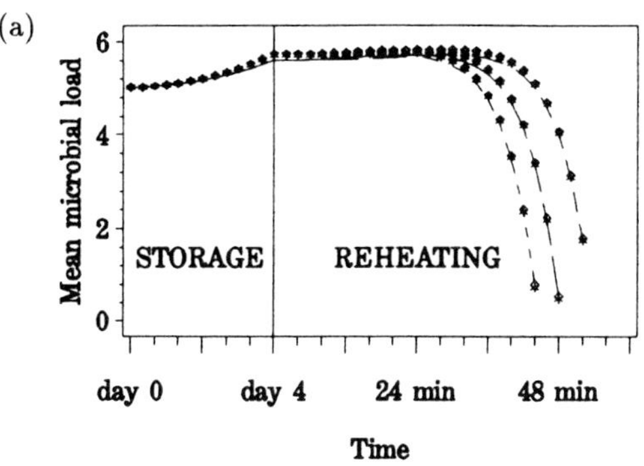

(b)

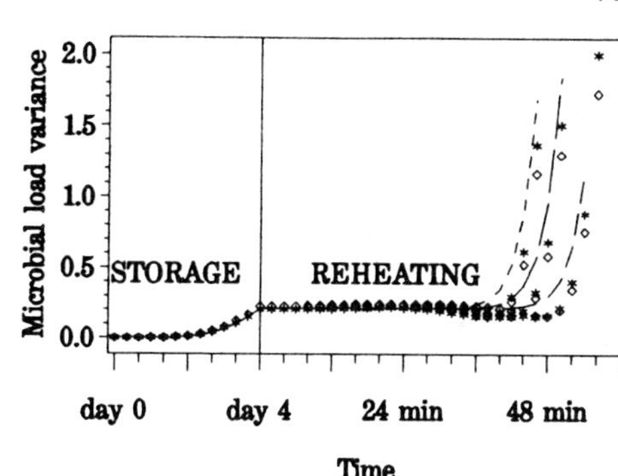

FIGURE 1. Mean (a) and variance (b) of the microbial load in a heated lasagna with random variable refrigerator and oven temperature at three different locations during cold storage and subsequent reheating of lasagna. Dashed lines: perturbation algorithm results at different places on the axis of revolution; ⋆ : Monte Carlo solution with 100 runs; ◇ : Monte Carlo solution with 1000 runs

agree well with these computed with the variance propagation algorithm. The perturbation algorithm fails to predict the apparent decrease of the variance at the onset of the thermal inactivation shown by the Monte Carlo simulations.

The variance of the microbial load increases sharply during the inactivation. Clearly small disturbances in the process parameters cause considerable uncertainties in predictions of the microbial load.

CONCLUSIONS

A first order perturbation algorithm has been derived for the computation of the mean and the variance of the microbial load during the thermal processing of foods in uncertain conditions. The algorithm has been applied to the growth and inactivation of *Lactobacillus* cells during cold storage and subsequent reheating of lasagna. The simulation results agree with the results obtained using the Monte Carlo method. The variance of the microbial load increases considerably as the thermal inactivation of the cells proceeds.

ACKNOWLEDGEMENTS

This study has been performed as a part of EEC–FLAIR (Food Linked Agro-Industrial Research) project AGRF–CT91–0047 (DTEE) and of FKFO–project (2.0095.92). The authors wish to thank the European Communities and the Belgian Fund for Collective Fundamental Research for their financial support.

REFERENCES

1 De Baerdemaeker, J., R.P. Singh and Segerlind L.J., Modelling heat transfer in foods using the finite element method, *Food Process Engineering*, 1977, **1**, 37–50.

2 Van Impe, J.F., Nicolaï, B.M., Martens, T., De Baerdemaeker J. and Vandewalle, J., Dynamic Mathematical Model to Predict Microbial Growth and Inactivation during Food Processing. *Applied and Environmental Microbiology*, 1992, **58**, 2901–2909.

3 Hayakawa, K., De Massaguer, P. and Trout, R.J., Statistical variability of thermal process lethality in conduction heating food — computerized simulation, *Journal of Food Science*, 1988, **53**, 1887–1893.

4 Nicolaï, B.M. and De Baerdemaeker, J., Computation of Heat transfer in Foods with Random Variable Thermophysical Properties, *International Journal for Numerical Methods in Engineering*, 1993, **36**, 523–536.

5 Nicolaï, B.M. and De Baerdemaeker, J., Probabilistic modeling and sensitivity analysis in heat transfer computations, In Singh R.P. and M.A. Wirakartakusumah (eds) *Advances in Food Engineering*. CRC Press, Boca Raton, Ann Arbor, London, Tokyo, 1992, 193–206.

6 De Baerdemaeker, J. (ed), Second year report on the EEC–FLAIR project 'Development of computer aided design procedures to improve the quality and safety of products with a limited shelf life', Agricultural Engineering Dept., K.U.Leuven, 1993.

7 Van den Broeck, P., Influence of non-stationary heat transport on the microbial quality of foods *Thesis K.U.Leuven* (in dutch), 1991.

VARIABILITY OF THERMAL PROCESS LETHALITY

K. Hayakawa, J. Wang[*], and P. de Massaguer[**]
Food Science Department, Cook College Rutgers University, P. O. Box 231, New Brunswick, NJ 08903 USA. [*]Process Development Division, Ross Laboratory, Columbus, OH, USA. [**]FEAA, University of Campinas, Campinas, SP, Brazil.

ABSTRACT

The variability of thermal process lethality should be predicted quantitatively for proper process design. Simple analytical and regression equations were derived for this prediction. According to sample applications of these equations, a high temperature-short time process required much tighter control of process parameter compared a conventional process and heating medium temperature influenced strongly lethality variability.

INTRODUCTION

Proper thermal process design requires quantitative data on the variability thermal processes. Therefore, several researches developed predictive methods for this variability. However, there was a need of a simple and generally applicable method for predicting the influence of all independent process parameters. Therefore, the present work was initiated to develop such methods.

PREDICTIVE EQUATIONS DEVELOPMENT

1. Analytical equation for lethality deviation

The following Eq. 1 estimates heating phase lethality (the equation obtained assuming the negligible value of $E_1\{ajI_o/z\}$, valid for virtually all thermal processes).

$$F_{ph} = (f/a)\exp\{a(T_1-T_r)/z\}\,E_1\{a(T_1-T_g)/z\} \qquad (1)$$

Deviation ΔF_{ph}, Eq. 2, was obtained from dF_{ph} when it was expressed in terms of multidimensional coordinates T_o, T_1, j, f, z, and t_b.

$$|\Delta F_{ph}| < [[\{jfL_g/(aI_o)\}\Delta T_o]^2 + [\{aF_{ph}/z + fL_g/(aI_o)\}\Delta T_1]^2$$

$$+[\{fL_g/(aj)\}\Delta j]^2 + [(t_b L_g/f)\Delta f]^2 + [\{a(T_r\text{-}T_1)F_{ph}/z^2 - fL_g/(az)\}\Delta z]^2 + [L_g \Delta t_b]^2]^{1/2} \quad (2)$$

2. Theoretical regression equation

A Monte Carlo simulation based screening and regression analyses were performed, using a computerized method (1), to develop a new theoretical regression equation for predicting the coefficient of variations of F_p, W_{fp}, for a z value of 10C°. The approach for this development is described in a previous paper (2).

$$W_{fp} = 37.7975 + 4.0029\, X_{fh} - 6.1851\, X_{t1} - 1.5010\, X_{wz} + 3.1002\, X_{wfh} + 1.3056\, X_{wjh} + 2.2739\, X_{fp} + 0.9157\, X_{fh}\, X_{wfh} - 1.2010\, X_{t1}\, X_{wfh} + 0.9075\, X^2_{t1} + 0.1\%\, 70\, X^3_{t1} \quad (3)$$

where, $X_{fh} = 2.6816\, \ell n\, (\bar{f}_h + 6.4516) - 10.293$; $X_{t1} = 15.77270\, \ell n\, (160.3036 - \bar{T}_1) - 59.4352$; $X_{wz} = 11.8880\, \ell n\, (27.5 - W_z) - 34.260$; $X_{wfh} = 8.5106\, \ell n\, (W_{jh} + 6.3333) - 22.9744$; $W_{wjh} = 3.2688\, \ell n\, (W_{jh} + 2.1667) - 8.6646$; $X_{fp} = 17.4793\, \ell n\, (18.6889 - F_p) - 45.7362$

(4)

APPLICATION OF PREDICTIVE EQUATIONS AND DISCUSSION

Equation 2 was applied to these parametric means and deviations representative of a conventional process: F_p = 300s, Δt_b = 15s, f_h = 1800s, Δf_h = 60s, j_h = 2.0, Δj_h = 0.2, T_1 = 120°C, ΔT_1 = 0.2C°, T_o = 70°C, ΔT_o = 2C°, z = 10C°, and Δz = 1.0C°. The estimated ΔF_p is 77s or 21.3% of the target F_p. The two most contributors to the F_p deviation were Δf_h (7.9% of F_p) and Δj_h (4.8%). The same equation was applied to a high temperature- short time (HTST) process of a condition heating food with the following conditions: Fp = 300s, Δt_b = 3s, f_h = 240s, Δf_h = 7.8s, j = 1.1, Δj = 0.1, T_1 = 145°C, ΔT_1 = 0.2C°, T_o = 85°C, ΔT_o = 0.2 C°, z = 1.0 C° and Δz = 0.1C°. The estimated ΔF_p was 364s (102% of the target F_p). The two most contributors to the lethality deviation were Δz(64.2%) and Δj_h (16.6%). Since it would be extremely difficult to reduce these parametric variations, a significantly larger target F_p is required for the HTST process.

Equation 3 was used to perform a sensitivity analysis. For this all parameter except one were kept at their frequent levels. The frequent values of f_h, $\bar{T}_1$, W_z, W_{fh}, W_{jh}, and F_p were 2400s, 117°C,. 10%, 7.0%, 12%, and 300s, respectively. The result shows that W_{fp} was influenced most strongly by $\bar{T}_1$, followed by W_z and F_p.

CONCLUSIONS

The new equations were obtained for predicting the variability of thermal process lethality. Sample application of these equations showed over 100% deviation in F_{ph} for an HTST process and strong influence of $\bar{T}_1$ on W_{fp}.

ACKNOWLEDGMENT

This is publication F-10209-1-93 and F-10575-1-93 of the New Jersey Agricultural Experiment Station supported by State Funds and the Center for Advanced Food Technology, a New Jersey Commission on Science and Technology Center; U. S. Army Research Office; U. S. Hatch Act Fund, and N. J. State Fund; Computer time support by Rutgers University Computing Services; and supercomputing time grant by Pittsburgh National Supercomputing Center.

REFERENCES

1. Hayakawa, K., de Massaguer, P., and Trout, R. J. Statistical variability of thermal process lethality in conduction heating food--computerized simulation. J. Food Sci., 1988, **53**, 1887-1893.

2. Wang, J., Wolfe, R. R., and Hayakawa, K. Thermal process lethality variability in conduction-heated foods. J. Food Sci., 1991, **56**, 1424-1428.

NOMENCLATURE

a = ℓn 10 (-); E_1 ( ) Exponential integral (-); F_p Sterilizing value (s); f Slope index of semilogarithmic food temperature history curve (s); I_o = T_1-T_o (C°); j Intercept constant of semilogarithmic food temperature history curve (-); L_g = exp[-a(T_g-T_r)/z] (-); T Temperature (°C); T_g & T_1 Food temperature at the end of heating phase of process and constant heating medium temperature, respectively (°C); t Time (s); t_b Time at the end of heating phase of process (s); W Coefficient of variation (standard deviation/mean) x (100) (%); X Statistical design variable used in Eq. 3. Appended subscript signifies associated process parametric mean or W (-); z Slope index of phantom thermal death time curve (C°).

Subscript (general): e ... End of process, h ... heating phase, o ... initial, r ... reference

Subscript appended to X and/or W: fh, fp, jh, t1, wfh, wjh, wz, z....X and/or W related to f_a, F_p, j_h, T_1, W_{fh}, W_{jh}, W_z, and z, respectively.

Superscript: - average

MICROWAVE HEATING OF LIQUID FOODS

M. KYEREME & R. C. ANANTHESWARAN
Department of Food Science
The Pennsylvania State University
University Park, PA 16802, USA

ABSTRACT

Time-temperature profiles in model Newtonian and non-Newtonian liquid foods in cylindrical containers were studied. Dimensional analysis was used to develop prediction models based on the variables that govern microwave heating of liquid foods. A flow visualization technique was used to investigate the flow patterns in the heated liquid samples.

INTRODUCTION

Microwave heating and its potential applications have received much attention in the food industry in recent years (1). Among microwave applications, microwave pasteurization and sterilization processes for liquid foods hold the greatest promise. Although, it is in the developmental stages, microwave sterilization has the potential to produce products of superior overall quality than those processed by conventional methods. The objectives of this study were to develop mathematical models for predicting time-temperature profiles during microwave sterilization of Newtonian and non-Newtonian liquid food products.

MATERIALS AND METHODS

Dimensional analysis was used to develop a model based on experimental data obtained for different concentrations of sucrose solutions (2). A second model, representing non-Newtonian liquids, was developed using data for aqueous solutions of sodium carboxymethylcellulose (CMC). All heating experiments for the model building stage were conducted in a GE (Model JE2800) household microwave oven (General Electric Co., Louisville, KY) with nominal output power of 700 W at 2450 MHz. A microwave-transparent Sani-Pro PS-300 (Sani-tech, Andover, NJ) container (7.2 cm dia. x 10.6 cm height) was used to contain the liquid samples. A quantity of 375 mL of each solution was heated and temperature-time data were recorded in intervals of 2 s using Luxtron (Model 750) fluoroptic probes placed at predetermined locations within the cans. An improvement of a technique developed by Liu (3) was used to visualize the flow patterns in heated liquid samples. A pouch containing crystals of potassium permanganate were used to trace the flow profiles of the liquid. The can was left to stand undisturbed until the crystals dissolved forming a colored solution in the immediate surrounding of the liquid before heating the can in the oven. A video camera was placed at an appropriate distance from the oven door and used to record the flow patterns in the heated sample.

RESULTS AND DISCUSSION

Based on the regression analysis of the different dimensionless terms for the Newtonian (sucrose) data, the Model I (equation 1) was chosen. This model had an R^2 of 0.994:

$$\left(\frac{k(T-T_i)}{P_o z^2}\right) = 1.14x10^{-10}(tf)^{0.6318}\left(\frac{\eta C_p}{k}\right)^{-0.0695}\left(\frac{C_p t^2 T_i}{R^2}\right)^{0.1855}$$

$$(\varepsilon'')^{-0.1025}\left(\frac{Z}{R}\right)^{-2.1689}\left(\frac{X}{Z}\right)^{0.0034} \qquad (1)$$

This equation describes the temperature rise as a function of the variables that govern microwave heating of Newtonian liquids.

A second dimensionless Model II (equation 2) was developed based on non-Newtonian data obtained for aqueous solutions of different concentration levels of sodium carboxymethylcellulose (CMC). The R^2 for this model was 0.967:

$$\left(\frac{k(T-T_i)}{P_o z^2}\right) = 0.2068\left(\frac{\eta_a C_p}{k}\right)^{-0.01689}\left(\frac{kt}{\rho C_p R^2}\right)^{0.8619}\left(\frac{C_p t^2 T_i}{R^2}\right)^{0.0833}$$

$$\left(\frac{Z}{R}\right)^{-2.1467}\left(\frac{X}{Z}\right)^{0.03349} \qquad (2)$$

The apparent viscosity was measured at a shear rate of 0.1 s^{-1}.

Each of the models was verified by comparing predicted temperatures with the experimentally measured temperatures. For 30% and 60% sucrose solutions, the Model I predictions and the experimental values were in good agreement. Comparisons of predicted and observed temperature profiles at selected points in 'Golden Delicious' apple juice were also made (figure 1a). The model generally under-predicted temperatures close to the sidewall and along the central axis. However, predictions at the sample surface were not very good.

Comparisons of temperatures predicted by Model II for 0.5% CMC and 0.8% CMC and experimentally measured temperature profiles showed close agreement. Model II was validated using data obtained for 0.65% CMC solution heated in a 2500 W pilot-scale microwave oven for 5 min. The predicted temperature profiles compared reasonably well with the observed values except at the sample surface (figure 1b). As in the case of Model I, predictions at points close to the surface deviated from the observed values. Extensive deviations were observed after 200 to 220 s of heating.

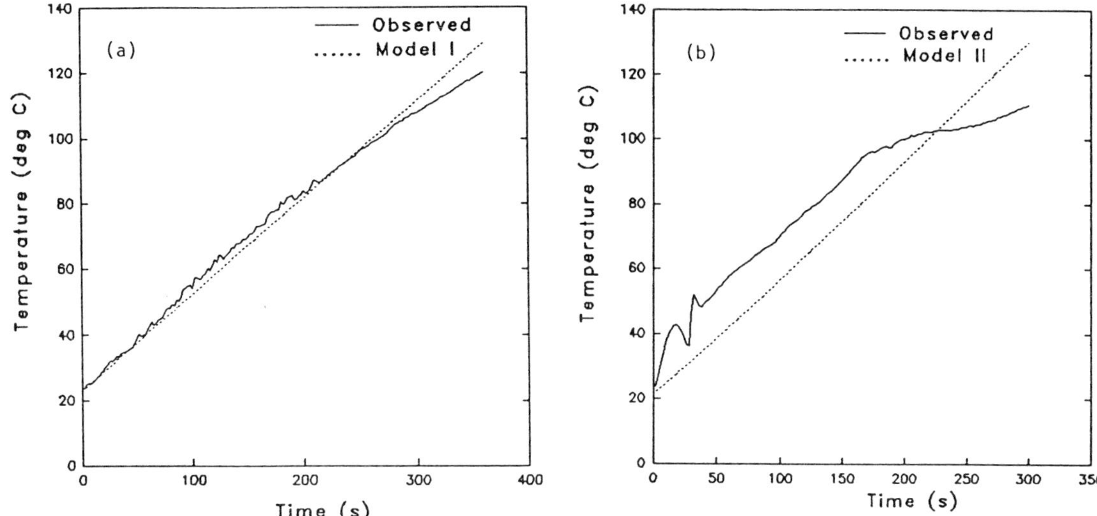

Figure 1 Predicted and observed time-temperature profiles along the central axis at a depth of 2.3 cm from the surface in (a) apple juice (b) 0.65% CMC.

The following observations were made based on the flow visualization studies. In the case of water, after 110--120 s, a plume of heated liquid started to form at the center of the can. As the heating continued, the plume of water gradually moved upwards along the central axis, up to the sample surface. The fluid at the surface then began to flow down along the sidewalls. Upon reaching the bottom of the can, the downward flow moved towards the center to join the central upward flow. The fluid in a zone located halfway between the center core and the sidewall was virtually stagnant. This was very apparent, especially, close to the bottom of the container. Complete mixing in the water samples occurred after 200 to 240 s heating. The patterns of flow in the CMC solutions were generally similar to those described for water.

CONCLUSIONS

Dimensional analysis was used to develop comprehensive models to describe microwave heating. Each model consisted of dimensionless terms that were significant in accounting for the variations in the temperature response term. The time-temperature profiles at different locations (except at the surface) within the liquids heated by microwave energy were predicted with a high correlation coefficient. An improved technique was developed to visualize the flow patterns generated in liquid products heated by microwave energy.

REFERENCES

1. Decareau, R.V. Microwaves in the Food Processing Industry. Academic Press, New York, 1985.

2. Komolprasert, V. and Ofoli, R.Y. Mathematical modeling of microwave heating by the method of dimensional analysis. J. Food Proc. Pres. 1989, 13: 87-106.

3. Liu, L.Z. Effect of viscosity and salt concentration on microwave heating of simulated liquid foods in cylindrical containers. M.S. thesis. The Pennsylvania State University, University Park, PA, 1990.

SELECTION OF HEATING MEDIA FOR THERMAL PROCESSING OF RETORT POUCHES

P.R. MASSAGUER, C.F. CARDELLI & H.G. AGUILERA
Depart. de Ciência de Alimentos, Fac. de Engenharia de Alimentos, Universidade Estadual de Campinas, C.P. 6121, 13081-970, Campinas, São Paulo, Brasil.

ABSTRACT

Heating media for retort pouches processing in a modified still vertical retort were selected through a heat distribution test at 105, 113, and 121°C. Nylon-6 bricks were used as transducer for pouches processed in the horizontal position. Steam, steam/air (90/10 and 70/30), and water with overpressure were tested. For all media, 121°C led to better heat distribution. Steam mixtures (90/10) and water (126 KPa) had heat distribution as good as pure steam at 121°C. No homogeneity was achieved with mixtures of 70% steam/30% air. No significant differences ($\alpha = 0.05$) among heating rates and lag factors were detected for transducers processed in pure steam or 90% steam/10% air at different levels of the rack.

INTRODUCTION

Since temperature distribution is greatly influenced by retort design and operation procedures, any new retort equipment must take into consideration the necessity of providing adequate temperature and heat distribution under fully loaded operation conditions. A steady and predictable temperature through the retort is important for the delivery of the desired lethality to all packages.

Steam-heated water with overpressure and steam/air mixtures are heating media commonly used to sterilize food in pouches but they are less efficient in the distribution and transfer of heat.

The NFPA (1) National Food Processors Assoc. has recently established guidelines for temperature distribution verification of retorts. They suggest that the temperature within the retort after 1 min following come up should have a maximum range of 3°F (1.7°C) and should be within 1.5°F (0.8°C) of the reference temperature device, and recommend verification of heating rate distribution throughout the loading area.

One method to systematically evaluate temperature distribution uses the mean temperature and standard deviation as a function of processing holding time and thermocouple location (2). The second method uses the range (maximum minus minimum) temperature as an expression of temperature uniformity (3).

The objective of this work was to select heating media, using steam as pattern, for retort pouches processed in horizontal position in a modified vertical retort which offer the best temperature distribution and heat penetration.

MATERIAL AND METHODS

A modified vertical retort (Dixie Inc., USA) for retort pouches processing was used (3). Thirty Nylon-6 bricks (15x110x150 mm) were used to fill a ten-level rack with 20 mm clearance between trays, simulating commercial size retort pouches filled with a conductive food. The rack was located 75 mm from the cross steam spreader (bottom).

1) Heating media testing and processing conditions: Pure steam, steam 90%/10% air, and 70%/30% air and water at the same overpressure as the mixtures were tested at 105, 113, and 121°C. For 121°C gauge pressure were 126 KPa and 191 KPa for water and steam mixtures. Venting time for pure steam was 7 min. For steam and its mixtures process time was 20 min and for water a 30 min process was set. The heat penetration test was run until the slowest heating brick reached 120°C from ambient temperature.

2) Temperature measurement: For temperature distribution studies, ten needle type, Ecklund thermocouples were fixed one at each level. Once secured, the thermocouples formed a descending spiral, in such a way that in level 10, the thermocouple was close to the retort wall, while in level one the thermocouple was close to the steam spreader. b) For heat penetration tests, 5 blocks with the measuring junction of the thermocouple positioned in the center were used and were located as follows: 2 in level 2, 1 in level 6, and 2 in level 10.

Heating and cooling medium temperatures were also recorded each 0.9 min with a digital temperature indicator (Jonhis Inst. Ltd., Brazil). The rest of the rack was filled with nylon transducers.

RESULTS AND DISCUSSION

For all heating media tested, setpoint 121°C led to better distributions. Results are summarized in Table 1.

TABLE 1
Comparison of heating media during temperature distribution and heat penetration test with Nylon-6 bricks, setpoint 121°C

Heating mean	$\overline{T}_1$(°C)	STD(°C)	DGT(°C)	Coldest Rack Level	Mean[a] Heating rate (min)	Mean[a] Lag Factor
100% Steam	121.8	0.11	0.6	1	9.1	0.77
90% Steam	121.4	0.18	0.9	4	9.2	0.71
70% Steam	121.4	0.31	1.8	10	–	—
water-LO	121.7	0.16	1.1	8	–	—
water-HO	126.6	0.18	1.4	10	–	—

$\overline{T}$ - Overall mean process temperature excluding come up time.
STD - Average standard deviation.
DGT - Maximum temperature difference.
a - no significant difference ($\alpha = 0.05$).

Most efficient media were the mixtures of steam 90% with a DGT of 0.9°C close to pure steam DGT = 0.6°C. Poor distribution was shown by steam mixture 70%, with the largest DGT (1.8) and STD = 0.31°C.

The coldest rack level was detected from a plot of the mean temperature and STD for each location, over the entire temperature maintenance period. Level 10 was identified as coldest position and was selected to locate pouches for future heat penetration tests. Mean heating rates and lag factors did not show significant differences (α = 0.05) when compared by rack position and heating media for pure steam and 90% mixtures. Figs. 1 and 2 show mean temperature and STD as a function of heating time for all thermocouples for water with 126 KPa (LO) and 191 KPa (HO).

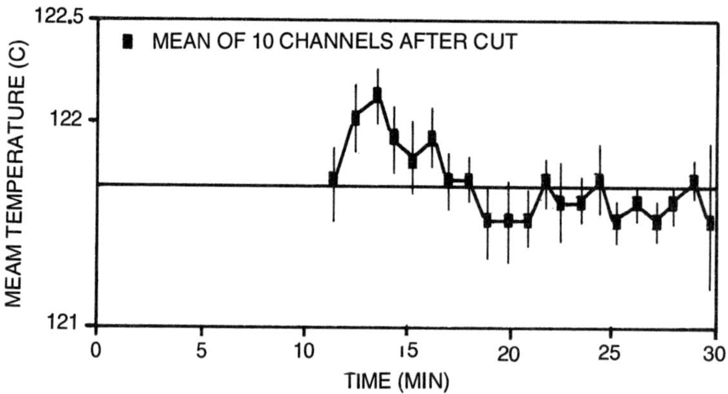

Figure 1. Temperature distribution for water 126 KPa - 121°C.

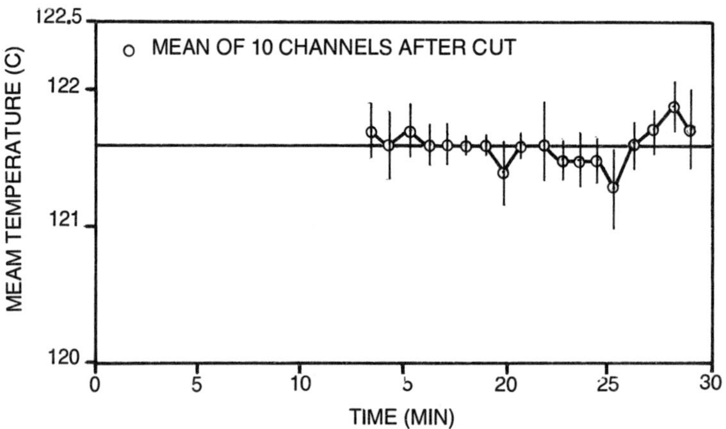

Figure 2. Temperature distribution for water 191 KPa - 121°C.

It is widely recognized that for retort pouches overpressured processes are necessary in order to counteract gases expansion within the packages during heating and protect their integrity during cooling. Therefore, among the heating media tested, we have selected steam/air mixtures (90/10) for processing in the modified Dixie retort. Such medium showed a maximum DGT of 0.9°C, not significantly different from those of pure steam in the transducers.

REFERENCES

1. NFA Guidelines for thermal process development for foods packages in flexible containers. Nat. Food Process. Ass., Washington, D.C., 1985.
2. Tung, M.A., Britt, I.J. & Ramaswamy, H.S. Food sterilization in steam/ air reports. Food Technol., 1990, 44, 105.
3. Cardelli, C.F., Aguilera, H.G. & Massaguer, P.R. Adaptation of a vertical retort for thermal processing of retort pouches containing food: Heat distribution tests. Proceeding of the 4th Brazilian Thermal Science Meeting, 1992, 149-152.

NUMERICAL ANALYSIS OF HEAT TRANSFER IN THE STERILIZING OF CANNED FOODS BY VACUUM PACKING

XIANG JUN HUANG, TAMOTSU HANZAWA and NOBORU SAKAI
Department of Food Science and Technology,
Tokyo University of Fisheries,
Tokyo 108, Japan

ABSTRACT

The heat transfer coefficients for steam condensation onto the surface of canned foods by vacuum packing were obtained by numerical calculation of fundamental equations and experimental results. A correlative equation was obtained for the heat transfer coefficient in a can under a vacuum pressure. By using these results, the temperature distributions in vacuum packed cans were calculated numerically. To check the calculated results, the temperature distributions were measured in a can under the same conditions as those of the calculations. The experimental temperature distributions were in close agreement with the calculated values.

INTRODUCTION

Canned foods by vacuum packing were developed in recent years and were packed under high vacuum conditions with little covering water instead of the traditional packing medium. These canned foods are not seasoned uniformly due to the lack of packing medium and add a peculiar flavour and a deep fragrance to the foods. The heat transfer in a sterilizing process for these canned foods is performed by steam which is converted by evaporation of a little water under conditions of high vacuum and heating. Therefore, it seems that a degree of vacuum and volume of water are important factors for the heat transfer in this can. However, there are few works concerning the heat transfer by steam in a can. Thus the mechanism of heat transfer is not yet established under these operating conditions. In this study, as basic research on the heat transfer in the can, the effect of degree of vacuum on the heat transfer coefficient was investigated and a correlative equation was obtained.

EXPERIMENTAL APPARATUS AND PROCEDURE

Figure 1 shows a schematic diagram of the experimental model can. Cans of type No. 5 and No. 6 were used. To measure the temperature in the can, Cu-constantan thermocouples were set through a small hole in the side wall of the can and were hardened in place with an adhesive paste. The sample was a paste of arum root which was similar to a food with the thermal physical property and surface condition. The shape and size were cylindrical and 20 mm (diameter) × 20 mm (height). The experimental data were given for the vacuum pressure of 180 mmHg abs. and for the heating temperature of 127°C.

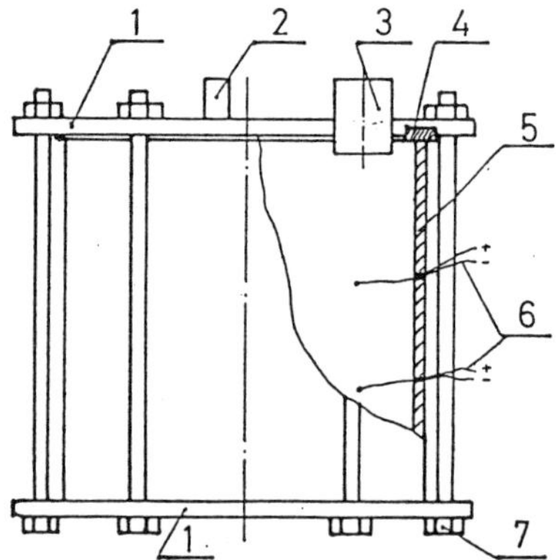

Figure 1 Model can. 1, Lid; 2, pipe for pressure measurement; 3, tap of silicon gum; 4, packing gum; 5, vessel of acrylic resin; 6, thermocouple; 7, bolt.

RESULTS AND CORRELATION

1) Effect of vacuum pressure on heat transfer: Figure 2 shows some typical experimental temperature variations at the center of the sample (the slowest heating point) in the can with the vacuum pressure, H_G, as a parameter. Can of type No. 5 was used and the number of samples as 24 pairs. From Figure 2, it seems that there is the most suitable vacuum pressure for the heat transfer to the solid food in the can. In Figure 2, the dotted line shows the temperature variation at the slowest heating point in the sample in the case of traditional packing medium.

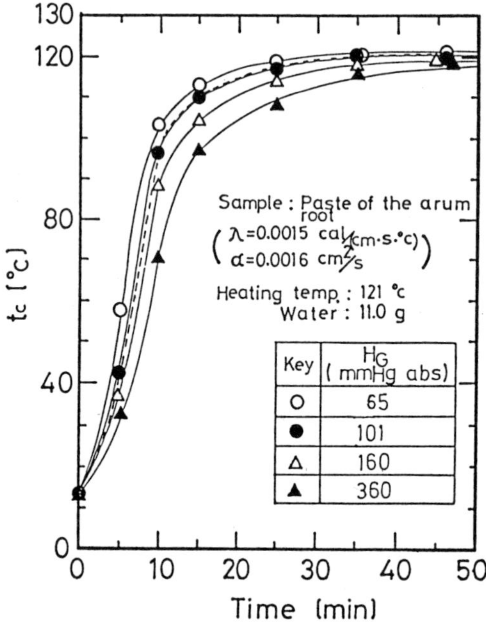

Figure 2 Effects of vacuum packing.

2) Correlation of heat transfer coefficient and vacuum pressure: in this study, the heat transfer coefficient, h, was obtained by comparison of the experimental temperature distributions in a sample in this can and the calculated ones from fundamental equations under the same conditions as those of this experiment. Figures 3 and 4 show the same typical case of coincidence with the calculated

temperature distribution on the central axis in the can with the experimental ones for the axial direction (Figure 3) and for the cross-sectional direction (Figure 4). From these figures, the calculated values were in sufficiently good agreement with the experimental ones and in this case h were given for Figure 3 of 0.0027 cal/cm^{2}s°C and for Figure 4 of 0.009 cal/cm^{2}s°C.

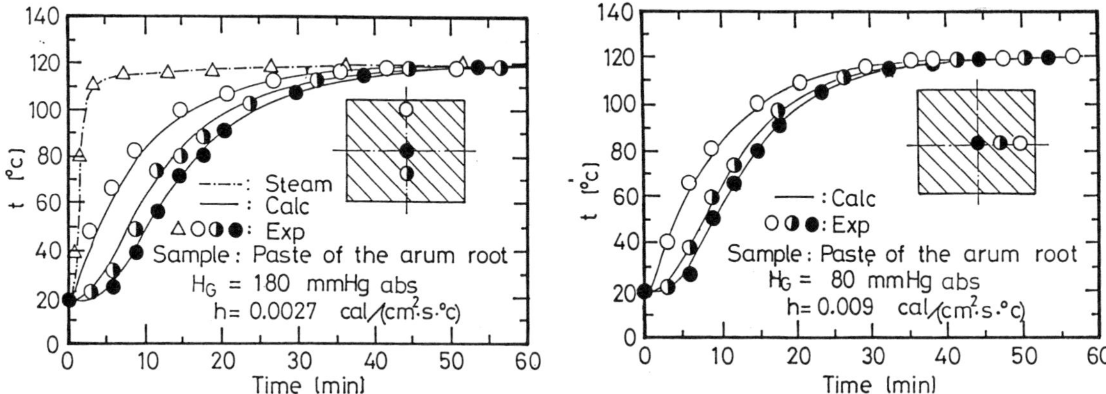

Figure 3 Comparison of calculated temperature profile with experimental ones.

Figure 4 Comparison of calculated temperature profile with experimental ones.

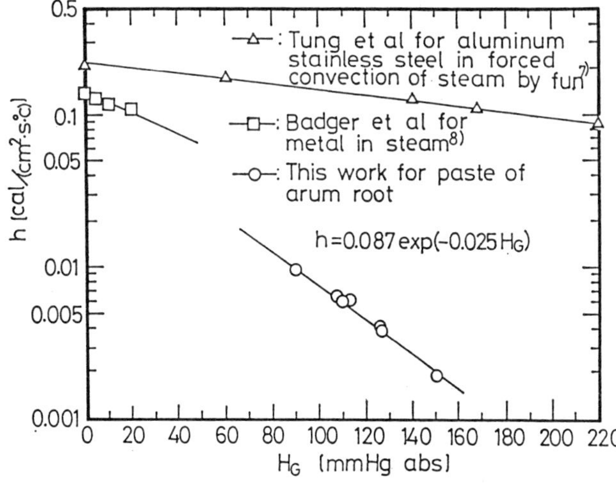

Figure 5 Relation between h and H_G.

Figure 5 shows the relationship between H_G and h. From Figure 5, the following correlative equation was obtained.

$$h = 0.087 \exp(-0.025H_G)$$

Figure 5 also shows some previous results for h[1,2]). These results are given for a polished surface and under forced convection of steam. By using this correlation, a check for h was performed with the calculation and experiment. The results indicated the validity of correlative equation.

REFERENCES

1. Badger *et al. Heat Transfer and Crystallization*, p. 24 (1945).
2. Tung, M. A. *et al. J. Food Sci.* **49**, 939 (1984).

HEAT TRANSFER IN LIQUID-FILLED CONTAINERS DURING END-OVER-END ROTATION

IAN J. BRITT, ALLAN T. PAULSON, ROBERT STARK* AND MARVIN A. TUNG
Department of Food Science and Technology, Technical University of Nova Scotia
Box 1000, Halifax, Nova Scotia, B3J 2X4 Canada
*Agriculture Canada Research Station, Kentville, Nova Scotia, B4N 1J5 Canada

ABSTRACT

Studies were carried out in a steam/air end-over-end (EOE) rotational batch retort using liquid-filled cans to assess the importance of radial position and processing conditions on the uniformity of product heating. Results of these experiments designed to simulate processing of fluid food materials, indicated a trend toward increased accumulated lethality with increased radius of rotation. However, differences were small in magnitude and not significantly different ($p > 0.05$) for some processes.

INTRODUCTION

Agitation of packages containing fluid foods during thermal processing can increase the effective heat transfer rate throughout the product during heating and cooling, and thereby reduce the time required to achieve a safe thermal process. One form of forced convection may be achieved by rotating entire product cars filled with food containers within a batch-type retort. In some applications it is necessary or desirable to use overpressure processes which provide a pressure higher than the steam pressure equivalent to the selected processing temperature. In such cases, this superimposed pressure can influence headspace gas volume and hence product agitation during rotation. In this work experiments were conducted in a 1300 mm diameter single-basket forced convection steam/air retort (J. Lagarde Autoclaves, Montélimar, France) to assess the importance of radial position and processing conditions on the uniformity of product heating in steam and steam/air mixtures.

MATERIALS AND METHODS

Cylindrical metal cans of size 300x409 (76.2 mm dia. x 115.9 mm high) were filled with 1% (w/v) aqueous bentonite suspensions, to a headspace equal to 5% of the can volume, and closed without vacuum. Test containers were fitted with calibrated needle-type thermocouples located 13 mm (0.5 in) from the bottom for vertically oriented cans in still processes and at the geometric center of cans in rotary processes. Temperature responses were assessed at various positions throughout the fully loaded retort by logging data at 10 s intervals. Experiments were conducted in triplicate at rotational speeds of 0, 10 and 20 rpm with target retort temperatures of 115 and 125°C, using "pure" steam as well as mixtures of 75% steam/25% air to provide overpressure conditions.

Heat penetration temperature history data from test cans in each position were compared based on accumulated lethality, F_o, (T_{ref}= 121.1°C, z=10 C°) during the heating phase calculated for each container using the Improved General Method for the first 50 or 20 min of the heating period for retort temperatures of 115 and 125°C, respectively.

RESULTS AND DISCUSSION

Tables 1 and 2 summarize the accumulated lethality results for processes at 115 and 125°C, respectively. Position #1 was at the center of a vertical section through the load, #2 and #4 were at the edge of the load extending horizontally and vertically, respectively, whereas #5 was at the outside corner of the plane and #3 was midway between #1 and #2. The thermocouple from Position #1 malfunctioned for some experiments; therefore, those data were discarded.

F_o-values from experiments at 115°C showed increased variability for steam/air compared to steam processes at 0 and 20 rpm; however, there were no significant differences ($p>0.05$) in the mean F_o-values at either steam fraction and there was a small decrease in variability for steam/air compared to steam processes at 10 rpm.

Process lethalities from experiments at 125°C were more variable for rotational processes as compared to still cooks (0 rpm) for steam processes and increased in variability for 0 and 20 rpm processes using steam/air compared to saturated steam. For still processes, there was a significant ($p<0.05$) decrease in mean F_o-values for steam/air processes which indicated slower heating, possibly due to decreased surface heat transfer associated with the steam/air heating medium. While mean F_o-values were not significantly different ($p>0.05$) between the steam and steam/air processes at 20 rpm, variability was nearly twice as large for steam/air compared to steam processes. However, only small changes in variability were observed between steam and steam/air processes at 10 rpm.

Overall, there was less variability in processes at 115°C compared to corresponding processes at 125°C. This was reasonable since low viscosity fluids heat quickly and tend to follow the retort temperature. Product temperatures approached the retort temperature asymptotically during heating, and unaccomplished temperature differences (retort minus product temperature) for the convection heating material approached zero during the heating period. Since the accumulation of lethality at 115°C calculated by the Improved General Method is 0.245 of equivalent time at 121.1°C compared with 2.45 times at 125°C, it follows that less variability would be expected among processes at the lower temperature where lethality accumulated at a lower rate. It is also noteworthy that while significant differences ($p<0.05$) in mean F_o-values among container positions were not identified for all process conditions, there was a trend toward increased lethality with increased radius of rotation at both retort temperatures.

CONCLUSIONS

While rotation resulted in increased heating rates within some containers, the effects were not consistent through the product load; moreover, under these experimental conditions, some containers did not heat more quickly when agitated than during still processes. These findings emphasize the need to determine the slowest heating region within a retort for a specific product and rotational process prior to completing heat penetration experiments, in order to ensure delivery of a minimum commercial process to all containers in a retort load.

Table 1
Individual and Overall Mean Values and Coefficients of Variation (%) for the Accumulated Lethality (F_o, min) During the First 50 min of the Heating Phase at 115°C.

Position	Steam			Steam/Air		
	0 rpm	10 rpm	20 rpm	0 rpm	10 rpm	20 rpm
1	8.4^{a}	--	9.0^{a}	7.7^{a}	--	--
2	8.7^{b}	8.4^{a}	9.3^{b}	8.4^{b}	8.8^{ab}	8.2^{a}
3	8.6^{ab}	8.3^{a}	9.5^{ab}	8.0^{a}	8.7^{a}	7.8^{a}
4	9.2^{c}	8.3^{a}	9.6^{c}	8.9^{c}	9.0^{b}	8.0^{a}
5	9.0^{c}	9.3^{b}	9.4^{ab}	8.8^{c}	9.3^{c}	9.4^{a}
Mean	8.8^{xy}	8.6^{xy}	9.4^{y}	8.4^{x}	9.0^{xy}	8.4^{x}
CV, %	3.32	5.15	2.25	6.14	2.92	10.39

Values in columns or means with the same letter are not significantly different ($p > 0.05$) based on 3 runs at each condition.

Table 2
Individual and Overall Mean Values and Coefficients of Variation (%) for the Accumulated Lethality (F_o, min) During the First 20 min of the Heating Phase at 125°C.

Position	Steam			Steam/Air		
	0 rpm	10 rpm	20 rpm	0 rpm	10 rpm	20 rpm
1	11.7^{a}	--	--	7.7^{a}	11.2^{a}	--
2	12.7^{a}	12.3^{a}	14.4^{a}	11.2^{c}	14.9^{ab}	14.8^{a}
3	12.1^{a}	11.5^{a}	13.5^{a}	9.0^{ab}	12.3^{ab}	12.7^{a}
4	13.2^{a}	11.8^{a}	13.7^{a}	10.9^{bc}	13.5^{ab}	13.4^{a}
5	13.1^{a}	15.0^{a}	17.7^{b}	11.4^{c}	16.6^{b}	18.3^{a}
Mean	12.6^{y}	12.7^{y}	14.8^{y}	10.1^{x}	13.7^{y}	14.8^{y}
CV, %	5.40	15.35	12.64	16.41	17.10	21.39

ACKNOWLEDGEMENTS

The authors acknowledge financial support from the Natural Sciences and Engineering Research Council of Canada and technical assistance from Mr. Gerry F. Morello.

DEVELOPMENT OF HIGH SPEED BATCHWISE STERILIZER BY STEAM FUSION

KOHJI AONO
Development Department, Iwai Kikai Kogyo Co.,Ltd.
17-10, 3-chome, Higashi-Kohjiya, Ohta-ku, Tokyo 144 ,Japan

ABSTRACT

A new batchwise sterilizer has been developed. Direct steam mixing and flush cooling are available in the vessel.

As a result, this system achieves a short cycle time (about 1 hr) and can be operated under UHT as well as plate heat exchanger and accomplishes an extremely high yield.

INTRODUCTION

This system can be applied to medium-viscous (～3000 cps) and containing particulate (～10 mm) food products. This system consists of :
(1) vessel (with agitator and sensors), (2) utility supply (including steam, air, chilled water) and, (3) control panel (consists of PLC, displays of sensors,etc.) A flow diagram of the steam fusion system is showm in Fig. 1.

OPERATING SEQUENCE

First, food is fed into the vessel, and they are gently mixed by a paddle. Then pure steam is directly injected onto the surface of the food through nozzles. Food near the nozzles is strongly agitated and steam condenses quickly and mixes with the product.

Thus, they are heated in a few minutes to ultra high temperature. After holding for the desired time, the pressure in the vessel is released to atmospheric pressure under sterile conditions. They are flush cooled in a few minutes to the boiling point. If further cooling is desired, vacuum flushing is available to 50℃. Schematic diagram of time-temperature in this system is showm in Fig. 2.

Finally, aseptically cooled food is pneumatically fed to the filler.

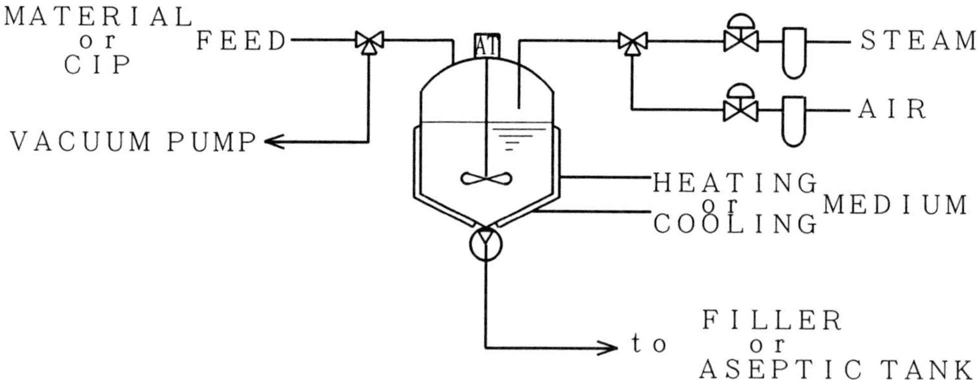

Fig. 1 FLOW DIAGRAM OF STEAM FUSION SYSTEM

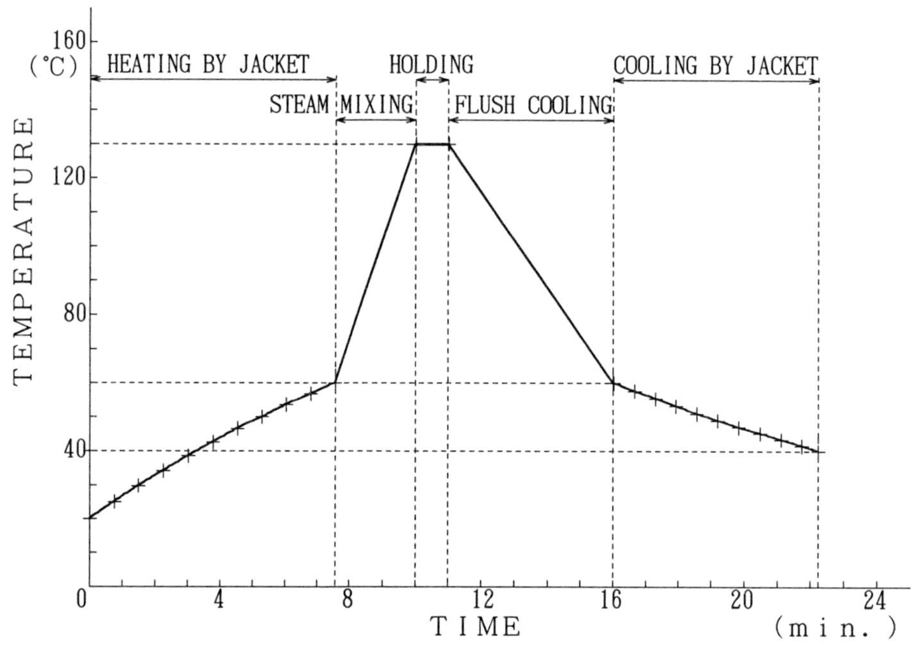

Fig. 2 SCHEMATIC DIAGRAM OF TIME-TEMPERATURE IN STEAM FUSION SYSTEM (100L)

COMPUTER AIDED DESIGN AND OPTIMIZATION OF STERILIZATION OF CANNED TUNA.

Julio R. Banga[1], Antonio A. Alonso[2], Jose M. Gallardo[2] and Ricardo I. Perez-Martín[2]

[1]Dept. Chem. Eng. (Fac. Ciencias), Universidad de Vigo. Apto. 874, 36200 Vigo, Spain.
[2]Instituto de Investigaciones Marinas (CSIC). Eduardo Cabello, 6. 36208 Vigo, Spain.

ABSTRACT

Methods for optimization of thermal processing have usually been experimentally tested using isotropic and homogeneous foods or analogs. We have developed a set of computational tools that allow the design and optimization of the thermal processing of canned foods and applied these tools to a real and complex food system: canned tuna, which is one of the leading products in the Spanish food industry. Suitable models were developed for heat transfer and thermal degradation of nutrients and quality factors.

Several optimization problems were solved and experimentally tested: maximization of the overall retention of thiamine, maximization of the overall retention of surface lightness and miminization of process time. The optimal processes thus calculated have very significant advantages over those used in the industry: thiamine retention may be increased as much as 150%, surface browning greatly reduced and process time reduced between 50-80%. Therefore, the use of these computational tools can significantly improve both the quality and the productivity of the canning industry.

INTRODUCTION

The canned tuna industry is of great economic importance in Spain, where the production was of 116,000 tons in 1991, worth $500 million [1]. The leadership of canned tuna is repeated in the U.S.A., where it is the most frequently eaten kind of fish [2]. In spite of all this, the thermal processes currently in use cause a significant overprocessing, which is detrimental to the final nutritional and organoleptic quality. Since the pioneer work of Teixeira et al. [3], several authors have used optimal control techniques to improve thermal processing of canned foods [4, 5, 6], but no experimental validation has been published. Here we use new advanced computer-aided simulation and optimization methods [7, 8] to design new processes which, guaranteeing a determined microbiological lethality, give a maximum retention of nutrients and/or quality factors. The retention of nutrients is evaluated as volume average retention, because the use of the C cook-value is not recommended [9].

METHODS

In order to optimize the process, a good mathematical model of the system is needed, together with values for all the kinetic and thermophysical parameters involved. We have used the modeling approach proposed by Banga et al [8], which allows the consideration of anisotropic and non-homogeneous conduction-heated canned foods. The thermal diffusivity and effective heat transfer coefficient were simultaneously determined from experimental heat penetration data by multi-variate least squares fitting [8]. The nutrient considered was thiamine, as this is the most thermolabile vitamin. With regard to the organoleptic quality factor, surface color (lightness) is the property most valued by the consumer [10]. The kinetics of thermal degradation of thiamine and surface lightness were determined using an unsteady-state experimental procedure [10].

The optimization problems (maximization of the final retention of thiamine or surface lightness, minimization of process time) were solved using a stochastic optimal control algorithm, ICRS/DS (Integrated Controlled Random Search for Dynamic Systems), which is discussed elsewhere [7]. Optimal constant retort temperature (CRT) and variable retort temperature (VRT) profiles were calculated. It should be noted that only CRT are currently used in industry. However, VRT profiles may offer significant advantages in several situations [7, 10], and the use of non-linear control techniques could ease its implementation [11].

RESULTS

All the optimization problems were successfully solved with small computing effort. Figures 1 and 2 show the retention of thiamine and surface lightness vs. retort temperature for different lethality values considering the constant retort temperature (CRT) policy. In figure 3 it can be seen that the optimal CRT that maximizes thiamine retention is always higher than that of surface lightness. These optimal CRT, with retort temperatures ranging from 125 to 128°C when Fc=7-12 min is considered, are significantly superior to the processes currently used in the industry, that are typically carried out at lower temperatures (110-115°C) and therefore, during longer times. Thiamine retention may be increased as much as 150%, surface lightness greatly preserved and process time reduced between 50-80%. In the same way, the use of optimal VRT may reduce process time even more (figure 4).

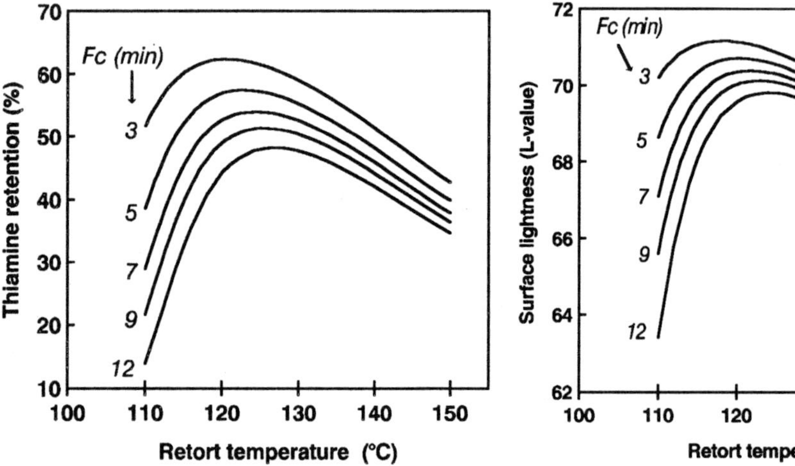

Fig. 1.- Thiamine retention versus (constant) retort temperature for different lethality (Fc) values.

Fig. 2.- Surface L-value versus (constant) retort temperature for different lethality (Fc) values.

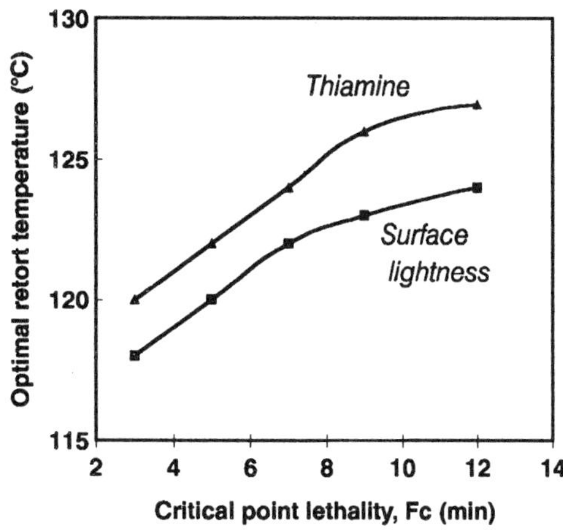

Fig. 3.- Optimal retort temperatures (max. thiamine and lightness) versus lethality (Fc).

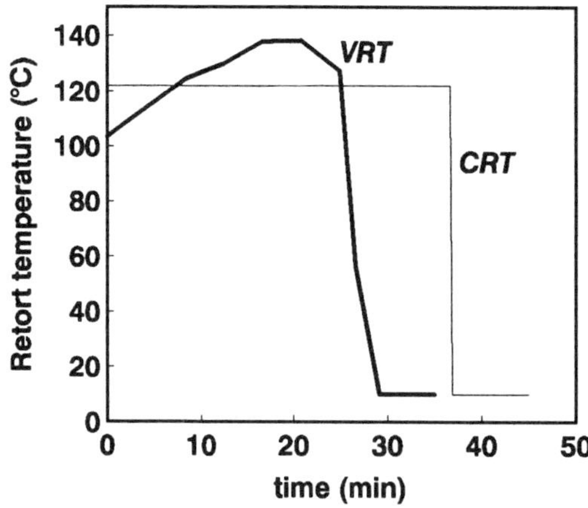

Fig. 4.- Optimal (minimum process time) variable retort temperature (VRT) process compared with the optimal constant retort temp. (CRT) profile that assures the same final surface lightness.

REFERENCES

1. Anonymous, Export directory of canned fish, seafood and salted fish. ANFACO,1992. Vigo, Spain.
2. Vondruska, J.W., W. Steven Otwell and R.E. Martin, Seafood consumption, availability and quality. Food Technol., 1988, 42(5):168-172.
3. Teixeira, A. A., G. E. Zinsmeister and J. W. Zahradnik, Computer simulation of variable retort control and container geometry as a possible means of improving thiamine retention in thermally processed foods. J. Food Sci., 1975, 40:656
4. Saguy, I. and Karel, M., Optimal retort temperature profile in optimizing thiamine retention in conduction-type heating of canned foods. J. Food Sci., 1979, 44:1485.
5. Hildebrand, P., An approach to solving the optimal temperature control problem for sterilization of conduction-heating foods. J. Food Proc. Eng., 1980, 3:123-142.
6. Nadkarni, M. M. and Hatton, T.A., Optimal nutrient retention during the thermal processing of conduction-heated canned foods: application of the distributed minimum principle. J. Food Sci., 1985, 50:1312-1321.
7. Banga, J.R., R.I. Perez-Martin, J.M. Gallardo and J.J. Casares, Optimization of the thermal processing of conduction-heated canned foods: study of several objective functions. J. Food Eng., 1991, 14(1):25-51.
8. Banga, J.R., A.A. Alonso, J.M. Gallardo and R.I. Perez-Martin, Mathematical modelling and simulation of the thermal processing of anisotropic and non-homogeneous conduction-heated canned foods: application to canned tuna. J. Food Eng., 1993, 18(4)369-387.
9. Silva C., Hendrickx, M. , Oliveira, F. and Tobback P., Critical evaluation of commonly used objective functions to optimize overall quality and nutrient retention of heat-preserved foods. J. Food Eng., 1992, 17(4):241-258.
10. Banga, J.R., Simulación y optimización del procesamiento térmico de conservas de alimentos. Ph.D. Thesis, 1991. Dept. of Chem. Eng., Universidad de Santiago de Compostela (Spain)
11. Alonso, A.A., R.I. Perez-Martin, N.V. Shukla and P.B. Deshpande, On-line quality control of (nonlinear) batch systems: application to thermal processing of canned foods. J. Food Eng., 1993, 19(3):275-289.

DIFFERENT STRATEGIES FOR CONTROLLING PRESSURE DURING THE COOLING STAGE IN BATCH RETORTS.

ANTONIO A. ALONSO, JULIO R. BANGA[1] AND RICARDO I. PEREZ-MARTIN

Food Eng. Group. Inst. de Investigacions Mariñas (CSIC), Eduardo Cabello,6. 36208 Vigo, Spain.
[1] Chem. Eng. Dept., Universidade de Vigo, Apto 874, 36200 Vigo, Spain.

ABSTRACT

A new strategy for controlling pressure and liquid level during the cooling stage of the sterilization process in steam retorts is presented. That situation, where valves dynamics are not negligiable is considered. In the two cases, performace is demonstrated to be superior to the on/off strategies. Besides this, controllers designed by IMC techniques permit the user to select just one parameter related to the speed of the response desired making control operation easier.

INTRODUCTION

During the sterilization process, the traditional mode of operation is carried out in several stages namely venting, heating and cooling. A complete discussion of all these stages is presented by Lopez[1]. To avoid sudden pressure drop which could result in bursting of cans, Steele[2] proposed a control strategy, executed on a microprocessor, where the final control elements were on/off valves. As the author reports, the main problem arises from the fact that quick actions are not possible by on/off valves and, as a result, sharp fluctuations of pressure occur.

The present work shows a new approach to the control of the cooling stage consisting of the separation of the process, esentially a multiinput-multioutput (MIMO) system, in a sequence of SISO (single input single output) subprocesses where pressure is mantained at the desired value using PID-type controllers designed by IMC techniques[3]. This strategy is compared with an on/off algorithm and its better performance demonstrated on a simulation example.

PROCESS DESCRIPTION

Water is used to cool the product while the pressure is mantained by means of air. To compensate for excessive pressure the bleeder is employed. Water level must reach a certain value in order to guarrantee that all the containers are inmersed in water. Then, it has to be mantained by acting on water flow and/or drain opening. The heat transferred to the containers is also considered, for the case of conduction-heated products. All these events may be well represented by a multivariable model consisting on a set of nonlinear ordinary and partial differential equations[4]. Parameters needed to solve the model have been taken from Mulvaney[5]. The different strategies employed in this work will be implemented on the basis of this model.

PID-TYPE CONTROLLER DESIGN UNDER IMC.

The 1991 International Process Control Conference states that 90% of industrial control requirements are still met by proportional-integral-derivative (PID) type controllers. Rivera et al[6] proposed the use of PID controllers adjusted on the basis of IMC techniques so the number of tunning parameters is reduced. This fact makes it very attractive from the industrial standpoint because the controller parameters K, τ_I and τ_D are reduced to just one related directly to the form and speed of the closed loop response we wish (filter parameter).

The IMC design method is resumed in the next two steps:

1-Given the transfer function of the process, calculate its inverse. In case of RHP poles or zeros and/or dead time, they must not appear in the control law.

2- Apply a low pass filter on the controller with an order at least equal or greater than that of the inverse.

In our case, plant and model will be conveniently described by first order equations and pure integrators estimated on the basis of step tests.

RESULTS AND DISCUSSION

ON/OFF Control and its implementation. An algorithm for the control of pressure during cooling stage with on/off type of actions has been proposed by Steele[2]. Pressure is trying to be mantained between 2 and 2.04 bar (Pmin and Pmax respectivily). The minimun level (Lm) required is 40 l. Water flow is 0.2 Kg/s and air flow 0.015 Kg/s. Initially, the system is at 120 ºC and 1.95 bar and the retort is filled halfway by RO-100 cans (radius=0.0325 m, height=0.0350 m). Figure 1 represents the evolution of pressure during the process for a valve action time of 0.5 s. These results agree with those reported by Steele[2] in the sense that actions of the order of 0.5 s result in poor control during the cooling cycle

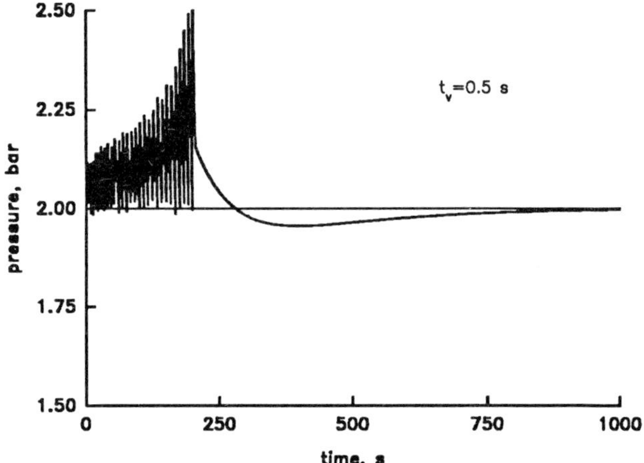

Figure 1. Pressure responses obtained when ON/OFF control is implemented.

PID type controller designed under IMC for the cooling stage. The proposed strategy consists of dividing the MIMO system in several SISO structures working in serial. Water flow will be mantained constant at the highest possible value (0.2 Kg/s) in order to guarantee fast cooling. The SISO subsystems covering the cooling cycle are corresponding to pressure control by acting on air flow, bleeder and drain opening respectivily.

The results of the simulations, in terms of pressure, are shown in figure 2 where a filter value of 10s has been employed to control pressure by acting on the drain opening. Figure 3 represents the manipulated variables movements. In case that valve dynamics must be considered, the alternative is to establish a sort of split control involving simultaneous actions on air flow and bleeder opening in response to the error in pressure. Once the minimun liquid level is achieved, pressure control is carried out by acting on drain opening.

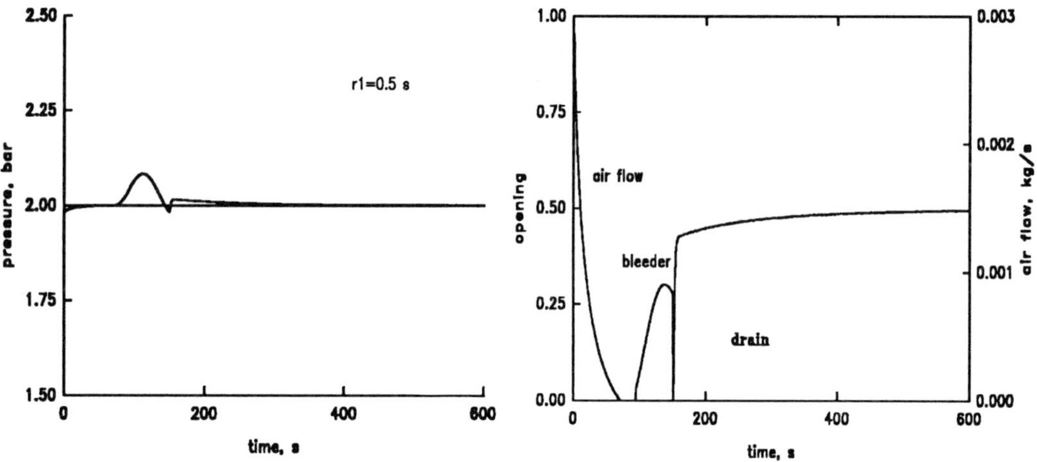

Figure 2. Pressure response obtained by serial IMC controllers.

Figure 3. Manipulated variable movements obtained with serial IMC controllers.

REFERENCES CITED.

1. Lopez, A. A complete course in canning and related processes. The Canning trade Inc, 1987, Baltimore.
2. Steele, D.J. Microprocessors and their application to the control of a horizontal batch retort. IFST Proc., 1983, **13(3)**, 183-193.
3. Morari, M. and Zafiriou, E. Robust process control. Prentice-Hall, 1989.
4. Alonso A.A., R. I. Perez-Martín, N.V. Shukla and P.B. Deshpande. On-line quality control of (nonlinear) batch systems: Aplication to thermal processing of canned foods. J. Food Eng, 1993, **19**, 275-289.
5. Mulvaney, S.J. Dynamic process modeling and evaluation of computer control of a retortfor thermal processing. Ph.D. thesis, 1987, Cornell University, Ithaca, NY.
6. Rivera, D.E., M. Morari, S. Skogestad. Internal model control. 4. PID controller design. Ind. Eng. Chem. Process Des. Dev., 1986, **25**, 252-265.

ACKNOWLEDGEMENTS.- This work was financially supported by a research grant fron the Comision Interministerial de Ciencia y Tecnologia, Spain (project ALI91-712).

THE INFLUENCE OF UNCERTAINTIES IN PROCESSING CONDITIONS ON THERMAL PROCESS CALCULATIONS

BART M. NICOLAI & JOSSE DE BAERDEMAEKER
Agricultural Engineering Department, K.U.Leuven, Kardinaal Mercierlaan 92, B-3001 Heverlee, Belgium

ABSTRACT

A numerical method for the computation of mean values and variances of the temperature and process lethality during thermal processing of foods under stochastic retort temperature conditions is derived. Simulations indicate that the stochastic fluctuations of the retort temperature may cause a considerable level of uncertainty in the predicted temperature and in the process lethality.

INTRODUCTION

The General Method for the design and evaluation of sterilization processes relies on the knowledge of the temperature history in the slowest heating zone of the container. During the last three decades there is a growing interest in the prediction of the temperature course in conduction–type foods. For this purpose the heat conduction equation is numerically solved under the assumption of accurately known process conditions such as the retort temperature [1-3]. However, in reality the retort temperature may fluctuate in an unpredictable way during the course of the thermal treatment. Recently a method has been proposed for the computation of the mean temperatures and temperature variances inside the food resulting from stochastic fluctuations of the retort temperatures [4]. In this paper this method is extended to calculate the mean and variance of the process lethality.

METHOD

Transient linear heat conduction in isotropic media in the absence of heat generation is governed by the Fourier equation with appropriate initial and boundary conditions. In the finite element method the continuum is subdivided in (finite) elements of variable size and shape which are interconnected in a finite number m of nodal points. In every element the unknown temperature can be approximated by a low order interpolating polynomial in such a way that the temperature is uniquely defined in terms of the (to be computed) temperatures $u_i(t)$ at the nodes. It is possible to show [5] that the unknown nodal temperature vector $\mathbf{u} = [u_1 \dots u_m]^T$ is the solution of the following linear differential system:

$$\mathbf{C}\dot{\mathbf{u}} + \mathbf{K}\mathbf{u} = \mathbf{f} \tag{1}$$

with $\mathbf{C}$ the capacity matrix and $\mathbf{K}$ the conductivity matrix, both $m \times m$ - matrices and $\mathbf{f}$ a $m \times 1$ vector which is a linear function of T_∞. In the DOT finite element code [5] equation (1) is solved using an implicit Euler finite difference method.

At every node i of the finite element grid the process lethality F_i (in D_{ref} units) can be described by the following *Bigelow* model [6]:

$$\dot{F}_i = g(u_i) = 2.303/(60D_{ref}) \times \exp(2.303(u_i - T_{ref})/z) \tag{2}$$

subject to the initial condition $F_i(t = 0) = 0$ and with D_{ref} the decimal reduction time (minutes) at the reference temperature T_{ref}, while z is the increase in temperature necessary to reduce this time requirement by a factor of 10.

A common formalism for describing time dependent random phenomena is that of a stochastic process [8]. It is assumed in this paper that the the mean value $\mathcal{E}(T_\infty)$ is constant and equal to $\overline{T_\infty}$. Further, the stochastic fluctuation of the retort temperature, $\Delta T_\infty \triangleq T_\infty(t) - \overline{T_\infty}$, is assumed to be a stationary first order Markov process with mean equal to zero, variance $\sigma^2_{T_\infty} = \sigma^2_{\Delta T_\infty}$ and correlation time α_{T_∞}. This means that ΔT_∞ can be considered as the output of the following linear system with a zero mean white noise input $w(t)$ with autocovariance σ^2_w equal to unity

$$\Delta\dot{T}_\infty(t) = -\Delta T_\infty(t)/\alpha_{T_\infty} + \sigma_{T_\infty}\sqrt{2/\alpha_{T_\infty}}\; w(t) \tag{3}$$

The smoothness of the stochastic process is determined by α_{T_∞}.

A conceptually simple method for the computation of heat transfer and microbial inactivation with stochastic parameters is the Monte Carlo approach (MC) [7]. In this method a set of stochastic retort temperature histories is generated and for each element of the set the heat transfer process is solved and the corresponding process lethality is computed. In the end the mean values and variances of the temperature for a given place and time are estimated by classical statistical means. A major drawback of the Monte Carlo method is the large amount of CPU time which is required when the number of Monte Carlo runs is increased to decrease the uncertainty of the estimates.

An alternative method is based on the stochastic systems theory [8]. According to this formalism the the mean temperature vector $\overline{\mathbf{u}}$ and the mean process lethality $\overline{F}_i$ at every node i can be found by solving (1) and (2) using the mean retort temperature. Further, (1) and (2) are linearized around their mean solutions $\overline{\mathbf{u}}$ and $\overline{F}_i$, respectively.

$$\Delta\dot{\mathbf{u}} = -\mathbf{C}^{-1}\mathbf{K}\Delta\mathbf{u} + \mathbf{C}^{-1}\frac{\partial\mathbf{f}}{\partial T_\infty}\Delta T_\infty \tag{4}$$

$$\Delta\dot{F}_i = (g(u_i)2.303/z)\Delta u_i \tag{5}$$

Equation (3), (4) and (5) can be combined into the following linear system

$$\dot{\mathbf{x}}(t) = \mathbf{F}\mathbf{x}(t) + \mathbf{G}w(t) \tag{6}$$

The variance matrix $\mathbf{V}_{\mathbf{x},\mathbf{x}} \triangleq \mathcal{E}(\mathbf{x}-\overline{\mathbf{x}})(\mathbf{x}-\overline{\mathbf{x}})^T$ of the solution of (6) then obeys the following *Lyapunov* matrix differential equation

$$\dot{\mathbf{V}}_{\mathbf{x},\mathbf{x}}(t) = \mathbf{F}\mathbf{V}_{\mathbf{x},\mathbf{x}}(t) + \mathbf{V}_{\mathbf{x},\mathbf{x}}(t)\mathbf{F}^T + \mathbf{G}\sigma^2_w\mathbf{G}^T \tag{7}$$

In this work the variance propagation algorithm (VP) (7) is solved using a first order explicit Euler algorithm, after reduction to a more convenient form [5]. The variances of the nodal temperatures and process values are then easily extracted from $\mathbf{V}_{\mathbf{x},\mathbf{x}}$.

The variance propagation algorithm requires in general only a fraction of the CPU time needed by the Monte Carlo method. A disadvantage of the method is that it generally provides only incomplete stochastic information (mean values, variances, no probability density functions) in contrast to the Monte Carlo method.

NUMERICAL RESULTS AND DISCUSSION

Simulations have been carried out in order to compare the results for the mean values and variances obtained by the numerical scheme which was presented above with those obtained by the Monte Carlo method.

The testproblem consisted of a A1-can (radius $r_0 = 3.41$ cm, height $L = 10.2$ cm) filled with a 30 % solids content tomato concentrate with k=0.542 W/m.K and ρc=3.89 10^6 J/m^{3o}C. The following process conditions were applied: $T_0 = 65°$C, $\overline{T_\infty} = 125°$C, $\sigma_{T_\infty} = 1°$C, $\alpha_{T_\infty} = 360$s, $h = 100$ W/m^{2o}C. The target microorganism was *Bacillus coagulans* for which $D_{ref} = 1.15$ min at 105°Cand $z = 9.2$ °C[9]. The target process lethality at the can center was F=3D. The finite element grid consisted of 100 elements with 4 nodes per element. In a first Monte Carlo experiment 100 runs were carried out, in a second experiment 1000 runs. From Fig. 1 it is clear that variances of the second Monte Carlo experiment agree very well with these computed with the variance propagation algorithm. However, the variances computed with the first Monte Carlo experiment are much more scattered indicating that the number of runs was not sufficient. The process lethality variance at the center of the can is about 0.1 D^2, indicating that the 95% confidence interval is equal to [2.4, 3.6]. Clearly small disturbances in the retort temperatures cause considerable uncertainties in the achieved process value.

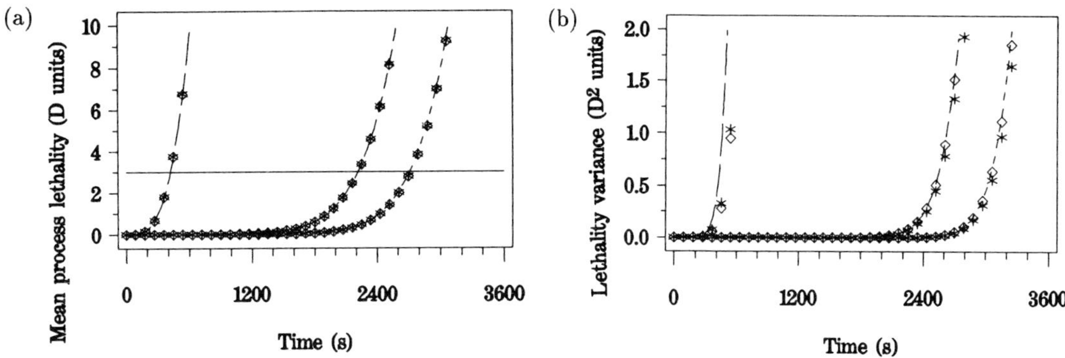

FIGURE 1. Mean (a) and variance (b) of the process lethality in a heated Al-can with random variable retort temperature at three different locations. ——— : VP at r=3.41 cm; — — : VP at r=1.71 cm; - - : VP at r=0 cm; ⋆ : MC solution with 100 runs; ◇ : MC solution with 1000 runs

CONCLUSIONS

A variance propagation algorithm has been derived for the computation of the mean and the variance of the process lethality during the thermal processing of foods with uncertain retort temperature. The mean values and variances of the temperature and process lethality within the container computed by this algorithm agreed well with those computed with the Monte Carlo method. A relatively large number of Monte Carlo runs (1000) was required in order to achieve a reasonable accuracy. It was shown that considerable uncertainties on the achieved process lethality may occur.

ACKNOWLEDGEMENTS

This study has been performed as a part of FLAIR (Food Linked Agro-Industrial Research) project AGRF-CT91-0047 (DTEE) and of FKFO–project 2.0095.92. The authors wish to thank the European Communities and the Belgian Fund for Collective Fundamental Research for the financial support.

REFERENCES

1 Teixeira, A.A. and Shoemaker C.F., *Computerized food processing operations*, Van Rostrand Reinhold, N.Y., 1989.

2 Tucker, G.S. and Holdsworth S.D., Mathematical modelling of sterilisation and cooking processes for heat preserved foods *Trans. IChemE, part C*, 1991, **69**, 5–12.

3 De Baerdemaeker, J., R.P. Singh and Segerlind L.J., Modelling heat transfer in foods using the finite element method, *Food Process Engineering*, 1977, **1** 37–50.

4 Nicolaï, B. M. and De Baerdemaeker J., Simulation of heat transfer in foods with stochastic initial and boundary conditions, *Transactions of the IChemE, part C*, 1992, **70**, 78–82.

5 Polivka R.M. and Wilson E.L., *Report no. UC SESM 76-2.* Dept. of Civil Engineering, University of California, Berkeley, California, 1976.

6 Stumbo, C.R., *Thermobacteriology in Food Processing*, 2nd Edition, Academic Press, N.Y., 1973.

7 Hayakawa, K., De Massaguer, P. and Trout, R.J., Statistical variability of thermal process lethality in conduction heating food — computerized simulation, *Journal of Food Science*, 1988, **53**, 1887–1893.

8 Melsa J.L. and Sage A.P., *An introduction to probability and stochastic Processes*, Prentice-Hall, Englewood Cliffs, New Jersey, 1973.

9 Mallidis, C.G., Frantzeskakis, P., Balatsouras, G. and C. Katsaboxakis, *International Journal of Food Science and Technology*, 1990, **25**, 442-448.

ICRS/DS: A COMPUTER PACKAGE FOR THE OPTIMIZATION OF BATCH PROCESSES AND ITS APPLICATIONS IN FOOD PROCESSING.

Julio R. Banga[1], Ricardo I. Pérez-Martín[2] and R. Paul Singh[3].

[1]Dept. Chem. Eng. (Fac. Ciencias), Universidad de Vigo. Apto. 874, 36200 Vigo, Spain.
[2]Instituto de Investigaciones Marinas. CSIC. Eduardo Cabello, 6. 36208 Vigo, Spain.
[3]Dept. Biological and Agric. Eng., Univ. of California, Davis. Davis, CA 95616. USA.

ABSTRACT

As the food industry is largely batch-oriented, many optimization problems in food engineering are actually optimal control problems. A computer package, ICRS/DS (Integrated Controlled Random Search for Dynamic Systems), for the optimal control of batch processes has been developed. It is based on the reduction of the optimal control problem to a constrained nonlinear optimization problem via an adequate transformation of the control function. This problem is then solved using an stochastic optimization algorithm which assures convergence with reasonable computation times. This package has been successfully used for the optimization of two important batch processes in the food industry: thermal processing of canned foods and air dehydration of foodstuffs. Different objective functions (overall nutrient retention, quality factor retention, process time, energy consumption, etc.) and constraints were considered for these processes. In all cases, ICRS/DS proved to be a reliable, useful and easy-to-use computational tool. Results show that, in some situations, the optimal non-linear control profiles have significant advantages over the traditional constant ones.

INTRODUCTION

Most processes in the food industry are batch operations that can be adequately modeled by heat and mass microscopic balances plus adequate kinetic and thermodynamic relationships. Therefore, these models usually are distributed parameter systems, where the transport equations are partial differential equations. Due to the dynamic nature of these systems, its optimization can only be achieved using an optimal control algorithm. The Maximum Principle of Pontryagin (MPP) is the classical approach to solve optimal control problems, and has been used to optimize thermal processing [1, 2] and drying of foods [3, 4, 5]. But the MPP is hard to implement and may not be efficient with highly constrained problems. It has been suggested in the literature [1, 6, 7, 8] that the so called Control Vector Iteration (CVI) methods have many benefits over the MPP. The basic idea of CVI is to transform the optimal control problem in a conventional non-linear programming problem (NLP) by suitable discretization of the control function(s).

OPTIMIZATION METHOD

We have developed a new fixed terminal time optimal control algorithm based on the CVI concept. This method, ICRS/DS (Integrated Controlled Random Search for Dynamic Systems), is based on the discretization of the control variable(s) using variable-length piecewise-linear polynomials. The transport phenomena equations are solved using finite differences with orthogonal collocation or finite elements methods. The resulting constrained NLP problem is then solved using a stochastic direct search optimization procedure based on the CRS method [9, 10]. ICRS/DS has proved to be a robust and easy-to-use method, and no convergence problems have been found even in complex problems.

OPTIMIZATION PROBLEMS

Two important food processing operations, thermal processing and air drying, have been optimized using ICRS/DS. The mathematical models used can be found elsewhere [11, 12, 13]. In order to validate our models and compare our optimization results with previous works, parameters values were taken from the literature [1, 3, 14]. In the case of thermal processing, the control variable is the retort temperature, and we have considered the following objective functions: maximization of the retention of a nutrient, maximization of the surface retention of a quality factor, minimization of process time and minimization of energy consumption. Each of these problems has its corresponding set of constraints, mainly related with the final microbiological lethality and quality retention. In the case of air drying, the air dry bulb temperature is the main control variable, but in some cases it was necessary to consider the relative humidity of the air as a second control variable. The objective functions considered were: maximization of nutrient or enzyme retention, minimization of process time, maximization of nutrient retention with a constraint on enzyme retention and maximization of energy efficiency. A more elaborated formulation of these optimization problems is given in Banga et al. [11] and Banga and Singh [13].

RESULTS AND DISCUSSION

In the case of thermal processing, the results for the maximization of nutrient retention are in agreement with the literature [1, 2]: the optimal control policy only provides a 5% improvement.

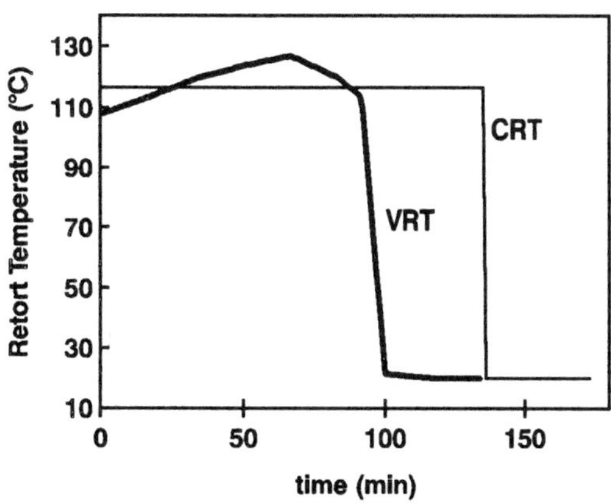

Fig. 1: Thermal processing: minimum process time variable retort temperature profile (VRT) compared with the optimal constant temperature process (CRT) that assures the same final retention of thiamin.

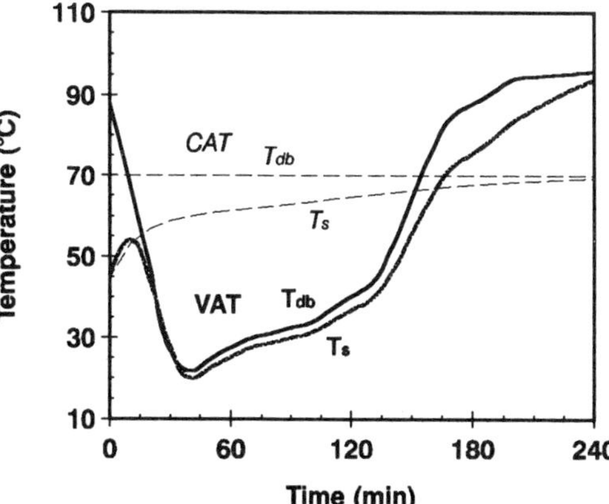

Fig. 2: Drying: optimal variable air temperature (VAT) profile that maximizes energy efficiency compared with the constant air temperature (CAT) profile of the same process time. The system temperatures (T_S) are also represented.

However, we have shown for the first time that the optimal control present significant advantages over current processes in the case of maximization of retention of a quality factor (over 20% increase) or minimization of process time (20-30% reduction, see Figure 1). This new optimal sterilization policies will soon be experimentally validated [15].

In the case of air drying, we have found that the optimal control profiles give significant increments of nutrient (13-50%) and enzyme retention (up to 70%). In the same way, process time can be reduced between 15-20% ,and energy efficiency increased up to 40% (see Figure 2). All these results suggest that advanced on-line optimal control of food processing operations is a promising area of research.

REFERENCES

1. Saguy, I. and Karel, M., Optimal retort temperature profile in optimizing thiamine retention in conduction-type heating of canned foods. J. Food Sci., 1979, 44:1485.
2. Nadkarni, M. M. and Hatton, T.A., Optimal nutrient retention during the thermal processing of conduction-heated canned foods: application of the distributed minimum principle. J. Food Sci., 1985, 50:1312-1321.
3. Mishkin, M., I. Saguy and M. Karel, Applications of optimization in food dehydration. Food Technol., 1982, 36(7):101-109.
4. Liapis, A.I. and R.J. Litchfiled, Optimal control of a freeze dryer- I. Theoretical development and quasi steady state analysis. Chem. Eng. Sci., 1979, 34:975-981.
5. Chang, T.N. and Y.H. Ma, Application of optimal control strategy to hybrid microwave and radiant heat freeze drying system. In Drying '85, eds. R. Toei and A.S. Mujumdar, Hemisphere Pub., New York, 1985, pp 249-53.
6. Evans, L.B., Optimization theory and its application in food processing. Food Technol., 1982, 36(7):88-96.
7. Edgar, T.F. and D.M. Himmelblau, Optimization of chemical processes. McGraw Hill, New York, 1988, pp. 364-371.
8. Cuthrell, J.E. and Biegler, L.T., Simultaneous optimization and solution methods for batch reactor control profiles. Comput. and Chem. Eng., 1989, 13(1/2):49-62.
9. Goulcher, R. and Casares, J.J., The solution of steady-state chemical engineering optimization problems using a random search technique. Comput. and Chem. Eng., 1978, 2:33-36.
10. Casares, J.J. and J.R. Banga, Analysis and evaluation of a wastewater treatment plant model by stochastic optimization. Appl. Math. Modelling, 1989, 13:420-424.
11. Banga, J.R., R.I. Perez-Martin, J.M. Gallardo and J.J. Casares, Optimization of the thermal processing of conduction-heated canned foods: study of several objective functions. J. Food Eng., 1991, 14(1):25-51.
12. Banga, J.R., A.A. Alonso, J.M. Gallardo and R.I. Perez-Martin, Mathematical modelling and simulation of the thermal processing of anisotropic and non-homogeneous conduction-heated canned foods: application to canned tuna. J. Food Eng., 1993, 18(4)369-387.
13. Banga, J.R. and R.P. Singh, Optimization of air drying of foods. J. Food Eng., 1993, accepted, in press.
14. Luyben, K.Ch.A.M., J.K. Liou and S. Bruin, Enzyme degradation during drying. Biotechnol. and Bioeng., 1982, 24: 533-552.
15. Alonso, A.A., R.I. Perez-Martin, N.V. Shukla and P.B. Deshpande, On-line quality control of (nonlinear) batch systems: application to thermal processing of canned foods. J. Food Eng., 1993, 19(3):275-289.

DISRUPTION OF MICROBIAL CELLS BY FLASH DISCHARGE OF HIGH PRESSURE GAS

KOZO NAKAMURA*, ATSUSHI ENOMOTO, MASARU HAKODA, HIDEO FUKUSHIMA, KIYOTAKA NAGAI, TAKAHIRO MUKAE AND YOICHI MASUDA
Department of Biological and Chemical Engineering, Faculty of Engineering, Gunma University, Tenjin, Kiryu, Gunma 376, Japan

* Department of Agricultural Chemistry, Faculty of Agriculture, University of Tokyo, Yayoi, Bunkyo-ku, Tokyo 113, Japan

ABSTRACT

To develop a novel sterilization method for heat-sensitive materials, we investigated the burst of microbial cells by rapid release of gas pressure under the various conditions. The wet cells of *S. cerevisiae* were well destroyed, when the organisms were saturated with CO_2 gas at 40 ℃ and 40 atm and then the pressure was suddenly released. On the other hand, the dry cells were poorly disrupted even under the same experimental condition. In particular, the gas with low solubility in water could provide no effects on the survival ratio of the yeast. The survival ratio for the spores of *B. megaterium QMB 1551* was also reduced, when the spores were applied to our system under the condition of 60 ℃ and 60 atm. From these findings, the mechanism of cell disruption is considered to be correlated with the gas absorption and desorption of the cells.

INTRODUCTION

Although heat sterilization is the most traditional and popular method to prevent foods from microbial spoilage, this process may commonly lead to undesirable changes in the taste and character of the products because of its high temperature. To resolve the problem, sterilization methods including no heat process have been widely studied. Among these methods, the use of ethylene oxide or radial rays is effective, but its application is limited to the special materials by the law in Japan. A high pressure treatment has been recently proposed as a new sterilization method, but its extremely high pressure (several thousand atmosphere) requires the expensive facilities.

In the present paper, we describe the disruption of microbial cells by flash discharge of high pressure gas under the various conditions of pressure, temperature, duration time and moisture content of the cells, using *Saccharomyces cerevisiae* and the spores of *Bacillus megaterium QMB 1551* as test microorganisms. Our new findings may be useful for the development of a simple, safe and inexpensive sterilization method for heat-sensitive materials.

MATERIALS AND METHODS

Microorganisms

S. cerevisiae (baker's yeast) was purchased from Oriental Yeast Co., Tokyo, Japan. The spores of *B. megaterium QMB 1551* were kindly provided by Dr. T. Nagai and Mrs. N. Amaya. These microorganisms were used in this study without precultivation.

Sterilization of the Microorganisms by Rapid Release of Gas Pressure

A glass cup containing 1.0 x 10^8 cells of *S. cerevisiae* or the spores of *B. megaterium QMB 1551* was placed in the slurry reservoir for preparative columns (40 ml capacity, GL Sciences, Tokyo, Japan), and then commercially available gas, CO_2 or N_2, was forced into the reservoir untill the pressure reached 10-100 atm. After standing for 30-300 min at the indicated temperature, the gas was expelled as rapidly as possible. The sterilization efficiency of the procedure was determined by agar plate (viable) counts.

RESULTS AND DISCUSSION

Disruption of *S. cerevisiae* by Explosive Decompression

As shown in Figure 1, the death of *S. cerevisiae* by rapid release of CO_2 gas pressure was found to be extensively affected by (1) pressure, (2) temperature, (3) duration time and (4) water content of the cells. Under the optimum condition (pressure, 40 atm; temperature, 40 ℃; duration time, 3 hr), the survival ratio of the wet cells could be reduced to $1/10^8$, while the dry cells were poorly killed even under the same experimental condition. Nitrogen gas with relatively low solubility in water, however, could provide no effects on the survival ratio of the yeast (data not shown). Similar phenomena have also been independently demonstrated by the other groups [1-2], using several kinds of gas (CO_2, N_2, N_2O and Ar) and microorganisms (*S. cerevisiae, E. coli,* and *B. subtilis*). From these results, the reduction of survival ratio is thought to be highly correlated with the gas absorption and desorption of the cells.

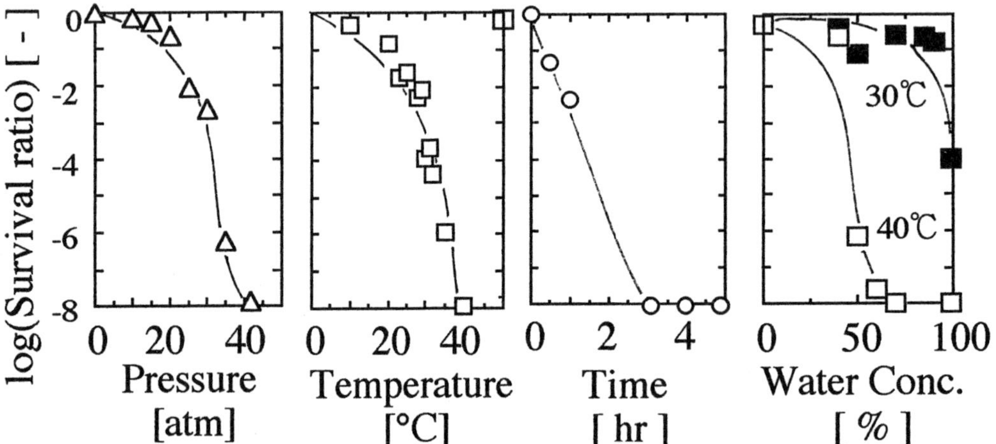

Figure 1. Effects of pressuring conditions by CO_2 gas on survival ratio of *S. cerevisiae*. (Experimental condition (if not indicated): pressure, 40 atm; temperature, 40 ℃; duration time, 5 hr; water content, 80 %)

Although the mechanism underlying the death of microorganisms is still not well understood, the rapid release of gas pressure seems to be responsible for the disruption of cells. In fact, the morphological changes of microbial cells such as appearance of "holes" in the cell surface were recognized by a scanning electron microscope (Figure 2). We are currently evaluating in detail relationship between the cell disruption and the discharge rate of the gas.

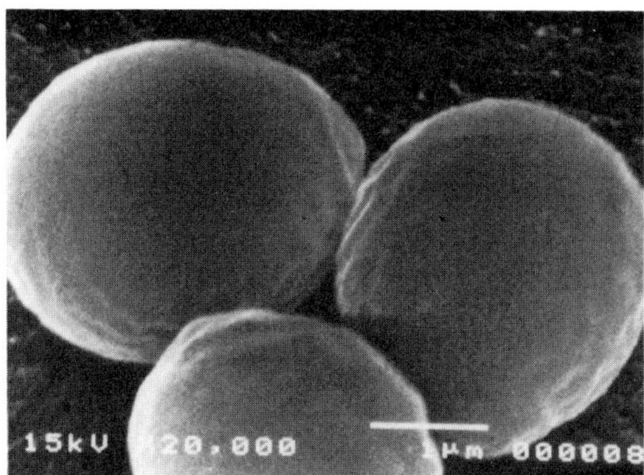

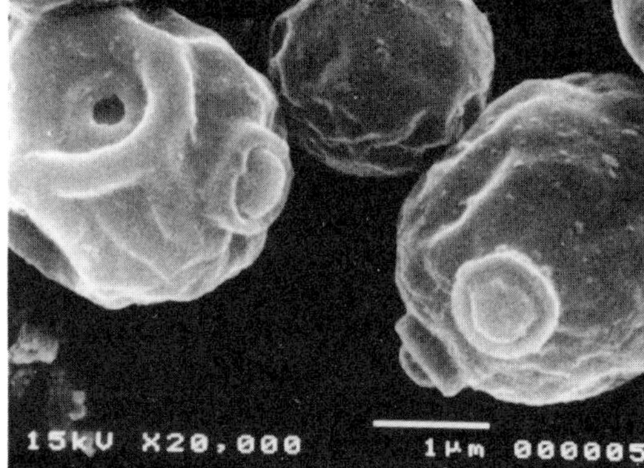

Figure 2. Electron micrographs of *S. cerevisiae* before (left) and after (right) the treatment. (pressure, 40 atm; temperature, 40 ℃; duration time, 4 hr; water content, 80 %)

Application of Our Sterilization Method to Bacterial Spores

Bacterial spores differ markedly from the vegetative cells in shape and structure, and are found to be extremely resistant to heat, radiation and most chemical disinfectants. The spores of *B. megaterium QMB 1551* were, therefore, treated by rapid release of CO_2 gas pressure. Within the limits of this experiment, the survival ratio of the spores could be reduced to 3.0×10^{-2} under the condition of 60 ℃ and 60 atm. This finding suggests that our system may be useful for the vegetative cells, and could also be applied to the bacterial spores.

ACKNOWLEDGEMENTS

The authors would like to thank Dr. Tadashi Nagai and Mrs. Noriko Amaya, Technological Development Center, SUNTORY Ltd., Mishima, Osaka, Japan for their kind gift of the spores of *B. megaterium QMB 1551* and helpful technical advice.

REFERENCES

1.Fraser, D., Bursting bacteria by release of gas pressure. *Nature,* 1951, **167,** 33-34.
2.Castor, T. P. and Hong, G. T. Critical fluid disruption of microbial cells. In *2nd International Symposium on Supercritical Fluids in Boston*, 1991, pp. 139-142.

STERILIZATION OF BEVERAGES UNDER NORMAL TEMPERATURE BY A HIGH-VOLTAGE, PULSED DISCHARGE

MASAYUKI SATO*,
KIYOSHI KIMURA[2]*, KEIKO IKEDA[3]*, TAKASHI OGIYAMA[4]* and KOICHI HATA[5]*
*Department of Biological and Chemical Engineering, Gunma University, Kiryu, Gunma 376, JAPAN, [2]*Yamatake Honeywell Co., [3]*Hitachi Co., [4]*Fuji Film Co., [5]*Honshu Seishi Co.

ABSTRACT

A high-speed, high-voltage pulsed discharge in water could be used to sterilize beverages or other liquid foods without heating and with no damage of flavor and taste. In the present study, a scaling-up from several to hundreds ml of sterilization vessel was tried experimentally. Survival ratio of *S. cerevisiae* was decreased with increasing electric field strength of the applied pulse. Effective sterilization was attained by a circulation of the sample liquid through the vessel using a peristaltic pump, in which the survival ratio decreased to about 10^{-6}. A scaling-up to 345 ml flow system was tried, and then a beer was treated and bottled. There was found no change in flavor and taste between the pulse-treated beer and the control.

INTRODUCTION

Recently, sterilization technique using electrical energy is widely recognized as an easily controllable and energy-saving method. For example, ohmic heating is now practically in use and has characteristics of direct heating without a heat conduction loss through walls. The application of a D.C. or A.C. voltage has a strong sterilization effect due to the compounds that are formed electrolytically on the electrode surface, such as chlorine, phosphorus, superoxide, and other toxic materials for human. Without formation of such compounds, effective sterilization is possible using high-speed, high-voltage pulsed discharge under ordinary temperatures without destruction of temperature-sensitive materials [1]. Some research works have been reported on pulsed discharge sterilization, where the volume of the sterilization vessel was only a few ml or less [2, 3, 4]. In the present study, the authors investigated a scaling-up from a few ml to hundreds ml for the purpose of practical applications.

EXPERIMENTS AND METHODS

Three kinds of vessels were used to treat the liquid sample containing *S. cerevisiae* as shown in Fig. 1. Fig. 1 - (a) shows a batch cell (sample liquid: 17 ml), where liquid is poured in and out through a hole located on the top of the vessel. A short tube of plexiglass tube (inner diameter: 46 mm; length: 12 mm) is sandwiched between two plexiglass plates (10 mm thick) with two electrodes glued on the inner surfaces of the plates. The diameter of the electrodes was 30 mm and the distance between the electrodes was 6 mm. Pulsed voltage was applied between the two electrodes. Fig. 1 - (b) shows a circulating system (50 ~ 100 ml of liquid including pump and flask volume) using the same vessel of (a), where liquid flows through holes located at the top and the bottom of the vessel by means of the peristaltic pump. Fig. 1 - (c) is a scaling-up flow system (345 ml), where liquid is supplied from the bottom and flows out from the top with no circulation.

The pulse power source consists of high voltage transformer, charge-discharge capacitor and spark gap, which is shown elsewhere [1]. An example of the pulse shape is as follows: pulse height, 5 ~ 25 kV; rising time, about 50 ns; and pulse width, about 500 ns.

S. cerevisiae was precultured and then suspended in a sample liquid. After applying the pulse, the sample liquid was diluted appropriately with water and inoculated onto malt agar medium. After culturing for 3 days at 30 °C, colonies were counted to calculate the survival ratio.

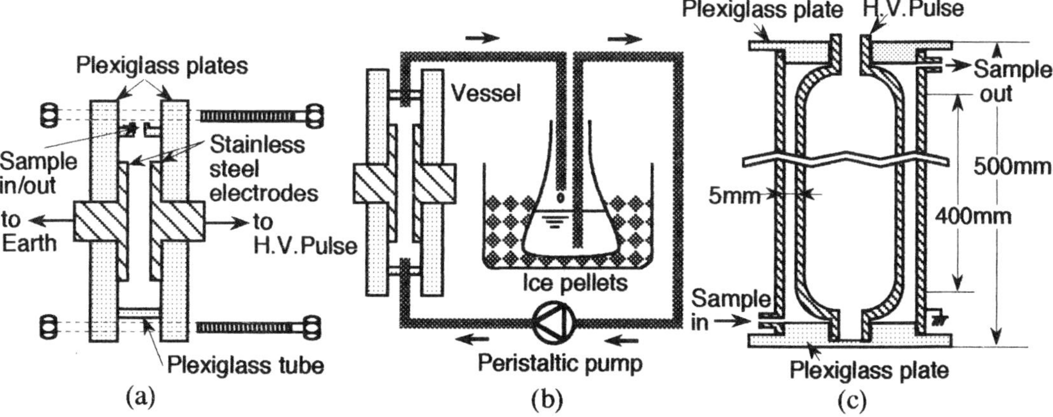

Fig. 1 Three kinds of vessels for pulsed sterilization.

RESULTS AND DISCUSSION

Using a small sterilization cell volume, it has been reported by some researchers that survival ratio (S/So) of the microorganisms decreased with increasing electric field strength of the applied pulse. In the present study (using 17 ml vessel, shown in Fig. 1 - (a)), however, it was difficult to reduce S/So to less than 10^{-2} because the electric field strength of each location in the vessel was radically different between the inner and outer parts of the electrode gap.

Fig. 2 shows S/So at several points along the vertical axis of the vessel. As is evident from the figure, the more effective sterilization was carried out between the electrodes, and the rest of the volume seemed to be a dead space. There are two possible ways to solve this problem: (1) to install a stirrer in the vessel or (2) to circulate the liquid by pumping.

As shown in Fig. 1 - (b), a modified apparatus was used with a circulation of the sample liquid by a peristaltic pump. It was found that a great sterilization effect by the circulation was attained, as shown with the line (c) in Fig. 3, compared to the curve (a) of no circulation, where the horizontal axis shows applied pulse energy per treated volume of the sample liquid. Reduction of S/So by the circulation (c) was less than 10^{-5} which means the flow rate of the circulation gives considerable influence on S/So.

Using a scaled-up flow system for pulse sterilization (Fig. 1 - (c)), de-gassed beer was treated with a flow rate of 80 ml/min (S/So was around 10^{-2} by this operating condition), and then bottled after recarbonation. There were no differences in flavor and taste between the pulse treated beer and the control.

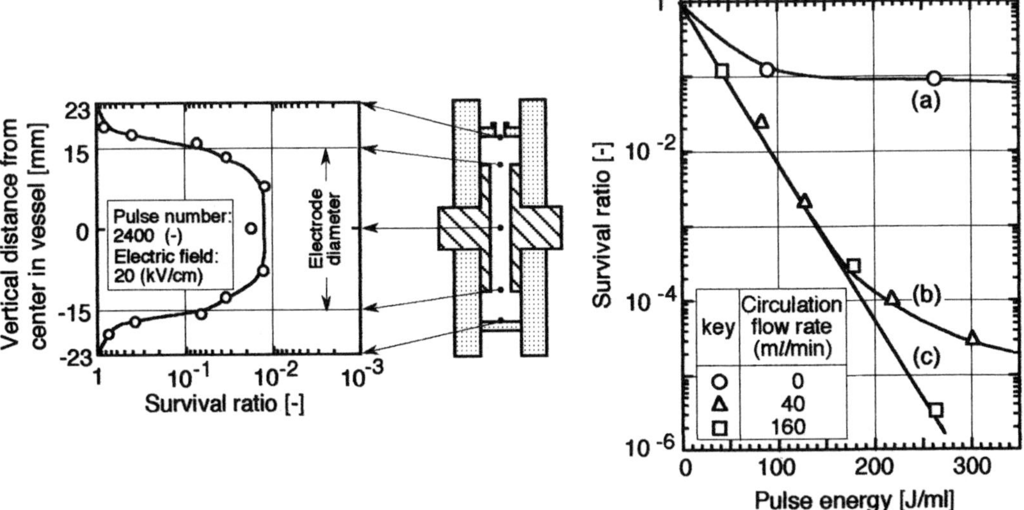

Fig. 2 Radial distribution of survival ratio using Fig. 1 - (a).

Fig. 3 Relation between applied pulse energy and survival ratio with and without circulation; Capacitor: 4000 pF, Pulse voltage: 12 kV, Pulse frequency: 50 Hz.

REFERENCES

1. Sato, M., Tokita, K., Sadakata, M., Sakai, T. and Nakanishi, K., Sterilization of microorganisms by a high-voltage, pulsed discharge under water., *Int. Chem. Eng.*, **30**, 695 - 698 (1990).
2. Sale, A. J. H. and Hamilton, W. A., Effects of high electric fields on microorganisms. I. Killing of bacteria and yeasts., *Biochim. Biophys. Acta*, **148**, 781 - 788 (1967).
3. Hülsheger, H. and Niemann, E. -G., Lethal effects of high-voltage pulses on *E. coli* K12., *Radiat. Environ. Biophys.*, **18**, 281 - 288 (1980).
4. Jayaram, S., Castle, G. S. P. and Margaritis, A., Kinetics of sterilization of *Lactobacillus brevis* cells by the application of high voltage pulses., *Biotech. Bioeng.*, **40**, 1412 - 1420 (1992).

COMPARATIVE EFFECTS OF GAMMA-RAYS AND ELECTRON BEAMS ON FILM DOSIMETERS AND BACTERIAL SPORES

T.HAYASHI, H.TAKIZAWA*, S.TODORIKI, T.SUZUKI*, M.FURUTA** AND K.TAKAMA*
National Food Research Institute, Ministry of Agriculture, Forestry & Fisheries, Tsukuba, Ibaraki, 305 Japan
* Faculty of Fisheries, Hokkaido University, Hakodate, Hokkaido, 041 Japan
**Research Institute for Advanced Science, University of Osaka Prefecture, Sakai, Osaka, 593 Japan

ABSTRACT

The responses of three film dosimeters, CTA, RCF and GAF, to gamma-rays were 30% larger than those to electron beams. Gamma-irradiation facilitated the germination of *B.pumilus* spores to a slightly higher degree than electron irradiation. Gamma-irradiation inhibited the outgrowth, growth and the synthesis of protein and RNA to a greater degree than electron beams. When the spores were irradiated with electron beams at a dose 30% higher than gamma-rays, the effects of the two types of radiation were same.

INTRODUCTION

While most of the commercial radiation sterilizations of foods, animal feeds, packaging materials and medical produces are conducted with gamma-rays from Co-60, electron beams have been increasingly utilized for treating these commodities. However, whether there is any difference in the bactericidal effect between the two types of radiation or not is controversial (1), which can be partly ascribed to inconsistent results on the dose-rate response of film dosimeters (2).

We comparatively studied the effects of gamma-rays and electron beams on *B.pumilus* spores as well as the responses of various film dosimeters to the two types of radiation.

MATERIALS AND METHODS

Irradiation

Dosimeters and spores were irradiated in a Gamma-cell 220 (2.1×10^{2}TBq of Co-60, 4.6×10^{3}Gy/hr, AECL) or in a Van de Graaff electron accelerator (2.5MeV, 147μA,

1.5×10^6Gy/hr, Nissin High Voltage Co.Ltd.).

Measurement of responses of film dosimeters to radiations
CTA (FTR-125, Fuji Photo Film Co.,Ltd.), RCF (FWT-60-00, Far West Technology Inc.) and GAF (GAFCHROMIC, GAF Chemicals Co.) were used as film dosimeter in this study. Several pieces of the film dosimeters were stacked together in layers and irradiated in the Gamma-cell 220 or the electron accelerator.

Observation of germination, outgrowth and growth of spores
The OD_{550} of spore suspension in Demain's Medium was measured during incubation at 37C to observe the germination of spores, and the OD_{550} of spore suspension in G-Medium was measured to observe the germination, outgrowth and growth.

Measurement of biosynthesis of protein and RNA
Incorporation of C-14 labeled leucine and uridine into TCA precipitates of the spores during incubation in G-Medium at 37C was used as a measure of the synthesis of protein and RNA, respectively.

RESULTS AND DISCUSSION

Relationships between OD_{280} of CTA and OD_{400} of GAF and between OD_{510} of RCF and OD_{400} of GAF
The curve obtained by plotting OD_{280} of CTA and OD_{400} of GAF for gamma-rays was coincident with that for electron beams, and the curve obtained by plotting OD_{510} of RCF and OD_{400} of GAF for gamma-rays was coincident with that for electron beams (Fig.1). These results indicate that CTA and GAF show the same response to gamma-rays and electron beams and RCF and GAF respond to the two types of radiation in the same manner. We have reported that both CTA and RCF show 30% greater responses to gamma-rays than electron beams (3), which together with the results shown in Fig.1 leads to the conclusion that CTA, RCF and GAF show 30% greater responses to gamma-rays than responses to electron beams.

Effects of gamma-rays and electron beams on *B.pumilus* spores
Irradiation facilitated the germination of *B.pumilus* spores and inhibited their outgrowth and growth. These effects of gamma-rays were larger than those of electron beams, when the spores were irradiated with gamma-rays at the same dose as electron beams. However, the effects of the two types of radiation were same, when the spores were irradiated with electron beams at a dose 30% higher than gamma-rays. The synthesis of protein and RNA was also inhibited by gamma-irradiation to a greater degree than electron irradiation, when the spores were irradiated at the same dose. The synthesis of the macromolecules was inhibited by electron beams to the same degree as gamma-rays, when the spores were irradiated with electron beams at doses 30% higher than gamma-rays (Fig.2).

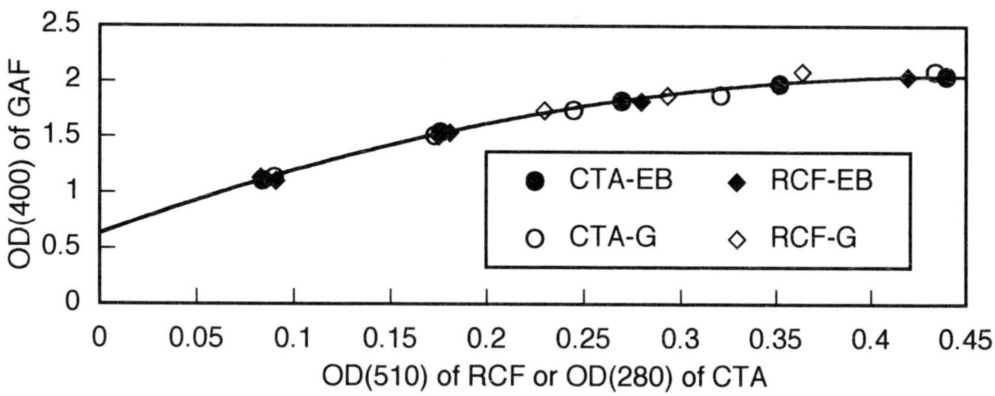

Fig.1 Relationship of responses of film dosimeters

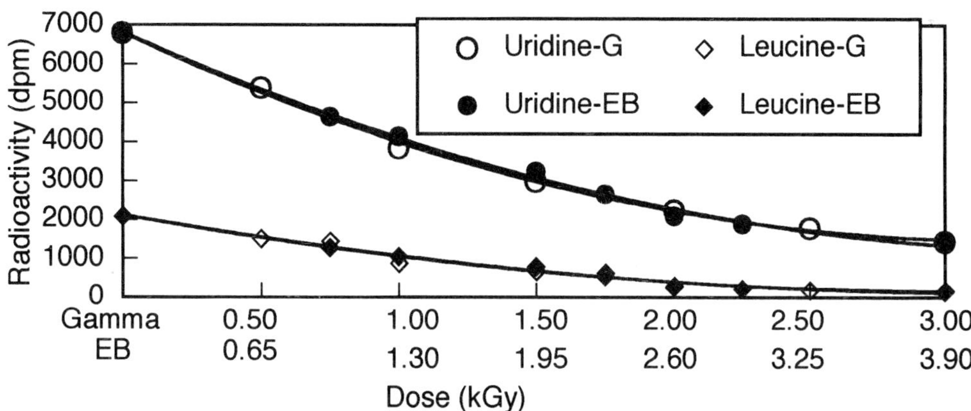

Fig.2. Incorporation of radioactivity of C-14 labeled uridine and leucine into irradiated spores after incubation for 2.5hr (uridine) or 4hr (leucine)

REFERENCES

1. Hayashi,T., Comparative effectiveness of gamma-rays and electron beams in food irradiation. In Food Irradiation, ed. S,Thorne, Elsevier Applied Science Publishers, London, 1991, pp.169-206.

2. Hayashi,T., Todoriki,S., Takizawa,H. and Furuta,M., Comparison of the cellulose triacetate (CTA) dosimeter and radiochromic film (RCF) for evaluating the bactericidal effects of gamma-rays and electron beams. Radiat.Phys.Chem., 1992, **40**, 593-595.

3. Hayashi,T., Todoriki,S. and Furuta,M., Dose rate dependence of cellulose triacetate dosimeter and radiochromic film dosimeter. Radiat.Phys.Chem., in press.

GROWTH-INHIBITORY EFFECT OF CERAMICS POWDER SLURRY ON BACTERIA

JUN SAWAI[1], HIDEO IGARASHI[2], ATSUSHI HASHIMOTO[3], and MASARU SHIMIZU[1]

[1]Department of Chemical Engineering, Faculty of Technology, Tokyo University of Agriculture & Technology, 2-24-16 Nakanachi, Koganei, Tokyo 184, JAPAN
[2]Department of Microbiology, Tokyo Metropolitan Research Laboratory of Public Helth, 3-24-1 Hyakunin-cho, Tokyo 169, JAPAN
[3]Department of Bioproduction & Machinery, Faculty of Bioresources, Mie University, 1515 Kamihama-cho, Tsu 514, JAPAN

ABSTRACT

Growth-inhibitory effects of ceramics powder slurry on *Esherichia coli* and *Staphylococcus aureus* were determined by measuring the conductance change of growth medium caused by the bacterial growth (conductance method). MgO, ZnO and α-Al_2O_3 powders were used. It was found that MgO and ZnO powders inhibited the growth of the bacteria. MgO and ZnO powders acted on bacteria in bactericidal and bacteriostatic manner , respectively. For evaluationg the growth-inhibitory effect, the conductance method is more useful than conventional methods (ie. Agar plate method).

INTRODUCTION

With development of the food industry, the relations between food preservation and food processing are getting closer and becoming more inseparable from each other, and new food protection technologies are expected. The growth-inhibitory effect of ceramics on bacteria holds considerable attention. However, there is no fundamental study on the growth-inhibitory effect of ceramics, and the method evaluationg the growth-inhibitory effect of ceramics on bacteria has not been established.

Recently, several automated methods for montoring the growth of bacteria, such as turbidometric method, conductance method, have been developed. Since the ceramics powder slurry is muddy, turbidometric method could not be used. Conductance method can deal with the muddy sample, therefore, the conductance method may be suitable for evaluating the growth-inhibitory effect of ceramics on bacteria. The conductance method relies on the fact that metabolizing microoganisms alters the chemical composition of the growth medium and that these chemical changes causes a change in the conductance of the medium [1]. The present study aims at getting a grasp of the growth-inhibitory effect of ceramics powder slurry on bacteria by conductance method.

MATERIALS AND METHODS

Test Organism

E. coli 745 and *S. aureus* 9779 stored at Tokyo Metropolitan Research Laboratory of Public Health were used. The test bacteria were cultured in Brain Heart Infusion broth (BHI broth; Difco) at 308K for 24 h on a reciprocal shaker. The culture was suspended in sterile saline to give a final bacterial concentration of about 10^3 CFU/ml.

Preparation for Ceramics Powder Slurry

As ceramics powder, we selected three kinds of metallic oxides, MgO, ZnO and α-Al_2O_3(Kishida Chemical Co. LTD., extra grade). The ceramics powder was heated at 453 K for 20 min and was steriled. The powder was suspended in sterile saline to give a specified concentration.

Measurement of the Conductance Change of Growth Medium

For measuring the conductance change caused by bacterial growth, BACTOMETER^R microbial monitoring system model 64(bio Mérieux VITEK) was used. Modified plate count agar(DIFCO) was poured into the well of module equipped with paired electrodes until it covered the electrodes. After the agar was solidified, both of the bacterial suspension and the ceramics powder slurry were added onto the agar. The modules were set in the incubation room of BACTOMETER^R , and the conductance change of the agar was monitored for 48 h, at 308 K.

Determination of Bactricidal or Bactariostatic Effect.

If the conductance change was not observed over the incubation, a little quantity of the slurry was cultured in BHI broth at 308 K for 24 h. Futhermore, for indentification of bacteria, Eosin Methylene Blue (EMB) agar and Mannitol Salt agar added 3% Egg Yolk (MSEY agar) were used for *E. coli* and *S. aureus* , respectively. From these experiments, it was determined whether the ceramics powder slurry acted on the test bacteria in bactricidal manner or bactariostatic one.

RESULTS & DISCUSSION

Figure 1 shows the growth-inhibitory effect of MgO powder slurry on *E. coli*. The conductance curve of the control(ceramics powder concentration is 0 mg/ml) changes markedly at the incubation time of about 6 h. The change is due to the bacterial growth and metabolism. This point is defined as "Detection Time(DT)". DT becomes higer with increase in the concentration of MgO powder. DT at the concentration of 2.7 mg/ml is 8 h, and DT at the concentration of 5.4 mg/ml is 26 h. From these results, it is found that MgO powder inhibits the growth of *E. coli*. At the concentration higher than 10.8mg/ml, the DT did not appeared, indicating no bacterial growth. In this region of MgO powder concentration, from the results of incubation in BHI broth, MgO powder acted on E.coli in a bactericidal manner. ZnO powder also showed the growth-inhibitory effect on *E. coli*. But, in the region where bacterial growth was not observed, ZnO powder acted on E.coli in bacteriostatic manner. Growth-inhibitory effect of MgO powder is stronger than that of ZnO powder. α-Al_2O_3 powder did not show the antibacterial activity. The results obtained for *S. aureus* are similar in tendency to those for *E. coli*. However, for *S. aureus*, ZnO powder showed the stronger growth-inhibitory effect than MgO powder.

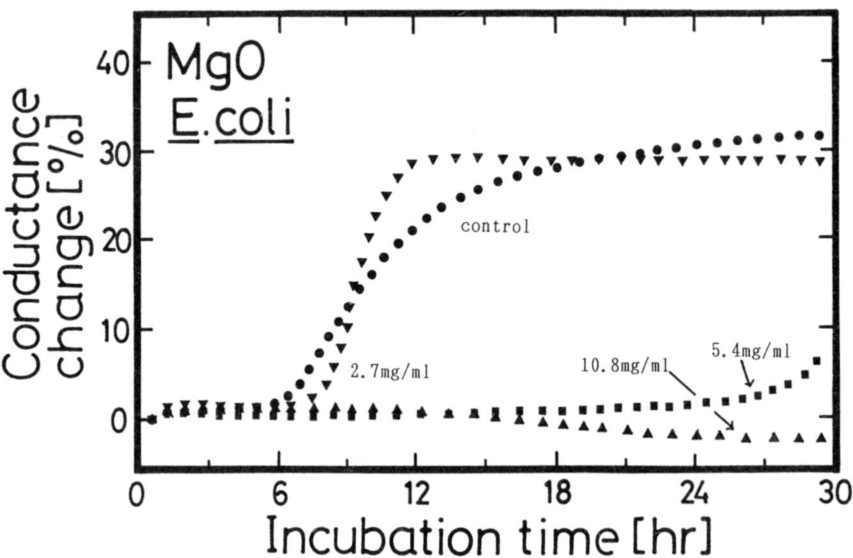

Fig.1 Growth inhibition of *E. coli* by using MgO powder slurry(conductance method)

By using the conductance method, it was found that MgO and ZnO powders showed the growth-inhibitory effect on *E. coli* and *S. aureus*. The conductance method is very useful for evaluating the growth-inhibitory effect of ceramics powder slurry on bacteria.

REFERENCE

1)Eden, R. and Eden, G., Impedance Microbiology, Reserch Studies Press, Hertfordshire, 1984, pp 11-18

RECENT STUDIES ON ASEPTIC PROCESSING OF PARTICULATE FOODS

NIKOLAOS G. STOFOROS
Thermopilon 5, 35100 Lamia, GREECE

ABSTRACT

Recent studies on mathematical modeling, liquid to particle heat transfer coefficient calculations, residence time distribution, and microbiological validation of aseptic processes of particulate foods are briefly presented. The equations coupling particle residence time distribution with the heat transfer problem are also outlined.

NOMENCLATURE

- A system wall heat transfer surface area, m^2
- C_p specific heat, J/kgK
- D time at a constant reference temperature required to achieve a decimal reduction of the initial concentration of a heat labile substance, s
- F_s integrated particle F value, s
- f_i fraction of exiting solid particles with equal residence times, dimensionless
- L total length of system, m
- m mass of product in the system, kg
- N final concentration of a heat labile substance in the solid product component, number of microorganisms/ml, g/ml, or any other appropriate unit
- N_o initial concentration of a heat labile substance in the solid product component, number of microorganisms/ml, g/ml, or any other appropriate unit
- n total number of classes of particles with equal residence times, dimensionless
- T temperature, °C
- t residence time, s
- U_o overall heat transfer coefficient, heating medium/system wall/internal liquid, W/m^2K
- z longitudinal distance measured from system entrance, m
- α thermal diffusivity, m^2/s
- θ time, $\theta=(z/L)\overline{t_p}$, s

Subscripts

- f liquid product component
- i index indicating a class of particles with the same residence time
- m heating medium
- p solid product component

Symbols

- $\overline{}$ appropriately averaged value

INTRODUCTION

For proper design of aseptic processes for low-acid particulate foods, the following elements must be considered: mathematical modeling, liquid to particle heat transfer coefficient, residence time distribution, and microbiological validation. In this overview, recent studies associated with the above topics are briefly presented. Space limits not only the discussion, but also the references included here. A more detailed discussion as well as a rather complete reference list can be found in [1].

MATHEMATICAL MODELING

Several mathematical models for particle temperature predictions, microbial destruction, and quality degradation, applicable to aseptic processing of particulate foods, have been reported in the literature, *e.g.*, [2-8]. The concept of particle residence time distribution, or the fact that the liquid and solid components might travel through the system with different velocities, have been incorporated in to the mathematical models by some investigators [2, 4-6, 8]. However, to our knowledge, the effect of the above on both fluid and particle temperatures, has not been reported. Equation (1), derived from an overall energy balance, is proposed here as the governing equation for each component of an aseptic unit (*e.g.*, the holding tube). There are two main assumptions associated with Eq. (1): first it is assumed that there is only longitudinal fluid temperature variation, and second that each particle travels with constant velocity throughout the whole system length.

$$U_o A (T_m - T_f) = \frac{\overline{t_p}}{t_f} m_f C_{pf} \frac{dT_f}{d\theta} + m_p C_{pp} \sum_{i=1}^{n} f_i \frac{d\overline{T_{pi}}}{d\theta} \tag{1}$$

Using the same time variable, θ, the equation governing the heat transfer, by conduction, to a solid particle can be written as

$$\nabla^2 T_{pi} = \frac{1}{\alpha_p} \frac{\overline{t_p}}{t_i} \frac{\partial T_{pi}}{\partial \theta} \tag{2}$$

Remaining microbial or quality factors concentration in each particle can be evaluated from the known temperature distribution within the particle and appropriate volume averaged techniques. Average integrated particle sterilization values can be then calculated from Eq. (3). Presumably, if a continuous probability density function is known, the summation in Eq. (1) and (3) should be replaced by the appropriate integral.

$$\overline{F_s} = D \log_{10} \left(N_o - \sum_{i=1}^{n} f_i \overline{N_i} \right) \tag{3}$$

LIQUID TO PARTICLE HEAT TRANSFER COEFFICIENT

The boundary condition at the particle surface requires the use of the liquid-particle film heat transfer coefficient, h_p. There have been basically four methods used to calculate h_p. The traditional one involves the use of thermocouples to monitor particle temperatures *e.g.*, [3, 9]. Restriction of particle motion by the use of thermocouples necessitates the knowledge of the relative fluid to particle velocities for a non conservative use of the results. The second method is based on calculating h_p from the change in concentration of a heat labile substance (bioindicator) as it goes through the system *e.g.*, [10-11]. Due to inherent difficulties in using bioindicators, this method should be used only if direct temperature measurements are impractical [10]. Furthermore, this method requires mathematical models that very accurately describes the process. This is also one of the main requirement of the third method which uses only fluid temperature data to calculate h_p, *e.g.*, [7]. The fourth method involves the use of liquid crystals as particle temperature sensors, *e.g.*, [7], thus alleviating the problem of particle motion restriction. This method is not yet of use under actual processing conditions. A critical review on h_p calculations is presented in [12].

PARTICLE RESIDENCE TIME DISTRIBUTION

Particle residence times have been experimentally measured by monitoring tracer particles as they were passing through the system (visually or by videotaping, or by using photo-sensors [13], or radioactive or magnetic tracers). Following the traditional approach used to design in-container processes, aseptic process schedules can be based on center point lethality of the fastest moving particle. The question to be answered by the thermal process authorities is what will be the acceptable probability level, for continuous distributions, or the appropriate sample size, for discrete distributions, for the minimum, critical, residence time to be set.

MICROBIOLOGICAL VALIDATION

Microbiological validation is an essential part of designing thermal processes. The main problem with aseptic processes is to design an experiment so as to exclude any lethality accumulated during the cooling portion of the process. Particle breakage during cooling might lead to considerably less lethality compared to intact particles. This was verified by [14] which introduced a homogenizing mixer that immediately disintegrated partially cooled particles. It must be noted here, that there is always some lethality accumulated in the particle center during the cooling process, the minimum being (assuming the fluid enters the cooler at higher than the particle temperature) if the particle breaks when particle center temperature equals the fluid temperature.

REFERENCES

1. Stoforos, N.G., An overview of aseptic processing of particulate foods. In Developments in Food Science, Vol. 29, Food Science and Human Nutrition, ed. G. Charalambous, Elsevier Science Publishers B.V., Amsterdam, The Netherlands, 1992, pp. 665-77.
2. Chandarana, D.I. and Gavin, A. III, Establishing thermal processes for heterogeneous foods to be processed aseptically: A theoretical comparison of process development methods. J. Food Sci., 1989, **54**(1), 198-204.
3. Chang, S.Y. and Toledo, R.T., Heat transfer and simulated sterilization of particulate solids in a continuously flowing system. J. Food Sci., 1989, **54**(4), 1017-23, 30.
4. Lee, J.H., Singh, R.K., and Larkin, J.W., Determination of lethality and processing time in a continuous sterilization system containing particulates. J. Food Eng., 1990, **11**, 67-92.
5. Manson, J.E. and Cullen, J.F., Thermal process simulation for aseptic processing of foods containing discrete particulate matter. J. Food Sci. 1974. **39**, 1084-89.
6. Sastry, S.K., Mathematical evaluation of process schedules for aseptic processing of low-acid foods containing discrete particulates. J. Food Sci., 1986, **51**(5), 1323-28, 32.
7. Sawada, H., An analytical heat transfer model for liquid/particle systems. Ph.D. thesis, Dept. of Agricultural Engineering, Univ. of California, Davis, CA, 1992.
8. Yang, B.B., Nunes, R.V., and Swartzel, K.R., Lethality distribution within particles in the holding section of an aseptic processing system. J. Food Sci., 1992, **57**(5), 1258-65.
9. Awuah, G.B., Ramaswamy, H.S., and Simpson, B.K., Surface heat transfer coefficients associated with heating of food particles in CMC solutions. J. Food Proc. Eng., 1993, **16**, 39-57.
10. Hunter, G.M., Continuous sterilization of liquid media containing suspended particles. Food Technol. in Australia, 1972, **4**, 158-65.
11. Weng, Z., Hendrickx, M., Maesmans, G., and Tobback, P., Immobilized Peroxidase: A potential bioindicator for evaluation of thermal processes. J. Food Sci., 1991, **56**(2), 567-70.
12. Maesmans, G., Hendrickx, M., DeCordt, S., Fransis, A., and Tobback, P., Fluid-to-particle heat transfer coefficient determination of heterogeneous foods: A review. J. Food Proc. Pres., 1992, **16**, 29-69.
13. Yang, B.B. and Swartzel, K.R., Particle residence time distributions in two-phase flow in straight round conduit. J. Food Sci., 1992, **57**(2), 497-502.
14. Unverferth, J.A., Chandarana, D.I., and Stoforos, N.G., Aseptic processing of particulate foods. A biological assay of hold tube-only lethality. Paper No. 45, Institute of Food Technologists Annual Meeting, New Orleans, LA, June 20-24, 1992.

LAMINAR TUBE FLOW OF A NEWTONIAN FLUID CONTAINING LARGE SPHERES - APPLICATION TO ASEPTIC PROCESSING

JOHAN FREGERT and CHRISTIAN TRÄGÅRDH
Food Engineering, Lund University
P.O.B 124, S-221 00 Lund, SWEDEN

ABSTRACT

The velocity profiles of a sucrose solution containing large alginate beads have been measured with hot-wire anemometry for the fluid and with pulsed ultrasonic velocimetry for the beads. The velocity profiles of two test cases are discussed. The hot-wire method compares well with expected velocity profiles while the ultrasonic Doppler methods implementation needs improvements to increase its precision.

INTRODUCTION

The introduction of continuous aseptic processing lines for low acid liquid foods containing large solid food particles, *e.g.* vegetable soup, resulted in increased interest in the mechanisms of momentum transfer in such a mixture [1].

The fluid residence time in a chemical reactor has been well explored for such defined vessels as tanks and tubes and also for more rheologically complex fluids the residence time distributions (RTD) are known [2]. The introduction of large particles ($D_{particle}/D_{tube} \approx 0.1$) in the mixture results in a more complicated distribution and furthermore to different RTDs for the particles and liquid.

The test of a residence time distribution model requires experimental data of the primitive variables such as local velocities of the phases. One objective of this study was to implement a set of methods for obtaining them.

MATERIALS AND METHODS

Experimental set up

The velocities of the two phases were measured in a horizontal plexiglass tube

(D_t=44 mm). Two measuring sites were situated 1255 mm apart. The particle velocity at the first site and the fluid velocity at the second site were measured simultaneously, and vice versa. The particles were made from a 2.0 % alginate solution (Kelco Manugel DMB) with 0.15 % colloidal iron and set in a 0.10 M $CaCl_2$ solution. The fluid was a sucrose solution which was deaerated in a vacuum chamber before being filled into the experimental loop.

Pulsed ultrasonic Doppler velocimetry

The particle velocity was measured with pulsed ultrasonic Doppler velocimetry developed for measurement in bubble reactors by the Institut für Technische Chemie, Universität Hannover [3]. The probe was traversed at an angle of 45 ° to the flow direction. The measurement duration was 3 sec. 10 measurements were made and repeated 3 times. With the help of a peak analysis software (Peakfit, Jandel) the velocity spectra were fitted to a Gaussian distribution. The mean velocity of all measurements were calculated for each location. The location of the measurement volume and the velocity were calibrated against a moving thread.

Hot-wire anemometry

The fluid velocity was measured with hot-wire anemometry (Dantec CTA 56C), using a conical film probe which was built to withstand mechanical stress. The velocity measurements were calibrated against a Poiseuillian velocity profile in a separate loop. The particle interactions were first eliminated and then the mean velocity over a period of 25 seconds (1 kHz) was calculated.

Physical properties

The fluid density was measured with a calculating digital density meter (Anton Paar DMA45). The particle density was measured by taking the time for spheres to free fall in the sucrose solution. The particle diameter was measured on a photograph.

RESULTS AND DISCUSSION

The physical properties and the flow conditions from the case study are presented in table 1.

TABLE 1

Physical properties and flow conditions of test case.

Case	ρ_f kg/dm^3	ρ_p/ρ_f	μ mPa s	D_p mm	*Re*	V_{mean} m/s	C_p vol. %
I	1.322	1.007	183	4.8	79	0.250	5.4
II	1.202	1.008	9.2	5.1	794	0.138	9.8

In figure 1 and test case I the measurement of the fluid velocity profile is in accordance with the calculated Poiseuillian flow. In test case II the velocity profiles are different from a Poiseuillian flow due to the different operating conditions. The precision of the measurement of particle velocity and the location of the measurement volume was not high. Therefore the particle velocity was calibrated against the fluid velocity at site II. The precision of the Pulsed ultrasonic Doppler velocimetry must be increased. The steep gradients of a laminar velocity profile makes a small error in the measurement of position a large error in the overall profile and in particular the slip velocity, if any.

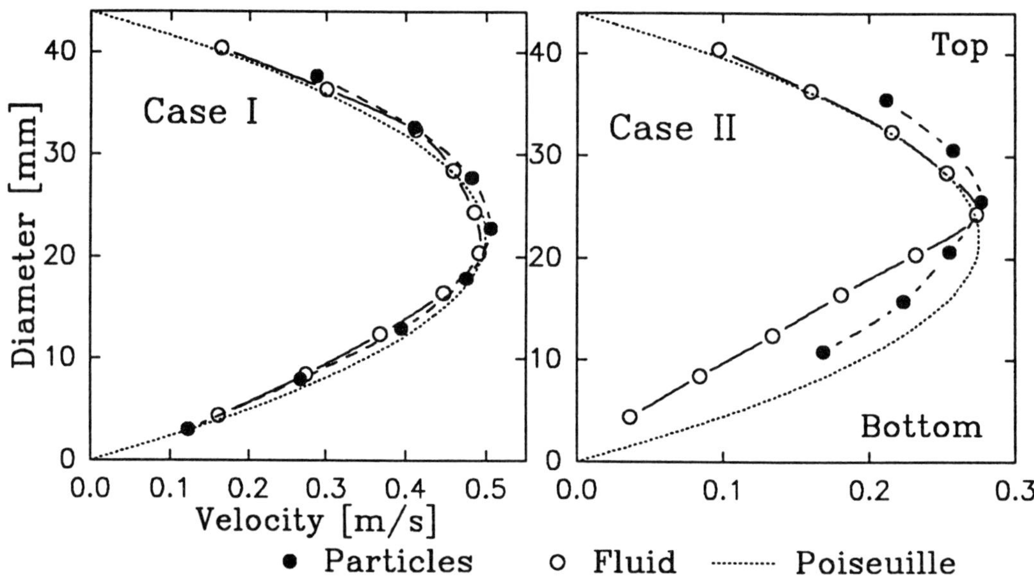

FIGURE 1. Velocities of particles and fluid in a horizontal tube of the test cases I and II. Both profiles at site 2.

CONCLUSIONS

The hot-wire anemometry is a useful method for the study of liquid velocity in solid-liquid flow. The pulsed ultrasonic Doppler velocimetry implementation needs further development for sensitive systems with steep velocity gradients.

REFERENCES

1. Sastry S.K. and Zuritz C.A., A review of particle behaviour in tube flow: Applications to aseptic processing. *J.Food Process Engng.*, 1987, **10**, 27-52
2. Wen C.Y. and Fan L.T., *Models for flow systems and chemical reactors*, Marcel Dekker, New York, 1975.
3. Lübbert A., Korte T. and Schügerl L.K., Ultrasonic Doppler measurements of bubble velocities in bubble columns. In *Measuring techniques in gas-liquid two-phase flows*, ed. J.M. Delaye and G. Cognet, Springer Verlag, Berlin, 1984, 479-494.

FLOW VELOCITIES OF PARTICLES IN HOLDING TUBE

Hideo Tozuka, Tadashi Fujiwara and Mitsukuni Mori
Research Laboratories, Japan Canners Association

INTRODUCTION

Flow velocity of a particle in a holding tube is one of the critical factors affecting microbial safety of continuous thermal processing of foods containing particles. Some have investigated the flow characteristics of the particles traveling through the tube (Taeymans et al., 1985; Dutta and Sastry, 1990; Hang et al., 1991; Yang and Swartzel, 1992; Bark, 1992). The objective of this work was to investigate flow rate of particles as affected by particle size, flow patterns and densities of fluids.

EXPERIMENTAL

Model of particles and carrier fluids
The test particles were plastic balls with a density of 1160 kg/m^3 and diameters of 10 mm and 15 mm. The carrier fluids used were tap water (1.0 mPaS) and a 2.0% carboxymethylcellulose (CMC) solution (640 mPaS); which showed various flow patterns in this experiment. The fluid density was adjusted by adding NaCl to study the influence of density on the flow rates of the particles.

Setup
Test particles and carrier fluids were circulated in the experimental circuit consisted of the transparent plastic tubes and a rotary pump. The tube was composed of three straight pipes having 1 m length and 34.4 mm inner diameter connected to 180° U bends. The tube was inclined 2.9° upward to simulate a holding tube as used in commercial aseptic processing systems. Motion of the particles was recorded using a video camera.

Operation
All experiments were conducted at room temperature and atmospheric pressure. Pump speed was adjusted to a desired flow rate (0.12, 0.16, 0.20, 0.24 m/sec). The flow rate was measured by an electromagnetic flowmeter. Under these conditions, it was assumed, based on

the Reynolds numbers, that a flow pattern of the tap water was a turbulent flow and that of the CMC solution was a laminar flow. First, to avoid any interaction with particles, a single particle was introduced into the setup to investigate the effects of the particle size, flow patterns and densities of fluids. Second, a large number of particles was added (particle to fluid ratio; 1.0% v/v) to examine the effect of the particle interaction.

RESULTS

Flow behavior of the particle moving in the straight tube showed a uniform motion. Effects of the flow patterns and densities of the fluid on the particle velocity are summarized in Table 1.

In turbulent flow
In water, the velocity of the single particle increased linearly as fluid average velocity increased. The smaller the differences in densities between the particle and fluid, the greater the particle velocity. There was no significant difference in the velocity between particle sizes, 10 mm and 15 mm.

In laminar flow
In the CMC solution with low density (1010 kg/cm^3), the particles flowed slower than the fluid average velocity. However, they flowed faster than the fluid average velocity in the CMC solution nearly as dense as particles. Flow rate of the 15 mm particle was greater than that of the 10 mm particle when rolling along the bottom of the tube. This phenomenon indicates that a greater velocity near the tube centerline under the laminar flow gave the larger force to the 15 mm particle than 10 mm particle.

Flow rate distribution of particles
In the CMC solution having a density equal to the particles, they flow spreads in the tube. Thus, the flow rates showed a wide distribution. Flow rate distribution of the particles having a heavier density than that of fluid was sharp. In fact, most of the particles moved slowly at the bottom of the tube. The maximum velocity of the 2.0% CMC solution was calculated as 1.9 times than average velocity of the fluid based on the flow behavior index. It was confirmed experimentally that no particles flowed faster than the maximum fluid velocity.

Mixture of various size particles
When the particles having diameters of 10 mm and 15 mm flowed simultaneously, the 15 mm particle sometimes collided with the 10 mm particle, hence the flow rate of the later particle was accelerated.

CONCLUSION

Flow rate of the particle used here was affected by the density and average velocity of the fluid. The fluid flow pattern was the critical factor affecting flow velocity of the particles. Since the maximum velocity of particle was smaller than the maximum rate of fluid, the process design for continuous thermal processing should be scheduled in accordance with the maximum velocity of fluid. However, the majority of the particles flowing slower than the fluid average velocity would be over-cooked.

Table 1
Flow velocities of single particle [a)]

Fluids	NaCl content	Density (kg/m^3)	Average fluid rate(m/s) 0.12	0.16	0.20	0.24
Water [b)]	0	1000	0.09	0.13	0.17	0.21
	20	1140	0.11	0.15	0.19	0.23
	23	1160	0.15	0.20	0.24	0.29
	25	1180	0.12	0.17	0.21	0.25
2.0%CMC [c)] solution	0	1010	0.07	0.10	0.12	0.17
	18	1140	0.11	0.18	0.24	0.29
	20	1160	0.16	0.21	0.28	0.34
	23	1180	0.19	0.26	0.30	0.33

a)10 mm plastic ball (density:1160 kg/m^3)
b)Viscosity:1.0 mPaS (turbulent flow)
c)Viscosity:620 mPaS (laminar flow)

REFERENCES

1 Taeymans, D., Roelans, E., Lenges, J, Residence time distribution in a horizontal SSHE used for UHT processing of liquids containing solids, Presented at the 4th. ICEF, 1985

2 Dutta, B. and Sastry, S. K., Velocity distribution of food particle suspensions in holding tube flow: Experimental and modeling studies on average particle velocities, J. Food Sci, 1990, 1448

3 Hong, C.W., Pan, B. S., Toledo R. T., and Chiou K.M., Measurement of residence time distribution of fluid and particles in turbulent flow, J. Food Sci, 1991, 255

4 Yang, B. B. and Swartzel, K. R., Particle residence time distributions in two-phase flow in straight round conduit, J. Food Sci, 1992, 497

5 Bark, G, Fluids and particles, SIK's Annual report 1991/1992, 1992, 11

FLOW OF SOLID-LIQUID FOOD MIXTURES

SHI LIU, J-P PAIN*, PJ FRYER.
Dept. of Chemical Engineering,
Pembroke St., Cambridge, UK.
* University of Compiegne, France

ABSTRACT

The design of efficient continuous sterilisation equipment for solid-liquid food mixtures requires that the temperature and the velocity distributions within the fluid are known. Experiments have been conducted to measure the variation in velocity in flows of carrots in water, and data given in terms of flows of single particles.

INTRODUCTION

Some commercial food sterilisation processes involve the transport and heating of solid-liquid food mixtures of high solids fractions, up to 50% solids of particle diameters up to 25 mm in non-Newtonian carrier fluids. Fluid velocities are restricted by the need not to damage the mixture. To design these processes, and confirm the sterility of the final product, it is vital to be able to predict the temperatures of solid-liquid mixtures during processing. This requires information on the flow patterns of food mixtures.

Most published work on the conveying of solid-liquid mixtures considers high-density solids in turbulent water flows [1]. In contrast, food flows can use non-Newtonian carrier fluids and be of low Reynolds number. In fully developed Newtonian laminar flow, particles can travel twice as fast as the mean flow; this assumption is commonly used in process plant to ensure sterility. The flow of single phase food liquids through process plant has been studied [2-4], and work on suspensions described [5]. Dutta and Sastry [6-7] studied velocity distributions in flows of up to 0.8% by volume; more realistic concentrations, of up to 40% solids, have been studied by [5] in a scraped surface heat exchanger. This paper summarises results on the flow of food solids in water carried out as part of a larger study [8-9].

FLOW OF FOOD SOLIDS

Details of equipment are given in [9]. The test section consisted of a 5 m of perspex pipe of 44 mm inside diameter. Three coils which detected the passage of metal-doped tracer particles were mounted along the test section of the pipe at 1m intervals. Tracers were constructed from shaped polythene wrapped with metal foil. Foods and tracers of similar densities had identical settling rates; tracer particles were representative of foods of the same density.

The flow in water of 6 mm carrot cubes of up to 35% delivered solids concentration was studied. Densities of particles ranged between 1010 and 1080 kg/m^3. After the test section the fluid was returned to the feed tank through a flexible pipe, from which samples could be taken to determine the flow rate and the delivered solids concentration. Recirculation of particles and tracers allowed data to be collected rapidly and efficiently.

Single particles. Preliminary experiments studied the flow of single particles [8]. If denser than the fluid, particles tend to sediment, whilst if lighter they tend to rise. Dimensional analysis suggests that the particle velocity is a function of a number of variables: particle and fluid densities, particle and pipe diameters, fluid viscosity and flow velocity. Conventionally, the modified Froude number

$$Fr_p = \frac{v_m}{\sqrt{gd(s-1)}} \qquad (1)$$

where s is the specific gravity of the particle, has been used to correlate particle flows. Several flow patterns were identified, ranging from particles sitting *stationary* on the pipe wall; which corresponds to the stationary deposit bed in pipe flows, which occurs for low flow velocities and/or high particle density, through *sliding* flows where the particle moves along the pipe wall with constant velocity, to fully suspended behaviour at high flow rates.

The ratio of particle to fluid velocity, $v_r = v_p/v_m$, was used to correlate data. Tube Re does not correlate the velocity ratio well; for the denser tracers, a Reynolds number of 8 000 was needed to start the particle moving, whilst for the majority of particles the velocity ratio was already in the region of unity at this point. An approximate analysis, based on a force balance between fluid drag and friction for a spherical particle on the wall of a pipe, is used by [9] to suggest that a plot of v_r versus Fr_p^{-1} will be a straight line. Although this analysis is simplistic, it can be used as a basis for plotting data. Data for single particles of various shapes and effective diameters between 6.35 mm and 15.29 mm have been obtained. Variation between data sets is seen, but data are correlated by:

$$v_r = 1.16 - 0.7234/Fr_p \qquad (2)$$

plotted as the continuous line in Figure 1, which also includes lines of the same slope but with intercepts 10% greater and less than 1.16 (i.e. 1.16 ± 0.116). Most single particle data lie within these boundaries. At high Froude number the particle becomes suspended, and thus travels faster than the mean velocity of the fluid. At high Froude numbers flow is turbulent; the fluid centre line velocity is ca. 1.2 times the mean velocity, explaining the intercept at $Fr_p^{-1} = 0$. Here, it would not be expected that an analysis based on a friction balance would apply.

Higher solids fractions. Delivered solids fractions of carrots of up to 35% have been examined. Characteristically, the flow was stratified; a slow moving bed forms below a region of low solids fraction. Dye tracer experiments and observation of particles indicated that the liquid velocity was significantly higher than that of the bed. At high velocity, the flow appeared well mixed with food particles dispersed over the cross section of the pipe. Some slow moving particles were seen even at high fluid velocities. Experiments were carried out with cubic tracers of s = 1.01, 1.03, and 1.05, representative of the heaviest, average and lightest particles. Three types of behaviour were seen:

(i) heavier tracers ($s > 1.04$) remained trapped among the bed of food particles and sometimes slid along the bottom of the pipe, giving low velocities,
(ii) medium tracers (s ca. 1.03) could flow in any positions in the pipe, although for the majority of the time the tracer formed part of the sedimented bed,
(iii) light particles ($s < 1.01$) tended to flow on top of the bed of food particles or in the upper portion of the pipe where food particles were fewer, giving high velocities.

Results are given as a function of solids fraction in Figure 1, together with equation (2) for comparison. Figure 1(a) shows data for the heavier particle; the particle velocity is less than that of a single particle, and the data are not widely scattered. Much greater scatter is seen in Figure 1(b) and 1(c); the particle took up a much more random range of positions in both the bed and the fluid above it. Highest particle velocities are found for the lightest particle, which can travel up to about 1.5 times the mean flow velocity. Highest velocities are found for solids fractions between 10 and 20%. At low solids fraction the effect of the particles is minimal, so that the velocity of the particles can be modelled by the single particle correlation. The bed becomes significant for fractions greater than ca. 10%, increasing the liquid phase velocity. At solids fractions above ca. 20%, the whole pipe is filled by the high solids fraction phase; the flow can again be predicted by the single particle equation. Observation suggests that the top of the bed continues to move at higher velocities than the base, but the difference decreases at higher solids fractions. More complex effects are seen for viscous carrier fluids; this will be reported later [10].

CONCLUSIONS

It is important to be able to predict velocity and temperature distributions in continuous food sterilisation plant. Experimental work has been carried out to characterise flows of solid-liquid mixtures in horizontal tubes. It has been shown that it is possible to obtain significant velocity differences between particles depending on their densities, but that at high solids fractions flows are more uniform. Work is continuing to improve the correlations produced.

Acknowledgements. This work was supported by Sous-Chef Ltd. The authors wish to express their gratitude to the late Mr Richard Sperring. SL wishes to acknowledge additional financial support from the British Council. J-P P's time at Cambridge was supported by the EC .

References

1. WILSON, KC, pp 103-124 in *Slurry handling design of solid-liquid systems* ed NP Brown and NI Heywood, Elsevier, 1991
2. RAO, MA, & LONCIN, M, *Lebesnm.Wiss. u-Techn.*, **7**, 4-17. 1974
3. HEPPEL, NJ, *J.Food Eng.*, **4**, 71-84. 1985.
4. SANCHO, MF & RAO, MA. *J.Food Engineering*, **15**, 1-20, 1992.
5. SINGH, RK, & LEE, JH. pp7-62 in *"Advances in Aseptic Processing Technologies"*, ed RK Singh and PE Nelson, Elsevier, 1992.
6. DUTTA, B & SASTRY, SK.*J. Food Sci.*, **55**, 1703-1710. 1990.
7. DUTTA, B & SASTRY, SK. *J. Food Sci.*, **55**, 1448. 1990
8. LIU, S, PAIN, J-P, & FRYER, PJ. *Entropie*, **28**(170), 50-58, 1992.
9. LIU, S, PAIN, J-P, PROCTOR, JM, DE ALWIS, AAP, & FRYER, PJ. in press, *Chem.Eng. Commun*, 1993.
10. LIU,S. PhD thesis, Cambridge University, in preparation.

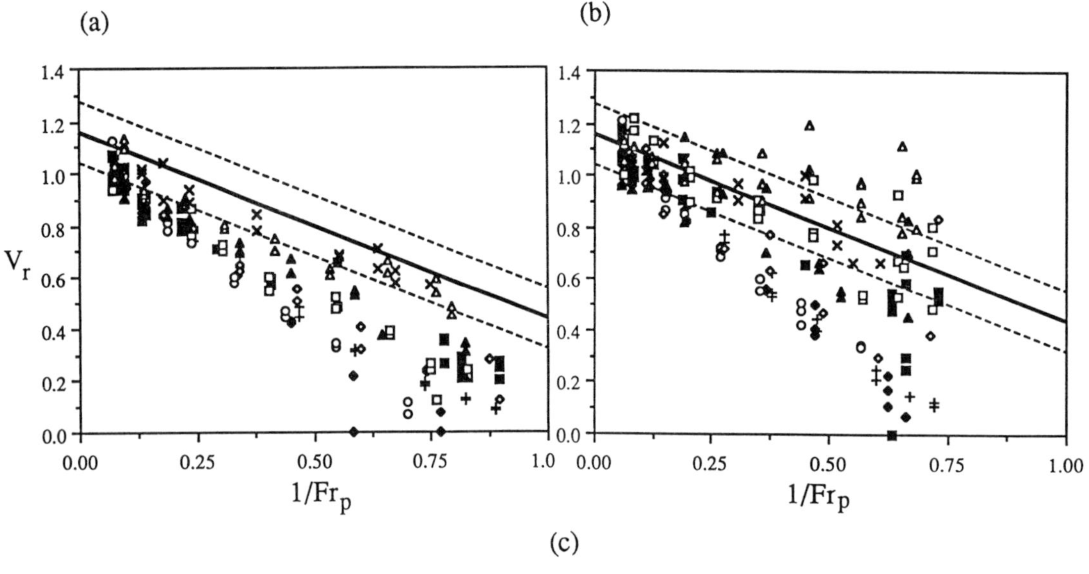

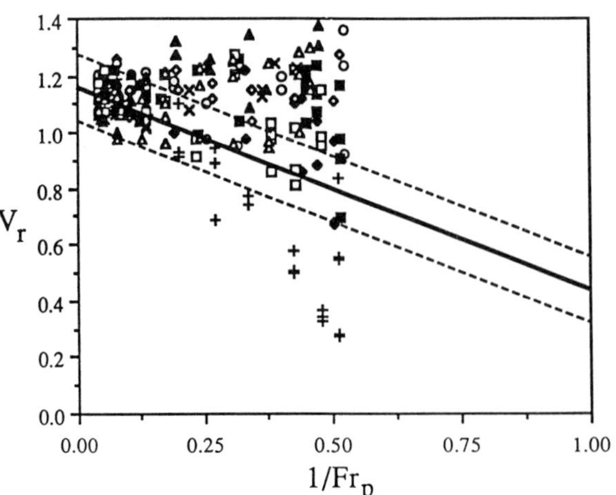

Figure 1. Plot of the velocity ratio in the flow of carrots of different concentrations. Heavy line shows equation (2); dotted lines show ± 10% limits.

(a) d = 7.6 mm, s = 1.04
(b) d = 7.4 mm, s = 1.03
(c) d = 8.3 mm, s = 1.01

Delivered solids fractions in all cases:

+: 4.3% ◆: 8.3%
○: 11.4% ◇: 11.5%
■: 14.6% □: 16%
▲: 18% △: 26%
✳: 35%

MODELING OF TIME-VARIANT HEAT TRANSFER IN A TWO-PHASE SYSTEM

K. H. PARK and R. L. MERSON
Department of Food Science and Technology
University of California, Davis, CA 95616, U. S. A.

ABSTRACT

A numerical analysis was performed to evaluate the convective heat transfer coefficients at the particle-liquid interface as a time-dependent function of system properties for liquid foods containing particulates. The transient energy equations for the sphere and fluid were solved by an alternating-direction implicit method to find the temperature distribution near the particle surface for near zero relative velocity, Re = 0.1. The instantaneous convective heat transfer coefficients varied over the sphere surface and heat flow reversal was observed near the rear stagnation region, which caused the displacement of the coldest point from the center to the particle surface. Lethality computations showed that the critical point was shifted from the particle center toward the rear stagnation point.

INTRODUCTION

The establishment of thermal processes for ensuring bacteriological safety of two-phase foods in continuous processes is not simple because it is difficult to measure the temperature of a moving food particle and determine the amount of heat which penetrates to the particle center. In the absence of experimental data, a mathematical model is one approach to estimating temperature distributions in the particles [1, 2, 3, 4, 5]. In these works, designs for aseptic processes have been developed assuming that the convective heat transfer coefficient is constant throughout the process. When the magnitude of the relative velocity is small, however, the coefficient is time dependent [6] and much smaller than usually assumed from steady state correlations.

Since the hot fluid loses heat as it moves along the particle surface, the heat flux into particle and the surface/fluid temperature difference depend on time and the spatial temperature variation inside the thermal boundary layer. Furthermore, with a variation in the local heat transfer coefficient on the particle surface, the geometric center of the particle will not be the coldest point at all times. This is quite important with respect to lethality because treating the center of the particle as the critical point may be wrong and yield an improper estimation of the lethality.

Therefore, we conducted an engineering analysis using numerical methods in order to obtain solutions for transient heat transfer from the fluid to the particle in a holding tube and explored the migration of the slowest heating point inside the particle as a function of system properties.

NUMERICAL METHOD

A single solid sphere which is suddenly immersed in an unbounded Stokes flow environment was considered as the idealization of the case of very small relative motion between the fluid and the particle. Unbounded means that the effects of the tube wall and other particles were ignored. The non-dimensional velocity components for creeping flow, which have been found analytically [7], are:

$$v_r = -\frac{1}{2}\cos\theta\left(\frac{1}{r^3} - \frac{3}{r} + 2\right), \quad v_\theta = \frac{1}{4}\sin\theta\left(-\frac{1}{r^3} - \frac{3}{r} + 4\right) \tag{1}$$

The velocity distribution satisfies the no-slip conditions; that is, $v_r = v_\theta = 0$ at the surface of the sphere.

The dimensionless unsteady energy equations for the spherical particle and fluid were solved simultaneously using the finite difference method of alternating direction to access the time history of transient heat transfer, with proper boundary conditions [8]. The time-dependent film coefficient expressed in the form of a dimensionless Nusselt number was evaluated at specific position and given time from Newton's law of cooling.

$$Nu_\theta = \frac{2a h_p(\theta,\tau_p)}{k_f} = -\frac{2}{T_s - 1}\left(\frac{\partial T_f}{\partial r}\right)_{r=1} \tag{2}$$

RESULTS AND DISCUSSION

The surface-averaged Nusselt numbers decreased with time and the values throughout most of the heating process were found to be significantly below the corresponding steady state values. The convective heat transfer coefficients decreased from the front of the sphere toward the rear and the values became negative in the vicinity of the rear stagnation point toward the end of process [9]. In this region, the heat is transferred not from the fluid to the sphere, but in the opposite direction. In addition to the local variation of heat transfer, the thermal wake behind the rear stagnation point caused by the heat flow reversal forced the coldest point within the sphere to migrate along the axis of symmetry, from the center toward the rear stagnation point [8].

TABLE 1

Comparison of the accumulated lethality on the axis of symmetry ($\theta = \pi$) of an acrylic sphere at Pr =1,000. Data used for computation; 1.27 cm for sphere radius, 25 °C for initial particle temperature, 121.1 °C for fluid temperature, z = 10 °C and F_o at 121.1 °C = 4 min.

τ	Time	Lethality						
	(min.)	r = 0.0	r = 0.2	r = 0.4	r = 0.5	r = 0.6	r = 0.8	r = 1.0
0.1	2.26	0.000	0.000	0.000	0.000	0.000	0.000	0.013
0.5	11.28	0.111	0.110	0.120	0.131	0.146	0.196	0.304
0.7	15.79	0.392	0.373	0.373	0.380	0.393	0.437	0.537
0.9	20.31	0.821	0.769	0.740	0.733	0.732	0.747	0.813
1.0	22.56	1.084	1.010	0.961	0.945	0.933	0.926	0.967
1.1	24.82	1.375	1.278	1.206	1.179	1.155	1.123	1.134

The migration of the coldest point significantly affects the lethality computation. Table 1 shows that the accumulated lethality varies along the axis of symmetry and the value at r = 0.8 toward the rear stagnation point is about 17 % less compared to the lethality at the particle center when the center achieves the required lethality of unity. This example shows that the migration of slowest heating point and the its temperature magnitude could significantly affect the lethality computation. Since the surface temperature at the rear stagnation point was high initially so that significant lethality was accumulated in that region at short time, the critical point having the least lethality was located inside the particle not at the coldest point.

CONCLUSION

Numerical solution of the transient energy equations showed that the temperature distribution near the particle surface and the convective heat transfer coefficients were time- and spatially dependent functions. The coldest point migrated on the axis of symmetry away from the particle center and at the end of the heating resided at the rear stagnation point. The lethality comparison showed that the particle center was not the critical point and the lethality at the critical point was found to be a function of particle and fluid properties.

REFERENCES

1. Chandarana, D.I., Gavin, III, A. and Wheaton, F.W., Particle/fluid interface heat transfer under UHT conditions at low particle/fluid relative velocities. J. Food Proc. Eng., 1990, **13**, 191-206.
2. Chang, S.Y. and Toledo, R.T., Simultaneous determination of thermal diffusivity and heat transfer coefficient during sterilization of carrot dices in a packed bed. J. Food Sci., 1990, **55**, 199-205.
3. Sastry, S.K., Heskitt, B.F. and Blaisdell, J.L., Experimental and modeling studies on convective heat transfer at the particle-liquid interface in aseptic processing systems. Food Technol., 1989, **43**, 132-36, 43.
4. Stoforos, N.G., Park, K.H. and Merson, R.L., Heat transfer in particulate foods during aseptic processing. Paper No. 545, presented at the Institute of Food Technologists Annual Meeting, Chicago, IL, June 25-29, 1989.
5. Zuritz, C.A., McCoy, S., and Sastry, S.K., Convective heat transfer coefficients for irregular particles immersed in non-Newtonian fluid during tube flow. J. Food Eng., 1990, **11**, 159-74.
6. Sun, X., Schmidt, S.J. and Litchfield, J.B., Convective heat transfer coefficient measurement using magnetic resonance imaging. Paper No. 82-6581, presented at International winter meeting of ASAE, Nashville, TN, December 15–18, 1992.
7. Bird, R.B., Stewart, W.E. and Lightfoot, E.N., Transport Phenomena, John Wiley and Sons, New York, 1960, pp. 133.
8. Park, K. H. and Merson, R. L., Coldest point location in convectively heated particles. Paper No. 92-6850, presented at International winter meeting of ASAE, Nashville, TN, December 15–18, 1992.
9. Park, K. H. and Merson, R. L., Numerical study of transient heat transfer between a particle and a fluid in a holding tube. Paper No. 299, presented at the Institute of Food Technologists Annual Meeting, New Orleans, LA, June 20-24, 1992.

HEAT GENERATION AND TRANSFER IN ELECTRIC HEATING OF A LAMINAR FLOW OF FOOD

L. ZHANG and P.J. FRYER
Department of Chemical Engineering,
University of Cambridge, Pembroke St.,
Cambridge, CB2 3RA, United Kingdom.

ABSTRACT

Electric heating allows even heating of food mixtures. Food flows including mixtures are often characterised in terms of laminar flows of Newtonian or power-law fluids. A model for the electrical heating of such flows has been developed. High temperatures are possible near the walls of the heater where food is slowest-moving; however, simulations of power-law flows where the viscosity is temperature dependent suggests that the region of high temperature is small.

INTRODUCTION

The continuous electrical (ohmic) heating of foods is a commercial process in which a food fluid is sterilised in flow due to the passage of electrical current. The sterilisation of the food is a function of the temperature-time history of the material, which must therefore be predictable. The electrical processing of food containing particles has been studied thoroughly at Cambridge [1-3]. The heating rate of a two-phase mixture is a complex function of the electrical conductivity of the two phases and the shape of the particles. Models for the process require solution of the field equations around the particles to take account of uneven heating effects.

Although solid-liquid flows are not homogeneous [4-5], the flow of food mixtures has often been modelled using continuum approaches such as power-low equations. The behaviour of such fluids under electric heating is worthy of study because it allows an estimation of the variation in temperature and flow velocity that will occur in a practical flow situation. This paper outlines a model for heat transfer and generation in laminar flows in the electrical heater, and demonstrates temperature and flow variations that might occur in practice.

FLOW AND HEAT GENERATION

Solution of the equations for the electrical heating of a two-phase mixture is described elsewhere [2-3]. The voltage distribution must be calculated by solving Laplace's equation for voltage V:

$$\nabla(\kappa \nabla V) = 0 \tag{1}$$

from which the electric field intensity, E = grad V, and the heat generation $Q = \kappa|E|^2$ can be calculated at each point. The Navier-Stokes equation for flow and the energy equation must be solved simultaneously with (1):

$$\frac{Du}{Dt} = -\nabla p + \mu(\nabla^2 u) + \rho F \qquad \rho c_p \frac{DT}{Dt} = \nabla(\lambda \nabla T) + Q \qquad (2\text{-}3)$$

In a pipe flow, approximations can be made; 2-D axial symmetric solution of (1) and a laminar Newtonian or power-law equation for the flow of axial velocity u.

$$\text{Newtonian:} \quad \tau = \mu \frac{du}{dr} = -\frac{rdp}{2dx}; \quad \text{Power-law:} \quad \tau = K\left(\frac{du}{dr}\right)^n \qquad (4)$$

The temperature field can be obtained by solving

$$u\frac{\partial T}{\partial x} = \alpha \frac{\partial(r\partial T/\partial r)}{r\partial r} + G \qquad (5)$$

where α is the thermal diffusivity and G is the inherent heating rate $G = \kappa|E|^2/\rho C_p$ [3]. If physical properties are temperature independent, and a parabolic velocity profile is assumed, α and G are the only parameters needed. Manipulation of (5) then gives:

$$U\frac{\partial \theta}{\partial \xi} = \Lambda \frac{\partial(\eta \partial\theta/\partial\eta)}{\eta\partial\eta} + \Gamma \qquad (6)$$

where $U = \frac{u}{u_0}$, u_0 is mean velocity, $\xi = \frac{x}{L}$, $\eta = \frac{r}{R}$, $\theta = \frac{T-T_0}{T_0}$, $\Lambda = \frac{\alpha L}{R^2 u_0}$ and $\Gamma = \frac{GL}{T_0 u_0}$.

In the linear case, Λ and Γ can be used in design.

Models for flow which include temperature-dependent physical properties have been developed. Equation (1) is solved to find the Q field, and an iterative procedure [8] used to find the resulting temperature and flow fields. Fluid viscosity μ will be a strong function of temperature, and will change with x and r, a problem analogous to that in a tubular polymeriser [6-7]. The solution method used is similar to that of [6]. The velocity profile for a known radial variation in μ can be found by integration of equation (4) to give

$$\frac{u(r)}{u_0} = \frac{R^2 \int_r^R r/\mu \, dr}{2\int_0^R r \int_r^R r/\mu \, dr dr} = \frac{R^2 \int_r^R r/\mu \, dr}{\int_0^R r^3/\mu \, dr}, \quad \frac{u(r)}{u_0} = \frac{R^2 \int_r^R (r/k)^{1/n} dr}{\int_0^R r^2 (r/k)^{1/n} dr} \text{ for power law liquid} \qquad (7)$$

In this 1-D model, u_0 is not a function of x.

Figures 1 and 2 show velocity and temperature profiles generated for a power law fluid and for a Newtonian liquid. The effect of the wall is to slow the flow near it; since the temperature of the material depend on how long it has spent in the heater, slow velocities result in high temperatures in the wall region. The reduction in viscosity which occurs as a result of the increase in temperature can flatten the velocity profile of the fluid; the effect of this is that the system approaches the ideal of plug flow assumed in many models of electrical heating. It may be that the food fluid contains components whose viscosity increases at high temperature, however, such as starches which undergo gelation; under these conditions different velocity profiles may result. Very high temperatures may lead to the formation of fouling deposit on the walls of the heater or, if the temperature exceeds the boiling point of the fluid at the pressure in the heater, to nucleate boiling.

Convective effects have not been simulated here, but have been seen even in stationary fluids [9]; it would thus be expected that the peaks in the temperature profile will be less pronounced than seen here.

REFERENCES

1. de Alwis, AAP, Halden, K and Fryer, PJ. *Chem.Eng.Res.Des.*, **67**, 159-168, (1989).
2. de Alwis, AAP, Zhang, L, and Fryer, PJ. pp 103-142 in "*Advances in Aseptic Processing Technologies*", ed RK Singh and PE Nelson, Elsevier, London. (1992)
3. Zhang L and Fryer, PJ.*Chem. Eng. Sci.* 48(4)633-642, (1993)
4. Liu, S, Pain, J-P, and Fryer, PJ. *Entropie*,**28**(170), 50-58, 1992.
5. Liu, S, Zhang, L, Pain, J-P and Fryer, PJ. *IChemE Symposium Ser.* **126**, 79-88, (1992).
6. McLaughlin, HS, Mallikarjun, R and Nauman, EB, *AIChE J*, 32,(3)419-425, (1986)
7. Lynn, S and Hoff, JE, AIChE J., 17, 475 (1971)
8. Zhang L and Fryer, PJ. submitted to *Chem. Eng. Sci.* (1993)
9. Fryer, PJ de Alwis, AAP, Koury, E, Stapley, AGF and Zhang, L, *J. of Fd Eng*, **18**,101-125 (1993)

ACKNOWLEDGEMENTS

Original financial support for work on electrical heating at Cambridge was provided by APV Baker. This work was supported by an AFRC grant to LZ.

NOMENCLATURE

c_p	Specific heat capacity	(J kg/K)	V	Voltage	(V)
E	Voltage gradient	(V/m)	x	distance	(m)
F	Force	(N)	α	Thermal diffusivity	(m^2/s)
G	Inherent heating rate	(K/s)	Γ	Dimensionless variable	(-)
K	Coefficient in power-law	(-)	Λ	Dimensionless variable	(-)
L	Length	(m)	κ	Electrical conductivity	(S/m)
n	power in power-law	(-)	λ	Thermal conductivity	(W/ mK)
p	Pressure	(N/m^2)	μ	viscosity	($N/m\ s^2$)
Q	Heat generation rate	(W/m^3)	ρ	Density	(kg/m^3)
R	Radius of heater	(m)	θ	Dimensionless temperature	(-)
r	Radius variable	(m)	τ	Shear stress	(N/m^2)
T	temperature	(K)	ξ	Dimensionless distance	(-)
t	time	(s)	η	Dimensionless radial variable	(-)
T_0	Initial temperature	(K)			
U	Dimensionless velocity	(-)			
u	Velocity	(m/s)			

Subscript and superscript

0 initial or mean

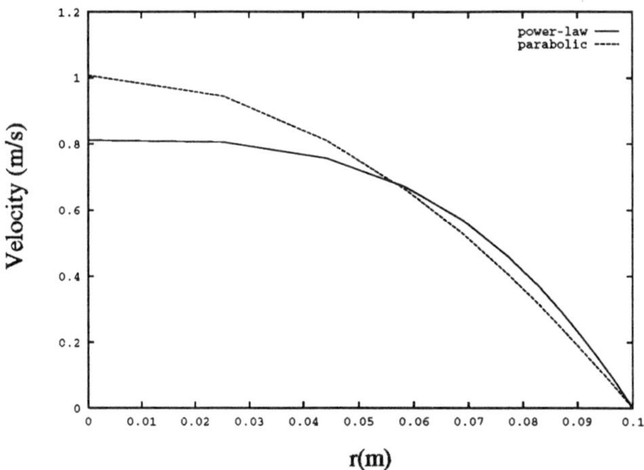

Figures 1 Velocity profile for the heating of tomato juice using a power law fluid with K = 0.728 exp(-0.058 T) and n = .453 exp (7.13×10^{-3} T) for the case where $\Gamma = 5$, $\Lambda = 0.02$, compared with parabolic flow, in a 0.2m diameter tube.

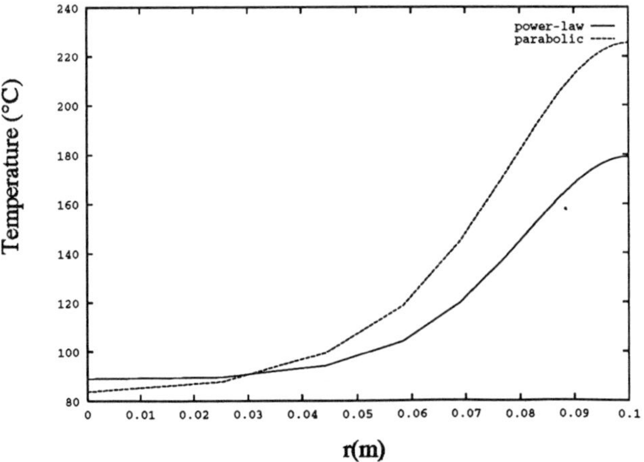

Figure 2 Exit temperature distribution for the same systems as figure 1.

A NEW NONDESTRUCTIVE SYSTEM TO EVALUATE QUANTITATIVELY THE EFFICIENCY OF FOOD ANTISEPTICS

Katsutada Takahashi
Laboratory of Biophysical Chemistry, College of Agriculture
University of Osaka Prefecture, Sakai, Osaka 593, Japan

ABSTRACT

A system to quantitatively measure the efficiency of food antiseptics based on the detection of metabolic heat evolved by microbial cells has been designed. The apparatus has 24 measuring units and the growth activities of microbes at various concentrations of an antiseptic are determined from the heat evolution processes during culture. When the measurement is completed, a drug potency curve is drawn so that the effects of the antiseptic can be quantitatively described. The method is nondestructive so that it is also very useful for the study of food putrefaction.

INTRODUCTION

Although biological activities of microbial cells are often compared by observation of their metabolic heat[1,2], little effort has been put forth to adapt this property to the quantitative study of food putrefaction where microbial propagation is essentially responsible. The author has developed a highly sensitive multiplex batch calorimeter useful for the detection of small heat effects arising from cellular metabolism, and used it to analyze the growth behavior of microbes in different culture media as well as for the putrefaction process of foods in order to obtain the efficiency of the antiseptic on them.

In this paper, the method that also includes the design of the apparatus is described together with some results obtained from the putrefaction experiment conducted on boiled soybeans containing different amounts of an antiseptic.

MATERIALS AND METHOD

Apparatus

The apparatus, a type of multiplex calorimeter having 24 calorimetric units , is schematically shown in Figure 1. (The design is now manufactured by Nippon Medical & Chemical

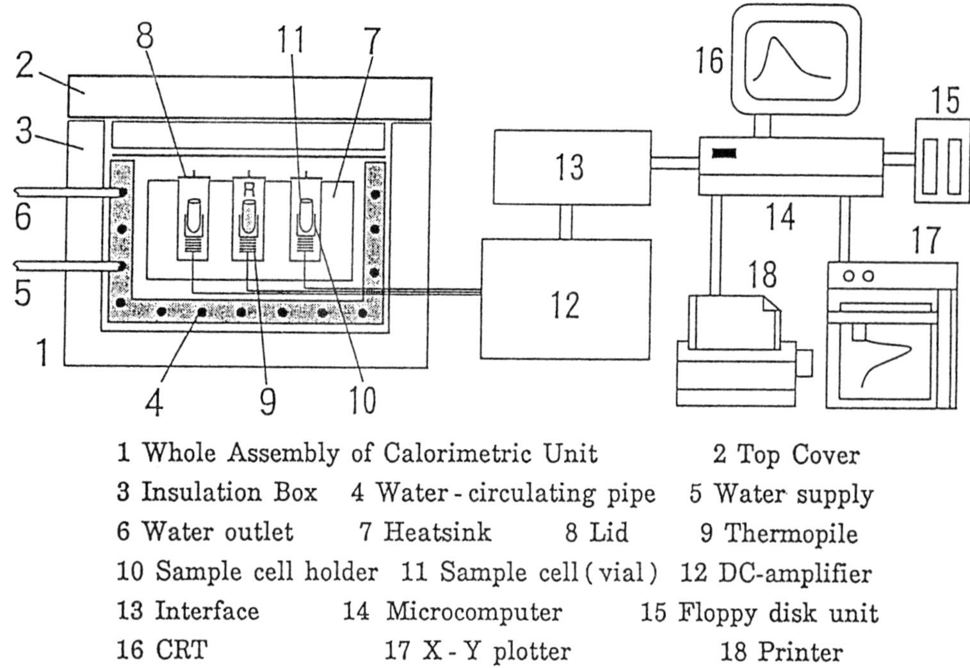

Figure 1. Schematic illustration of Bio Thermo Analyzer H-201.

Instruments Co. Ltd., Osaka, Japan (Fax: 81-6-445-7641) and is commercially available under the name "Bio Thermo Analyzer H-201").

The measurement is made by setting up the sample vial (30 cm^3)(11) containing samples in the calorimetric units of the heatsink(7) that is maintained at a constant temperature. The thermopile plate (9) placed at the bottom of sample cell holder(10) detects any temperature differences between the sample cell and the heatsink. The differential voltage between the thermopile plates of the sample and the reference units is led out, then digitalized by an A/D-converter, and stored on floppy disks(15) through a microcomputer(14) for further computational analysis.

RESULTS AND DISCUSSION

In Figure 2, the heat evolution processes ($f(t)$ curves) obtained for the putrefaction of boiled soybeans containing various amounts of sodium benzoate are shown. As the concentration of the antiseptic increased, the $f(t)$ curve shifted toward a longer incubation period with a decrease in the initial slope. Thus, the change in putrefying ability was correlated with the concentration of sodium benzoate.

In order to characterize the above effect more quantitatively, the extent of the delay in putrefaction in the presence of the antiseptic was analyzed by a simple mathematical model. If, for a given sample containing the antiseptic at concentration i, an incubation time to attain the putrefaction to a definite level is defined as $t_\alpha(i)$, the extent of delay in putrefaction will conveniently be expressed by $t_\alpha(0)/t_\alpha(i)$, where $t_\alpha(0)$ is the incubation time required for attaining the same level of putrefaction when no antiseptic is added. In Figure 3 the drug

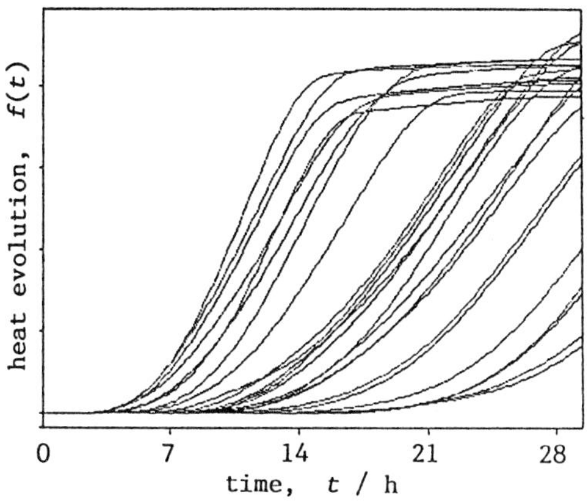

Figure 2. Putrefaction of boiled soybeans containing various amounts of sodium benzoate at 30°C as measured by Bio Thermo Analyzer.

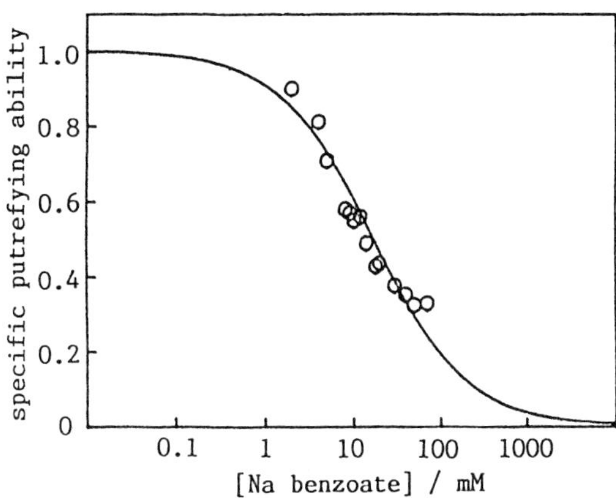

Figure 3. Drug potency curve for the action of sodium benzoate on boiled soybean as determined from the experiment shown in Figure 2.

potency curve of sodium benzoate against the putrefaction of boiled soybeans obtained from the above experiment is shown in the normalized form of $t_{\alpha}(0)/t_{\alpha}(i)$ *versus* i. By a regression analysis a parameter characterizing the efficiency of the antiseptic was obtained to be Ki = 13.3 mM, the concentration of sodium benzoate at which the putrefying ability of the boiled soybeans is repressed by 50%.

CONCLUSIONS

From the result given above together with those obtained from the same series of experiments conducted on the other foods and the antiseptics, it may be concluded that the method described here is very useful and that it will contribute to further advances in the quantitative evaluation of food antiseptics.

REFERENCES

1. Takahashi, K. Application of calorimetric methods to cellular processes. *Thermochimica Acta*, 1990, **163**, 71-80.
2. Kimura, t., and Takahashi, K., Calorimetric studies on soil microbes. *J. Gen. Microbiol. (London)*, 1985 , **131**, 3083-89.

DEVELOPMENT AND APPLICATION OF ASEPTIC NEW MATERIALS(ASEPLA)

S.KUNISAKI, K.NODA, T.SAEKI, and T.AMACHI
Institute for Fundamental Research, SUNTORY Ltd.
1023-1, Yamazaki, Shimamoto-cho, Mishima-gun, Osaka 618, Japan.

ABSTRACT

ASEPLA(Aseptic Plastic) is new material with high antimicrobial activity, prepared by a mixing resin with a powdered antimicrobial zeolite containing Ag. It was confirmed that ASEPLA has higher antimicrobial activity in raw water containing mineral components than in the deionized water. Thus, it was expected that mineral components in raw water would play an important role for its antimicrobial activity. It was recognized that ASEPLA can maintain high antimicrobial activity even in the presence of Cl^-.

INTRODUCTION

In a water-purifying device, residual chlorine is eliminated prior to the demineralizing process. However, the elimination of chlorine necessarily results in undesirable propagation of microorganisms in the water-transporting pipes or water-reserving tanks. In order to prevent the microorganism contamination, we have developed and applied a system using an antimicrobial plastic, ASEPLA. ASEPLA is a new resin material with a superior antimicrobial activity, prepared by mixing an antimicrobial zeolite containing Ag (ZEOMIC) made by Shinanen Co. Ltd. with plastic materials. In this paper, fundamental characteristics of ASEPLA will be discussed.

MATERIALS AND METHODS

Antimicrobial Activity of ASEPLA

Microorganisms isolated from raw water was suspended in sterilized water to at the concentration of about 10^1 to 10^5 cells/ml. The ASEPLA sample(5x5cm) was soaked in 150 ml of the microbial suspension, and allowed to stand at 28°C for 96 hrs with stirring. One ml portion was taken from the suspension every 24 hours to count the living cell numbers at each sampling times.

RESULTS

Effect of Initial Microbial Concentration on the Antimicrobial Activity

Three types of microbial suspension at the initial concentration of about 10^1, 10^3 and 10^5 cells/ml were prepared to test the antimicrobial activity of ASEPLA.As shown in Figure 1,the antimicrobial activity is dependent on the initial cell number,that is,the increase of its activity by decrease of the cell concentration. Concentration of Ag in each suspension was about 3 ppb.

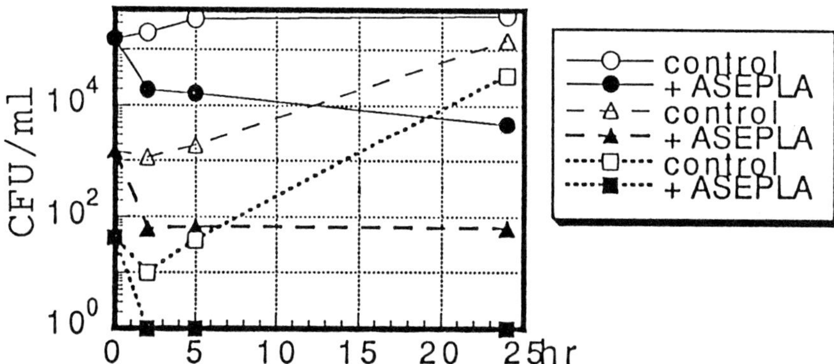

Figure 1. Effect of Initial Microbial Concentration on the Antimicrobial Activity

Effect of Mineral Components of Water

Antimicrobial activity of ASEPLA was examined both in raw water with mineral components (total hardness:98) and in deionized water.In the presence of ASEPLA,living cell in raw water could hardly be detected within 2 hrs,while in deionized water living cells were observed for 5 hrs but not after 24 hrs(Figure 2).The concentration of silver in raw water used for this experiment reached to 25 µg/L, which was almost similar to that of deionized water being used,28 µg/L.Thus,the effect of some ionic components can be expected.

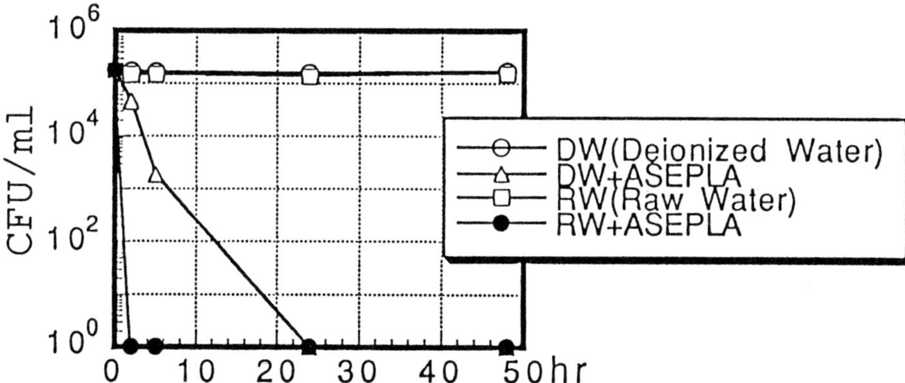

Figure 2. Effect of Mineral Components of Water

Effect of Silver Concentration Released from ASEPLA

Test solution containing 20 µg/L of silver was prepared by dipping of ASEPLA samples in deionized water for 1 week. Diluted solution was used to investigate the minimum concentration of silver required for the antimicrobial activity. Living cell in microbial suspension containing more than 10 µg/L of silver could hardly be detected after 2 hrs, while in the suspension containing only 1 µg/L of silver propagation of the microorganisms were observed as shown in Figure 3.

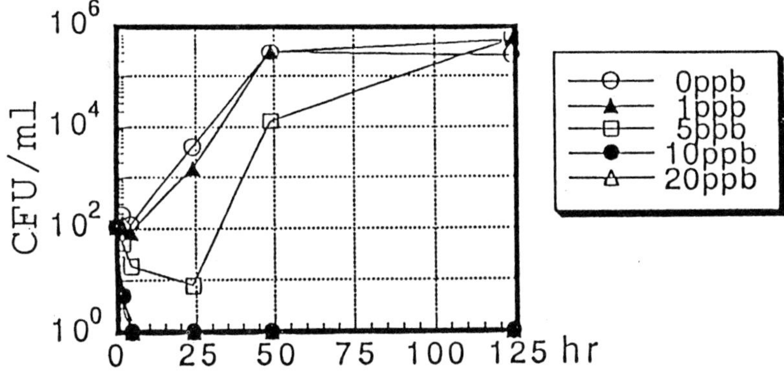

Figure 3. Effect of Silver Concentration Released from ASEPLA

Comparison of Antimicrobial Activity between ASEPLA and $AgNO_3$ in the Presence of NaCl

The antimicrobial activity of ASEPLA was not affected by the presence of NaCl at 0.8 w/w% concentration(Figure 4).However,in the case of $AgNO_3$(100 ppb),the effect of silver,which can terminate microorganisms within 2 hrs,was extremely weakened by the co-existence of NaCl (0.8%). This would be attributed to the formation of insoluble AgCl.

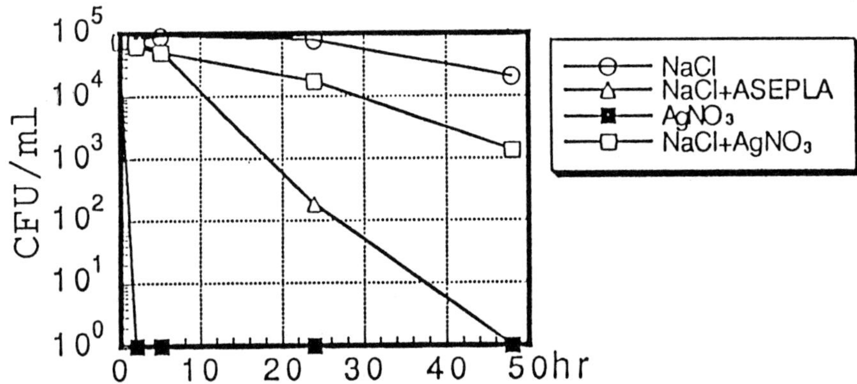

Figure 4. Comparison of Antimicrobial Activity between ASEPLA and $AgNO_3$ in the Presence of NaCl

DISCUSSION

ASEPLA has the antimicrobial property when soaked into aqueous solution. It was confirmed that ASEPLA shows higher antimicrobial activity in raw water containing mineral components than in the deionized water.Thus,it was expected that mineral component in raw water would play an important role for the antimicrobial activity. It was suggested that concentration of Ag was correlated with the ASEPLA activity,and the inactivation of microbes was caused by silver released into liquid from ASEPLA.It is well known that $AgNO_3$ has high antimicrobial activity. However, $AgNO_3$ cannot be used for the water containing Cl^- such as sea water because of formation of insoluble salt.On the other hand,ASEPLA can maintain its activity even in the presence of Cl^-. Consequently,it was suggested that ASEPLA differed somewhat from $AgNO_3$ in existing its antimicrobial activity.

CONTINUOUS STERILIZATION OF PARTICULATE FOODS BY OHMIC HEATING: CRITICAL PROCESS DESIGN CONSIDERATIONS

SUDHIR K. SASTRY
The Ohio State University
Department of Agricultural Engineering
590 Woody Hayes Drive
Columbus, OH 43210, USA

ABSTRACT

Research at this laboratory for the past several years has involved determination of electrical conductivities of foods, microbial death kinetics, process modeling and experimental verification. Finite element models developed for heating of solid-liquid mixtures in a continuous flow ohmic heater indicate that if all particles and liquid are of equal electrical conductivities, the particle cold spots heat slightly faster than the liquid. For high concentration mixtures, if all particles are of low electrical conductivity, the mixture heats slowly due to high effective resistance, but the particles still heat faster than the fluid. However, if a single particle of unusually low electrical conductivity enters the heater, it will thermally lag the fluid since the current has alternate low-resistance pathways around it. Under these conditions, the potential for underprocessing exists. Particle concentration has been found to be important in determining whether or not particles heat faster than fluids. Under low concentrations, particles will typically lag fluids, while high concentrations favor faster particle heating. These findings have been verified experimentally in a static ohmic heater. Conditions involving a radial velocity profile are discussed.

INTRODUCTION

Recent industry interest in continuous sterilization of particulate foods has focused much attention on the technology of ohmic heating, in which liquid-particle mixtures are heated by passing an electrical current through them. The resulting internal generation has been reported to cause rapid and uniform heating of foods, and the technology holds promise for continuous sterilization. However, fundamental understanding of the process is limited.

MATHEMATICAL MODEL FOR OHMIC HEATING

The energy generation rate during ohmic heating depends on the field strength (∇V) and a temperature-dependent electrical conductivity.

$$\dot{u} = |\nabla V|^2 \sigma_0 [1 + mT]$$

where σ_0 is the electrical conductivity at 0°C, m a temperature coefficient, and T the temperature. When a solid-liquid mixture is heated, the relative rates of heating depend on the electrical conductivities of the respective phases. Fluid temperatures depend on the energy generation rate, extent of fluid mixing and heat transfer from solids. Solid temperatures can be calculated from the conduction heat transfer equation for a solid with internal energy generation. The voltage field is subject to two types of variations: large scale, due to the medium being heated along the heater length, and smaller scale variations due to differences in phase conductivities. Large scale effects result in the major temperature changes over the heater geometry. Small-scale effects depend on the extent of fluid mixing in the particle vicinity, and are typically transient in character, depending on the local particle/fluid structure, temperatures and relative movement. Formulations for prediction of fluid and solid temperatures in continuous flow heaters in plug flow have been published [1]. A more recent model analyzes the situation where a radial velocity profile exists in the flow.

IMPORTANT RESULTS

Results of simulations have shown that the fluid and particle heating rates depend on the electrical conductivities (σ) and volume fractions of the phases. If particles are less electrically conductive than the fluid, they can thermally lag the fluid, unless the volume fractions are sufficiently high. Under this condition, the particles can heat faster than the fluid. One critical condition involves the presence of an isolated low electrical conductivity particle in an environment of high electrical conductivity. Under these conditions, substantial thermal lags can occur. In addition, (within a static heater) the isolated low-conductivity particle can alter the electrical field around it to create localized hot and cold zones [2]. The

thermal nonuniformity is minimal if sufficient fluid motion occurs.
From the standpoint of process safety, it is necessary to consider the case of the particle of the lowest possible σ, within a medium of the highest possible σ for a given mixture. If a residence time distribution exists, the worst-case might be that of the isolated low-conductivity particle that is also the fastest in the mixture. Orientation effects will of course have effects, with long-thin particles being significantly affected by being oriented either parallel to or perpendicular to the field.

Studies within static and continuous flow heaters indicate that some key differences exist between the two types of heaters, particularly from the standpoint of fluid motions. Consequently results obtained from one type of heater should not be extended into another type without adequate accounting of these differences. Additionally, it is important to note that the behavior of multiple particles is substantially different from those of single particles, hence care needs to be exercised in extending work with limited particle populations to conditions of high solids concentrations.

CONCLUSIONS

Critical factors associated with process design involve the condition of the lowest electrical conductivity phase being surrounded by that of highest electrical conductivity. Additional safety considerations arise if this particle is also the fastest moving. Particle volume fraction has significant effects both from the standpoint of internal energy generation and flow patterns about the particles.

REFERENCES

1. Sastry, S.K. A model for heating of liquid-particle mixtures in a continuous flow ohmic heater. J. Food Proc. Engr., 1992. **15**, 263-278.

2. Fryer, P.J., de Alwis, A.A.P., Koury, E., Stapley, A.G.F., and Zhang, L. Ohmic processing of solid-liquid mixtures: heat generation and convection effects. J. Food Engr., 1993, **18**, 101-125.

QUALITY CHANGES OF ASEPTICALLY PACKED APPLE JUICE DURING STORAGE AND THE PREDICTION OF ITS SHELF-LIFE

Shin-Hwei Yuo, Sun-San Lin, Sue-Ywe Chen, Ching-Chuang Chen, and Chu-Chin Chen
Food Industry R & D Institute
P.O. BOX 246, Hsinchu, Taiwan, Republic of China

ABSTRACT

The effects of storage time and temperature on the quality of aseptically-packed single-strength apple juice were investigated by estimating the changes of sensory characters and chemical components. The concentrations of 5-hydroxymethyl-furfural and furfural increased with storage temperature and time, the rate of formation followed zero order kinetics with activation energies of 25.49 and 14.94 Kcal/mol, respectively. The Hunter L, a, b values did not change significantly when stored at low temperatures, i.e.. 4, 20 °C. The Hunter L value decreased while the Hunter a value and absorbance at 420nm increased along with storage time when stored at 37 °C. The concentration of sucrose decreased along with storage time when stored at 37 °C, while the concentrations of glucose and fructose increased. Storage time exhibited a significant effect on the reduction of malic acid; however, tartaric acid did not change. Sensory scores decreased during storage and changes in the scores of samples stored at 37 °C were greater than those under other storage temperatures. The Q_{10} values of flavor scores of samples stored at 4-14 °C, 14-24 °C, and 25-35 °C were 1.652, 1.597, and 1.549 accordingly. The predicted shelf-life of aseptically packed apple juice stored at 37 °C, 25 °C, 20 °C, and 4 °C was 30.8, 52.0, 68.2, and 161.4 weeks, respectively.

INTRODUCTION

Recently, the consumption of fruit juice in Taiwan has been growing rapidly and most of the juices were diluted products with 30% juice contents or less. Fruit juices supplied by local manufacturers include guava, carambola, and passion fruit, and the popular imported juice consists of orange, apple, grape, and peach concentrates. Non-enzymatic browning (NEB) reaction is one of the primary cause of quality change in juice during thermal processing and storage , which is a combined effect of the Maillard browning reactions, degradation of ascorbic acid, and caramelization of sugars. Furfural and 5-hydroxy-methyl furfural (HMF) are two known intermediates of NEB reactions. As a consequence, the accumulation of furfural or HMF is a good indicator for the extent of NEB reactions in juice (4). The concentrations of these compounds can also be established for estimating the shelf-life of juice products. Therefore, the purpose of the present study was to analyze the quality changes in aseptically packed apple juice during storage, coupled with the adequateness of various chemical and physical parameters to estimate the quality and the shelf-life of aseptically packed apple juice.

MATERIALS AND METHODS

All chemicals were purchased from E. Merck. All solvents for HPLC were HPLC grade and were purchased from Fisher Scientific. Apple juice concentrate (70° Brix) was obtained from a local supplier (Chou Chin Industrial Corp., Changhwa, Taiwan). Aseptic packages (ca. 400mL) were acquired from PKL Corp. Apple juice (13.3 Brix) was diluted from concentrate with suitable amount of added apple flavor (0.03%, Haarmann & Reimer Ltd., Holzminden, Germany). Condition for aseptic processing (85°C, 45s) and packaging (ca. 350mL per package) of apple juices were cited elsewhere (2). Aseptically packed apple juice was divided into four groups and stored immediately at 4 °C, 20 °C, 25 °C, and 37 °C for 24 weeks.

Absorbance of apple juice at 420nm was determined on a Hitachi U-200 UV-Visible Sepectrophotometer. The Hunter L, a, and b values were determined with transmittance measurement on a Color and Color Difference Meter ("Color Ace " model TC-1). HPLC analyses were conducted on a Jasco 800 HPLC system equipped with an 830-RI detector and an 875-UV detector. The procedures for the analysis for furfural and HMF were as described (4). Mobile phases used for sugars and organic acid were 80% acetonitrile and 1% $(NH_4)H_2PO_3$ solution (pH = 2.4), respectively.

Sensory evaluations using 9 point hedonic scale were conducted by 12 panelists. Duplicate juice samples were evaluated at each time.

RESULTS AND DISCUSSION

The concentrations of HMF and furfural in aseptically packed apple juice were gradually increased during storage (Fig. 1), which agreed with a previous report (1). Kinetic studies indicated that the formation of both HMF and furfural followed a zero order reaction, with activation energy (Ea) of 25.49 and 14.94 kcal/mole, respectively. The result agreed with a previous study (5). According to the result of statistical analysis, the concentrations of both HMF and furfural were inversely related to the sensory scores of apple juice. The Hunter L value of apple juice stored at 37 °C was significantly different from those stored at lower temperatures (Table. 1). For apple juice stored at 37 °C, sucrose concentration decreased gradually with concomitant increase of fructose and glucose. The result agreed well with a previous report (1). It was also found that the concentration of malic acid decreased gradually during storage; however, tartaric acid did not change significantly.

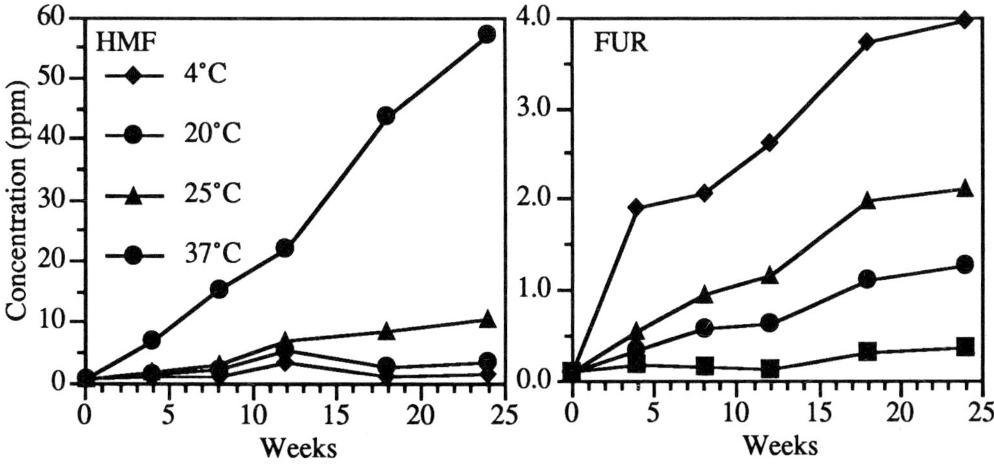

Figure 1. The change of 5-hydroxymethyl furfural (HMF) and furfural (FUR) concentrations in aseptically packed apple juice during storage at various temperatures.

After 24 weeks of storage, the sensory scores of samples stored at 37 °C were closer to the acceptable level (i.e. 4). Juice stored at lower temperatures after 24 weeks of storage appeared to be more acceptable to panelists than those stored at 37°C. Flavor scores were selected as the parameters for the estimation of the shelf-life of aseptically packed apple juice. With the Arrhenius equation, the activation energy was estimated to be 7.169 kcal/mole and the Q_{10} values from 4 °C to 37 °C ranged from 1.652 to 1.549.

A sensory flavor score of 4.0 was chosen as the minimum score of acceptability, thus the estimated shelf-life of aseptically packed apple juice stored at 37 °C was 30.8 weeks. The shelf-life of other samples stored at lower temperatures was calculated using the same shelf-life value of those stored at 37°C and applied to the previously described equation (3). The estimated shelf-life of aseptically packed apple juice stored at 25°C, 20°C, and 4°C was 52, 68.2 and 161.4 weeks, respectively.

TABLE 1

Effects of storage temperatures on Hunter L, a, and b value and absorbance value at 420nm of aseptically packed apple juice

Temperature	L	a	b	A420
4°C	81.232a	-1.572a	29.033a	0.487a
20°C	80.817a	-1.187ab	28.834a	0.461a
25°C	80.159a	-0.940b	28.954a	0.470a
37°C	77.984b	0.100c	29.414a	0.519b

NOTE : Duncan's multiple range test, data within the same column with the same letter are not significant different from each other for $P < 0.05$.

ACKNOWLEDGMENTS

This research was supported by Ministry of Economy (No. 91T8112B3), Republic of China.

REFERENCES

1. Babsky, N.E., Toribio, J.L. and Lozano, J.E. Influence of storage on the composition of clarified apple juice concentrate. J.Food Sci., 1986, 51, 564-567.

2. Chen, C.C., Lin, S.J., Chen, S.Y., Chen, C.C., Zen, S.M. and Jeng, J.G., Development and evaluation of a laboratory scale aseptic processing and packaging system. Research Report 633-3, Food Industry R&D Institute, Hsinchu, Taiwan, ROC., 1991.

3. Labuza, T.P. and Schmidl, M.K. Accelerated shelf-life testing of food. Food Tech., 1985, 39, 57 - 64.

4. Nagy, S. and Dinsmore, H.L., Relationship of furfural to temperature abuse and flavor change in commercial canned single strength orange juice. J. Food Sci., 1974, 39, 1116 - 1119 .

5. Resnik,S.and Chirife,J. Effect of moisture content and temperature on some aspects of nonenzymic browning in dehydrated apple. J. Food Sci., 1979, 44, 601 - 605.

COMPARISON OF THE IMMERSION BIOTEST AND THE SPORE-TEST METHOD FOR EVALUATING THE INTEGRITY OF SEALS OF FLEXIBLE RETORT PACKAGES

GUN WIRTANEN, EERO HURME, RAIJA AHVENAINEN,
LENA AXELSON-LARSSON* AND TIINA MATTILA-SANDHOLM
VTT Food Research Laboratory, P.O. Box 203, SF-02151 Espoo, Finland
* Packforsk, Swedish Packaging Research Institute, P.O. Box 9, S-16493 Kista, Sweden

ABSTRACT

Penetration of vegetative cells of *Enterobacter aerogenes* and spores of *Bacillus subtilis* was investigated through microholes in the seal area of retort packages. The packages were filled with mashed potatoes. Defects in the packages used in the biotest evaluation were made before sealing by pushing a wolfram thread of either 50 µm or 100 µm diameter through the seal. In the biotest with *E. aerogenes* the packages were immersed for 1 h and in the spore test with *B. subtilis* the solution was left on the seal area for 30 min. The packages were then dried and stored at 30 °C for 6 and 20 days. The results demonstrated that the storage time did not influence the outcome of the positive results. The effective diameter of the microholes in the seal was measured by an electrolytic test, Microhole Tester. When the microhole size was less than 100 µm the spore test gave more positive samples than the test with vegetative cells and when the size was greater than 100 µm all samples were positive with both test methods.

INTRODUCTION

The destructive or non-destructive testing of containers during and after processing is one way of demonstrating the integrity of the packages. In order to select and develop non-destructive leak-testing methods for flexible and semi-flexible packages it is essential to obtain information, e.g. by biotesting, concerning the critical microhole size for bacterial penetration [1]. The aim of this study was to compare the spore test and the immersion test in order to determine which biotest method gives more accurate information about the integrity of packages and which is less time consuming in practice.

MATERIALS AND METHODS

Packaging of the Model Foodstuff

Mashed potatoes were packed in 200 ml polypropylene-based containers with ethylene vinyl alcohol as a barrier layer (Bebo Plastik, Germany). Tungsten thread of either 50 µm or 100 µm in diameter was placed on the seal area before sealing and then the filled containers were sealed in a commercial processing line (Lieder Maschienenbau, Germany) and autoclaved. After autoclaving the threads were withdrawn from the seals and the containers were tested using either the immersion test or the spore test. Some of the containers were tested with blocked microholes, the blocking being performed with agar. Intact containers with no holes in the seal area were used as reference packages.

Immersion Biotest with *Enterobacter aerogenes*

A modification of the biotest method proposed by the National Food Processors Association and ASTM in the USA was used. In the test the containers were placed in a bath of fresh *Enterobacter aerogenes* (ATCC 13048) suspension containing approximately 10^7 cfu/ml and incubated at 24 ± 1 °C for 60 min. Thereafter the containers were removed from the bath, wiped and cleaned and transferred to the incubator, where they were incubated upside down at 30 °C. The samples were withdrawn after incubation of either 6 or 20 days for investigation of bacterial growth. The containers were emptied and washed before seal defect testing with the Microhole tester (Packforsk, Sweden).

Spore Biotest with *Bacillus subtilis*

In the spore test the containers were placed with the longer seal facing up. A suspension containing spores of *Bacillus subtilis* (Merck 10649) was pipetted into the seal area and left there for 30 min at 24 ± 1 °C. A solution containing about 10^6 spores/ml sterile distilled water was used. The containers were further treated as described in the section on the immersion test above.

Confirmation of Bacterial Growth

The bacterial growth of *E. aerogenes* was confirmed in tubes containing Brilliant Green Bile (Difco, US) solution by gas production in Durham tubes. Positive samples were plated on Levine EMB agar (Difco, US), where the test organism forms colonies with grey-brown centres. *E. aerogenes* is a Gram negative rod, which was confirmed by Gram staining.

The samples taken from containers tested in the spore test with *B. subtilis* were cultivated on blood agar plates (Orion Diagnostica, Finland). Hemolysis of the blood was observed around the colonies on the plates from positive samples. Cells of *B. subtilis* were also tested by Gram staining and identified as Gram positive rods.

Measurement and Confirmation of Microholes

The effective diameter of microholes was measured by an electrolytic test (Microhole tester). The method is based on the fact that the electrical conductance of a package of insulating material is drastically altered by a small hole [2]. The seals were also tested with a dye test (Ageless Seal check).

RESULTS AND DISCUSSION

The results obtained with the samples tested with the two methods and the two incubation times are presented in Table 1. Prolongation of the incubation time from 6 to 20 days had no effect on the outcome of the case of positive samples. When the effective diameter of the microhole measured by the Microhole tester was less than 100 µm the spore test gave more positive results. A comparison of the two biotest methods with open and blocked microholes is presented in Table 2. When the size of the microholes was less than 100 µm the number of positive samples was higher in the spore test than in the immersion test. Blocking of the microholes had no effect on the number of positive samples obtained with the two methods. In fact, it is possible that the blocking agar enabled the motile *E. aerogenes* to enter the package. Containers with very small microholes were not tested with *E. aerogenes*. When the effective diameter was greater than 100 µm all samples were positive with both test methods. Microholes were also detected in the control containers with "intact" seals. The spore test appears to be very sensitive, because some packages with no microholes detected in the measurement using the Microhole tester showed positive reactions in this test. According to this study, testing of the integrity of packages should be based on more than one method in order to confirm the product quality depending on seal integrity.

TABLE 1

The occurrence of bacterial growth in autoclaved plastic containers filled with mashed potatoes and biotested using two methods, the spore test with spores of *Bacillus subtilis* and the immersion test with *Enterobacter aerogenes*. The microholes in containers both with intact seals and leaking seals made using tungsten thread of diameter 50 μm were measured with the Microhole tester.

Effective diameter of microhole (μm)	Number of contaminated / biotested packages in biotest with *B. subtilis*		biotest with *E. aerogenes*	
	Storage after biotesting		Storage after biotesting	
	6 days	20 days	6 days	20 days
0	6/25	2/22	1/26	0/25
1-25	6/8	3/6	2/4	2/7
26-50	5/5	2/5	1/3	6/7
51-100	8/9	12/12	3/6	6/7
101-150	2/2	1/1	2/2	1/1
151-200	1/1	nd	4/4	nd
>201	nd	nd	6/6	5/5
total	28/50	20/46	19/51	20/52

TABLE 2

As in Table 1 except that the microholes were made using tungsten thread of diameter 100 μm and the microholes in half of the samples were blocked with agar. The incubation time was 6 days.

Effective diameter of microhole (μm)	Number of contaminated / biotested packages in biotest with *B. subtilis*		biotest with *E. aerogenes*	
	Open microholes	Blocked microholes	Open microholes	Blocked microholes
0	1/2	nd	0/11	nd
1-25	9/14	nd	0/11	nd
26-50	1/9	4/4	0/2	nd
51-100	3/4	4/4	1/3	nd
101-200	6/6	7/7	1/1	nd
201-300	8/8	5/5	12/12	5/5
301-400	9/9	4/4	6/6	7/7
> 400	nd	1/1	6/6	13/13
total	37/52	25/25	26/52	25/25

REFERENCES

1. Ahvenainen, R., Mattila-Sandholm, T., Axelson, L. and Wirtanen, G., The effect of microhole size and foodstuff on the microbial integrity of aseptic plastic cups. Pack. Technol. Sci., 1992, **5**, 101-107.

2. Axelson, L., Cavlin, S. and Nordström, J., Aseptic integrity and microhole determination of packages by electrolytic conductance measurement. Pack. Technol. Sci., 1990, **3**, 141-162.

FOOD PACKAGING AND SHELF-LIFE

IVÅN VARSÅNYI
Canning Technology Department,
University of Horticulture and Food Industry,
Ménesi ut 45, H-1118 Budapest, Hungary

ABSTRACT

The results of our research call attention to the relationship between the shelf-life of packed food items and packaging. Mathematical models of deterioration allow prediction of the shelf-life of foods when the packaging and the circumstances of storage do not change. The packaging can protect the product from outside effects but the modification of consequences on the hygienic state and on the reactivity of the food matrix is limited by packaging techniques.

INTRODUCTION

One of the most important tasks of food packaging is to ensure the quality – nutritive and sensory values – and quantity of packed food items from production (harvesting) until consumption. Quality assurance is a very complex task because the foods are mainly bio-products and consequently they are unstable, 'living' and changing materials.

To inhibit or reduce the quality changing rate of packed food, with reference to the shelf-life declared, we have to choose the most appropriate packaging material and technique (e.g. modified atmosphere packaging). We investigated various products of the food industry to determine the main factors that change quality and the effect of packaging on shelf-life.

The selection of the most appropriate packaging material depends on the food matrix. Deterioration of a chemical nature may be caused by internal and external factors, for example oxygen molecule uptake, double bond changing, or molecule chain breaking.

Food deterioration of physical origin is mainly caused by temperature and by micro- and macroclimate differences between the package and storage room. This causes water adsorption or water desorption; rheological changing, or modification of physical state.

Very frequently food deterioration is of biological origin. The quality change is caused by multiplication of microorganisms, by toxin production or by enzymes or enzyme systems. It is important however, to emphasize that the activity of enzymes and microorganisms depends on the water activity of the food and the temperature of the storage room, which belong to the physical effects.

MATERIALS AND METHODS

The shelf-life is defined as the time during which the most rapidly changing (deteriorating) important property or properties of the food change until the well defined value limits of the characteristic property, which are described in the standards in given package under certain storage conditions, are exceeded.

A continuous mathematical model was set up to follow the quality change of foods and in this way to determine the shelf-life of packed products. Various physical, chemical and microbiological methods were used to select the most rapidly changing property(ies) of packed food. For that selection we used comprehensive mathematical statistics, and two-way variance analyses were applied to determine the effects of packaging, storage time, temperature, etc. Of course the homogeneity of variances was also investigated. The number of measurements was 3–5 parallel for 2–3 repetitions.

RESULTS

Analysing the reaction kinetics and mechanism of food deterioration, we found that they may be followed with five different mathematical models. The value of regression constant (a) is identical with the critical characteristic value measured at the start of storage and the regression coefficient (b) defines the rate of deterioration.

A first order polynomial (linear type function) described the rate of deterioration, where the changing of the critical property (y) is constant in the function of storage time (x). A second order polynomial (quadratic type function) is suitable for following the quality change of foods stored mainly between −10°C and +5°C. The rate of deterioration, described by exponential function, changes rapidly as a function of storage time. The origin of deterioration is chemical and/or biochemical. The sigmoid type deterioration is characteristic in about one third of processed foods. The quality generally changes as a combined effect of several factors (physical, chemical, microbiological). A relatively small group of food items change as a function of storage time using a hyperbolic function.

CONCLUSIONS

Quality protection of foods needs a suitable and marketable packaging. Quality protection of foods and the barrier properties of packaging have a tight correlation, with special regard to the structure of polymers. Therefore an organic part of our research programme was to investigate the correlations between the kinetics and mechanisms of food deterioration and the packaging quality. Surveying the results of our investigations it may be stated that the mechanism of deterioration and the packaging quality. Surveying the results of our investigations it may be stated that the mechanism of deterioration does not change significantly as a function of packaging but the rate of deterioration is variable. This means that we can modify the shelf-life of foods (reduce or extend) by packaging methods and by the quality of packaging materials and by the auxiliaries. The value of permeability constant (P) and the shelf-life of packed foods have a stochastic relationship which depends on the food matrix and the barrier properties of packaging.

MODIFIED ATMOSPHERE AND MODIFIED HUMIDITY PACKAGING TO EXTEND THE SHELF LIFE OF FRESH MUSHROOMS (Agaricus bisporus)

S. ROY, R. C. ANANTHESWARAN & R. B. BEELMAN
Department of Food Science
The Pennsylvania State University
University Park, PA 16802, U.S.A

ABSTRACT

Modified atmosphere packaging (MAP) increased the shelf life of fresh mushrooms up to 9 days at 12°C. Sorbitol, a food grade moisture absorber, was used to control the relative humidity within packages. Less sorbitol was required in MAP than in conventional packages. Optimum O_2 concentration and optimum relative humidity in MAP was found to be 6% and 90%, respectively.

INTRODUCTION

The shelf life of fresh mushrooms (Agaricus bisporus) is limited to 1-3 days at ambient temperature. Retardation of metabolic processes using modified atmosphere packaging (MAP) will increase the shelf life of mushrooms. The mushrooms stored in lower oxygen concentration was found to have better color, lower maturity index, and lower weight loss and disease incidence than those stored in higher oxygen concentration (1). Condensation of water in MAP was successfully avoided by using moisture absorbents to lower the In-package relative humidity (IPRH) of packages containing mature green tomatoes (2). In this present study, optimum O_2 concentration in MAP was determined and the effect of sorbitol as a moisture absorber on the shelf life of mushrooms in packages was evaluated.

MATERIALS AND METHODS

Mushrooms (Agaricus bisporus) were packaged in conventional packages (CP) and MAP. In CP, the mushrooms were placed in 600 ml polystyrene trays overwrapped with PVC film. The film was punctured with two 3 mm holes. In MAP, mushrooms were placed in 1000 ml trays. A 60-gauge polyethylene film (Cryovac Inc., Duncan, SC) was used to make a pouch. The polystyrene trays containing mushrooms were inserted into the pouch and heat sealed. The moisture absorbers were sealed in a Tyvak paper pouch which in turn were placed in the tray underneath the mushrooms. The packaged mushrooms were stored at 12°C chamber with 70% RH and were analyzed for quality after 3, 6 and 9 days of storage. The maturity index was determined according to a 7 point scale (1= least mature and 7= most mature) (3). Surface color of mushroom caps was measured using a Minolta Chroma Meter (Minolta CR-200). The surface moisture content of mushrooms were measured using a near infrared (NIR) spectroscope (NIRSystems model 6500) (4). The oxygen and carbon dioxide concentrations in the packages were monitored every day, for a period of 7 days, using a Gas Chromatograph (Hewlett Packard 5890 Series II). A Vaishala HMP 23 UT RH probe (Campbell Scientific, Logan, UT) was inserted in packages containing 5, 10 and 15 g of sorbitol and IPRH was monitored at 12°C for a period of 7 days.

RESULTS AND DISCUSSION

The oxygen concentration of MAP containing different amount of mushrooms are shown in figure 1a. The packages containing 100 and 120 g of mushrooms (which reached a steady state O_2 concentration of approximately 6% and 2% respectively) did not mature during storage ($p>0.1$) (figure 1b).

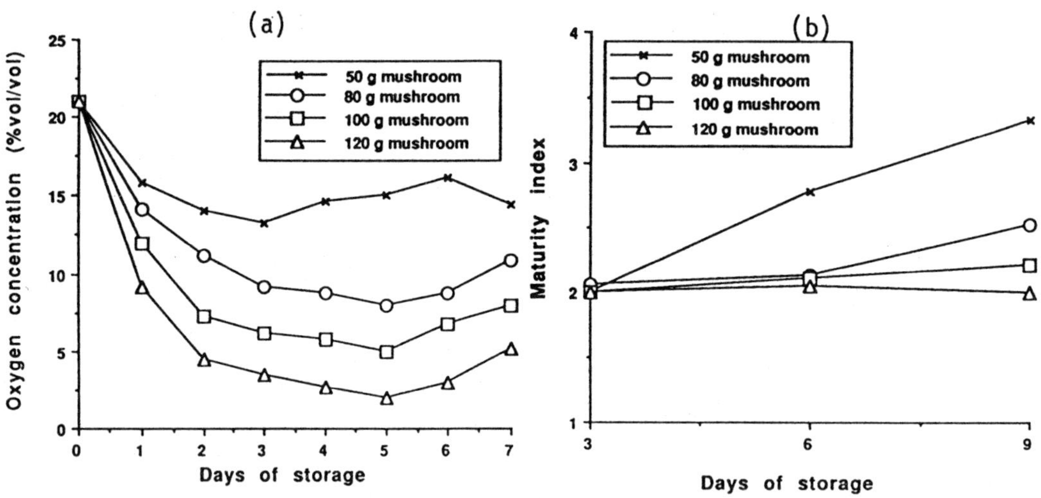

Figure 1. (a) O_2 concentration during MAP (b) Effect of days of storage (O_2 concentration) on maturity

Mushrooms in steady state O_2 concentrations of 15% and 10% had higher maturity after 6 and 9 days of storage, respectively. Mushrooms at 6% steady state O_2 concentration had the best color values at the end of 6 days of storage.

Mushrooms in CP containing 15 g of sorbitol had best color throughout the storage period. There were no differences in maturity index or color between mushrooms with or without sorbitol during MAP. Mushrooms packaged without sorbitol and with 5 g of sorbitol had higher surface moisture content after 9 days of storage than after 3 days of storage. The surface moisture content of mushrooms packaged with 10 and 15 g of sorbitol remained constant during storage. The IPRH with different amount of moisture absorbers is shown in figure 2a.
The packages containing 10 g of sorbitol had an IPRH of 88-90% which was therefore considered optimum to store mushrooms in MAP.

Shelf life of mushrooms in MAP was compared with that in CP, both with and without sorbitol in the packages. MAP mushrooms, both with and without sorbitol, had significantly higher

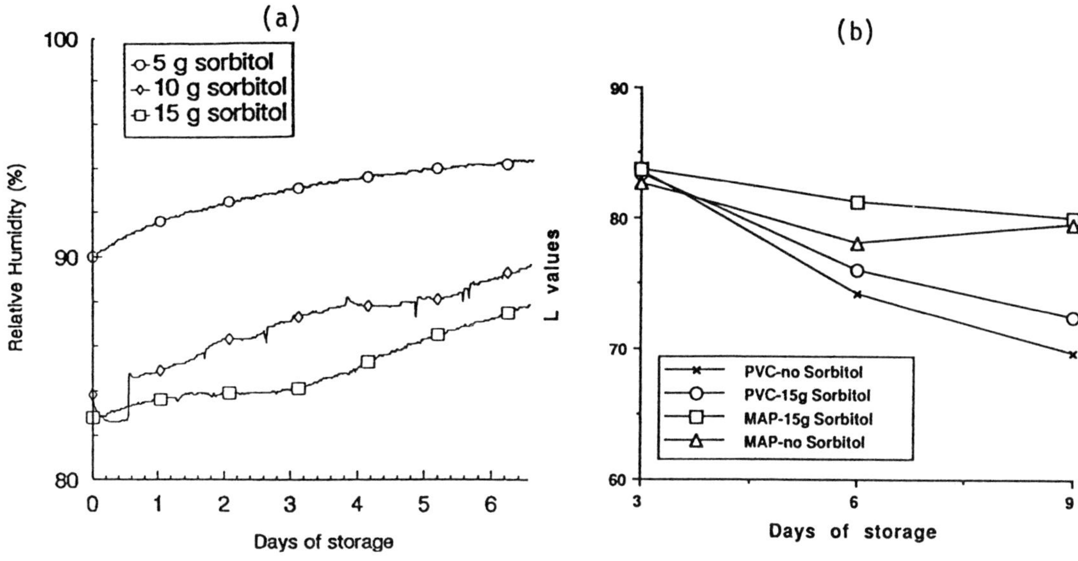

Figure 2(a) Effect of sorbitol on IPRH in MAP (b) Effect of packaging and sorbitol on L value

L values (figure 2b) than those in CP ($p<0.1$). After 6 days of storage, the maturity index of mushrooms in MAP was significantly lower than those in CP ($P<0.1$). MAP mushrooms with 15 g sorbitol had higher L values and significantly lower ΔE values than the rest of the treatments after 6 days of storage ($p<0.1$).

CONCLUSIONS

Modified atmosphere and modified humidity packaging improved the shelf life of mushrooms to 9 days at 12°C. The color of mushrooms were better when sorbitol was used in conventional PVC overwrap packages. Lowering the O_2 concentration to 6% in MAP mushrooms retarded the rate of maturation, but lowering the O_2 concentration to 2% increased the rate of browning. Less amount of sorbitol was necessary in MAP than in conventional packages.

REFERENCES

1. Burton, K. S. Frost, C. E., and Nichols, R. A combination of plastic permeable film system for controlling post-harvest mushroom quality. Biotechnology Letters. 1987. 9(8):529-534.

2. Shirazi, A. Modified Humidity Packaging of Fresh Produce. Ph.D. Dissertation. Michigan State University, East Lansing, MI. 1989.

3. Guthrie, B. D. Studies on the control of bacterial deterioration of fresh, washed mushrooms (Agaricus bisporus/brunescens). M.S. Thesis, The Pennsylvania State University, University park, PA. 1984.

4. Roy, S., Anantheswaran, R. C., Shenk, J. S., Westerhaus, M. O and Beelman, R. B. Determination of moisture content of mushrooms using VIS-NIR spectroscopy. J. Sci. Food Agric. 1993 (in press).

DESIGN OF PERFORATED POLYMERIC PACKAGES FOR THE MODIFIED ATMOSPHERE STORAGE OF BROCCOLI

JATAL D. MANNAPPERUMA AND R. PAUL SINGH
Department of Biological and Agricultural Engineering
University of California, Davis, CA, USA

ABSTRACT

The optimum modified atmospheres required for the extension of shelf life of fresh produce span a wide range of CO_2 and O_2 compositions. However, the atmospheres that can be established inside intact polymeric packages are limited by the narrow range of the permeability ratio of polymeric films. This limitation can be overcome by using perforations. A mathematical model to aid in the design of perforated polymeric packages was developed. The predictions of the model were verified by an experimental study using broccoli.

INTRODUCTION

The commercially available polymeric films span a wide range of gas permeabilities. However, the ratio of permeability of CO_2 and O_2 for most of the polymers fall within a narrow range around 3 to 6. This limits the atmospheric compositions obtainable through the use of intact polymeric packages also to a narrow range. This range covers the complete range of O_2 but does not allow CO_2 contents above 5%.

It is possible to establish high CO_2 - low O_2 atmospheres through the introduction of perforation in the polymeric packages. Although this possibility was known for sometime [4], package design methodology incorporating this concept were never fully developed. The objective of this study was to develop a method for the design of perforated polymeric packages to establish high CO_2 - low O_2 atmospheres.

Literature Review

Tomkins [4] investigated the effect of several design parameters on the atmosphere inside a polymeric package of fresh produce. These parameters included, weight of fruits, type of film, temperature and number of perforations. The perforations lowered the equilibrium CO_2 level and raised the O_2 level relative to intact packages. The drop in CO_2 level was approximately equal to the rise in O_2 level. Use of two films with widely different permeability ratios has been suggested [5]. A simple mathematical model based on steady state mass balances was used to evaluate package design parameters. Polyethylene and vegetable parchment films (permeability ratios, 3.9 and 0.9) were used to construct experimental packages for Macintosh apples with target package atmosphere 7.6% O_2 and 14.0% CO_2. The experiment verified the design through close agreement of the steady state package atmosphere [5]. A combination of relatively impermeable oriented polypropylene film and a microporous film has been used to package mushrooms [1]

The basic principles involved in the design of modified atmosphere packages have been previously illustrated by Mannapperuma and Singh [2] using a plot of the recommended modified atmospheric windows on a two dimensional chart with O_2 concentration as the abscissa and CO_2 concentration as the ordinate. A simple mathematical model was used in conjunction with this plot to illustrate the effect of permeability ratio, respiratory quotient, temperature and perforations on the package atmosphere. The present study is a logical extension of this work.

Mathematical Model

When fresh produce is stored in a package, the respiration activity lowers the O_2 concentration and elevates the CO_2 concentration in the package relative to the ambient atmosphere, establishing a concentration difference. This concentration difference creates a flux of oxygen into the package and a flux of carbon dioxide out of the package. In a properly designed package under steady state conditions, these two fluxes should be equal to the O_2 consumption and the CO_2 generation by the produce in the package, respectively. The design parameters of the package are determined using these equalities. An appropriate mathematical model was developed using perforations and fluxes of gases through the perforations. The solution of this mathematical model provided following solutions for the weight of fresh produce and area of perforations required to create the known optimum atmosphere in polymeric package using a film with known area and thickness.

$$W = \frac{P_{1f} A_f}{R_1 b_f} \frac{\beta_f - \beta_p}{\beta - \beta_p} (c_1 - x_1) \qquad 1$$

$$A_p = A_f \frac{P_{1f}}{P_{1p}} \frac{\beta_f - \beta_p}{\beta - \beta_p} \qquad 2$$

Where: A is Area of the film (m^2); β is permeability ratio (ratio of permeability of CO_2 to O_2, dimensionless); b is thickness of the film (μm); c is gas concentrations in ambient (% atm); P is permeability (ml-μm/m^2-h-atm); R is respiration rate (ml/kg-h);W is weight of produce in the package (kg); x is gas concentrations in package atmosphere (% atm); suffixes 1 and 2 denote oxygen and carbon dioxide; suffixes f and p denote film and perforations.

EXPERIMENTAL STUDY

The experimental study involved the design of a perforated package for broccoli. The recommended modified atmosphere for storage of broccoli [3] provided values for x_1, x_2 and β. $x_1 = 2.0\%$; $x_2 = 9.0\%$; and $\beta = 2.11$. The respiration rate of broccoli near optimum conditions from the literature was used to calculate values for R_1 and R_2 . $R_1 = 9.5$ ml/kg-h; and $R_2 = 9.5$ ml/kg-h

A low density polyethylene film of 100 μm thickness, and a package size of 200 mm x 200 mm was selected resulting in the values for A_f and b_f. $A_f = 0.08$ m^2; and $b_f = 100$ μm. The permeability of this film at 2.5 C was determined experimentally. $P_{1f} = 2120$ ml-μm/m^2-h-atm, $P_{2f} = 10470$ ml-μm/m^2-h-atm, and $\beta_f = 4.94$

Substitution of these values in equations 1 and 2 results in a design weight of 120 grams of broccoli and a perforation area of 6700 μm^2. The experimental study was planned based on these calculations.

A metal wire heated by discharging a capacitor was used to make the perforations with an average diameter of 75 µm (Area=4500 μm^2). Replicate packages of 120g of broccoli with 1, 2 and 3 holes were prepared and stored in an incubator controlled at 2-3 C temperature. The package atmosphere was monitored at weekly intervals until the steady conditions were reached.

RESULTS AND DISCUSSION

The equilibrium atmospheres inside packages with 1, 2 and 3 holes were as follows. 1 hole: 4500 μm^2, x_1=1.2%, x_2=5.6%; 2 holes: 9000 μm^2, x_1=3.3%, x_2=8.8%; 3 holes: 13500 μm^2; x_1=4.9%, x_2=8.2%

The atmospheres inside packages with 1 and 2 holes were close to the design target atmosphere of x_1=2.0% and x_2=9.0%. The deviation from the design values and the high variation among experimental values were attributed to non uniformity of perforations.

Improved techniques for making perforations should result in better agreement with the predictions by the model. The method described in this study can be used successfully to achieve optimum modified atmospheres of a wide range of fresh fruits and vegetables.

REFERENCES

1. Burton, K. F., Frost, C. E. and Nichols. R. 1987. A combination plastic permeable film system for controlling post-harvest mushroom quality. Biotechnology Letters, 9(8):529

2. Mannapperuma, J. D. and Singh, R. P. 1993. Modeling of gas exchange in polymeric packages of fresh fruits and vegetables. Technical Report, University of California, Davis, CA.

3. Saltveit, M. E. 1989. A summary of requirements and recommendations for the controlled and modified atmosphere storage of harvested vegetables. Fifth Proceedings of The International Controlled Atmosphere Research Conference. June 14-16, Wenatchee, WA.

4. Tomkins, R. G. 1962. The conditions produced in film packages by fresh fruits and vegetables and the effect of these conditions on storage life. Journal of Applied Bacteriology. 25(2):290.

5. Veeraju, P. and Karel, M. 1966. Controlling atmosphere in a fresh fruit package. Modern Packaging, 40(2);166.

PROTECTION AGAINST PERISHABLENESS – NEW PACKAGING MATERIALS AND PRINCIPALS OF PERMEATION MEASUREMENT

JOCHEN HERTLEIN and HORST WEISSER
Institute for Brewery Installations and Food Packaging Technology,
Technical University of Munich, 85350 Freising–Weihenstephan, Fed. Rep. of Germany

ABSTRACT

To ensure food quality and food safety, a large number of plastic barrier materials has been developed. Good barrier properties are reached by combining properties of different plastic films and aluminium. Recently a new generation of packaging material is available where a plastic layer is vacuum coated with a thin layer of either metal or silicium oxide.

The measurement of gas permeation is usually carried out under steady state conditions. The transient state, before equilibrium is reached, can last for more than 24 hours for high barrier films. Therefore a numerical model describing the transient state was used to predict steady state permeability. The duration of experiments can be shortened and more profound knowledge of the permeation process is won.

INTRODUCTION

One important property of plastics is their permeability to gases and vapors. Because many food products are sensitive to attack by oxygen and water vapor, a lot of work is done to improve the barrier properties of plastic films using different techniques: coextrusion of different plastics, lamination techniques, and vacuum deposition of metals and oxides. Especially the vacuum deposition gaines importance in food packaging technology as a replacement of aluminium-plastic laminates [1]. Beside the very good barrier properties, vacuum coated films have several advantages: They are decorative and the coating is exceedingly thin (around 50 nm) compared to a 7-12 μm aluminium layer in laminates. The base for the deposited material is PET, BOPP, OPA, PVC, PE, cellulose, and paper. Some metals that can be used for vacuum coating are aluminium, nickel, chromium, copper, their alloys, and oxides. Silicium oxide is used in different ratios of SiO and SiO_2 and is often denoted as SiO_x. Good barrier properties are achieved for x = 1.4-1.8. The wide range of possible polymer-metal combinations and the growing importance of vacuum coated films in food packaging technology require extensive permeation measurements. The feasibility of the differential pulse method in combination with the oxygen permeability test system Mocon Oxtran 100 (ASTM D 3985-81) for plastic films and barrier plastic films was examined to shorten the duration of permeation measurements.

PERMEATION MEASUREMENT

All test arrangements for permeation measurements can be classified in one of three large groups of measurement techniques: integral permeation methods, differential permeation methods, and sorption/desorption methods. Typical response curves with a transient and steady state portion are recorded. The permeability is usually determined from the steady state section of these response curves. One interesting alternative is the differential pulse method. Instead of measuring the steady state gas flow through a polymer membrane, the permeability is determined in a non-steady state experiment [2].

Differential Pulse Method

The most important difference between the traditional techniques and the differential pulse method is that instead of maintaining a defined partial pressure difference during the entire experiment, a rectangular partial pressure pulse is sent along one side of the sample. The pulse is so short that steady state permeation is not reached. Instead, a typical response curve (marked *difference* in fig. 1) is recorded. The permeation parameters (diffusion, solubility, and permeability coefficients) can be calculated by evaluating the peak height F_{max} and half width $t_{0.5}$. The response curve to the partial pressure pulse can be interpreted as the mathematical sum of two response curves. The first response curve is the result of a change in the partial pressure difference from p_1 to p_2, the second of a change from p_2 to p_1 ($p_2 > p_1$) (fig. 1). The latter one is a virtual curve needed to explain the resulting response curve. The time delay between both curves is identical to the pulse length ϑ. By introducing dimensionless quantities, the response signal Y can be written as:

$$Y = 2 \sum_{n=1}^{\infty} (-1)^n \left(exp(-n^2 \pi^2 t_q) - exp(-n^2 \pi^2 (t_q - \vartheta_q)) \right) \tag{1}$$

$$\text{with} \quad t_q = \frac{D\,t}{L^2} \quad \text{and} \quad \vartheta_q = \frac{D\,\vartheta}{L^2} \tag{2}$$

The maximum response signal Y_{max} and the half width $t_{q,0.5}$ can be calculated using equation (1). These theoretical values are compared with results from experiments (F_{max}, $t_{0.5}$, and ϑ). The permeation coefficient P and the diffusion coefficient D can be calculated with:

$$D = L^2 \frac{t_{q,0.5}}{t_{0.5}} \quad \text{and} \quad P = \frac{L}{A\,\Delta p\,\eta} \frac{F_{max}}{Y_{max}} \tag{3}$$

A denotes the sample area and η the sensor's sensitivity.

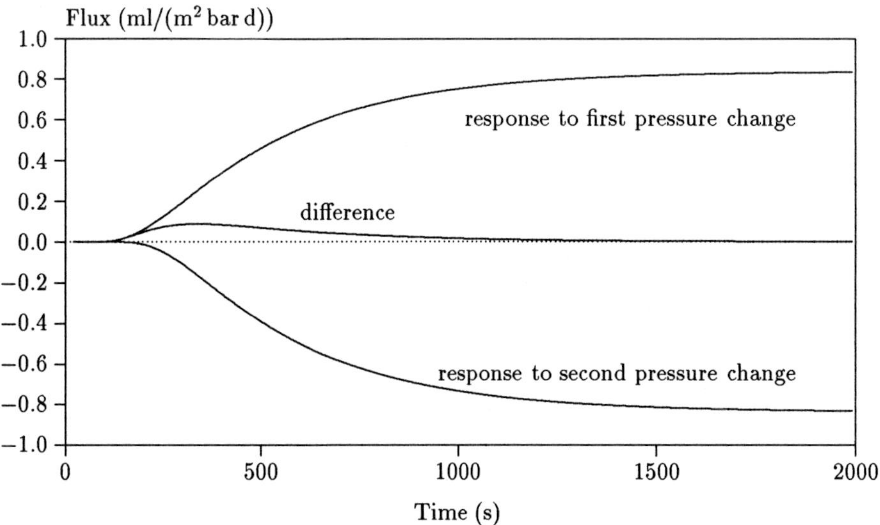

Figure 1. Response signal versus time for a PET_{met}/PE (12 μm/80 μm) sample with pulse time ϑ = 60 s, D = 2.53 10^{-12} m^2/s, P = 8.92 10^{-15} ml m/(m^2 Pa s).

RESULTS AND CONCLUSIONS

The most important postulate for the feasibility of the differential pulse method is, that a rectangular partial pressure pulse can built up at one side of the tested plastic film. It could be shown in experiments that the Mocon Oxtran 100 A fullfills this important requirement when the pulse lasts for 60 s and more. The maximum oxygen concentrations measured at the test chamber entry were 85 to 90% for shorter pulses and the curve was bell-shaped. When the measured pulse area was evaluated it showed that the area was larger than that for the ideal pulse by 14 s · bar, regardless of the pulse duration. This means that 14 s had to be added to all pulses. Other time delays such as time delays in tubes or the response time of the Sensor did not influence the response curves. Response curves of films with a permeability $< 0.5\,\mathrm{ml/(m^2\,d\,bar)}$ were too flat for interpretation. Table 1 lists the permeation, diffusion, and solubility coefficients for several plastic films.

Table 1 shows good agreement for the permeability coefficients determined with the pulse method and under steady state conditions for the monolayer films. This difference is larger for the barrier films due to the flatness of the response curve and difficulties in evaluating peak height and half width. The transport parameters D and P could not be calculated for the silicium oxide coated PET sample. Still, for a lot of applications the calculated parameters are precise enough and helpful – especially when a large number of samples has to be compared. The results shown here were all determined with a 60 s pulse. Longer pulses did not influence the calculated parameters significantly, shorter pulses should not be used because of the above mentioned behavior of the test equipment.

The Mocon Oxtran 100 can be used to determine the transport parameters and the oxygen flux through barriers with the differential pulse method. The time needed to perform traditional permeation experiments can be reduced at least by a factor of two.

TABLE 1
Values of D, S, and P for different monolayers and vacuum coated films calculated for a 74 s pulse. Test conditions: 23°C and 50% r. h., 10 samples.

	pulse method			steady state
sample	D $\left(\frac{m^2}{s}\right)$	S $\left(\frac{ml}{m^3\,Pa}\right)$	P $\left(\frac{ml\,m}{m^2\,s\,Pa}\right)$	P $\left(\frac{ml\,m}{m^2\,s\,Pa}\right)$
PVC	$(4.5 \pm 0.1)\ 10^{-13}$	$(8.1 \pm 0.5)\ 10^{-1}$	$(3.6 \pm 0.3)\ 10^{-13}$	$2.5\ 10^{-13}$
PA 6	$(1.1 \pm 0.1)\ 10^{-11}$	$(7.7 \pm 0.7)\ 10^{-2}$	$(8.7 \pm 0.5)\ 10^{-13}$	$6.8\ 10^{-13}$
OPP	$(9.5 \pm 0.3)\ 10^{-13}$	(2.4 ± 0.2)	$(2.3 \pm 0.1)\ 10^{-12}$	$2.6\ 10^{-12}$
OPA_{met}/PE	$(2.6 \pm 0.6)\ 10^{-12}$	$(3.0 \pm 2.0)\ 10^{-3}$	$(7.0 \pm 3.5)\ 10^{-15}$	$2.2\ 10^{-14}$
PET_{met}/PE	$(2.5 \pm 0.5)\ 10^{-12}$	$(3.9 \pm 2.4)\ 10^{-3}$	$(8.9 \pm 4.2)\ 10^{-15}$	$1.1\ 10^{-14}$
PET_{SiO_x}/PE	-	-	-	$5.1\ 10^{-15}$

REFERENCES

1. Jamieson, E. H. H. and Windle, A. H., Structure and oxygen-barrier properties of metallized polymer film. *Journal of Materials Science,* 1983, **18**, 64-80. *Die Verpackung,* 1992, **33**, No. 5, 197-200.
2. Pálmai, G. and Oláh, K., New differential permeation rate method for determination of membrane transport parameters of gases. *Journal of Membrane Science,* 1984, **21**, No. 2, 161-83.

EDIBLE WHEAT GLUTEN FILMS: OPTIMIZATION OF THE MAIN PROCESS VARIABLES AND IMPROVEMENT OF WATER VAPOR BARRIER PROPERTIES BY COMBINING GLUTEN PROTEINS WITH LIPIDS

Nathalie GONTARD, Stephane GUILBERT, Sylvie MARCHESSEAU and Jean-Louis CUQ
Laboratory of Génie et Technologie Agro-Alimentaire, CIRAD-SAR, BP 5035, 73 rue J. F. Breton, 34032 Montpellier, France.
and LGBSA, Université de Montpellier II, place E. Bataillon, 34095 Montpellier, France.

ABSTRACT

An edible wheat gluten film was developed and relationships between film-formation conditions and properties were studied using Response Surface Methodology. pH and ethanol concentration of film-forming solution had strong interactive effects on film water solubility and moisture permeability. Mechanical properties were affected by gluten concentration and pH. Various formulations and methods of fabricating films consisting of gluten and lipids were investigated.

INTRODUCTION

Edible films and coatings have been used to protect pharmaceuticals, improve shelf-life and properties of food products (3, 4). Proteins, especially wheat gluten, as edible film-forming agents have been studied less extensively than lipids or polysaccharides. The objective of the presented work was to develop an edible wheat gluten film, to gain a better understanding of relationships between gluten film-formation conditions and film properties, and to improve moisture resistance by combining wheat glutenwith lipid materials.

MATERIALS AND METHODS

The film was prepared from a film-forming solution which was obtained by dispersing gluten proteins in ethanol, acetic acid (to adjust the pH) and water. Glycerol (20 % p/p gluten) was added as plasticizer (2). The film-forming solution was used for casting film and allowed to dry. Puncture test, opacity and water vapor permeability measurements (standard procedures) were made on films with a controlled constant thickness (0,05mm) and equilibrated at 56% relative humidity at 25°C (1).

RESULTS AND DISCUSSION

Response surface methodology was used to determine the influence of some film-formation conditions on film properties. From the statistical results (1), pH and ethanol concentration (ET) of the film-forming solution had strong interactive effects on film opacity, water solubility and water vapor permeability (WVP). Mechanical properties seemed to be particularly affected by the concentration of gluten (GL) and pH.

Figure 1 shows that high GL (12,5%) and pH above 5 induced high puncture strength which involve a high number and/or a better localization of bonds between proteins chains. During the drying of the film solution, ethanol and acetic acid, which are responsible for proteins dispersion, were first evaporated, allowing the formation of bonds between protein chains. During this stage, the proximity of protein chains induced by high GL could facilitate and improve the formation of such cross-bond.

The film heterogeneity induced by high ET led to complete and rapid disintegration and dispersion in water (figure 2). All of the effects and interactions resulted in the lowest film solubilities with a minimum value of about 40%, with ET and pH simultaneously varying between ET= 40% - pH 2 and ET= 20% - pH 5. Such conditions allowed sufficient unfolding of the gluten protein chains in the solution and resulted in a water resistant film.

The shape of the WVP response surface (figure 3) is characteristic of the strong interaction between ET and pH. At high ET, heterogeneity of the film could explain the ease of water transport. Films formed under low pH and low ET conditions had very high WVP. Low pH was certainly responsible for the unfolding of proteins with exposure of hydrophilic residues on the protein surface. The individual negative effects of ethanol and acetic acid on WVP appeared to be limited by the simultaneous variation of these two factors (diagonal). The effect of hydrophilic groups at low pH seemed to be counterbalanced by the effect of ethanol, thus improving the exposure of less polar groups,consequently limiting the WVP increase.

Rather than determining the optimum for all responses, it would be preferable to choose particular film-formation combinations based on specific uses of film.

To improve the water vapor barrier properties of wheat gluten film, gluten proteins have been combined with lipids materials. Edible composite films constituting of wheat gluten as structural matrix and lipids as moisture barrier were developped using emulsion technique. The effect of lipids on the physical properties of composite films depended on the lipids characteristics and on the interactions between the lipid and the protein structural matrix.

Beeswax was the most effective lipid for decreasing WVP but these films were opaque, weak and disintegrated easily in water.

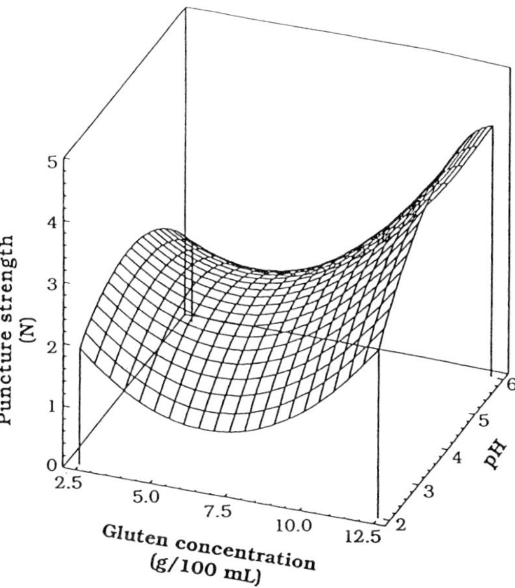

Figure 1. Response surface for the effect of gluten concentration and pH of the film-forming solution on film puncture strength at a constant ethanol concentration of 45 mL/100 mL.

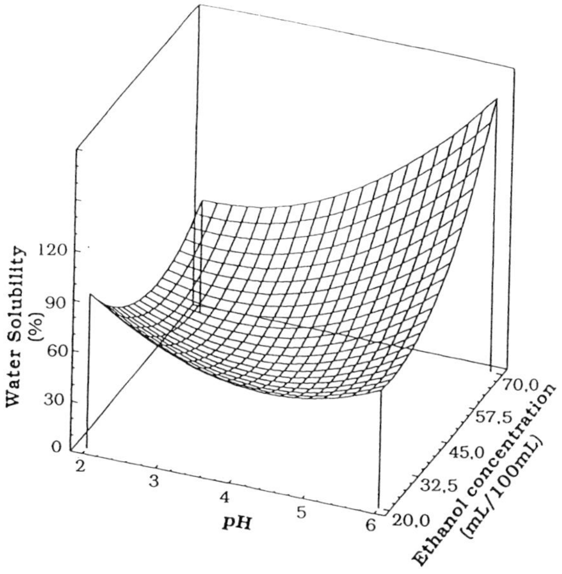

Figure 2. Response surface for the effect of pH and ethanol concentration of the film-forming solution on film water solubility at a constant gluten concentration of 7.5 g/100mL.

Combining wheat gluten proteins with a diacetyl tartaric ester of monoglycerides increased resistance and maintained transparency. Using the coating technique, solid lipids such as beeswax or paraffin wax, deposited in a molten state onto the gluten base film, were effective water vapor barriers. A film consisting of wheat gluten, glycerol and diacetyl tartaric ester of monoglyceride as one layer, and beeswax as the other yielded a WVP of 0,0048 g.mm/m2.mmHg.24h. (table 1) which was less than that obtained with low density polyethylene.

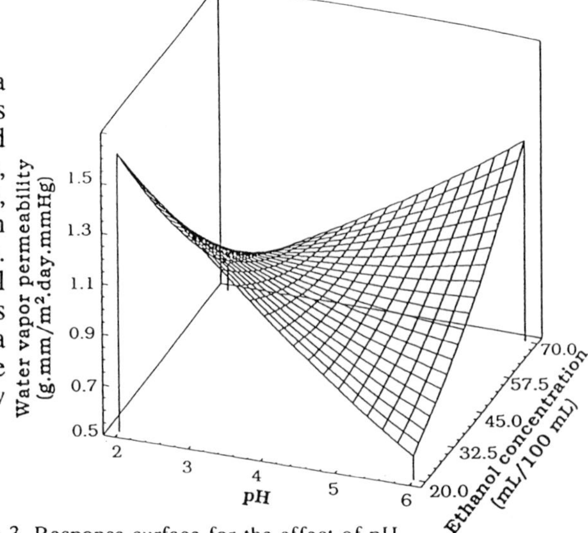

Figure 3. Response surface for the effect of pH and ethanol concentration of the film-forming solution on film water vapor permeability at a constant gluten concentration of 7.5 g/100mL.

TABLE 1

Water vapor permeability of edible and non edible films

Film	Temperature (°C)	Δp (mmHg)	Thickness (mm)	Water vapor permeability (g.mm/ m².mmHg.24hr.)
Starch	25	19.2-7.3	0.036	29.3
Casein-gelatin	30	28.9-18.5	0.25	7.1
Gluten, Glycerol (16.6%)	30	32.2-0	0.05	1.05
Gluten, DATEM (20%), Glycerol (13%)	30	32.2-0	0.05	0.55
Plain cellophane	37.8	46.7	-	0.83-0.166
Polyethylene (low density)	37.7	44.3-0	0.025	0.010
C16-C18 MC/HPMC, Beeswax (4mg/cm²)	25	23,0-0	0.056	0.0076
Gluten, Beeswax (5.3 mg/cm²)	30	32.2-0	0.09	0.0048
Waxed paper	37.8	46.7-0	-	0.0016-0.125
Aluminium foil	37.7	44.3-0	0.025	0.00006

CONCLUSION

The feasability of using wheat gluten as a film-forming agent appeared promising considering the overall film properties and the possible improvement of this properties, especially moisture barrier property, through introduction of lipids materials.

REFERENCES

1. Gontard N., Guilbert S., Cuq J.L., Edible wheat gluten films: influence of the main process variables on film properties using response surface methodology. ,1992, J. Food Sci., 57 (1), 190-195.

2. Gontard N., Guilbert S., Cuq J.L. . Water and glycerol as plasticizers affect mechanical and water vapor barrier properties of an edible wheat gluten film., J. Food Sci., 1993, 58 (1), 206-211.

3. Guilbert S. . Technology and application of edible protective films. In: Food packaging and preservation, 1986, Edr. M. Mathlouthi, Elsevier Applied Science Publishers, New York, 371-394.

4. Kester J.J., Fennema O.R. Edible films and coatings: a review., 1986, Food Technol., 40, (12), 47-59.

SELECTION OF LAMINATED FILM FOR A FOOD PACKAGING

MITSUYA SHIMODA AND YUTAKA OSAJIMA
Department of Food Science and Technology, Faculty of Agriculture, Kyushu University, Hakozaki, Higashi-ku, Fukuoka 812, Japan

ABSTRACT

A method of selecting a laminated film for pouch was proposed by showing the influence of oxygen permeability [a:cm^3/cm^2 day atm], storage period [day], the ratio of surface area [S:cm^2] to volume [V:cm^3] of pouch. In order to correlate flavor deterioration to the amount of oxygen permeated, the following equation was introduced; α=p a T S/V, where p is a partial pressure of oxygen in atmosphere and α [dimensionless] was defined as a specific volume of oxygen permeated. As a result, the flavor of packaged soup significantly deteriorated above α=0.0034.

INTRODUCTION

Flavor deterioration of food packaged in a pouch can be attributed to sorption of flavor compounds into a liner 1,2), and an oxidation induced by oxygen permeated 3). Flavor compounds in Japanese style soup packaged in a plastic pouch lined with linear low density polyethylene (LLDPE) film were not sorbed into the liner, but the flavor of soup in the pouch made of nylon(15μm)/LLDPE(60μm) film deteriorated during storage. However, the flavor scarcely deteriorated in the pouch laminated with aluminum foil. The influence of the oxygen on flavor deterioration was investigated quantitatively using pouches made of films with different oxygen permeabilities.

MATERIALS AND METHODS

Table 1 shows six kinds of laminated films. The oxygen permeability was ranged from 0.3 to 98 x 10^{-4} [cm^3/cm^2 day atm], and the film laminated with aluminum foil was used as a control pouch. The liner of every pouch was LLDPE film. The soup, which was manufactured from soy sauce, dried bonito, and seaweed by Ichiban Food Co.,LTD, was packaged in the pouches (9x15cm, 200ml) made of the films in Table 1. The sterilization was done by heating at 90℃ for 20 min after packaging and the samples were stored under dark at 25 ℃. Preparation of odor concentrate from the soup was carried out by adsorptive method using the column packed with Porapack Q. Sample was passed through the column (2x10cm) and adsorbed volatile compounds were eluted with 50ml of diethylether after the column was washed with 50ml of water. Cyclohexanol as an internal standard was added to the ether solution, and then ether was evaporated under a nitrogen stream. A Shimadzu GC-14A gas chromatograph equipped with a 60m x 0.25mm i.d. DB-WAX column was used. The oven temperature was programmed at 60 to 230 C at 3 ℃/min. For identification of GC components, a Shimadzu GCMS-9020DF gas chromatograph-mass spectrometer was used. Sensory test on the soup flavor was done with the samples stored for 3 weeks at 25 ℃. Sample 1 (pouch 1) and sample 6 were presented to panelists. The remaining 4 samples were arranged in order of the flavor deterioration.

Table 1 Laminated films and oxygen permeabilities

Film contruction 1)	Oxygen permeability 2)
1. PET^{12}/LLDPE^{60}	98 x 10^{-4}
2. NY^{15}/LLDPE^{60}	28 x 10^{-4}
3. KNY^{15}/LLDPE^{60}	4.5 x 10^{-4}
4. *KNY^{15}/LLDPE^{60}	1.5 x 10^{-4}
5. EVOH^{15}/LLDPE^{60}	0.3 x 10^{-4}
6. NY^{15}/AL^{7}/LLDPE^{60}	0

1) PET;polyethylene terephthalate (thickness, 12 μm), NY;nylon, KNY;nylon coated with vinylidene chloride, *KNY;nylon coated with high barrier vinylidene chloride, EVOH;ethylene vinylalcohol copolymer, AL;aluminum foil, LLDPE;linear low density polyethylene
2) [cm^3/cm^2 day atm]

RESULTS

Forty seven volatile compounds in the soup were determined and 36 of them were identified. The similarity of GC pattern before and after storage was used as an index of the flavor deterioration, since the flavor deterioration could not be attributed to any specific components. Fig. 1 showed sample 6 was maintained higher than 0.98 even after 4 weeks. On the other hand sample 1 decreased significantly after a week. The decreasing rate in the similarity of other samples were compatible with their oxygen permeabilities. As shown in Table 2, 5 persons of the panelists arranged samples 2 and 3 correctly, while samples 4 and 5 were confused with each other. This showed that flavor deterioration progressed with an

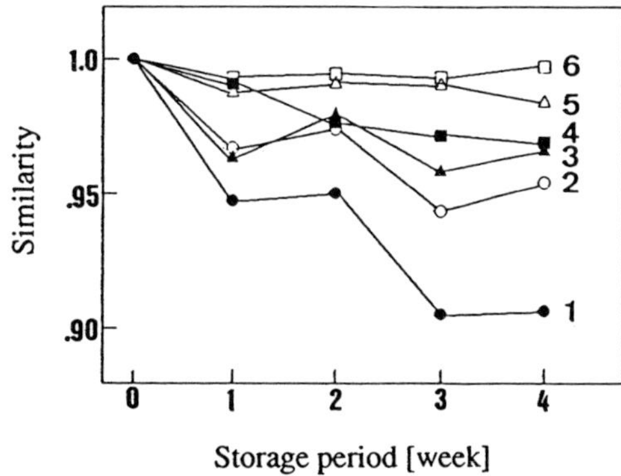

Storage period [week]

Fig. 1 Change in the simirality of gas chromatogram during storage
Symbols refer to the samples packaged in the pouch (1-6) in Table 1.

Table 2 Results of sensory test

Panelist	Sample					
P1	1	2	3	5	4	6
P2	1	2	3	5	4	6
P3	1	2	3	4	5	6
P4	1	2	3	4	5	6
P5	1	2	3	4	5	6
P6	1	2	5	3	4	6
P7	1	2	4	5	3	6

1)Samples 1 and 6 were presented as a pouch of maximum or minimum oxygen permeability and samples 2-5 were arranged between 1 and 6.

increase in oxygen permeability in the case of oxygen permeability being higher than 4.5×10^{-4} [cm^3/cm^2 day atm], but was not appreciable with the permeability lower than 1.5×10^{-4} [cm^3/cm^2 day atm]. Every panelist mentioned that sample 6 preserved well its flavor, and sample 1 smelled stable flavor. As a result, the soups in the pouches with oxygen permeability higher than 4.5×10^{-4} [cm^3/cm^2 day atm] deteriorated its flavor within 3 weeks.

DISCUSSION

In order to select the most appropriate film for the flavor of Japanese style soup packaged in a pouch, the following consideration was given with variables of oxygen permeability [a:cm^3/cm^2 day atm], storage period [T:day], pouch volume [V:cm^3], pouch surface area [S:cm^2]. α=p a T S/V, where α [cm^3/cm^3; dimensionless] was defined as a specific volume of oxygen permeated, and p is a partial pressure of atmosphere. Fig. 2 shows the plots of α *vs.* TS/V with the pouch 1 - 6.

The limiting α value was estimated from p=0.2, $1.5 \times 10^{-4} \leqq a \leqq 4.5 \times 10^{-4}$, T=21 days, S=270 cm^2, and V=200 ml, and shown as a region hatched. Fig. 2 shows the flavor deteriorates within 4 days in pouch 2, *etc.*, and it enables to select a most appropriate film.

Above consideration was given under the following assumptions: 1) flavor deterioration is attributed to oxygen permeated, 2) rate of reactions including oxygen molecule is much faster than the rate of oxygen permeation.

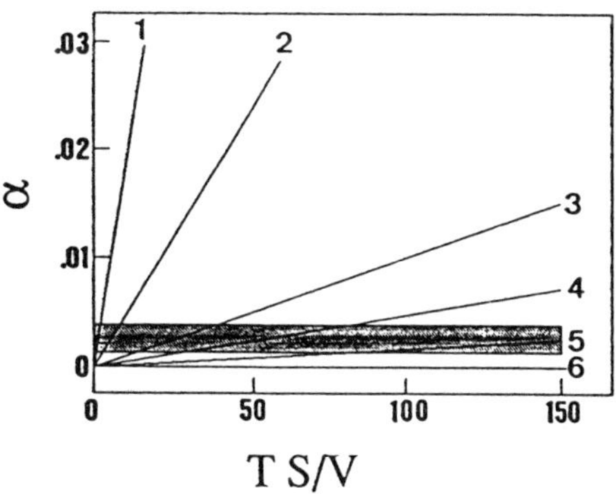

Fig. 2 Relationship between flavor deterioration and a specific volume of oxygen permeated. α=p a T S/V where α is a specific volume of oxygen permeated [dimensionless], p;partial pressure of oxygen in atmosphere, a;oxygen permeability [cm^3/cm^2 day atm], T;storage period [day], S;surface area of pouch [cm^2], V;volume of pouch [cm^3].

REFERENCES

1. M. Shimoda, T. Ikegami and Y. Osajima, Sorption of flavor compounds in aqueous solution into polyethylene film. J. Sci. Food Agric., 1988, **42**, 157-163.

2. T. Ikegami, K. Nagashima, M. Shimoda, Y. Tanaka, and Y. Osajima, Sorption of volatile compounds in aqueous solution by ethylene-vinyl alcohol copolymer films. J. Food Sci., 1991, **56**, 500-503.

3. J. R. Mclellan, L. R. Lind, and R. W. Kime, A shelflife evaluation of an oriented Polyethylene Terephthalate package for use with hot filled apple juice. J. Food Sci., 1987, **52**, 365- 369.

CONTINUOUS MICROWAVE DRYING OF POLYETHYLENE TEREPHTHALATE (PET)

C.A.R. ANJOS, J.A.F. FARIA, and A. MARSAIOLI, JR.
Dept. of Food Engineering (DEA), Faculty of Food Eng. (FEA), UNICAMP
C.P.6121 - 13081-970, Campinas, SP, BRAZIL

ABSTRACT

Microwave heating techniques have been applied to the drying of polyethylene terephthalate (PET) resin, aiming to improve the critical operational conditions prevailing in the conventional air drying of that polymer and which may affect the final quality of molded beverage bottles if not carefully controlled. Moisture adversely affects intrinsic viscosity of PET, a high value of it being required to produce quality preforms. PET is very hygroscopic, which makes drying a mandatory operation prior to the production of preforms. Considering the critical variables dew point, air temperature, air volume and product holding time, it became evident from this study that the last two might be substantially reduced. That would cause certain advantages to be gained as concerns to the speed of drying, quality control, and the overall capitalized cost, which may render microwave heating feasible, in spite of the high cost of microwave equipment per unit of capacity.

INTRODUCTION

Polyethylene Terephthalate (PET) is a polyester resin that has been applied to food packaging, particularly into soft drinks bottles by way of the injection-stretch-blow-molded process. PET as a raw material in the solid form absorbs moisture from the atmosphere, i.e. it acts like a desiccant. Moisture is absorbed during storage to a value as high as 0.6 % (w/w). In practice moisture level of commercial PET is likely to range from 0.1 to 0.3 % after storage under cover for relatively short periods. Prior to injection molding of preforms, PET has to be carefully controlled dried to less than 0.004 % (40 ppm), as an essential prerequisite to attain maximum product performance in the final processing. The conventional drying method for PET has been one of using hot air with desiccation. This can be done on either a batch or continuous basis under a normal holding time of 4 to 6 hours, carried out in the range of 170-180°C. Similarly, many other industrial plastics have to be dried to some acceptable level of residual water content, usually less than 1.0 % (w/w), prior to injection molding, using hot air with desiccation as well. All these methods have in common the use of relatively high temperatures, ranging from 80 to 120^{0}C depending on the type of plastic, with a normal dwell time of 4 to 5 hours [1]. This time can be significantly reduced by introducing a microwave/RF drying cycle of from 10 to 15 minutes before the hot air cycle, rendering the total drying time now reduced to 30-60 minutes [2]. It was thought in this study that the application of microwave heating techniques to PET could also be less rigorous than in the conventional heating, bringing the benefit of time or energy saving as well as increasing product quality. As information on this type of process has been scarce as far as the industrial applications are concerned, the reported preliminary experiments were conducted either in a laboratory oven or in a pilot plant, by various exposures to dielectric heating at a frequency of 2.45 GHz, taking advantage of one already existing continuous microwave drying installation at the Dept. of Food Engineering of State University of Campinas in the area of microwave processing [3].

MATERIALS AND METHODS

Commercial PET from ICI (Melinar) was used as raw material for the experiments, with an average water content of 3500 ppm, a little higher than the originally delivered product due to a long period of storage. Microwave heating on a laboratory scale was carried out at 2.45 GHz in a modified domestic oven (40 liters x 850 watts). Due to the lack of a continuous microwave power control, a procedure was adopted based on an empirical model developed by Mudgett [4] applied to the energy coupled inside the oven cavity as a function of load volume : a spirally wound flexible polyethylene tube was introduced into the oven, provided with a suitable hole through which the tube ends crossed the top wall and were outside connected to the tap water and drain pipe. Water volume was controlled by the length of the spiralled tube, whereas the flow was monitored by a precision flowmeter of scale 30-245 liters/hour. Water inlet and outlet temperatures were also monitored by the use of suitable thermocouples and a digital indicator. For each test, 50 g PET samples were weighed into a single spiralled glass tube, which was introduced into the center of the oven cavity, supported by a special plastic round tripod and having its extremities also crossing the top oven wall. A double stage system for filtering and desiccating air, resulting from a combination of a compressed air dryer by refrigeration in series with a molecular sieve, was used to supply cleaned air at a -30 to -35^{0}C dew point, which was measured by means of a precision flowmeter of scale 40-350 liters/hour and made to flow inside the glass tube packed with the PET sample. Air moisture was controlled by the use of an automatic Shaw dew point meter, provided with a precision analog scale from 0 to -80^{0}C. As an alternative for certain tests, an electrical heater was also used to heat the air before passing through the PET sample. The temperature of the PET pellets was recorded as an average taken by inserting a thermocouple midway inside the glass tube immediately after the microwave treatment. For tests which approximate industrial conditions, a novel microwave rotary dryer as described in [3] was adapted through the removal of the original blower which was substituted by the same double stage system to generate the filtered and desiccated air as applied to the laboratory experiments, but operating at a flow rate in the range of from 4000 to 8500 liters/hour. After adjusting the desired PET mass flow rate at an average 5.5 kg/hour, experiments were carried out to establish the residence time distribution of the pellets inside the rotating cavity, as a function of rotation speed and inclination of the drum. The moisture contents of PET samples for both laboratory and pilot plant tests were determined in a Aquastar moisture titrator.

RESULTS AND CONCLUSIONS

Tests were first run in the laboratory oven in order to establish the best combination of energy level and exposure time for the moisture removal of samples. Processing times ranged from 30 to 60 minutes. Some difficulties arose because of lack of homogeneity of the electrical field distribution, causing deviations of the samples temperatures as great as 10^{0}C. Also energy level evaluations based on calorimetric measurements produced errors on the order of 20-30 %. Even so samples were partially dried, ranging from 21.9 % moisture removal for 230 W in 30 min. up to 38.6 % removal for 700 W in 40 min. The pilot scale treatments were carried out under the conditions : 4250 liters/hour of air at -35^{0}C dew point, with and without preheating, energy levels from 120 to 360 Wh/kg and average PET mass flow rate of 5.5 kg/hour. Under the maximum energy level moisture removal was 81.1 % of total water content, not sufficient yet in relation to the recommended moisture of less than 40 ppm. From the observed results it can be concluded that the microwave treatment of PET has a good potential as compared to conventional drying method, even though a considerable improvement of the techniques related to the energy modulation of microwave is still necessary in order to have a better control of the new process.

REFERENCES

1. Hasan, M. and Mujumdar, A.S., Drying of Polymers, in Handbook of Industrial Drying. Edited by Arun S. Mujumdar, Marcel Dekker, Inc., N.York, 1987.

2. Lightsey, G.R. and Russell, L.D., TVA Report (Low Temperature Processing of Plastics Utilizing Microwave Heating Techniques) November, 1985, (Unpublished), cited in R.W. Bruce and M.W.Mccurdy, Dielectric Measurements of Particulate Organic Polymers, in Symposium Proceedings of the Materials Research Society : Microwave Processing of Materials, volume 124, Pittsburgh, PA, 1988.

3. Marsaioli Jr., A., Conforti, E. and Kieckbush, T.G., A Prototype of a Combined Hot Air and Microwave Rotary Cylindrical Oven for Continuous Drying of Granular Products, Proceedings of the Fifth International Congress on Engineering and Food, vol. 2, edited by W.E.L. Spiess and H. Shubert, Elsevier Applied Science, 1990.

4. Mudgett, R.E., Electrical Properties of Foods, in Properties of Foods, edited by M.A. Rao and S.S.H. Rizvi, Marcel Dekker, New York, 1986, pp. 329-390.

STERILIZATION OF PACKAGING CONTAINERS BY GASIFIED HYDROGEN PEROXIDE AND THE APPLICATION TO FLEXIBLE ASEPTIC PACKAGING MACHINE

Y.SHIBAUCHI, T.TANAKA, K.HATANAKA
Technical Research Institute, Snow Brand Milk Products Co.,Ltd.
Minami-dai 1-1-2, Kawagoe, Saitama 350, Japan

ABSTRACT

Gasified hydrogen peroxide (H_2O_2) was studied to sterilize preformed packaging containers in the combination of irradiation of ultraviolet rays (UV), and it was successfully applied to a flexible aseptic packaging machine that is applicable to a variety of preformed containers with different shapes and sizes. An effective and efficient H_2O_2 vaporizer was developed based on the study of its vaporization and decomposition characteristics.

The sterilization test with inoculated containers showed a synergistic effect of the H_2O_2 gas and the subsequent UV-rays treatment, inactivating more than 10^6 spores of *B.subtilis* and *B.stearothermophilus* per a 100ml-container. Residual H_2O_2 of the containers after drying treatment by heated sterile air was confirmed to be low enough, less than 0.1ppm under the drying condition of 80°C for five seconds.

INTRODUCTION

The liquid H_2O_2 has been studied as means of sterilization of packaging materials such as plastic films and sheets since it has a strong inactivation effect for most microorganisms[1),2),3),4),5),6)]. However the application to preformed containers has been limited because of the difficulty in reducing residual H_2O_2 on the surface of containers after the sterilization treatment. Usually the more complex in the shape of containers are, the more serious the difficulty in reducing the H_2O_2 residue. Accordingly, in our development of the versatile aseptic packaging machine, an effective sterilization method applicable to the containers with complicated surface shapes was required. In addition, since the softening temperature of common plastics used for food containers such as polyethylene or polystyrene is about 80 to 90°C, it should be noted that the higher temperature beyond that cannot be applied to the containers.

EXPERIMENTAL RESULTS AND DISCUSSION

In advance of the development of H_2O_2 vaporizer, vaporizing characteristics were studied. The H_2O_2 droplet of about 0.01g was gently put on the surface of a heated stainless steel plate to measure the vaporizing time. Fig.1 shows the vaporizing time in comparison with water. In case of water, the vaporizing time decreased as the plate temperature was increased and came to the minimum point at about 140°C, then increased again due to the so-called spheroidal phenomenon that declines heat transfer from the plate to the droplet because of a vapor film formed between them.

The H_2O_2 droplet showed, however, a bit different trend. After reaching the minimum point at about 170°C, the vaporizing time did not increase so much because it became unstable due to oxygen bubbles formed in the droplet by the thermal decomposition of H_2O_2. An additional test showed that the decomposition rate became smallest when the vaporizing time was shortest.

Fig.2 shows the residual H_2O_2 characteristics per 100ml-container when the drying air temperature was changed. The drying time was fixed to five seconds for all measurements. The amount of residual H_2O_2 decreased exponentially as the drying temperature increased. The figure indicates that the drying treatment at about 80°C may reduce the residual H_2O_2 to 0.1ppm or lower.

Tab.1 shows a part of results of the sterilization experiments using H_2O_2 gas, UV-rays and their combination for the 100ml-containers inoculated with spores of *A.niger, B.subtilis* and *B.stearothermophilus.* The concentration of liquid H_2O_2 submitted to gasification was varied from 3% to 35%, and the intensity of applied UV-rays was about 160mW·s/cm^2 for each container. Independent use of H_2O_2 gas or UV-rays resulted in some residual spores, but their combination sterilized all samples completely even when 3%-H_2O_2 was used.

Fig.3 shows the schematic picture of the developed H_2O_2 vaporizer. The H_2O_2 gas is generated in the heated vaporizing pocket with the help of the heated air, and passes through the entrainment strainer, in turn get to the gas-outlet after the temperature being raised properly.

Fig.4 shows the preformed container sterilizer comprising the H_2O_2 gas generator, a UV-lamp and an H_2O_2 dryer employing heated air. Each container is suspended by its flange between two parallel rails and conveyed forward by a shoving plate. The shoving plate and the walls of the both sides and the top and bottom form a closed chamber which helps to sterilize whole surface of the container. The distance between the rails is adjustable so that it would be applicable to different containers in shapes and sizes.

A conformation similar to Fig.3 is applied to the lid sterilization (Fig.5) except its rotational conveying system. Each lid is held by vacuum heads and submitted to the similar sterilization process. This lid sterilizer is applicable to any shapes of lids as far as their sizes are in the range from 60×60mm to 120×120mm.

These container and lid sterilizing system were successfully applied to the flexible aseptic packaging machine, and the aseptic filling test showed the enough commercial sterility of products and the stability in operation.

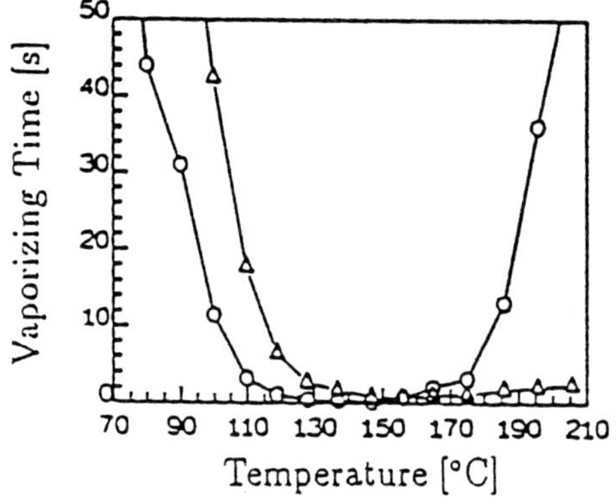

Fig.1 Vaporizing Characteristics of A Droplet

o : Water, △ : H_2O_2

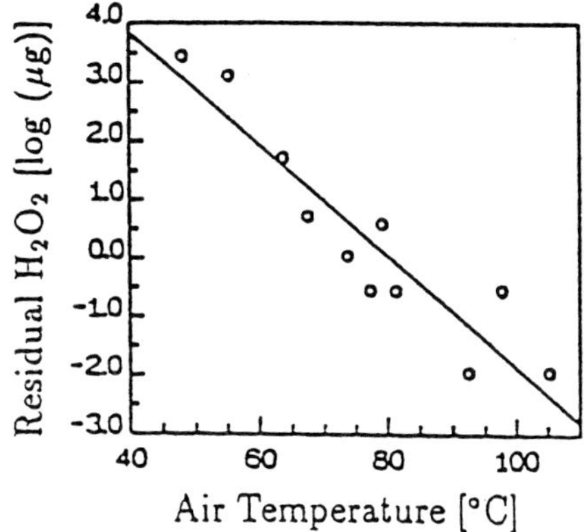

Fig.2 H_2O_2-removal Characteristics by Heated Air

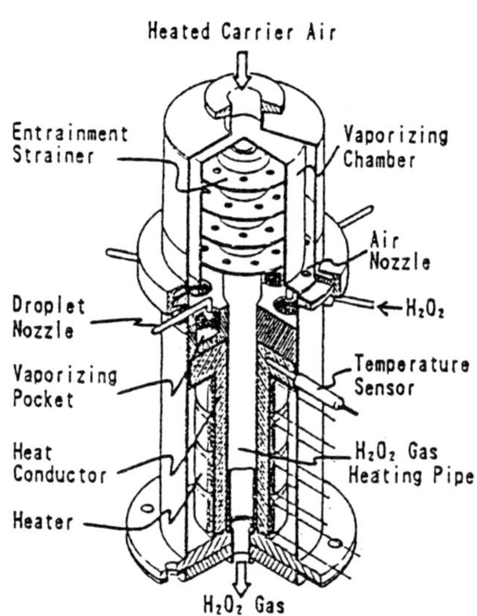

Fig. 3 H_2O_2 Vaporizer

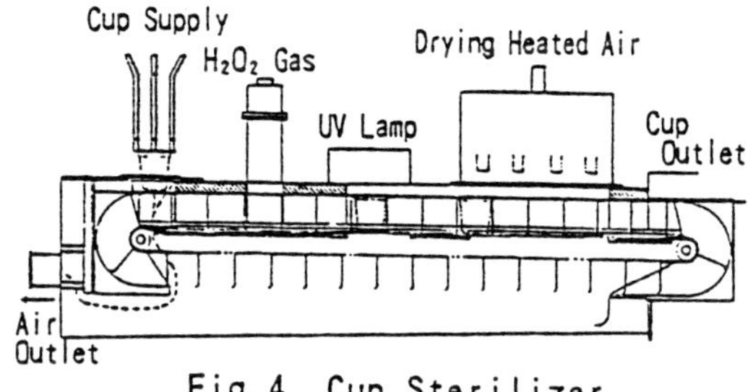

Fig. 4 Cup Sterilizer

Tab. 1 Results of Sterilization Experiments

H_2O_2 [%]	U V	*A. niger*	*B. subtilis*	*B. stearoth.*
–	○	1700	120	172
3	–	0	2800	12000
5	–	0	1886	3920
10	–	0	87	368
35	–	0	72	4
3	○	0	0	0
5	○	0	0	0
10	○	0	0	0
35	○	0	0	0

H_2O_2 [%]:Concentration of hydrogen peroxide-water solution to be vaporized.

U V :Ultraviolet rays were applied(○), not applied(–).

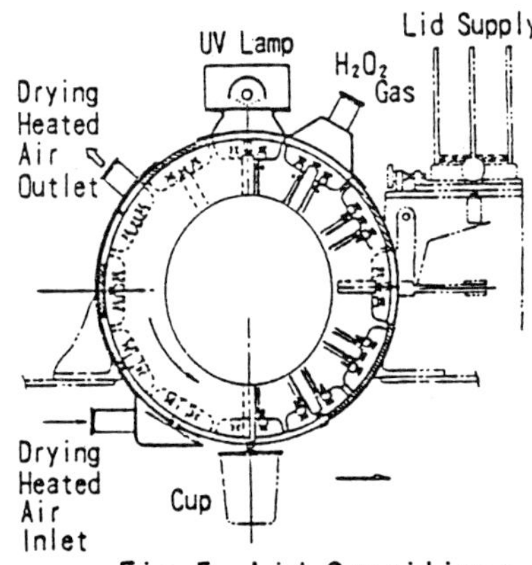

Fig. 5 Lid Sterilizer

REFERENCES

1)Smith,Q.J. and Brown,K.L., Food Technol.,15,169(1980)
2)Stevenson,K.E. and Shafer,B.D., Food Technol.,11,111(1983)
3)Baldry,M.G.C., J.Appl.Bacteriol.,54,417(1983)
4)Stannard,C. and others, J.Food Protect.,46(12),1060(1983)
5)Waites,W.M. and others, Appl.Microbiol.,7,139(1988)
6)Wang,J. and Toledo,R.T., Food Technol.,11,60(1986)

ASEPTIC FILLING OF BEVERAGES IN GLASS BOTTLES - NEW DEVELOPED PROCESSES IN GERMANY

HORST WEISSER and MICHAEL ZUBER
Institute for Brewery Installations and Food Packaging Technology.
Technical University of Munich, 85350 Freising-Weihenstephan, Federal Republic of Germany

ABSTRACT

The aseptic filling of beverages in glass bottles to achieve commercial sterility requests a high standard of the packaging material, the filling line, and on the microbiological hygiene and sanitation.

As the consumer prefers food products from sterilization processes using physical rather than chemical treatment the use of saturated steam is an interesting method to sterilize the glass and the filling line. Three examples for recently built continuous filling lines of the German companies Bosch (for milk), KHS (for juices) and Krones (for beer) are shown. The different filling systems are compared with respect to constructive details, operating conditions, and examples of installation.

INTRODUCTION AND DEFINITIONS

As the consumer prefers food products from sterilization processes using physical rather than chemical treatment the use of saturated steam is an interesting method to sterilize the glass and the filling system.

Since such concepts as *aseptic, ultra-clean, sterile, beverage-sterile* cannot be used unambiguously in the field of sterile packaging, a definition of aseptic packaging is useful as an introduction to the following examples.

Aseptic packaging represents the packaging of a sterilized product into sterile packages in a sterile environment by sterile means whereby neither the product nor the packaging material nor the internal atmosphere is exposed to a further sterilization process after the packaging stage, and the product does not suffer microbial growth in the context of the environment and storage conditions.

As with sterilized canned food, no absolute sterility is achievable in all cases with aseptically filled foods. However, the microbial contamination must be reduced drastically, food-poisoning by pathogenic microorganisms must be avoided with reliability.

With *beverage-sterile filling* commercial sterility is achieved. During the filling process the entering of product-spoilage microorganisms through filler, empty bottles, closures, and environment must be prevented. The product has to be free of beverage-spoilage microorganisms when it is fed to the filler. The decision whether to use sterile filters or a flash pasteurizer is not so much a question of food engineering but of marketing (draft beer). It is also determined by the capital investment and operating costs involved.

Conventional aseptic filling systems

Aseptic filling systems for the filling of soft beverages into

- *carton-laminate systems*, e.g. Tetra Pak, Pure-Pak and PKL Combibloc, and
- *Composite can systems*, e.g. Bosch Hypa-S.

now exist for several years [1].

RECENTLY DEVELOPED GLASS ASEPTIC SYSTEMS

In the last years, several companies developed aseptic or at least beverage-sterile packaging processes to fill premium quality beverages like milk, juices, and beer into glass bottles.

Bosch

A pioneer in this field was the *Bosch* company which developed the GlaSeptik®-process in co-operation with a glass manufacturer (*Oberland Glas*) and the manufacturer of the Twist-Off® closures (*Continental Can Europe, Schmalbach Lubeca*). The filling line for non-acid and low acid products consists of a rinser, a bottle sterilizer, an aseptic filling machine, and a sealing machine designed especially for the aseptic closing of glass jars [2]. The different process steps are:

- Rinsing and pre-sterilizing of the jars with steam.
- Final sterilisation with hydrogen peroxide (H_2O_2) and drying with sterile air in a bottle sterilizer.
- Transportation of the sterile jars to the filler via a sterile connection tunnel.
- Filling with a longitudinal filler with a underlevel filling system (long filling pipe) to avoid or reduce the formation of foam or the intrusion of oxygen. If necessary, the oxygen content in the headspace can be reduced either by steam-exhaustion or by flushing the headspace with an inert gas.
- Sealing the jar with Twist-Off® or PT- caps in a two-lane linear closing machine with 10 closing heads. Saturated steam is continuously fed into the top of the machine. This guarantees aseptic conditions during machine operation and generates a partial vacuum in the package after sealing.

The entire filling line is hermetically closed and a positive pressure of several mbar is applied to prevent microbial infections during the operation. Whenever it comes to machine standstills during the filling process, operators can use attached rubber hand gloves at critical points. No direct contact is necessary to remedy the interruption. The exhausted H_2O_2-vapor is reduced by using a catalyst. The whole line can be cleaned with CIP and sterilized using the SIP process.

KHS

In 1989 a new aseptic filling system, *Aseptronic GRA* , was presented at the *Drinktec*-fair by *KHS* (Klöckner Holstein Seitz, machinery for the production of beverages) and *Granini* (juice producer) as a joint venture project. This aseptic filling line consists of a combined sterilizer/filler and a modified closing machine [3]. The filling process can be separated into the following steps:

- Warm bottles (60 - 70°C) are supplied from a bottle washer.
- The bottles are pressed against the filling valve by the bottle platform. Bottle platform, bottle, and filling valve form a unit. They are enclosed into the so-called valve-bell (Figure 1).
- Each bottle is rinsed with steam continuously emerging from within the bell. A reduced steam flow enters the bottle via a long filling tube, flushes the internal bottle surface and the bottle mouth and after the bottle is full of steam also the external bottle walls. A uniform bottle temperature is reached.
- The bell is sealed hermetically.
- Bell and bottle are sterilized for 4 s at about 135°C.
- Sterile air or inert gases replace the steam in the bottle. The bottle is counter-pressurized either with sterile air or an inert gas.
- A low turbulence filling process using traditional techniques follows.
- The filling tube is emptied and the bottles are removed from the filling valves.
- The bottles are transported to a *Twist-Off* bottle closing machine in a steam atmosphere.

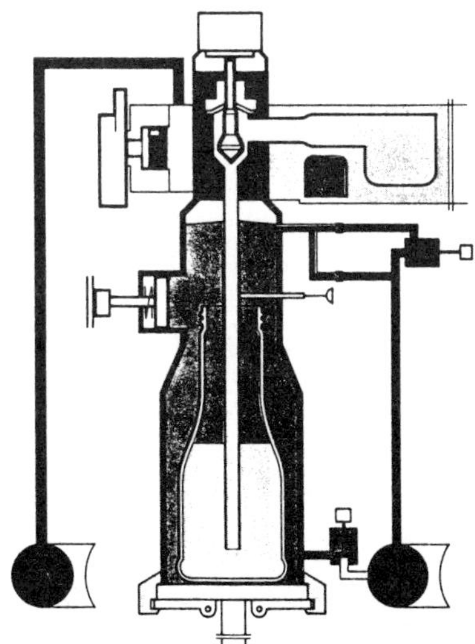

Figure 1. Aseptronic GRA filling system.

Krones

A sterile filling and closing system for beer was developed by *Krones* (Hermann Kronseder Maschinenfabrik) under the name *BSF-System* (Beverage-Sterile-Filling). The filling line consists of a combined sterilisation and filling and a conventional, CIP cleanable cap crowning machine [4]. The process steps are:

- Warm bottles (50–60°C) are supplied from a bottle washer or a rinser.
- The bottle's inner surface is sterilized with 110°C hot saturated steam (Figure 2). The bottle internal wall temperature rises to about 105°C. The oxygen content in this steam atmosphere is very low.
- The bottle is now CO_2-flushed to remove steam and air from the bottle.
- The bottle is pressurized with carbon dioxide. The bottle is brought up to filling pressure.
- The low turbulence filling process starts, when the product and the return gas valves open. The product streams along the internal walls into the bottle. The carbon dioxide is collected in the return gas/snift channel (Figure 2).
- The filling process ends as soon as the pre-set filling height is reached. The level probe sends an electric impulse to close the product and return gas valve.
- After an adequate settling time the bottles are snifted and fed to a CIP cleanable crowner under aseptic conditions.

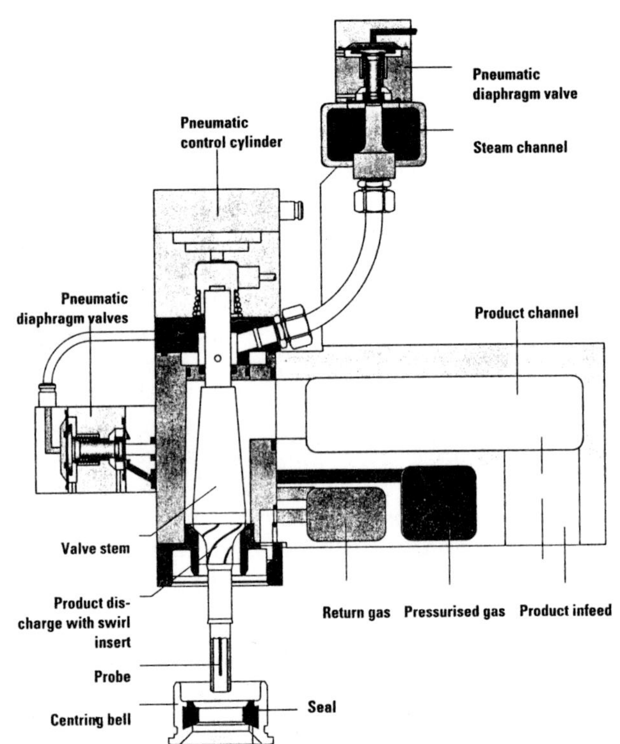

Figure 2. Krones pneumatic filler for beverage-sterile filling.

CONCLUSIONS

The *Bosch GlaSeptik* process is an interesting example of aseptic filling being used for microbiologically very sensitive low-acid products like milk, milk products, and dietetic products.

Until today the *KHS Aseptronic GRA* system, as well as the newly developed *KHS DDF-filler* (Dampfdruckfüllverfahren = vapor pressure filling process) with long filling tubes, and the *Krones BSF-filler* can only be used for low-acid beverages like fruit juices, nectars, and beer. With this type of beverage-sterile rotary only a commercial sterilization can be achieved.

Aseptic packaging into glass is as well a challenge to the food industry as it is an opportunity for cutting costs and improving the quality of the end products.

REFERENCES

1. Weisser, H., Fortschritte beim keimarmen und aseptischen Verpacken von Lebensmitteln. *Der Weihenstephaner*, 1991, **59**, No. 1, 48-52.
2. Buchner, N., Aseptisches Füllen von Behältern aus Glas und Kunststoffen. *ZFL - Intern. J. Food. Technol., Marketing, Packaging, and Analysis*, 1990, **41**, No. 5, 295-300.
3. Aseptische Abfüllung auch für Bier. *Neue Verpackung* 1992, **45**, No. 10, 42-46.
4. Kronseder, Hermann, Getränkesteriles Füllen. *Getränketechnik*, 1992, No. 5, 173-176.

A KINETIC MODEL FOR FOULING IN MILK PROCESSING

P. J. R. SCHREIER, I. TOYODA *, M. T. BELMAR-BEINY, and P. J. FRYER,
Department of Chemical Engineering,
University of Cambridge, Pembroke Street,
Cambridge CB2 3RA, United Kingdom.
*: Meiji Milk Company, Japan.

ABSTRACT

Fouling from milk proteins at pasteurisation temperatures has been examined experimentally in a simple counter-current heat exchanger, and a model based on the method of characteristics used to model the results. Protein reactions both at the surface and in the bulk are necessary to model the process.

INTRODUCTION

Fouling from milk at temperatures below 100°C results from a complex mixture of reactions on the heat exchange surface and in the bulk of the fluid [1]. When heated, the structure of the native milk protein ß-lactoglobulin is irreversibly altered by denaturation (intramolecular) and aggregation (intermolecular) reactions. Early work on fouling models concentrated on surface reactions [2], however [3] has shown that bulk reactions can be important in milk protein fouling. Simple models have been produced in which the amount of fouling was considered proportional to the volume of fluid which is hot enough to produce denatured and aggregated protein [3,4]. Recently, a mathematical model has been produced [5] which considers both surface and bulk reactions. The denaturation reaction is modelled as first order in concentration of native protein with the aggregation reaction as second order in concentration of denatured protein. The model of [5] describes a complex plate heat exchanger, in which a range of velocities and temperatures is found. Experiments in tubes, where the temperatures and flows are more easily known, are easier to interpret. This paper describes the basis of a theoretical study of a tubular pasteuriser to examine a model for fouling. The model is compared with experimental data on fouling produced in a laboratory-scale apparatus.

MODELLING EQUATIONS

Heat transfer

Experiments have been carried out in a simple shell-and-tube configuration [6]. This system can be modelled [7,8] using enthalpy balances on differential volume elements of fluid on the shell and tube sides of the generalised exchanger configuration. These lead to the following equations for fluid temperature distributions:

(a) Tube side fluid temperature, $T_f(t,x)$:

$$\frac{\partial T_f}{\partial t} + v_f \frac{\partial T_f}{\partial x} = \frac{U a_t}{(\rho C_p)_f}(T_s - T_f) \qquad (1)$$

(b) Shell side fluid temperature, $T_s(t,x)$:

$$\frac{\partial T_s}{\partial t} + v_s \frac{\partial T_s}{\partial x} = -\frac{U a_s}{(\rho C_p)_s}(T_s - T_f) \qquad (2)$$

where a is the heat transfer area per unit volume, v the fluid velocity and ρC_p the thermal capacity. Equations (1) and (2) assume that the fluids pass along the shell and tube sides in plug flow and that entrance and exit effects may be neglected, with the deposit sufficiently thin for its enthalpy to be neglected and for the tube side Reynolds number to remain unchanged. To study fouling, equations (1) and (2) must be solved simultaneously with any fouling model. Together with the fouling model, equations (1) and (2) may be transformed from partial differential equations into ordinary differential equations defined along characteristic lines in the x-t plane using the method of characteristics [9]. In effect this method allows the equations to follow a fluid element down the tube; it is thus intuitively sensible as well as mathematically rigorous.

Fouling

Two reactions are assumed, $N \rightarrow^{k_1} D \rightarrow^{k_2} A$, by which native protein, N, is transformed to aggregates, A. Reaction (1) is taken as first order and reaction (2) as second order. Fig. 1 gives a schematic diagram of the system. The fluid mechanics are simplified by assuming that there are two regions in the flow, one of uniform bulk temperature, T_b, and a thermal boundary layer of thickness δ, at the deposit-fluid interface temperature, T_{wd}. Protein concentration in the bulk and the thermal boundary layer are modelled separately. Mass transfer between the two regions is given by some mass transfer coefficient:

$$\text{Rate of transfer of aggregates between bulk and wall layer} = k_m \left(C_A^* - C_A\right) \qquad (3)$$

where the concentration of aggregate in the wall region is C_A^*.

$$\text{The rate of fouling is modelled as an adhesion reaction} = k_w\, C_A^* \qquad (4)$$

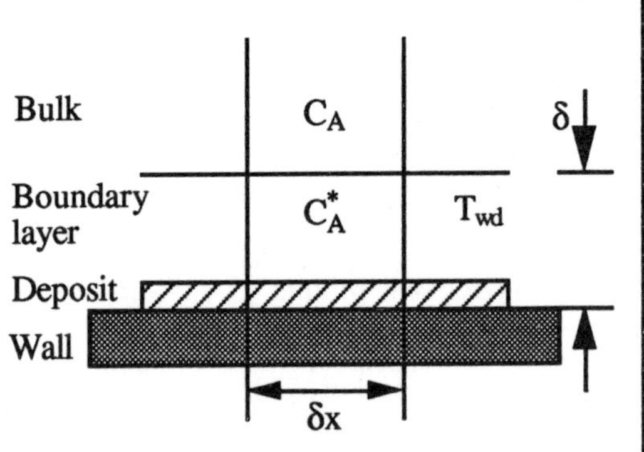

Fig. 1: Thermal boundary layer for material balance of equations (3) and (4).

Figs. 2 and 3:
$E_N = 261$ kJ/mol, $A_N = 3.37 \times 10^{31}$: 70°C-90°C
$E_D = 312$ kJ/mol, $A_D = 1.36 \times 10^{43}$: 70°C-90°C
$E_D = 56$ kJ/mol, $A_D = 1.86 \times 10^{6}$: 90°C-150°C
Protein concentration = 3.5 wt %
Fig. 2:
$k_w = 5 \times 10^{-7}$ m/s
Fig. 3:
$k_w = 3 \times 10^{-7}$ m/s
Initial denaturation = 30 %

TABLE I: Activation energies and reaction constants [5].

RESULTS

The model demonstrates the effect that bulk processes can have on fouling. Clearly, in any heating process, the temperature of the wall region will exceed that of the bulk, and so reaction will be faster in the wall region. If the concentration of aggregates in the bulk is low, mass transfer *into* the bulk will lower the aggregate concentration in the wall layer and the fouling rate; however, if the bulk is hot enough for protein reactions to take place then the mass transfer rate will fall, and C_A^* and the fouling rate will increase.

Two types of simulation have been carried out. The sensitivity of the model to variations in a range of process parameters has been examined. Two important factors are the concentration of native protein in the milk and the proportion of protein in the milk that has been denatured before it enters the heat exchanger tube. The latter is important since much experimental work has used reconstituted whey protein concentrate powders which will be denatured to some degree. Fig. 2 shows the effect of varying the degree of protein denaturation in the inlet protein, for the constants in Table I; clear variation in the inlet region is seen.

Experiments have been performed [6] in a heat exchanger tube using whey protein concentrate. The simple geometry of the single pass tubular heat exchanger may be modelled using equations (1) to (4). Fig. 3 shows a comparison of experimental data with a curve produced by the model, which shows good agreement with experimental data. This model demonstrates the importance of bulk reactions; further development is underway to fit a wider range of experimental data.

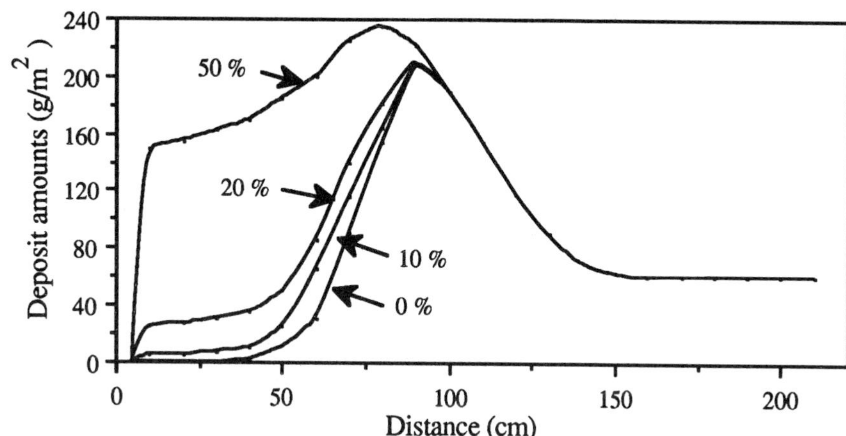

Fig. 2: Effect on fouling of varying degree of inlet protein denaturation.

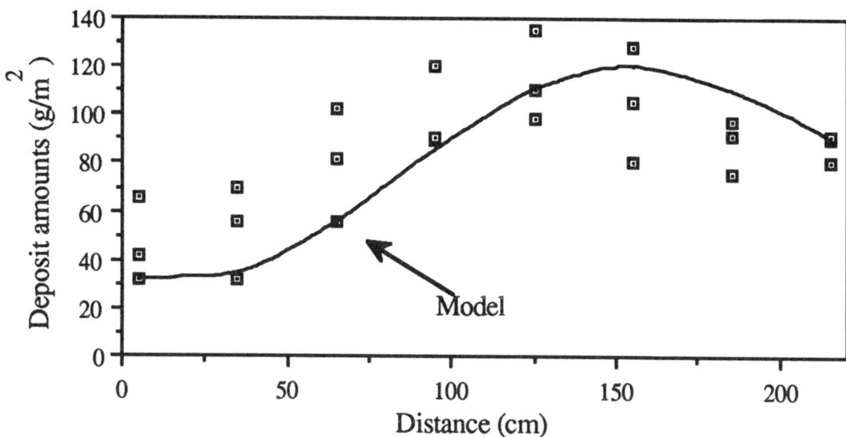

Fig. 3: Comparison of experimental data with the model.

REFERENCES

1. Fryer, P.J., Belmar-Beiny, M.T. and Schreier, P.J.R. (1993), presented at this conf.
2. Lalande, M., Tissier, J.-P. and Corrieu, G. (1984), *J. Dairy Res.*, **51**, 557-568.
3. Belmar-Beiny, M.T., Gotham, S.M., Fryer, P.J. and Pritchard, A.M. (1993), *J. Food Eng.*, **19**, 119-139.
4. Paterson, W.R. and Fryer, P.J. (1988), *Chem. Eng. Sci.*, **43**, 1714-1717.
5. de Jong, P., Bouman, S., van der Linden, H.J.L.J. (1992), *J. Soc. Dairy Tech.*, **45**, 3-8.
6. Gotham., S.M. (1990), *Ph.D. Thesis*, University of Cambridge.
7. Fryer, P.J. and Slater, N.K.H. (1985), *Chem. Eng. J.*, **31**, 97-107.
8. Fryer, P.J. and Slater, N.K.H. (1986), *Chem. Eng. Sci.*, **41**, 2365-2372.
9. Acrivos, A. (1956), *Ind. Eng. Chem.*, **48**, 703-709.

INITIAL EVENTS IN SURFACE FOULING

MT BELMAR-BEINY, PJR SCHREIER and PJ FRYER
Dept. of Chemical Engineering,
Pembroke Street, Cambridge, UK, CB2 3RA

ABSTRACT

Deposit formation (fouling) from dairy fluids is a severe industrial problem. Experiments were conducted to identify the species that adhere first into stainless steel surfaces at 96°C from turbulent flows of whey (inlet fluid temperatures of 68 and 73°C). The heating contact times were varied between 4 and 210 s. At 4 s, XPS analysis showed the first layer being made of proteinaceous material. For a fluid inlet temperature of 73°C, a lag phase of 150 s occurred before deposit aggregates were observed under SEM. These aggregates were identified with X ray elemental mapping as being made of protein and calcium. No phosphorus was found in any experiments. At 60 minutes contact time, both calcium and phosphorus were found at the interface deposit-stainless steel.

INTRODUCTION

Fouling lowers the efficiency of a food process and can endanger product sterility of the product. The problems of fouling are discussed in more detail elsewhere [1, 2]. The literature so far is not conclusive as to which species (proteins or minerals) adsorbs to the surface first [3, 4]. If it were possible to extend the induction period prior to fouling then plant operation might be improved and cleaning costs reduced. Here, experiments to study the initial stages of fouling are described.

MATERIALS AND METHODS

Whey protein concentrates (WPC) (35 and 75% protein) were dissolved to 1% protein concentration in distilled water. The test fluid was preheated before passing through the heat exchanger. For contact times below 3 minutes, a rig was constructed to allow turbulent flow by gravity at a Reynolds number ca. 15000. For longer contact times a second rig was built operating at a Reynolds number for the fouling fluid of 6250. The inlet fluid temperature was varied from 68°C to 73°C, and the wall temperature kept constant at 96°C. Fouled tubes were removed, dried and cut into small sections for Scanning Electron Microscopy (SEM), X ray microanalysis and XPS. Further details can be found in [3, 5].

RESULTS

Up to 120 s (73°C inlet temperature) the surface of the material became smoother. Through naked eye the surface initially looked yellow, then became green and blue as a result of an interference film. At 150 s the surface looked whitish to the naked eye and under the SEM, aggregates appeared on the surface in a random way (size range 0.2-0.7 μm). From this point the surface became completely covered. For a fluid inlet temperature of 68°C, however no aggregates were found on the surface even after 210s. XPS studies showed that, as the contact time increased (up to 210 s for 68°C, and 150 s for 73°C) the peaks corresponding to the metallic elements decreased and the carbon and nitrogen peaks increased. The presence of protein was followed through the carbon and nitrogen peaks. Figure 1 shows the atomic percentages for carbon, nitrogen and oxygen obtained for WPC75 adsorbed at 68°C and 73°C for different contact times. At both 68°C and 73°C the surface was already covered after 4 s. As the contact time increased the percentages of the three elements approached the theoretical values for ß-lg: carbon 65%, oxygen 19% and nitrogen 16%, becoming almost identical at 150 s, 73°C. Sulphur and calcium were detected in trace amounts (0.8 and 0.6% respectively) at 150 s, 73°C. No phosphorus was found.

Figures 2a and 2b show elemental maps obtained for calcium and phosphorus for WPC35 after 60 minutes contact time, 73°C. Both elements are concentrated at the interface deposit-stainless steel surface. Figure 3 shows an elemental map for iron obtained for 150s 73°C. The detection of iron is a good indicator of how the surface had become covered with deposit: the higher concentration of iron, the thinner the adsorbed foulant layer. Elemental maps for calcium and sulphur coincided with XPS results for the same specimen. Phosphorus was not found at 150s.

DISCUSSION

SEM analysis showed that for an inlet temperature of 73°C, there was a lag phase up to 150 s, in which no aggregates appear on the surface. This lag phase probably corresponds to the induction period. During this lag phase the surface became covered with a proteinaceous film, which has been seen under SEM [4]. The presence of whey proteins on the surface was successfully identified with XPS through the organic carbon and organic nitrogen. No sulphur, calcium or phosphorus were detected during the lag phase observed at 73°C or up to 210 s at 68°C. X ray elemental mapping prove to be a very useful technique in showing the composition and distribution of elements on deposits formed at the end of the induction period.

The surface analysis results indicate that proteins are most likely to be the first species to adhere to the surface. The heterogeneous distribution of minerals along the deposit, in which minerals eventually become concentrated at the interface deposit-surface is still a unanswered question. Our work showed conclusively that if calcium or phosphorus formed part of the first layers, their amount was insignificant in comparison to the amount of protein. Thus, protein adsorption controls the first stages of fouling. Any surface modification to the heat exchanger most affect protein binding sites.

ACKNOWLEDGEMENTS

MTBB wishes to acknowledge financial support from AFRC.

REFERENCES

1. Fryer, P.J., Schreier, P.J.R., Belmar-Beiny, M.T. (1993). This conference
2. Schreier, P.J.R., Toyoda, I., Belmar-Beiny, M.T. and Fryer, P.J. (1993). This conference.
3. Belmar-Beiny, M. T., & Fryer, P. J. (1992a). Bulk surface effects on the initial stages of whey fouling. Transactions of the Institute of Chemical Engineers, 70(part C), 193-199.
4. Belmar-Beiny, M. T., & Fryer, P. J. (1992b). A study of the initial stages of deposition from whey protein solutions. *Entropie*, 28(169), 51-58.
5. Belmar-Beiny, M., Gotham, S., Paterson, W., Fryer, P., & Pritchard, A. (1993). The effect of reynolds number and fluid temperature in whey protein fouling. Journal of Food Engineering, 19, 119-139.

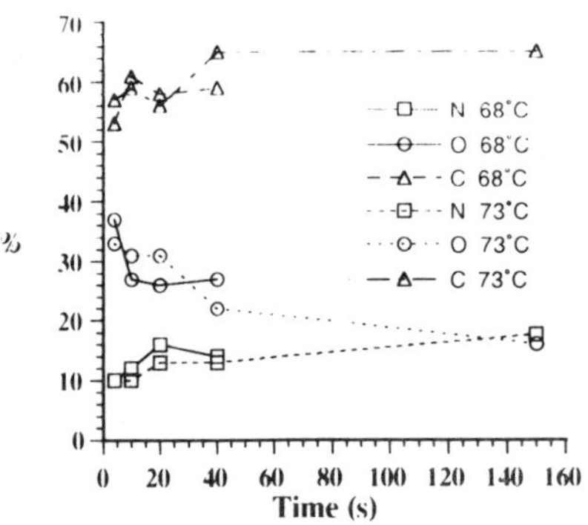

Figure 1. XPS atomic percentages for carbon, oxygen and nitrogen on fouled stainless steel surface (AISI 321) with WPC 75 at different contact times and inlet temperatures; wall temperature 96°C. Photoelectron angle of emission of 30° with respect to the surface.

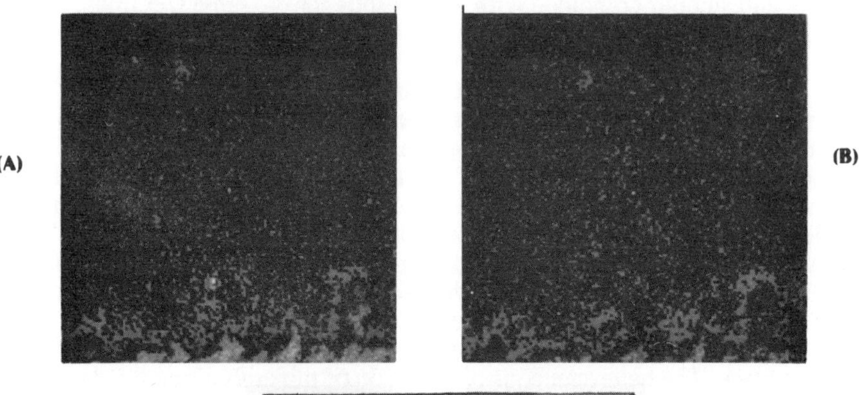

Figure 2. X ray elemental maps for (a) calcium and (b) phosphorus on fouled stainless steel surface (AISI 321) with WPC35 for 60 minutes contact time. (Reynolds 6250, temperatures: inlet 73°C and outlet 83°C, wall temperature 97°C). Dwell time 200 ms, magnification 1000 X. The grey scale is given in the figure, it reads from left to right.

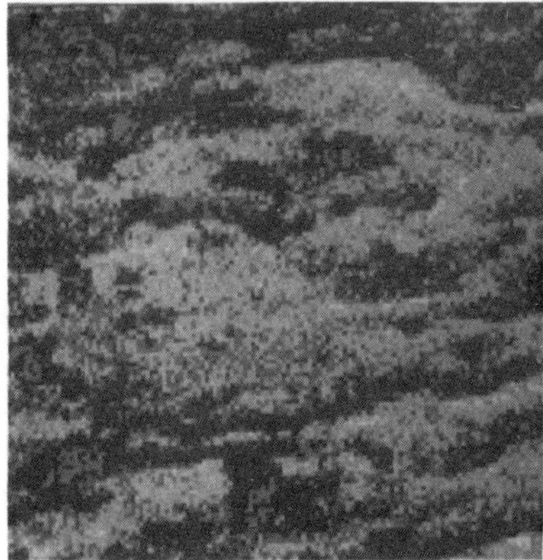

Figure 3. X ray elemental map for iron on fouled stainless steel surface (AISI 321) with WPC 75 for 150 s contact time. (Temperatures: inlet 73°C and outlet 75°C, wall temperature 96°C). Dwell time 200 ms, magnification 1000 X. Grey scale as in figure 2.

ADSORPTION OF PROTEIN ONTO STAINLESS STEEL PARTICLE SURFACE AND ITS DESORPTION BEHAVIOR

HIROSHI ITOH, TADASHI NAGAI*, TAKESHI SAEKI*,
TAKAHARU SAKIYAMA, and KAZUHIRO NAKANISHI
Department of Biotechnology, Faculty of Engineering, Okayama University,
3-1-1, Tsushima-naka, Okayama 700, Japan, *Institute for Fundamental Research,
Suntory Co., 1023-1, Yamazaki, Shimamoto-honmachi, Mishima-gun, Osaka 618, Japan

ABSTRACT

The adsorption and desorption of β-lactoglobulin were studied using stainless steel particles. The amount of β-lactoglobulin adsorbed was influenced by various factors such as temperature, pH, incubation time, and the protein concentration. The desorption behavior was analyzed by feeding NaOH or proteolytic enzyme solution into a glass column packed with the stainless steel particles on which the protein had been adsorbed and subsequently measuring its elution profile at the outlet. The amount of the protein remaiming on the stainless particles decreased accoring to a first-order kinetics at the initial stage of desorption, and then gradually reached a constant value. The desorption rate constant and the remaining amount were studied under various conditions.

INTRODUCTION

Cleaning is quite an important operation in food processing particularly from a viewpoint of sanitation of manufacturing processes. To optimize cleaning conditions, the mechanism for adsorption of soils onto the surface of the equipment and its desorption behavior should be clarified. We studied here various factors affecting the adsorption of β-lactoglobulin and its desorption behavior using NaOH or proteolytic enzyme solution was analyzed under various conditions. β-Lactoglobulin is one of the main soil components when milk products are heat-treated.

MATERIALS AND METHODS

In most cases, β-lactoglobulin (3x crystallized, Sigma Chemical Co.) dissolved in 2 ml of 35 mM phosphate buffer, pH 6.85, containing 100 mM NaCl was incubated with 2 g of stainless steel particles (100 meshes in mean size) with vigorous shaking under various conditions. The amount of protein adsorbed was measured from the difference in concentration of the solution before and after incubation. The stainless steel particles have a large surface area (0.17 m^2/g) to measure the amount of protein adsorbed with accuracy.

About 1.5 g-portion of the stainless steel particles on which β-lactoglobulin had been adsorbed was packed into a glass column (1.5 cm ϕ x 5 cm). The phosphate buffer was allowed to flow for 2 h, and then NaOH or proteolytic enzyme (Protin AC, Daiwa Kasei Co. Ltd., Osaka) solution was fed usually at 50°C. The protein concentration in the eluent was measured by the Lowry-Folin method. After 2-h feeding, the stainless steel particles were withdrawn from the column and the remaining amount of protein was directly measured. The amount of protein remaining on the surface at any time during the desorption process was calculated by cumulative summation of the protein eluted.

RESULTS AND DISCUSSION

Adsorption of β-lactoglobulin

The adsorption equilibrium of β-lactoglobulin at 25°C was similar to the Langmuir type adsorption isotherm, and the maximum amount of protein was adsorbed in the thickness of monolayer. When the temperature exceeded 65°C, the amount of protein adsorbed increased by several times. At 70°C the amount of protein adsorbed was 1.3 mg/g (7.6 mg/m^2) at the protein concentration of 10 mg/ml. The increase in the adsorbed amount with temperature is closely related to the denaturation of β-lactoglobulin [1]. When the pH of the solution was lower than pH 6, the amount of protein adsorbed considerably increased, indicating the effect of an electrostatic interaction between the protein and the stainless steel surface. The β-lactoglobulin which had been denatured at 70°C showed an adsorption behavior similar to that for obtained without pre-incubation, which might imply that the adsorption takes place at an instance when the protein suffers from denaturation.

Desorption behavior of β-lactoglobulin

The elution behavior of β-lactoglobulin with NaOH or enzyme solution was measured after feeding the phosphate buffer for 2 h. The amount of protein remaining on the stainless steel particle decreased accoring to a first-order kinetics at the initial stage of desorption, and then gradually reached a constant value as shown in Fig. 1.

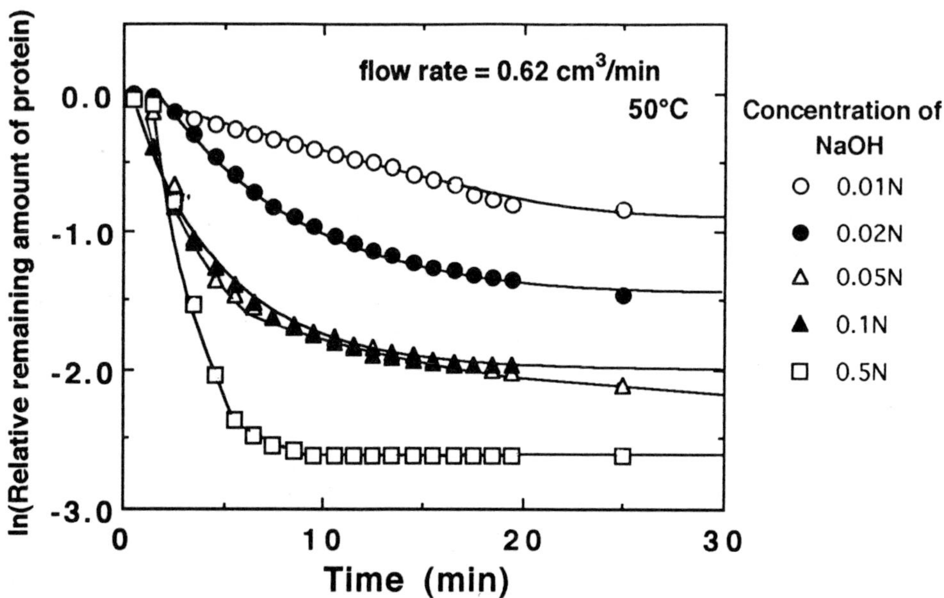

Figure 1. Relative amount of protein remaining on the stainless steel surface during desorption with NaOH.

With NaOH solution, the first-order rate constant for desorption, k_d, was dependent on the NaOH concentration, temperature and flow rate. From the temperature dependence of k_d the activation energy was estimated to be 210 kJ/mol. The k_d increased with 0.5th power of the linear flow rate, indicating the effect of mass transfer. The desorption behavior with the enzyme solution was different in some points from that using NaOH solution. The activation energy was 30 kJ/mol, which was much lower than that for the desorption with NaOH solution. The remaining amount of the protein after 2-h treatment was much less in the case of desorption with enzyme solution than in the case with NaOH solution at the same pH as shown in Fig 2.

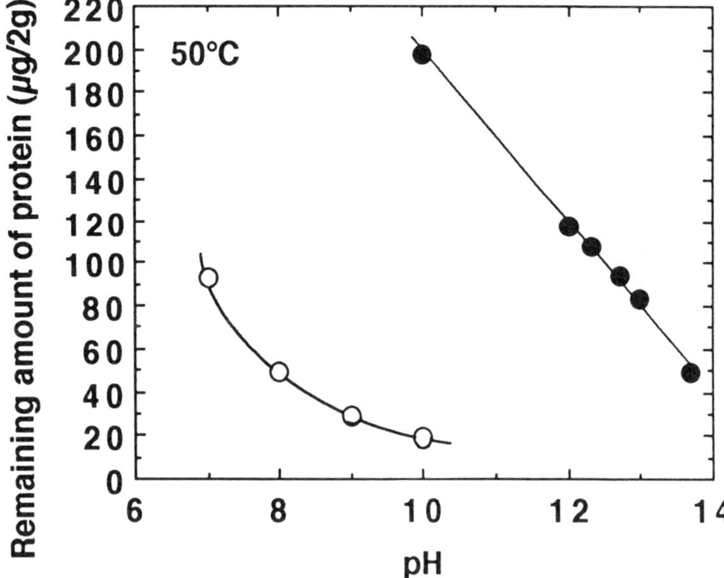

Figure 2. Amount of protein remaining after 2-h treatment with NaOH (○) or proteolytic enzyme solution (●).

CONCLUSIONS

The adsorption and desorption behaviors of β-lactoglobulin were studied using stainless steel particles of 100 meshes in mean size. In particular, the rate of desorption was studied by analyzing the elution profile of protein desorbed from the column packed with the stainless steel particles. Furthermore, the different behaviors in desorption with NaOH and proteolytic enzyme solution were shown.

REFERENCES

1. Arnebrant, T., Barton, K. and Nylander, T., Adsorption of α-lactoalbumin and β-lactoglobulin on metal surfaces versus temperature. J. Colloid Interface Sci., 1987, 119, pp. 383-390.

SUPERCRITICAL FLUID EXTRUSION - A NEW PROCESS

S.S.H. RIZVI and S.J. MULVANEY
Institute of Food Science
Cornell University, Ithaca, NY 14853

ABSTRACT

A new hybrid process which combines the attributes of supercritical fluids with extrusion processes is described. Based on the commonality of high pressure operation for both processes, supercritical fluid extrusion (SCFX) offers numerous added advantages over the conventional process. Major features and unique system configurations for the process along with microstructures of resulting products are presented.

INTRODUCTION

Food extrusion is a well established unit operation. It is used extensively in the manufacture of a large number of products worldwide. The process provides a unique means of utilizing mechanical and/or thermal energy to obtain uniformly processed products of defined characteristics. Developed primarily for the purpose of making puffed snacks from relatively dry (<20% water) granular cereal ingredients, the use of this technology now covers a wide spectrum of consumer foods and related products (Harper 1979). Examples of such products include ready-to-eat breakfast cereal, soup bases, pregelatinized starch, textured vegetable proteins, pasta and pet foods, among many others.

In general, the manufacture of extruded direct expanded products involves conversion of mechanical energy into heat by the turning screws. The temperature of the extrudate rises over 150°C under high shear, causing plasticization of the feed ingredient. The molten and plasticized mass is then forced through a die and the extrudate is puffed upon exit because of the phase change of liquid water into steam. Product attributes are controlled by manipulation of the specific mechanical or thermal energy inputs and residence time, as adjusted by such variables as moisture content, barrel temperature, screw speed, feet rate, degree of barrel fill, etc.

In view of the above, typical cooking extrusion processes for direct expanded products, operating under rather harsh environments of temperature and pressure, impose the following limitations. 1. Heat sensitive ingredients (proteins, flavorings, etc.) are excluded from the formulations. 2. Water's role is undesirably coupled. It is both the plasticizer for the dough and blowing agent for expansion. 3. Low in-barrel moisture causes extensive barrel and screw wear and produces undesirable dextrins as a result of high energy input. 4. Expansion is controlled by phase change of water, causing uncontrolled expansion with open cell structure. 5. The driving force for expansion is limited and cannot be independently varied.

The commonality of high pressure between supercritical fluids and extrusion processing and their own specific attributes provide a means to overcome some of the basic limitations of the conventional processes indicated above. Termed supercritical fluid extrusion (SCFX), the process involves introduction of supercritical carbon dioxide (SC-CO_2) carrying micronutrients, flavorants and colorants soluble in the fluid phase into a cooked and cooled dough contained within an extruder of modified configurations. The details of the process have been reported elsewhere (3,4) and a system schematic is shown in Figure 1.

While various other supercritical fluids can be advantageously exploited as well in this process, SC-CO_2 is generally the fluid of choice because of its greater solvency, moderate critical temperature and pressure, nontoxicity and availability (2). A few typical pressure profiles obtained with a TX 52 corotating twin-screw extruder consisting of 14 heads are shown in Figure 2. The low shear (pressure) configuration permits the process to occur at subcritical conditions for CO_2 while the medium shear configuration is designed for above critical conditions for CO_2 to prevail within the extruder barrel.

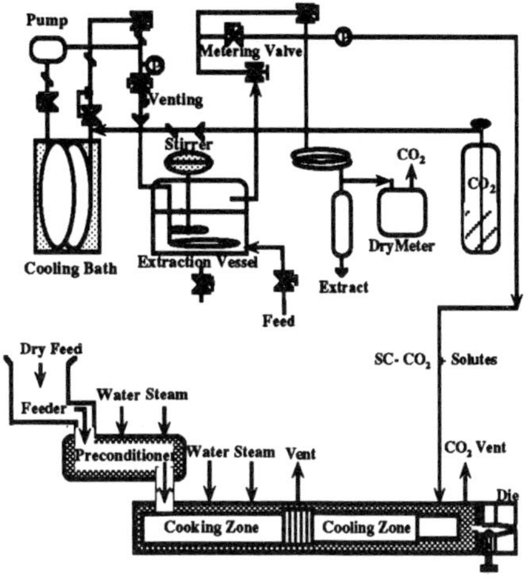

Schematic of Supercritical Fluid Extrusion System

Figure 1. Schematic diagram of the SCFX process showing a supercritical fluid system coupled to a modified extrusion system.

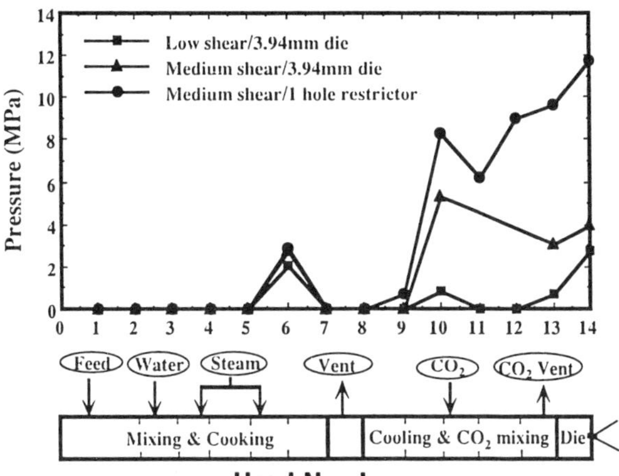

Figure 2. Typical pressure profiles and the corresponding zones for the SCFX process.

A scanning electron micrograph of both the crosssection and surface of SCFX produced corn curl is shown in Figure 3. The most noticeable features are the smooth surface and the closed thick-walled cells. These unique features can be effectively utilized to create a new generation of surface coated, coextruded or honey-comb structured products of superior quality.

1. Supercritical fluids can be used to simultaneously add solutes, lower the in-barrel dough viscosity and puff the extrudate. 2. Heat-sensitive proteins and other ingredients can be successfully utilized to make expanded products. 3. High moisture (low viscosity) extrudates reduce barrel and screw wear. 4. The role of water as a plasticizer and as a blowing agent is decoupled. Expansion is controlled independently of moisture content. 5. Control of the expansion process resides in two-phase flow of supercritical fluid-dough system. 6. Die shaping characteristics become similar to injection molding process because of variable forces for product expansion. 7. Formation of carbonic acid when SC-CO_2 dissolves in the aqueous phase of dough can be exploited to change rheological properties and taste of the extrudate. 8. The extruder can be operated as a controlled bioreactor to induce desirable reactions such as hydrolysis of starch without any residual acidity in the final product. 9. The use of supercritical fluids also offers the possibility of in-barrel extraction of undesirable components prior to puffing.

CONCLUSIONS

The use of a supercritical fluid in conjunction with conventional extrusion processes is likely to further enhance the extensive capability and range of application of extruders. Very special and innovative extruded products can be obtained using SCFX which can lead to new trends in the manufacture of snack foods, breakfast cereals, pasta and baked goods. Further research is needed both in engineering solutions to the many challenging problems this new technology poses and in pursuing the science involved.

(a)
Cross-section

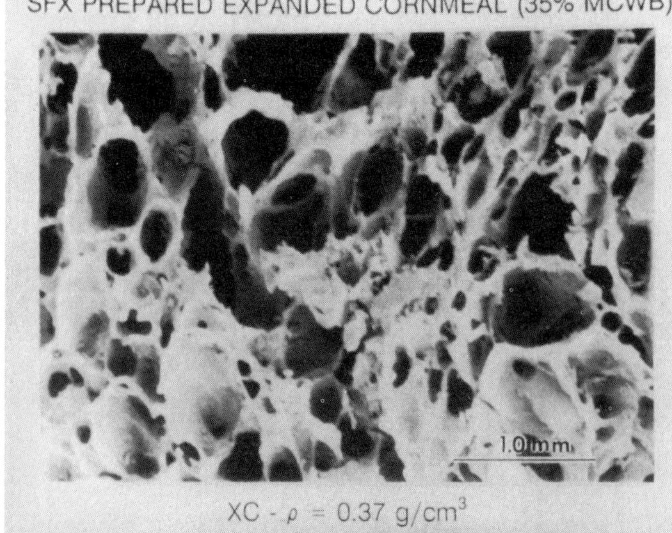

(b)
Surface

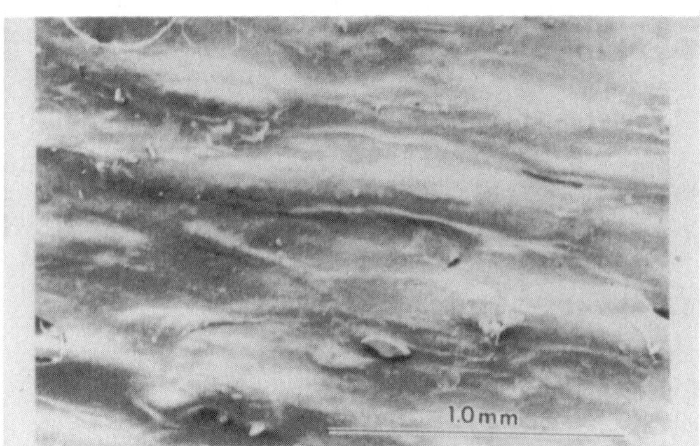

Figure 3. **Scanning electron micrograph of the SCFX produced corn extrudates.**

The SCFX process is envisioned to offer the following specific advantages over the conventional extrusion operations.

REFERENCES

1. Harper, J.M. 1979. Food Extrusion. Critical Reviews in Food Science and Nutrition 11:155.
2. Rizvi, S.S.H., Benado, A.L., Zollweg, J.A. and Daniels, J.A. 1986. Supercritical fluid extraction: Fundamental principles and modeling methods. Food Technology 40:55-64.
3. Rizvi, S.S.H. and Mulvaney, S.J. 1992. Extrusion processing with supercritical fluids. U.S. Patent 5, 120, 559.
4. Mulvaney, S.J. and Rizvi, S.S.H. 1993. Extrusion processing with supercritical fluids. Food Technology (in press).

RATE PROCESSES IN SUPERCRITICAL FLUID EXTRACTION

BEN J. MCCOY
Department of Chemical Engineering,
University of California, Davis, CA 95616

ABSTRACT

Rates of supercritical fluid (SCF) extraction with CO_2 depend on (a) partitioning of the extract between the natural matrix and the fluid, (b) rate of mass transfer of extract from the interior of the matrix to the fluid, and possibly (c) solubility of the extract in the supercritical fluid. These issues are discussed and illustrated by calculations. Investigations of the decaffeination of coffee beans demonstrate the concepts. Decaffeination was measured as a function of CO_2 flow rate, temperature and pressure. Soaking the raw beans in water prior to decaffeination enhanced the rate of extraction. The rate of decaffeination increased with both pressure and temperature. The mathematical model describes the external and intraparticle diffusion resistances and the distribution of caffein between water and supercritical CO_2. The partition coefficient for caffeine distributed between water and supercritical CO_2 depends on temperature and pressure.

INTRODUCTION

Supercritical fluid extraction, especially with CO_2, is the process of choice in many extractions of natural products, e.g., flavor from hops, essential oils from spices, and caffeine from coffee beans or tea leaves. Peker et al. (1992) presented a theoretical model with the capability to describe quantitatively a range of such physical extraction processes (notation is shown in Table 1). The general case is based on porous spherical particles whose pores are filled with a liquid. For a thin, gradientless bed of particles (or well-stirred vessel), the dimensionless differential equations with their initial conditions are

$$dx/d\theta + x/\alpha = -\phi(x - my)(1 - \alpha)/\alpha, \qquad x(\theta=0) = 0$$

$$dy/d\theta = \phi(x - my)(\beta + (1 - \beta)K), \qquad y(\theta=0) = 1/[\beta+(1-\beta)K$$

$$dF/d\theta = x/(1 - \alpha), \qquad F(\theta=0) = 0$$

Since the concentrations were quite low for the the flow system used by Peker et al. (1992), the solubility of the solute in the SCF CO_2 was not a limiting factor. For an enclosed batch system, the second term in the first equation, which represents the flow of extract out of the vessel, is set to zero, i.e., $x/\alpha = 0$. Peker et al. (1992) gave analytical solutions for x, y, and F, but the system of differential equations is easily solved numerically, e.g., by a Runge-Kutta method. By means of the coefficient m, the model accounts for the partitioning of the extracted solute between the matrix medium and the SCF CO_2. The equilibrium coefficient K allows for adsorption-desorption effects. The mass transfer coefficient ϕ accounts for intraparticle diffusion and external mass transfer resistances.

Coffee beans are always saturated with water during SCF decaffeination. This allows the caffeine crystals to dissolve completely in the water (K << 1), and transfer more easily to the CO_2. For <u>dry</u> coffee beans the extraction process is extremely slow, since the solid caffeine must desorb from the plant tissue (Peker et al., 1992). For caffeine distributed between CO_2 and water the partition coefficient m increases with both temperature and pressure, thus increasing the rate of decaffeination. A correlation that describes the data presented by Peker et al. (1992) is

$$m = m_o \exp(b_1T + b_2P), \qquad m_o=2.75\times10^{-14},\ b_1=0.0738,\ b_2=0.105$$

In principle, a coefficient m can also represent the partitioning of a solute between the SCF CO_2 and a nonpolar liquid.

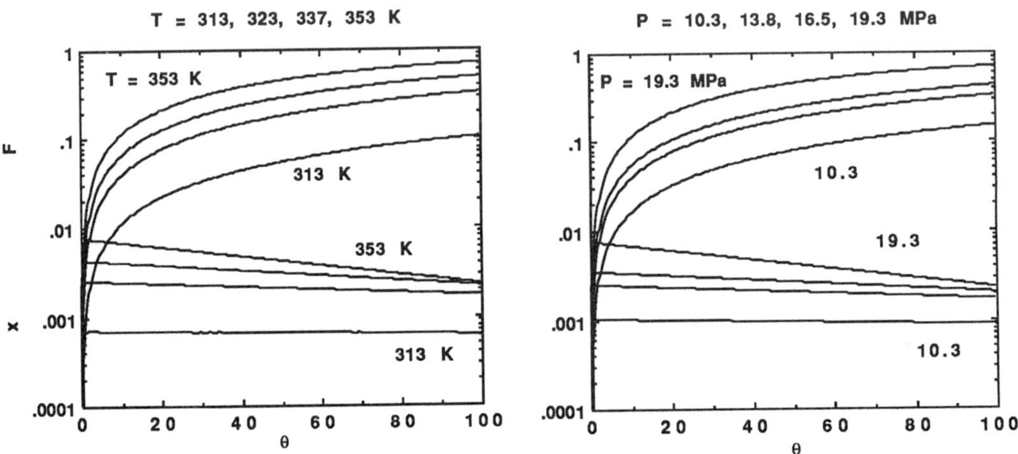

Figure 1. Effluent concentration, x, and cumulative fraction, F, for SCF CO_2 decaffeination; effect of changing T and P when α=0.371 and β=0.515.

Figure 1, based on parameter values given in Table 2, shows the effects of increasing temperature and pressure, respectively. The mass transfer coefficient ϕ decreases with P and T. The two sets of curves in Figure 1 are similar since the values of ϕ and m are of similar magnitude for the two sets. The effluent concentration x of caffeine immediately jumps to a value that slowly declines with time as the caffeine is depleted. Coffee-bean decaffeination usually requires about ten hours. The cumulative fraction approaches unity as caffeine is exhausted.

TABLE 1. Notation

Bi = Biot number (k_fR/D_e)
c = extractable solute concentration in the CO_2, g/ml
c_i = solute concentration in the pore volume of the matrix, g/ml
c_o = initial total solute concentration, g/ml
D_e = effective intraparticle diffusion coefficient for solute in the matrix, cm^2/s
F = cumulative fraction of solute extracted
k_f = external mass-transfer coefficient, cm/s
k_p = combined mass-transfer coefficient $(15k_f/R)/(5+mBi)$, cm/s
K = equilibrium adsorption coefficient
m = equilibrium partition coefficient of solute distributed between CO_2 and the matrix: (conc of solute in CO_2)/(conc of solute in matrix)
R = radius of spherical particles, cm
t = time
x = dimensionless solute concentration in CO_2 (c/c_o)
y = dimensionless solute concentration in matrix (c_i/c_o)
α = void fraction in bed
β = porosity of the particles
ϕ = dimensionless mass-transfer coefficient ($k_p\tau$)
θ = dimensionless time (t/τ)
τ = total bed volume/volumetric flow rate, s

TABLE 2. Values of parameters for Figures 1 and 2.

T, K	P, MPa	ϕ	m
313	13.8	1.07	0.00089
323	13.8	0.88	0.0036
337	13.8	0.68	0.0073
353	13.8	0.36	0.021
323	10.3	1.08	0.0013
323	16.5	0.79	0.0054
323	19.3	0.35	0.0215

REFERENCES

1. Peker, H., Srinivasan, M.P., Smith, J.M., McCoy, B.J., Caffeine extraction rates from coffee beans with supercritical carbon dioxide. AIChE J., 1992, **38**, 761-770.

MEMBRANE SEPARATION OF SUPERCRITICAL FLUID MIXTURE

KOZO NAKAMURA, TERUHIKO HOSHINO, AKITOMO MORITA
MASANORI HATTORI AND RYUJI OKAMOTO
Department of Biological and Chemical Engineering, Gunma Univ.
Kiryu, Gunma 376, Japan

ABSTRACT

The ceramic membrane as well as the polymer membrane were used to examine the possibility of membrane separation of the supercritical fluid mixture. The permeation rate of carbon dioxide exhibited a maximum peak near the critical pressure when it was measured at a constant temperature with change of pressure. Polyethylene glycol 400(PEG 400), PEG 600 and triolein were used as the model solutes with the maximum concentration of 1%. The rejection was negative in the composite polymer membrane(NTGS-2100) and approached zero with increase in the permeation flux. The rejection of PEG 400 was from 0.5 to 0.8 at 40℃ and 20 MPa in the thin porous silica membrane with the average pore size less than 0.6 nm.

INTRODUCTION

The supercritical carbon dioxide(SCCO_2) is a safe solvent with several excellent properties such as low viscosity and high diffusivity. The solubility of solutes in SCCO_2 is, however, much lower than that in appropriately chosen organic solvents, and the solvent ratio usually becomes large in SCCO_2-extraction. It means that the energy-saving recovery of SCCO_2 is necessary in a large scale extraction. The membrane could be useful for this purpose, and it will be also applied to the reactor of enzymatic reaction in SCCO_2. The research works on the membrane processing of supercritical fluid(SCF) were scarcely reported[1,2], and it is the purpose of this study to pursue the possibility of the membrane separation of SCF mixture using the specially prepared ceramic membrane as well as the commercial polymer membrane.

MATERIALS AND METHODS

(1)Membranes and modules: The polymer membrane was the composite membrane of the

polyimide ultrafiltration membrane with the skin layer of cross-linked silicone. This membrane(NTGS-2100) was developed by Nitto Denko Corp. for separation of organic gases. The circular flat membrane with diameter of 72 mm was used in the experiments. The thin porous silica membranes were prepared by the sol-gel technique using a cylindycal porous, α-alumina substrate(average pore diameter: 1 μ m, outside diameter: 1 cm, porosity: about 50%)[3]. Three membrane with the average pore size of 1.5 nm(A), 0.6 nm(B) and less than 0.6 nm(C) were used in the experiments, and the length was 51 mm. The modules were designed by ourselves and prepared by the manufacturing company of high pressure vessels.
(2)Materials: The carbon dioxide in a high pressure cylinder was of food grade and purchased from the local supplier. Polyethylene glycol 400 and 600 were the reagents of WAKO PURE CHEMICAL INDUSTRIES, Ltd., and triolein was the product of SIGMA with approximate purity of 99%.
(3)Methods: The experimental apparatus and the method were almost as same as those explained in the reference[2].

RESULTS AND DISCUSSION

(1)Permeation rate: Fig.1(a) is an example of the permeation rate when the membrane is ceramic. The permeation rate takes the maximum value near the critical pressure and then dropped to the almost constant value. This change is remarkable in the membrane with the larger pore size. The permeation rate is proportional to the reciprocal of kinematic viscosity as shown in Fig.1(b), which is expectative for the permeation of fluid through the porous membrane. Its dependence on the kinematic viscosity, however, becomes weaker beyond the critical pressure. The appearance of the peak followed by the sharp fall was also observed in the

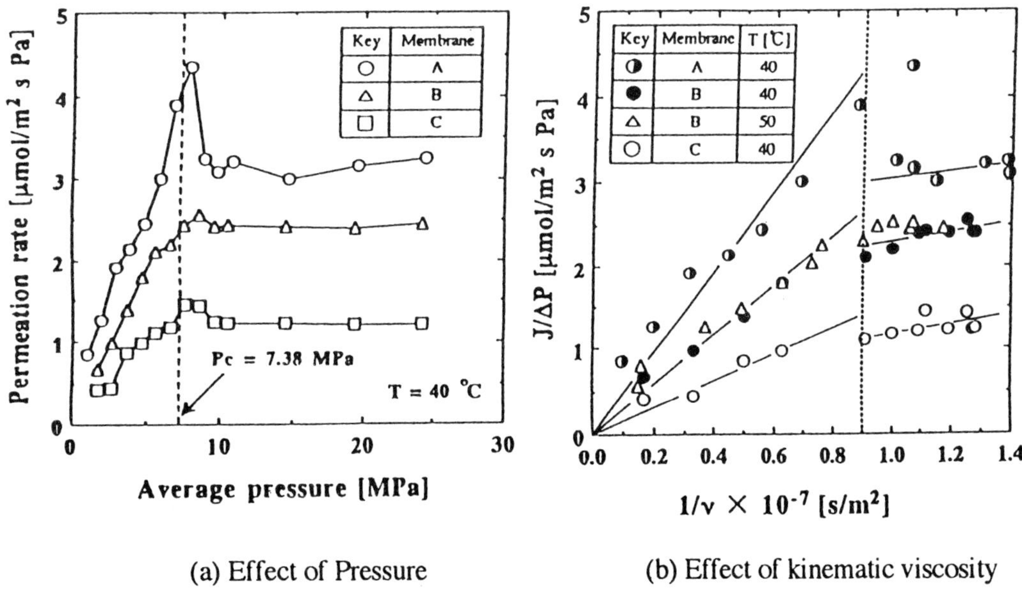

(a) Effect of Pressure (b) Effect of kinematic viscosity

Fig.1 Effect of pressure and kinematic viscosity on permeation rate of CO_2 in ceramic membranes.

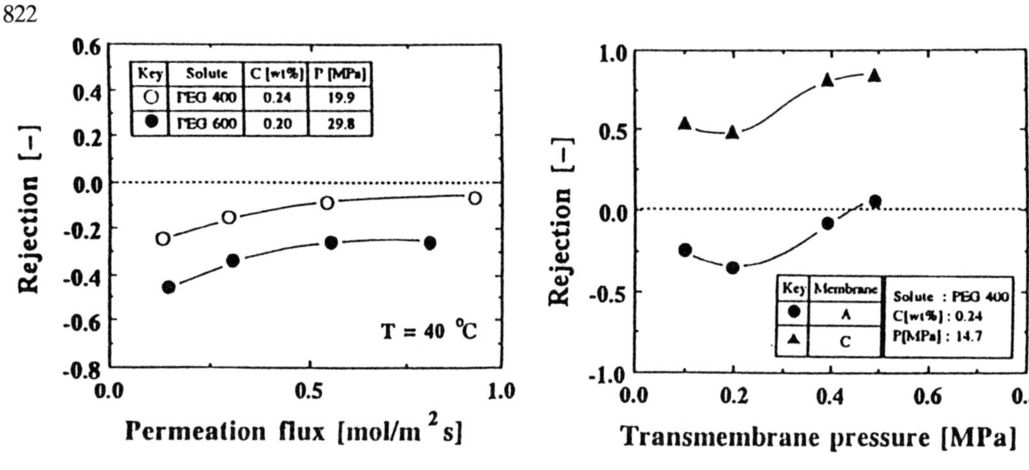

(a) Polymer membrane (b) Ceramic membrane

Fig.2 Rejection of polyethylene glycol in Polymer membrane(a) and ceramic membrane(b).

polymer membrane and could be related to the condensation of $SCCO_2$ to narrow the pores in the membrane.

Fig.2(a) and (b) show the rejection of PEG 400 and 600 measured using the polymer membrane and the ceramic membrane respectively. The rejection was negative in the polymer membrane probably due to the solubility of PEG in the polymer membrane being better than that of $SCCO_2$. The sieving effect of the membrane became remarkable to retain the solute well as shown in Fig.2(b) when one lot of the ceramic membrane with the average pore size probably less than 0.6 nm (not determined) was used. The effect of temperature, pressure, trans-membrane pressure and solute concentration was also measured, but the explanation of the results have to be omitted as space is limited.

CONCLUSIONS

1. The permeation rate of $SCCO_2$ became maximum near the critical pressure and fell sharply to the almost constant level in the polymer membrane and in the ceramic membrane with large pore size.
2. The rejection of PEG 400 and 600 was negative in the polymer membrane, while PEG 400 was well retained by the ceramic membrane the average pore size of which was supposed to be less than 0.6 nm.

REFERENCES

1. J.M.Martinet et.al.: *Proc. Int. Symp. on SCFs*, Tome **1**, 1988, p.343-347
2. K.Nakamura: *Fouling and Cleaning in Food Processing*(ed. by H.G. Kessler and D.B. Lund), 1989, p.258-267
3. M.Asaeda et.al.: *Proc. IFMEST'92 on Porous Materials* (to be published in 1993)

ADSORPTION ISOTHERMS FOR SUPERCRITICAL CARBON DIOXIDE ON PROTEINS AND POLYSACCHARIDES

KOZO NAKAMURA*, TERUHIKO HOSHINO, YOSHITSUGU SUZUKI AND NORIHIKO YOSIZAWA
Department of Biological and Chemical Engineering,
Gunma University , Kiryu, Gunma 376, Japan
* Department of Agricultural Chemistry, Faculty of Agriculture,
University of Tokyo, Yayoi, Bunkyo-ku, Tokyo 113, Japan

ABSTRACT

The adsorption isotherms for CO_2 on several proteins and polysaccharides were measured at pressures of 0-29.4 MPa by two different microbalances, gravimetric and piezoelectric. The adsorption isotherms measured gravimetrically all exhibited the maximum peak near the critical pressure Pc and remained constant at the higher pressure, the level differing each others. The piezoelectric microbalance using the quartz crystal (9 MHz, At-cut) could measure the "absolute" amount of fluid sorbed in the adsorbent which was larger especially at the high pressure than the "surface excess" adsorption measured gravimetrically. The density of adsorbed layer was calculated using the data of the surface excess and the absolute adsorption isotherms.

INTRODUCTION

The adsorption phenomenon of high pressure gas or supercritical fluid is fundamental to the application of expansion of biomaterials or sterilization of the microbial cells, and it is also related to enzymatic reactions and chromatographic or membrane separation. But the adsorption of supercritical fluid on the biological materials has not been well studied. In this study, the adsorption isotherms for carbon dioxide on several proteins and polysaccharides were measured at the pressures of 0-29.4 MPa by the different microbalances, gravimetric and piezoelectric.

MATERIALS AND METHODS

All materials used in this study were dry proteins and polysaccharides. The apparatus used for the adsorption experiments by a piezoelectric quartz crystal is shown in Fig. 1. The apparatus

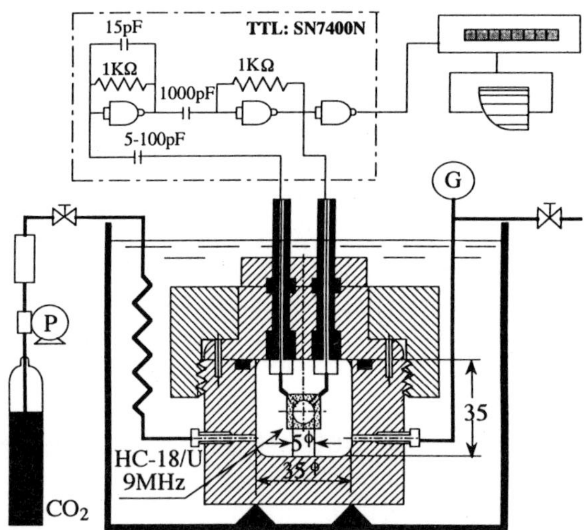

Figure 1. Experimental apparatus for piezoelectric microbalance.

by the quartz spring has been described in the literatures [1, 2]. The sample was coated on the surface of a 9 MHz At-cut quartz crystal and the frequency change was measured by a frequency counter to follow the time course of the weight change caused by sorption. The principle of piezoelectric quartz crystal microbalance is referred to elsewhere [3].

RESULTS AND DISCUSSION

Figure 2 shows the adsorption isotherms for CO_2 on polysaccharides (corn, potato, dextrin and cellulose) measured by the quartz spring at 313 K. In Fig. 2, the adsorption became maximum just above Pc and constant at the higher pressure. This profiles are similar to those of proteins reported previously [1, 2]. The high level adsorption of cellulose can be explained by its micro-structure.

Figure 3 shows the adsorption isotherms for CO_2 on proteins measured by the piezoelectric microbalance and the quartz spring at 313 K. The piezoelectric microbalance could

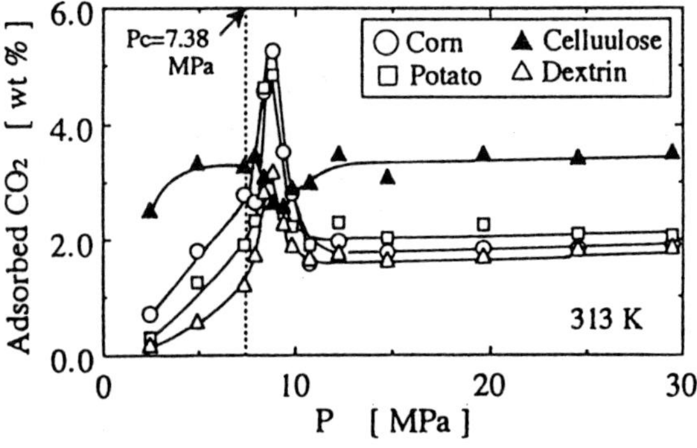

Figure 2. Adsorption isotherms for CO_2 on polysaccharides.

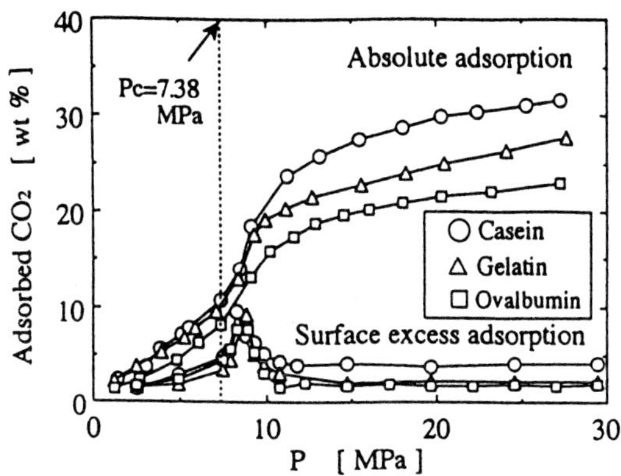

Figure 3. Adsorption isotherms for CO2 on proteins at 313 K.

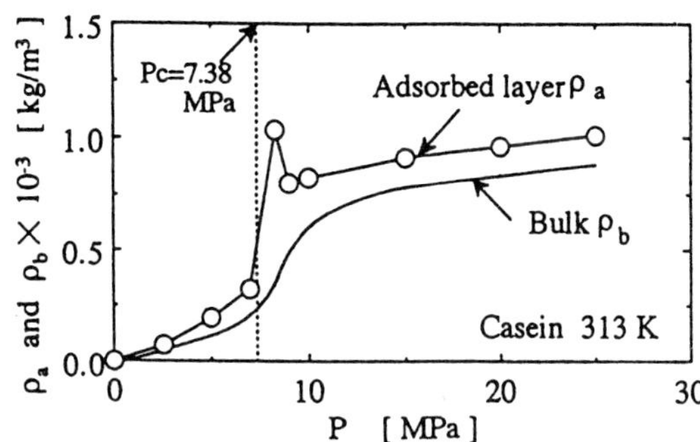

Figure 4. Density of adsorbed layer for casein at 313 K.

measure the "absolute" amount of fluid in the adsorbent which was larger especially at the high pressure than the "excess" adsorption measured by the quartz spring. The relation between the absolute adsorption ΔW and the excess adsorption $\Delta W'$ is expressed as $\Delta W = \Delta W' + \rho_b (\Delta W/ \rho_a)$. The density of the adsorbed layer ρ_a can be calculated from this equation, that is, $\rho_a = \rho_b \Delta W / (\Delta W - \Delta W')$ where ρ_b is the bulk density of CO2. The density of the adsorbed layer for casein was calculated at 313 K using the data of Fig. 3, and it is shown together with the bulk density in Fig. 4. There still appeared the maximum peak in the density of the adsorbed layer at the pressure where the adsorption became maximum. The peak density of 1.03×10^3 kg/m^3 was approximately equal to the hypothetical density estimated from the Dubinin-Nikolaev's equation which was recommended for estimation of the adsorbed layer above the critical temperature [4]. At the near-critical point, there will be occurring the clusters of CO_2 molecules which may be more condensable than the molecules. Therefore, the formation of the adsorption maximum at the near-critical point may possibly correspond to the formation and the disappepearance of cluster.

REFERENCES

1. Nakamura, K., Hoshino, T. and Suzuki, Y., *Biosci. Biotech. Biochem.*, (submitted).
2 Nakamura, K., Hoshino, T. and Ariyama, H., *Agric. Biol. Chem.*, 1991, **55**, 2341-2347.
3. Stockbridge, C.D., In Vacuum Microbalance Techniques, ed. H.K.Behendt, Plenum Press, NY, 1966, vol.5, pp. 147-190.
4. Suzuki, M., Adsorption Engineering, Elsvier, Oxford-NY-Tokyo, 1990, p. 44-46.

APPLICATION OF SUPERCRITICAL CO_2 FOR FOOD PROCESSING

Ž. KNEZ, M. ŠKERGET
University of Maribor, Faculty of Technical Sciences,
Dept. of Chemical Engineering, SLO-62000 Maribor,
Smetanova 17, Slovenia.

ABSTRACT

The influence of the process parameters to the composition of the extracts and mass transfer coefficients for extraction of paprika are presented.

The equilibrium solubilities of capsaicin (pungent component of paprika) in liquid and supercritical CO_2 were measured by the static analytic method. The data were correlated by use of the thermodynamic model relating the solubility to the solvent density and a model for the dependence of the enhancement factor on the density of the solvent. The data were also correlated with the Peng-Robinson equation of state.

INTRODUCTION

High pressure technology offers the industry an enormous opportunity to develop novel products of high value. High isostatic pressure is able to inactivate microorganisms and enzymes and thus can be applied for preservation of some foodstuffs which is usually performed by high temperature treatment (1).

Recently supercritical fluids have been applied as a solvent for non extractive applications in high pressure micronisation, in chromatography and as a chemical and biochemical reactions media (2). On the other hand, supercritical fluids have been used as a solvent for a wide variety of extractive applications.

For the design of a sub- or supercritical fluid extraction system, data are required on the pressure and temperature required for extraction and separation, the type and quantity of the solvent, the recirculation rate, and energy consumption. This information can be obtained from phase equilibria, mass transfer measurements, and from the thermodynamic analysis of the conditions in the extraction and separation step (3).

In the literature (4) the process for production of paprika oleoresin with supercritical CO_2 are relatively well described, however, there are practically no data on the solubility of capsaicin and no mass transfer data for extraction of paprika.

EXPERIMENTAL

Apparatus and experimental procedure for determination of mass transfer coefficients and of solubility of capsaicin CO_2 was already described (5,6).

RESULTS AND DISCUSSION

Mass transfer coefficients

Paprika was extracted in the temperature range 20-60°C and in the pressure range 80 to 450 bar and the kinetics of extraction were followed.

Mass transfer coefficients were determined for the constant rate periods where steady state mass transfer prevails(3), and are presented in Table 1.

TABLE 1

Parameters for the extraction of paprika and mass transfer coefficients.

Diameter of equivalent sphere	$d = 0.15x10^{-3}m$				
Specific area	$A_s = 27200\ m^2/m^3$				
Solvent	CO_2				
Temperature	42°C				
Pressure (bar)	90	90	150	300	400
Solvent flow rate (kg/h)	23.5	32.8	25.78	30.45	32.7
Mass transfer coeff. (10^7 m/s)	0.53	1.47	0.56	0.37	0.30

Equilibrium data

The experimental equilibrium solubility data for capsaicin in carbon dioxide were determined at the temperatures of 25, 40 and 60°C and in the pressure range from 70 to 400 bar. The equilibrium mole fraction (y_i) of the solute in the supercritical CO_2 as a function of system pressure is presented in Figure 1. The data were correlated with thermodynamic correlations such as ln y vs. ln ρ_r and log E vs. ρ_r (5), and with the Peng-Robinson equation of state. The physical data for capsaicin were determined by the Lydersen group contribution method.

The constants Ω_a,Ω_b and acentric factor in the Peng-Robinson equation of state were for capsaicin not evaluated from critical data. They were obtained by regressing the solubility data measured at 25 and 40°C. On this way the PREOS was modified. The omega-parameters were »set free«, so they have different values from those of PR. The results are summarised in Table 2. It can be seen that the temperature does not much influence the coefficient k_{ij}, which could be the consequence of the way of calculation.

Fig. 1 shows that the approximation of the experimental data is good for 25°C, where the average absolute relative deviation is 4.63% but it is worse for 40°C, where AARD is 30.07%.

TABLE 3
Properties of Capsaicin estimated by regressing the experimental data with the modified PREOS

Property	Value of property
Ω_a	0.13797
Ω_b	0.15968
χ	5.89741
$v^s(m^3/kmol)$	0.89142
$k_{ij}(298K)$	-0.27900
$k_{ij}(313K)$	-0.27700

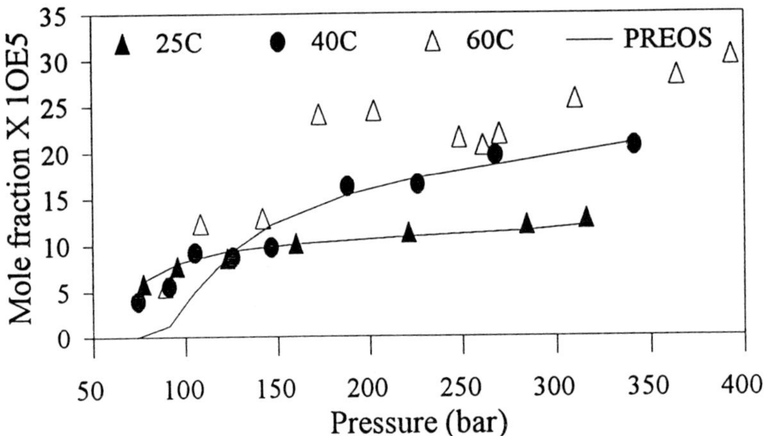

Fig. 1. Solubility isotherms of Capsaicin as a function of pressure and PREOS corr.

CONCLUSION

The measured mass transfer coefficients for extraction of paprika and equilibrium solubilities of capsaicin in CO_2 can be useful in design of paprika extraction plant.

REFERENCES

1. Hayashi, R., Utilization of pressure in addition to temperature in food science and technology, In High pressure and Biotechnology, Eds. C. Balny, R. Hayashi, K. Hermans, P. Masson, John Libertey, Grand Mote, 1993, pp 185-92.
2. Nakamura, K., Hoshino, T., Novel Utilization of Supercritical Carbon Dioxide for Enzymatic Reactions and other processings. Int. Workshop, Jakarta, Sept. 1991
3. Brunner, G., Mass separation with supercrit. gases, Int. Chem. Eng., 1990, **30**, 191-205.
4. Coenen, H., Hagen, R., Knuth, M., German Pat. DE 3114593 C1, 1982.
5. Knez, Ž., Steiner, R., Solubility of Capsaicin in dense CO_2, The Journal of Supercritical Fluids, 1992, **5**, 251-5.
6. Knez, Ž., Posel, F., Hunek, J., Golob, J., Extraction of plant materials with supercritical CO_2. In Proc. II. Int. Conf. on Sup. Fluids,ed.M.Mc Hugh, Boston, 1990, pp. 101-4.

SUPERCRITICAL FLUID EXTRACTION OF AROMA COMPOUNDS FROM AROMATIC HERBS (*Thymus zygis* and *Coriandrum sativum*).

A. Lopes Cardoso, M. Moldão-Martins, G. Bernardo-Gil*, M.L. Beirão da Costa
Dept. of Food Sci. & Tech. Inst. Sup. Agronomia, Tapada da Ajuda, 1399 Lisboa Codex.
*Dept. of Chem., Inst. Sup. Técnico, Av. Rovisco Pais, 1096 Lisboa Codex, PORTUGAL

ABSTRACT

Supercritical fluid (SCF) extraction was compared to steam and Clevenger distillation, in terms of yield and composition of extracts, when applied to thyme and coriander leaves. In thyme extracts, higher production yields were always obtained by the Clevenger extraction method. Using SCF extraction high yields were also obtained but the extracts included other kinds of compounds. In respect of chemical composition of steam and Clevenger distillation extracts show similar profiles. Supercritical extracts present, besides the same components as that of steam distillation, non aromatic compounds. All the extraction methods tested produced very low yields of extracts when applied to coriander leaves, nevertheless the patterns of the chromatograms are quite different among those methods.

INTRODUCTION

Aromatic herbs are quite abundant as flavoring ingredients in the productions of food. Many of these plants only produce large quantities of flavor compounds in a very short period of the vegetative cycle. So, an important task is to find the most suitable extraction procedure and to produce an extract of interesting composition, with good yield.

Thyme and coriander leaves are the herbs commonly produced in Portugal. In a previous work [1] the evolution of thyme aromatic compounds throughout the vegetative cycle were studied and it was found that the richer extracts were produced at the flowering and post-flowering stages.

CO_2 supercritical fluid extraction is an interesting process to be applied in the food industry as it produces solvent-free extracts and is non-toxic [2]. On the other hand it should be applied only when high added values products are expected.

The aim of present work is to study whether the SCF extraction used under several experimental conditions can produce aroma from coriander and thyme. The yields and the composition of the extracts were compared to those obtained by steam and Clevenger distillation.

MATERIAL AND METHODS

Raw materials
Thymus zygis spp silvestris was collected in the north of Portugal during the flowering period, and was stored in the dark at the room temperature for a week.
Coriandrum sativum leaves were purchased from commercial sources.

Extraction methods
SCF extraction was conducted under the following experimental conditions:
Thyme: T=313 K at 10, 15, and 20 MPa.
Coriander leaves: T=308 K at 9.5, 12.5 and 20 MPa.
Steam and Clevenger distillation were carried out at the atmospheric pressure for 30 min.

Methods of analysis
GC analysis was performed with HP 5890 instrument equipped with a HP-5 (5% diphenil, 95% dimethylpolysyloxane (50 m x 0.32 mm; coating thickness 0.17 μm) and a FID.

Analytical conditions: injector temperature 473 K; detector temperature 523 K; oven temperature 333 K/10 min; oven temperature programmed 337 K- 453 K at 2 K/min, then 453 K - 473 K at 10 K/min, then isothermal at 473 K for 50 min; flow rate of carrier gas 2 ml/min.

Relative concentration were evaluated using the peak areas, without correction for the response factor.

GC-MS analysis were performed with HP-5890 instrument equipped with a WAX capillary column (30 m x 0.32 mm; coating thickness 0.1 μm) and 7.00 eMvolts mass selective detector.

Analytical conditions: injector temperature 503 K; oven temperature 333 K/15 min, oven temperature programmed 333 K - 493 K at 2 K/min, then isothermal for 15 min; flow rate of carrier gas 2 ml He/min; source 70 eVolts.

Identification of compounds: standard of most of components were used. When the standards are not available mass spectrum of unknown compound were compared to the library.

RESULTS AND DISCUSSION

SCF extraction of thyme showed that the maximum yield was obtained for 313 K at 15 MPa, althougth the other two yields were quite similar. This yield (6.36%) is considerably higher than the one obtained by steam and Clevenger distillation (0.6 % and 0.7 %, respectively). This difference is explained by a high extraction of other than aroma compounds like waxes, carotenoids and chlorophyls.

The difference in composition of the extracts obtained by the different methods is shown in Figure 1. It can be seen that geranyol and geranyl acetate are almost exclusively extracted in the distillation procedures.

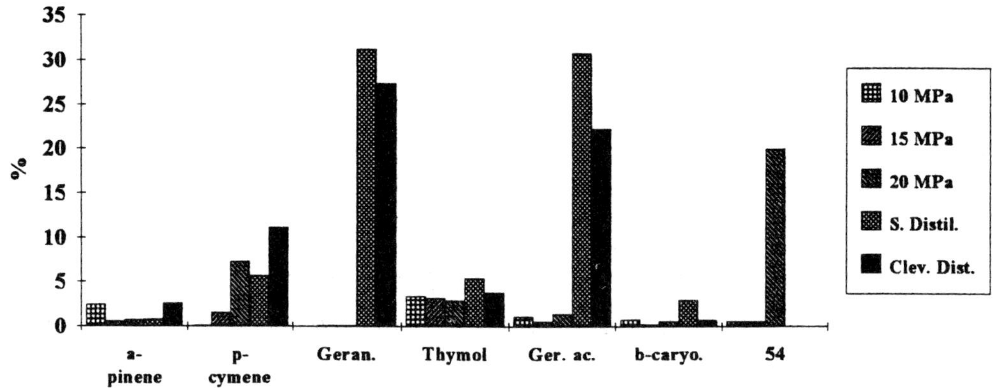

Figure 1. Comparative extract composition of thyme produced by different methods

α-Pinene is present to a similar extent in the SCF extracts obtained at 313 K at 10 MPa and in the product of Clevenger distillation. p-Cymene is almost absent in the SCF extract obtained at 10 MPa (0.07%) and is very low in the product obtained at 15 MPa (1.5%) although it is present in a high content on the other extraction procedures (11% for Clevenger, 5.6% for steam distillation and 7.2% for 20 MPa). Thymol is extracted at similar levels by all the tested methods.

Based on the results obtained in the extraction of coriander leaves, it must be stated that the yields were too low for all the methodologies to be quantified. SCF extraction showed that the maximum yield was obtained for 308 K at 12.5 MPa, althoagh the other two yields were quite similar. The difference in composition of the extracts obtained by the different methods is shown in Figure 2. Steam distillation produced a richer extract, mainly composed by monoterpenes, aldehydes and alcohols. Monoterpenes appear at a high composition in the supercritical fluid extract obtained for 308 K at 9.5 MPa. The SCF extracts are rich for higher molecular weight compounds, which have not yet been identified.

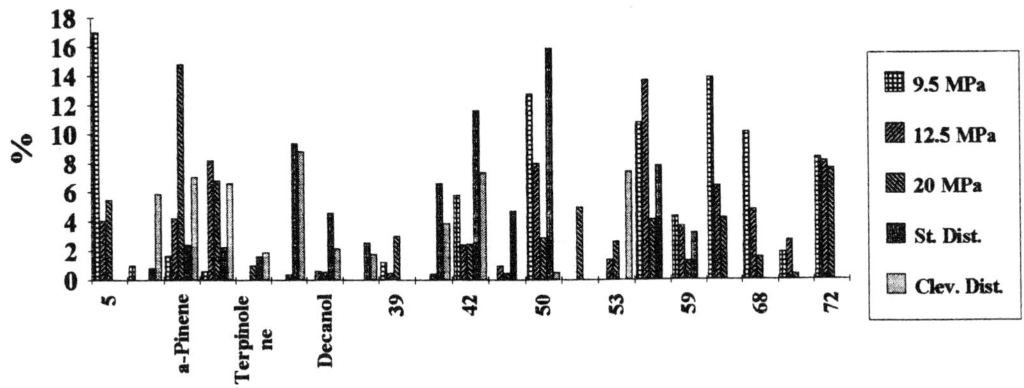

Figure 2. Comparative extract composition of coriander produced by different methods

REFERENCES

1. Moldão-Martins, M.M., Beirão-da-Costa, M.L.and Bernardo-Gil, M.G.,Comparative extraction procedures of volatile compounds from *Thymus Sigis*. 8th world Cong. of Food Sci. and Techn., Toronto, September 1991.
2. Caragay, A.B.and Little, A.D., Supercritical fluids for extraction of flavours and fragrances from natural products. Perfumer & Flavourist, 1981, **6**, 43-55.

Separation of aroma components from soy sauce by continuous supercritical CO_2 extraction

Yoshihisa KITAKURA*, Hitoshi IMAMURA**, Saburo HAYAKAWA**
Mitsutoshi HAMANO*, Hikotaka HASHIMOTO*

* Kikkoman Corporation, 339 Noda, Noda-shi, Chiba 278, JAPAN
** Nippon Sanso Corporation, 4-320 Tsukagoshi, Saiwai-ku, Kanagawa 210, JAPAN

ABSTRACT

Traditional soy sauce making process have not been able to separate aroma compounds from body component, which are regarded as useful food resources.

In this investigation, soy sauce aroma components were successfully separated from soy sauce by using a continuous supercritical CO_2 extraction system and the concentrate of soy sauce aroma was extracted at the same time. These new high quality food resources have not been available using traditional processes.

INTRODUCTION

Soy sauce is one of the Japanese traditional seasonings and is indispensable for Japanese-style cooking. Improvement of the separating technologies enables us to make new types of soy sauce. For example, electrical dialysis makes it possible to provide lower-salt soy sauce, which is suitable for preventing excess sodium intake; spray-drying makes it possible to provide powdered soy sauce, which is used for instant noodle soup, and other dried foods [1]; membrane makes it possible to provide non-pasteurized soy sauce, which has neither cooking flavor nor color increase [2].

This investigation is a part of an effort to develop a new type of soy sauce. A continuous supercritical CO_2 extraction system was expected to provide new soy sauce which has less aroma but has delicious taste and soy sauce aroma concentrate at the same time. Those new types of soy sauce are regarded as useful food resources.

MATERIALS AND METHODS

Materials

In this investigation, a soy sauce model system simulating real soy sauce was applied. It is comprised of a solution of salts, acids, amino acids, and several soy sauce aroma components in water. Each aroma component is added at the concentration of 100 ppm. This soy sauce model system is easy to analyze.

Extracting conditions and aroma concentrating conditions were investigated using this soy sauce model system and the examination using real soy sauce followed.

Extraction

A continuous supercritical CO_2 extracting pilot plant with a counter flow extracting tower was used. To judge the extracting conditions, the elimination of isoamylalcohol, furfurylalcohol, methionol, 2-acetylpyrrole was caliculated;

$$\text{Elimination} = 1 - \frac{\text{Final aroma concentration of solution}}{\text{Initial aroma concentration of solution}}$$

Quantitative analyses were carried out by gas chromatography.

Concentration

Since aroma components are very thin, it was decided to concentrate soy sauce aroma components into solvent. An absorbing column was attached to the continuous supercritical CO_2 extracting pilot plant.

We investigated the recovery of aroma components from aromatic CO_2 into absorbing solvent. First, absorbing pressure and temperature were examined. Next, various solvents were examined. The concentrating conditions were judged by aroma concentration in absorbing solvent after experiments with the same absorbing period.

RESULTS AND DISCUSSION

Extraction conditions

A Raschig ring was decided to be used because of its high elimination data. The longer extracting tower achieved higher elimination. As for extracting temperature, 40° C and 60° C gave higher elimination than 20° C. Extracting pressure was responsible for the elimination of furfurylalcohol and methionol. These components are more easily taken away under higher pressure.

Lower CO_2 flow rate and lower food material flow rate show higher elimination and furfurylalcohol and methionol are mostly eliminated under these conditions.

Concentration conditions

Atmospheric pressure, and 50, 80, and 200 kgf/cm^2 were examined to trap aroma concentrate. Among those pressures, 50 kgf/cm^2 was most suitable for our purpose. As for temperature, 40°C is more effective for aroma recovery than 30°C.

Water, 50% (w/w), and 99.4% (w/w) ethanol solutions were examined as solvents. Water is not suitable for solvent, because it does not absorb aroma components efficiently. On the other hand, the volume of 99.4% ethanol reduced during concentration. It appeared that CO_2 gas blows out ethanol, so 99.4% ethanol is not suitable for our purpose. It was decided to use 50% ethanol solution as the solvent.

Production

An extracting tower 1500 mm long with a Raschig ring in it was prepared and real soy sauce aroma was extracted under the condition of 200 kgf/cm^2, 40°C, and CO_2 flow rate of 12Nl/min. Soy sauce was fed under flow rates of 15 ml/min and 5 ml/min.

Aroma concentration of soy sauce becomes less than 1 ppm. Very weak soy sauce aroma was smelled when soy sauce was fed at 15 ml/min but less aroma at 5 ml/min.

Aroma components were absorbed into 100 ml of 50% ethanol solution under the condition of 50 kgf/cm^2, 30°C. Almost all aroma components in solvent are concentrated up to 10 or 30 times as dense as those in soy sauce. Loss of solvent was only 2 or 7 ml, so we expected this condition may be practicable.

CONCLUSIONS

From the above experiments, it is concluded that less aroma soy sauce and soy sauce aroma concentrate can be obtained using this pilot plant under the conditions suggested by the present investigation using a soy sauce model system.

ACKNOWLEDGMENT

This investigation was carried out under the governmental project of "The Japanese Research and Development Association for High Separation System in Food Industry" and was mentioned in its report published in book form [3].

REFERENCES

1. Hamano,K.,Behaviors of Water and Quality on the Dehydrated foods. Nippon Shokuhin Kogyo Gakkaishi, 1982, **30**, 125-132.

2. Hamano,K. and Suzuki,K., Nama-shoyu no Tokusei to Sono Seizo-gijutu. Japan Food Science, 1991, **30**, 53-57.

3. Renzoku-chorinkai CO_2 Chusyutu-ho ni yoru Shoyu no Ko-fuka-kachi-ka Gijutu no Kaihatsu. In Kinousei Shokuhin Sozai no Koudo Bunri Seisei to Kaihatsu, Syokuhin Sangyo High Separation System Gijutsu Kenkyu Kumiai, Tokyo, 1992, pp.453-478.

SUPERCRITICAL CARBON DIOXIDE EXTRACTION OF CAROTENOIDS FROM CARROTS

MOTONOBU GOTO, MASAKI SATO AND TSUTOMU HIROSE
Department of Applied Chemistry, Faculty of Engineering, Kumamoto University, 2-39-1 Kurokami Kumamoto 860, JAPAN

ABSTRACT

Carotenoids were extracted from freeze-dried and raw carrots with supercritical carbon dioxide. For the freeze-dried carrots carotenoid existing only about 10μm in the outer shell of each carrot particle was rapidly extracted and the rest was extracted very slowly. However for the raw carrots carotenoid was completely extracted and the extraction rate increased with pressure and ethanol as an entrainer. HPLC analysis of the extracts indicated that extracted carotenoids are mainly α- and β-carotene.

INTRODUCTION

Carotenoids are one of the major natural pigments which are widely distributed in nature. Carotenoids, in particular β-carotene, are important for food and pharmaceutical industries, since they are a precursor to vitamin A and a coloring material. The extraction of carotenoids from natural materials by supercritical carbon dioxide extraction process has been studied [1-3]. The objectives of this paper are to evaluate the effect of water present in the carrot cells on the extraction process and to study the extraction kinetics of carotenoids from carrots with SC-CO_2.

EXPERIMENT

Carrots were diced into 5mm cubes and freeze-dried. Freeze-dried carrots were ground and sieved into average size of diameter of 0.26mm, 0.47mm and 1.12mm. Fresh ground carrots were used as raw carrot sample. The carotenoid content in the carrots was determined by absorbance measurement at 448nm of ethanol extracts with cell disruption.

Freeze-dried or raw carrots were placed in an extractor (10ϕ x13mm). Pure CO_2 or CO_2 with 1, 3 and 5wt% ethanol was used as solvents. Extraction was carried out at 313, 323 and 333K and at pressures of 7.8 - 29.4MPa. The effluent concentration of carotenoids in SC-CO_2 was monitored continuously with a high-pressure UV detector at 436nm for dried carrots / pure CO_2 system. During the extraction run, the extracts were dissolved in ethanol periodically and the amounts of carotenoids were measured by the absorbance. The constituents of the extracts were determined by HPLC with YMC C-18 column with ethanol/acetonitril (40/60v/v).

RESULTS AND DISCUSSION

Initial carotenoids contents were about 7.0×10^{-3} and 2.0×10^{-2} kg/kg-dry for freeze-dried and raw carrots, respectively. Carotenoids must be partially denatured during the freeze-drying process. β-carotene is unstable, e.g., cis isomerization accelerated by heat and light [3]. The absorbance may decrease with the denaturation to reduce carotenoids contents.

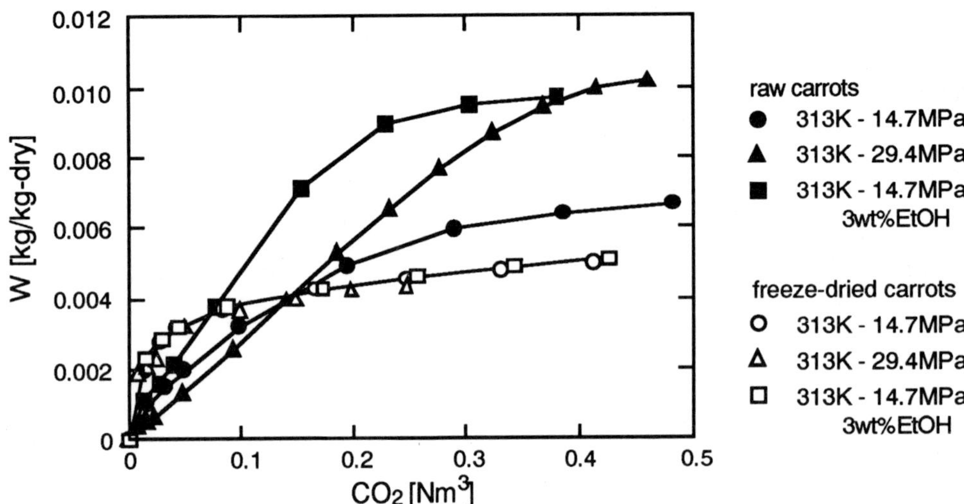

Figure 1. Effect of pressure and entrainer on the cumulative extraction histories from raw and freeze-dried carrots (Dp < 0.1mm, flow rate = 1.1×10^{-7} m^3/s at pump).

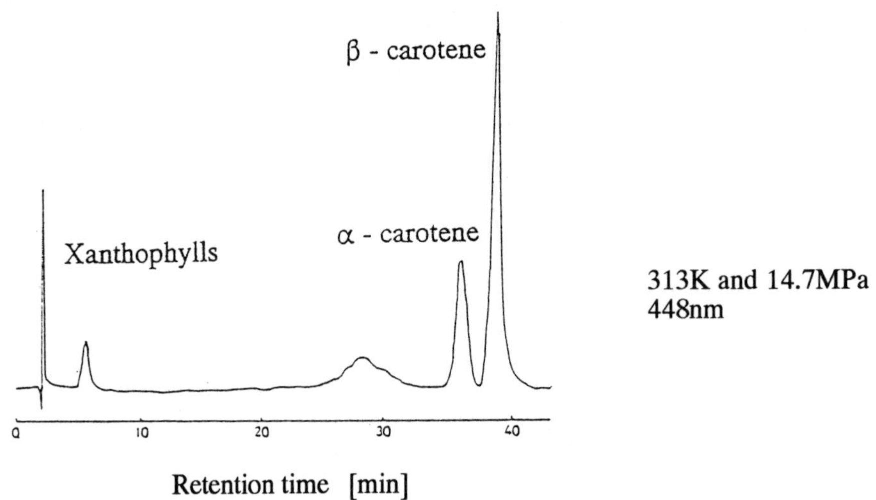

Figure 2. HPLC chromatogram of the extracts from raw carrot with SC-CO_2.

The influences of pressure and entrainer are shown in Figure 1 for both freeze-dried and raw carrots. The extracted residue was colorless for raw carrots, while it was still colored for freeze-dried carrots. For extraction from raw carrots, carotenoids were extracted completely and the increase in pressure and the addition of entrainer enhanced the extraction rate, because of increased solubility of carotenoids [3]. On the other hand, for extraction from freeze-dried carrots, about 70% of carotenoids were extracted and neither pressure nor entrainer affected the extraction histories. The discrepancy between the amounts extracted after the complete extraction and the original contents for raw carrots may be owing to denaturation of carotenoids during the extraction or the measuring process. Although the initial extraction rate for freeze-dried carrots was faster than that for raw carrots, the extraction rate for freeze-dried carrots decreased soon.

HPLC chromatograms showed that pigments extracted were mainly α- and β-carotene (Figure 2), and the relative amount of β-carotene was decreased by the freeze-drying process.

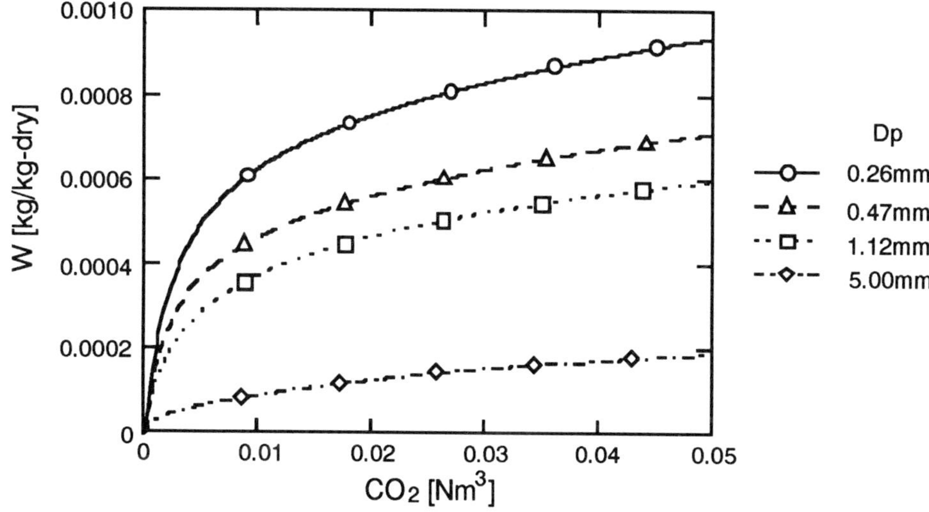

Figure 3. Effect of particle size on the cumulative extraction histories for freeze-dried carrots (313K, 14.7MPa, flow rate = $1.1x10^{-7}$ m^3/s at pump).

For the extraction from freeze-dried carrots, the influence of particle size on the extraction histories is shown in Figure 3. In each run, only the carotenoids existing in about 10μm of the outer shell of each particle, which corresponds approximately to the size of a broken carrot cell, could be extracted rapidly. The rest part was extracted very slowly. To extract carotenoids completely the carrot cells should be broken for freeze-dried carrots.

CONCLUSION

The cell membranes of freeze-dried carrots prevented the extraction of carotenoids with SC-CO_2. For complete extraction of carotenoid from freeze-dried carrots it is required disruption of the carrot cells. For raw carrots cell disruption was not essential.

ACKNOWLEDGMENT

This work was supported by a Grant-in-Aid for Scientific Research (No. 04238106) from the Ministry of Education, Science and Culture, Japan.

REFERENCES

1. Lorenzo, T.V., Schwartz, S.J. and Kilpartric, P.K., Supercritical fluid extraction of carotenoids from *Dunaliella algae*, Proc. of 2nd Int. Symp. on Supercritical Fluids, 1991, 297-298.

2. Favati, F., King, J.W., Friedrich, J.P. and Eskins, K., Supercritical CO_2 extraction of carotene and lutein from leaf protein concentrates., J.Food Sci., 1988, **53**, 1532-1536.

3. Cygnarowicz, M.L., Maxwell, R.J. and Seider, W.D., Equilibrium solubilities of b-carotene in supercritical carbon dioxide, Fluid Phase Equilibria, 1990, **59**, 57-71.

THE PERFORMANCE OF PREPARATIVE SUPERCRITICAL-FLUID CHROMATOGRAPHY OF LIPIDS AND RELATED MATERIALS

SHUICHI YAMAMOTO AND YUJI SANO
Department of Chemical Engineering,
Yamaguchi University, Tokiwadai, Ube 755, JAPAN

ABSTRACT

The plate height($HETP$) and the mobile phase velocity(u) relationships on ODS-silica gel columns of various particle diameters ($5 \sim 30\mu$m) were measured. Both the liquid (methanol-water) and the supercritical carbon dioxide were employed as the mobile phase. Naphtalene and its derivatives as well as unsaturated fatty acids were used as test samples(tracers). The stationary phase diffusion coefficient, the axial dispersion coefficient and the distribution coefficient were determined from the $HETP - u$ data for various experimental conditions and discussed.

INTRODUCTION

Supercritical-fluid chromatography(SFC) has several advantages against conventional liquid chromatography(LC) for the preparative and process-scale separation of lipids and related materials[1]. However, it is not easy to optimize the chromatographic separation process as a large number of operating and column variables must be tuned to obtain the maximum productivity[2]. In addition, SFC has two more important operating variables, temperature T and pressure p, which are usually not important in LC. It is therefore needed to investigate the fundamental separation mechanism in SFC over a wide range of experimental conditions in detail.

In this paper, the plate height(HETP) and the mobile phase velocity(u) relationships on ODS-silica gel columns of various particle diameters ($d_p = 5 \sim 30\mu$m) were measured. Both the liquid (methanol-water) and the supercritical carbon dioxide (SC-CO_2) were employed as the mobile phase. Naphtalene and its derivatives as well as unsaturated fatty acids were used as test samples(tracers). The stationary phase diffusion coefficient, the axial dispersion coefficient and the distribution coefficient were determined for various experimental conditions and discussed.

RESULTS AND DISCUSSION

$HETP - u$ curves obtained for SFC and LC are shown in **Figs.1** and **2**. These data were analyzed on the basis of the following equation[3,4].

$$HETP = 2(D_L/u) + d_p^2 HKu/[30D_s(1 + HK)^2] \quad (1)$$

where D_L=axial dispersion coefficient, D_s=stationary phase diffusion coefficient, K = distribution coefficient and H = volumetric phase ratio. The results are shown in **Table 1**. It is seen that the KD_s/D_m values and the $Pe(= ud_p/D_L)$ values for SFC are comparable to those for LC. When the $HETP - u$ values of SFC and LC are re-plotted as the reduced $HETP$, $h(= HETP/d_p)$ - the reduced velocity ν=(ud_p/D_m), they were expressed by a single curve(data not shown).

From these results it may be concluded that high separation efficiency at high flow velocities can be expected for SFC because the molecular diffusion coefficient is high and the viscosity is low(pressure drop is low).

These data were obtained at low sample loadings where the effect of sample volume or concentration to the peak width is negligible. Further investigation is needed on the elution behavior at high (or overloaded) sample loadings.

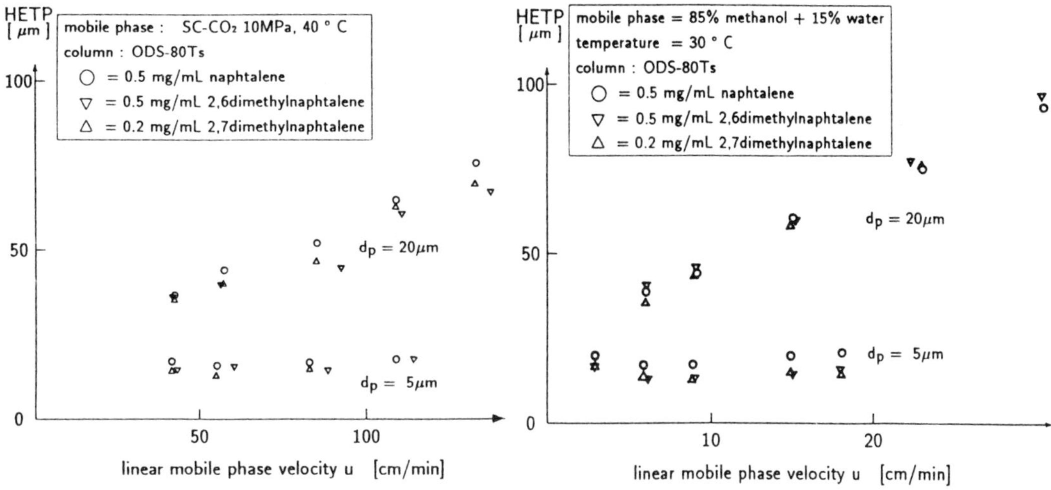

Fig.1 $HETP$ *vs.* mobile phase velocity u(SFC). Fig.2 $HETP$ *vs.* mobile phase velocity u(LC).

Table 1 Values of the parameters obtained from the $HETP - u$ data

solute	K [-]	$D_s \times 10^6$ [cm^2/s]	$D_m \times 10^6$ [cm^2/s]	KD_s/D_m [-]	Pe [-]	mobile phase
naphthalene	1.6	2.0	12	0.28	1.6	85%methyl-
2,6 dimethyl naphthalene	2.8	1.4	10	0.41	1.8	alcohol
2,7 dimethyl naphthalene	2.8	1.5	10	0.42	1.5	30°C
naphthalene	3.9	6.6	160	0.16	2.1	SC-CO2
2,6 dimethyl naphthalene	5.3	5.5	137	0.21	2.2	10MPa
2,7 dimethyl naphthalene	5.3	6.3	137	0.24	1.8	40°C

Molecular diffusion coefficient(D_m) values were calculated by the Wilke-Chang Equation.
Column:ODS-80Ts (d_p = 20μm, 4.6mm diameter and 15 cm length, Tosoh, JAPAN)

REFERENCES

1. Engelhardt,H., Gross, A., Mertens, R. and Petersen,R.,*J.Chromatogr.*, 1989, **477**,169.
2. Yamamoto,S., Nomura,M. and Sano,Y.,*J.Chromatogr.*, 1990, **512**, 77.
3. Yamamoto, S., Nakanishi, K. and Matsuno,R., *Ion Exchange Chromatography of Proteins*, Marcel Dekker, New York, 1988.
4. Yamamoto,S., Nomura,M. and Sano,Y.,*J.Chromatogr.*, 1987, **394**, 363.

CONTINUOUS PROCESSING OF MILK FAT WITH SUPERCRITICAL CARBON DIOXIDE

S.S.H. RIZVI and A.R. BHASKAR
Institute of Food Science
Cornell University, Ithaca, NY 14853

INTRODUCTION

Milk fat fractionation is an area of current research which holds potential for development of novel ingredients with varied functional properties based on their unique physical and chemical characteristics. As a first approximation, the ability to manipulate solvating power of supercritical fluids by varying density, offers a feasible approach to customized fractionation of milk fat by the phenomena of selective distillation and extraction. Research on milk fat fractionation with supercritical carbon dioxide (SC-CO_2), as with most other commercial uses of this technology, has thus far concentrated mostly on batch systems. With fluids, such as milk fat, which can be continuously pumped at high pressures, it is reasonable to anticipate that the processing time can be minimized and the economics made more favorable (1).

PHASE EQUILIBRIA

Studies on SC-CO_2 phase equilibria are needed to provide a deeper understanding of the processing conditions so that separation can be influenced on a more selective basis. This information can then be used to obtain new products (fractions) in a most cost-effective manner.

An extensive study of fluid-liquid equilibria for milk fat using a static method has been done by Yu et al. (2). They correlated experimental fluid liquid equilibria data of milk fat with SC-CO_2 using the Peng-Robinson (PR) equation of state (EOS) with Panagiotopoulos and Reid (PR) mixing rule.

In order to understand the separation of cholesterol from triglycerides, Yu et al. (2) used selectivity as an index. They found that the highest selectivity of cholesterol occurred at pressures between 8-12 MPa at 313.15 and 333.15 K. Yet, the highest selectivity value obtained is not sufficient enough for clean removal of cholesterol. Use of in-line adsorbents has been shown to remove over 80% cholesterol (3).

FRACTIONATION

Using a continuous countercurrent SC-CO_2 system (Figure 1), we have fractionated milk fat into six fractions, S1-S5 and a flavor fraction (S6) (Table 1). The short-chain (C4 - C8) and medium-chain (C10 - C12) fatty acids increased from S1 - S5. The long-chain (C14 - C18) and the unsaturated fatty acids decreased gradually from S1 - S5. The unsaturated to saturated fatty acid ratio decreased from S1 - S5, with a range of 0.75 - 0.47 as compared to 0.57 for milk fat.

The triglycerides followed the same trend as the fatty acids. The low-melting (LMT) and medium-melting (MMT) triglyceride concentration increased from S1 - S5, while the high-melting (HMT) triglyceride decreased from S1 - S5. The last fraction (S6) is a flavor fraction enriched with lactones which are four times the concentration in milk fat. The cholesterol content decreased in the raffinate (S1) by 51% but increased in the fractions S3,S4,S5.

The solid fat content is an indication of hardness for fats and oils. The percentage of solid and liquid fat at different temperatures calculated from thermograms of milk fat and two of its fractions are shown in Figure 2. The solid fat content increased in the order of S5-S4-S3-milk fat-S2-S1 fractions.

ECONOMICS

Singh and Rizvi (4) conducted a detailed economic analysis for separation of milk fat into four fractions. The calculated conversion cost for such a process was only 14¢/kg. This cost is considerably less when compared to other commercial uses of SC-CO_2 (Table 2).

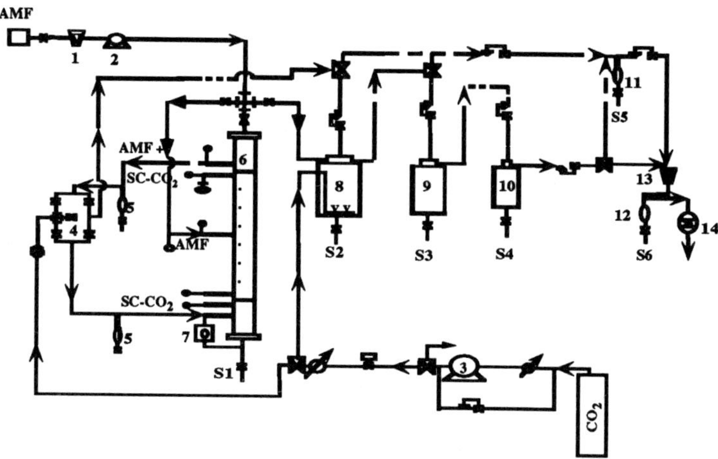

Figure 1. Schematic diagram of pilot-scale continuous $SC\text{-}CO_2$ processing system.

TABLE 1
Chemical composition of milk fat and its fractions with $SC\text{-}CO_2$.

	Feed	Raffinate	Extract				Flavor
	Milk fat	S1	S2	S3	S4	S5	S6
Temp(°C)	40	40	60	75	60	60	4
Press(psi)	3500	3500	3500	2500	1000	500	7
Yield(wt%)	100	21.0	15.0	48.0	4.0	11.0	1.0
FA(s)[1]:							
C 4-C 8	8.55	1.22	5.97	10.14	10.67	12.42	N/A
C10-C12	4.60	1.95	4.17	5.34	5.91	5.88	N/A
C14-C18	86.85	96.83	89.86	84.52	83.42	81.69	N/A
Unsat	31.28	41.57	34.37	28.19	27.23	26.32	-
Sat	55.01	55.26	55.49	56.33	56.19	55.59	-
Unsat/Sat	0.57	0.75	0.62	0.50	0.48	0.47	-
TRG(s)[2]:							
C24-C34	16.72	Traces	10.18	18.82	24.30	26.39	N/A
C36-C40	50.85	17.07	49.94	56.19	53.62	54.22	N/A
C42-C54	32.93	82.93	39.88	24.99	22.08	19.39	N/A
Chol[3] (mg/ 100g)	240.6	117.6	234.6	251.8	363.6	353.7	N/A
Chol[3] change (%)	-	-51.1	-2.5	+4.7	+50.7	+47.0	-
Carotenoids[4] (IU/100 g)	314	768	N/A[6]	N/A	N/A	N/A	N/A
Lactones[5] (μg/g)	61.74	11.90	N/A	N/A	N/A	N/A	217.7

[1]Fatty acids; [2]Triglycerides; [3]Cholesterol compared to original milk fat; [4]Carotenoids determined as ß-carotene; [5]Lactone determined as summation of d-10, d-12, d-14 & d-16 lactones; [6]Not analyzed

TABLE 2

Comparitive economics of various processes utilizing SC-CO_2

Reference	System studied	Capacity (tons/yr)	Convers. cost (¢/kg)	Capital cost (M $)	Major costs item
Passey (1991)	Batch system, defatting of peanuts (oil as byproduct)	1000	70	6.1	Labor & electricity
Leyers (1991)	Semi-continuous, coffee decaffeination	11,000	41	22.0	Utilities (steam)
Singh and Rizvi (1993)	Continuous, fractionation of milk fat into 4 streams	10,000	14	6.2	Utilities

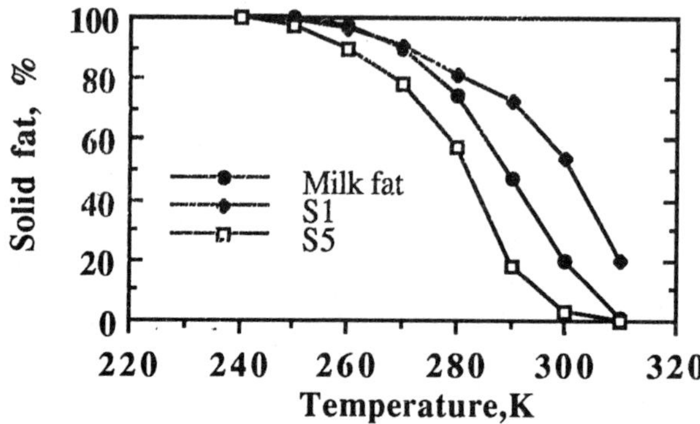

Figure 2. Variation in solid fat content with temperature of milk fat and its fractions.

CONCLUSIONS

The quality and not necessarily the quantity of fat is today's health issue. The very nature of milk fat, its unique triglyceride profile and flavor constituents coupled with SFE, can insure milk fat a long and economically viable future as a raw material to the world market. Additionally, SFE of milk fat can supply the dairy industry itself with modified fats needed to offer alternative product lines formulated in response to the demands of the scrutinizing consumer. Further, it can be seen that the generally held view of SFE being expensive is valid for batch/semi-continuous operations and not for continuous processing of biomaterials.

REFERENCES

1. Rizvi, S.S.H. Supercritical fluid processing of milk fat. Newsletter #6. Northeast Dairy Foods Research Center. Cornell University, Ithaca, NY. 1991, **11**, 1-4.
2. Yu, Z.R., Rizvi, S.S.H., and Zollweg, J.A. Fluid-liquid equilibria of anhydrous milk fat in supercritical carbon dioxide. J. Supercritical Fluids, 1992b, 123-129.
3. Rizvi, S. S. H., Lim, S., Nikoopour, H., Singh, M., and Yu, Z., Supercritical fluid processing of milk fat. In Engineering and Food, Vol 3, eds. W.E.L. Spiess and H. Schubert, Elsevier Applied Science Publishers, New York, 1989, p.145-160.
4. Singh, B. and Rizvi, S. S. H. Design and economic analysis for continuous countercurrent processing of milk fat with SC-CO_2. J. Dairy Sci. (In review), 1993, 00-00.

SEPARATION OF ANTIOXIDATIVE FERRUGINOL FROM JAPANESE CEDAR BARK BY SUPERCRITICAL CARBON DIOXIDE

HAJIME ŌHINATA, NAOKO INOUE, YOSHIO YONEI
Tobacco Science Research Laboratory
Japan Tobacco Inc.
6-2, Umegaoka, Midori-ku, Kanagawa 227, Japan

ABSTRACT

Extraction and separation of ferruginol as an antioxidant from Japanese cedar bark using supercritical carbon dioxide ($SCCO_2$) was studied. After being ground, the cedar bark was extracted with $SCCO_2$ and the ferruginol was then fractionated from the extracts using preparative-scale-supercritical fluid chromatography (P-SFC) with a 50mm i.d. column (TMS,15μm). Ferruginol, 0.4% in content in the cedar bark, was concentrated to 16% by $SCCO_2$ extraction and to 66% by P-SFC. 50-70% of ferruginol in the cedar bark was recovered as ferruginol fractions by a composite procedure of $SCCO_2$ extraction and P-SFC.

INTRODUCTION

Lipid peroxidation is known as one of the major factors causing deterioration of food during storage and processing. To inhibit lipid autoxidation of foods, natural antioxidants are being used commercially, but there remain some problems in their use as food additives. They are usually extracted with organic solvents such as ethanol, acetone and hexane, but these organic solvents are inflammable and present problems of physiological safety.

$SCCO_2$ as a solvent is very suitable for use in the food industries because CO_2 is an inert, odorless and tasteless gas with low toxicity, and it can be removed completely from extracts at room temperature.

Ferruginol(8,11,13-abietatriene-12-ol), a natural antioxidant, was isolated and identified from Japanese cedar bark. Its antioxidative activity was similar to that of α-tocopherol.

In order to develop an extraction and separation technique for ferruginol from cedar bark, a composite procedure of $SCCO_2$ extraction and SFC was studied.

MATERIALS AND METHODS

Sample preparation
The cedar bark which had been stripped from a tree was air-dried at room temperature and was then ground to particles of 0.4mm mean diameter by a single-disk refiner. The content of ferruginol in it was 0.37% and its moisture content was 10%.

Apparatus and procedure
$SCCO_2$ extraction experiments of the ground cedar bark were carried out under various pressures and temperatures in the ranges of 9-30MPa and 15-60°C in the type of semi-batch apparatus which is commonly used. Figure 1 shows a schematic diagram of the P-SFC used in this study. It was designed and assembled by the author's group and was equipped with a 1000mm x 50mm i.d. stainless steel column packed with a tri-methyl-silane modified silica(TMS) and heated electrically. The pore size of the packing materials was 120Å and the particle size was 15μm. The TMS was selected from among HPLC's based on the preliminary experiments.

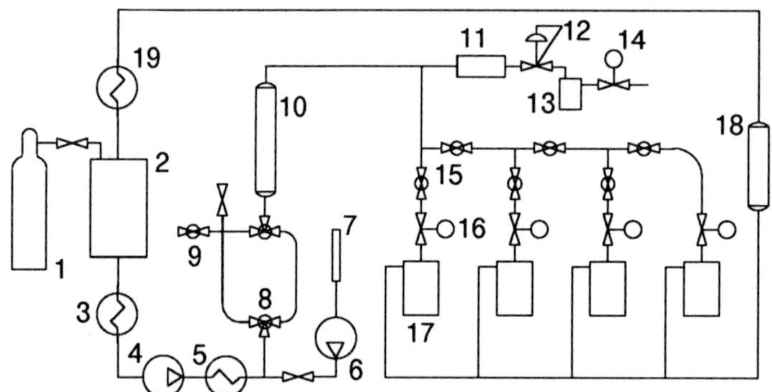

1.CO_2 cylinder, 2.CO_2 reservior, 3,5,19.Heat exchanger, 4.CO_2 pump, 6.Feed pump, 7.Feed, 8.Air-actuated 3-way ball valve, 9.2-way ball valve, 10.Column, 11.UV detector, 12.Pressure reducing regulator, 13.Separator, 14.Flow controlled valve, 15.Air-actuated 2-way ball valve, 16.Pressure controlled valve, 17.Fraction collector, 18.Gas cleaner

Figure 1. A schematic diagram of preparative-scale supercritical fluid chromatograph.

The sample was injected directly as an ethanol solution into the CO_2 stream by a piston pump or a sample injector consisting of stainless tubes and valves. At the column outlet line, a small fraction of the CO_2 (1ℓ/min) was split, one portion of which went to a UV detector for a chromatogram. Based on this chromatogram, four fractions were collected by the operation of 3 pairs of air-actuated 2-way ball valves.

RESULTS AND DISCUSSION

The extract containing 16-19% of ferruginol from the cedar bark was obtained by $SCCO_2$ extraction. The recovery of ferruginol showed maxima with increasing pressure at each of the temperatures tested and these values were about 80% at any temperature tested. In consideration of the SFC conditions described below, the $SCCO_2$ extraction conditions were determined at 15MPa and 40°C. Under these conditions, the extract yield was 1.8g/100g-material and the content of ferruginol in the extract was 16%. The recovery of ferruginol from the cedar bark was about 80%.

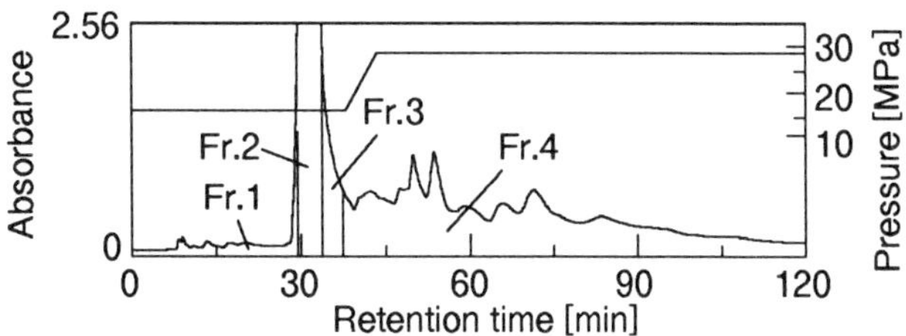

Figure 2. A typical chromatogram and cut point arrangements. (injection 3.4g, UV at 280nm, linear velocity 0.3cm/s)

Figure 2 shows a typical chromatogram obtained from the extract mentioned above and cut point arrangements. The optimal SFC conditions, a pressure at the middle of the column of 15MPa and the temperature of 60°C, had been previously determined by an analytical-scale SFC on a 250mm x 4.6mm i.d. column(TMS, 5μm). However, because injected sample could not be completely eluted at 15MPa, after the ferruginol fractions had been collected, the pressure was increased to 27MPa for 10 min and held there for 2 hours as the injected sample could be completely eluted at this pressure . Table 1 shows the analytical results of the four fractions shown in figure 2 with a capillary gas chromatography.

Table 1
The analytical results of the fractions(injection 3.4g)

Fraction No.	Recovery of solute[%]	Content of ferruginol[%]	Recovery of ferruginol[%]
Fr.1	4.8	11.2	3.0
Fr.2	20.0	65.6	71.6
Fr.3	5.1	48.9	13.7
Fr.4	70.1	3.1	11.7

In Fr.2 shown in Table 1, a fraction containing 66% of ferruginol could be fractionated from the extract and the recovery of ferruginol was then 72%. Its recovery from the cedar bark was finally 57%. By using a longer column or smaller packing materials, it is thought that it will be possible to further concentrate ferruginol in the fractions.

ACKNOWLEDGMENT

We would like to thank the Japanese Research and Development Association for High Separation System in Food Industry for its support and approval of this presentation.

MOISTURE ADSORPTION ISOTHERMS AND REACTION CHARACTERISTICS OF IMMOBILIZED LIPASE IN SUPERCRITICAL CARBON DIOXIDE

KOZO NAKAMURA*, TOSHIO SUGIYAMA AND HIDEKI SUDO
Department of Biological and Chemical Engineering,
Gunma University
Kiryu, Gunma 376, Japan

* Department of Agricultural Chemistry, Faculty of Agriculture,
University of Tokyo, Yayoi, Bunkyo-ku, Tokyo 113, Japan

ABSTRACT

The water adsorption characteristics of several carrier particles as well as the immobilized lipase were measured in supercritical carbon dioxide ($SCCO_2$) by the flow method. The maximum sharp peak appeared near the critical pressure when the adsorption of water was measured at 50℃ at various pressures. The data of the adsorption experiments could be correlated with a single curve for each material when the concentration of water in $SCCO_2$ was normalized by the solubility. The lipase of *Rhizopus delemar* adsorbed less water than CPG 75 and Duolite A-568 with a specific surface area as large as 150 to 200m^2/g.

INTRODUCTION

The acidolysis reaction catalyzed by lipase is crucially influenced by the water concentration in $SCCO_2$,an excellent reaction medium and it was reported that the water content in the immobilized lipase could be directly related to the activity of enzymatic ester synthesis in a limited amount of water. In this study, the water adsorption characteristics of several carrier particles, as well as the immobilized lipase, were measured to elucidate the effect of water on the reactions of esterification and also to identity good carrier particles for the preparation of immobilized lipase with high activity.

MATERIALS AND METHODS

(1) Materials: The adsorbents used were anionic exchange resin (Duolite A-568), porous glasses (CPG-75, -1000, -3000), ceramics (SM-10, NGK Insulators Ltd), lipase of *Rhizopus delemar* (Seikagaku Kogyo, 600 U/mg) and immobilized lipases (Lipozyme IM20, IM60, IM, Novo Industry A/S). The physical properties of those adsorbents, except the enzyme, are shown in Table 1.

(2) Measurement of water adsorption: The flow method was used to measure the amount of water adsorbed in the materials. Moistened carbon dioxide was fed into the column packed with an adsorbent, and the water concentration was continuously measured at the exit of the pressure regulator by the moisture meter (AQUAMATIC+, MEECO Inc.). The blank test was done with the empty column to measure the amount of water adsorbed on the walls of the pipes and the column, and this amount was substracted from the total amount adsorbed. The dimensions of the column were 4.6 mm in inner diameter and 50 mm in length.
(3) Immobilization of lipase: The lipase was immobilized on CPG by the silane coupling method and on SM-10 by the adsorption method with glutaraldehyde cross-linking. The activity was assayed by the acidolysis of triolein (5.6 mM) with stearic acid (17.6 mM) under the condition : 50℃, 29.4 MPa and total water concentration 28 mM.

RESULTS AND DISCUSSION

Water Adsorption

When the amount of water adsorbed on Lipozyme IM20 was measured in the $SCCO_2$ at almost constant water concentration (0.027 to 0.036 mM), the maximum adsorption peak (0.06 wt%) appeared near the critical pressure and was followed by a sharp decline to zero adsorption. The water adsorption soon returned to a level of 0.02 wt% and then gradually decreased with increase in pressure. This gradual decrease is supposed to occur by the solubility of water in $SCCO_2$ increasing with pressure. The water adsorption isotherms measured at various temperatures and pressures were well correlated using the concentration normalized by the solubility, as in **Fig.1**, although the data taken at the near-critical pressure deviated from those taken at higher pressures.

Table 1
Physical properties of immobilizing materials

Materials	Mean pore diameter [Å]	Pore volume [cm^3/g]	Specific surface area [m^2/g]	Mean diameter [μm]
Duolite A-568	180	0.80~0.85	200	460
CPG 75	74	0.47	152.7	146
CPG 1000	1010	0.79	21.8	122
CPG 3000	2869	1.06	8.9	120
SM-10	365	0.56~0.80	77.7	180

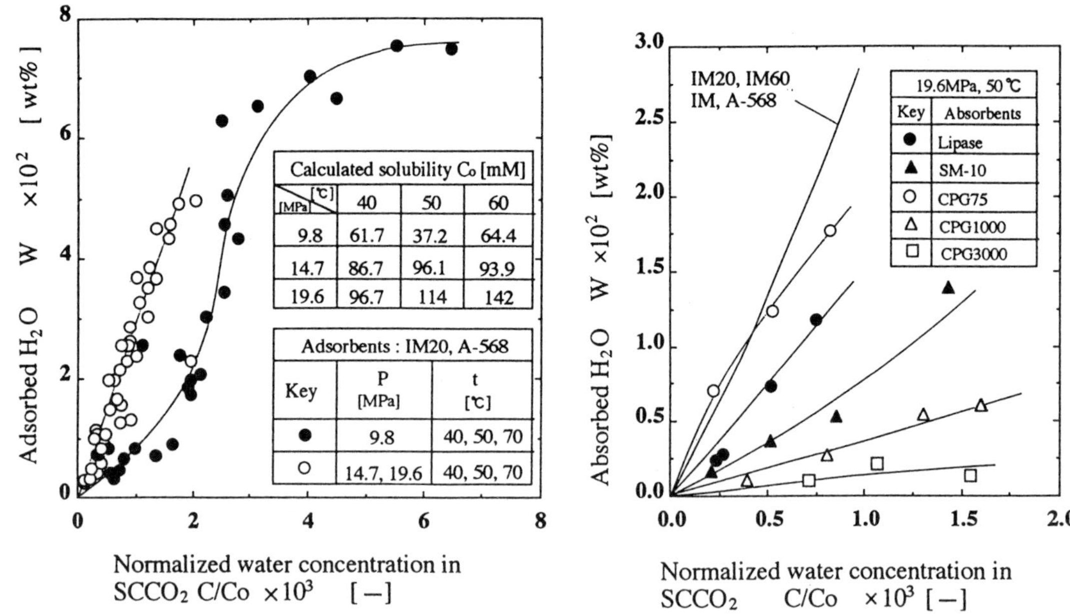

Fig. 1 Water adsorption on several carrier particles and immobilized lipase

Table 2
Specific activity of immobilized enzymes

Carriers	Immobilization efficiency[-]	Amount of immobilized protein		Specific activity			
				Protein base [U/mg-protein]		IME base [U/g-IME]	
		[mg/g-support]	[mg/m²-support]	Hydrolysis	Esterification	Hydrolysis	Esterification
CPG-75	0.20	21.9	0.14	1.47	0.14	32.2	3.14
CPG-1000	0.48	52.8	2.42	0.54	0.28	28.3	14.8
CPG-3000	0.29	30.7	3.45	1.49	0.50	45.7	15.3
Duolite A-568	0.20	22.9	0.30	1.03	0.60	23.7	13.8
IM	—	—	—	—	—	43.4	60.2

Activity of immobilized enzyme (Table 2)

The high esterification activity for IM could mainly be ascribed to the lipase of *Mucor miehei*, and the water adsorption per unit surface area might be also an important factor for specific activity, although the water adsorption isotherms were measured at too low concentrations of water to estimate the amount of water adsorbed on each immobilized enzyme under the reaction condition.

CONCLUSIONS

(1) A sharp change of the isothermal water adsorption was observed near the critical pressure.
(2) The water adsorption isotherms in $SCCO_2$ were correlated with a single curve for each adsorbent using the normalized water concentration.

ENZYME REACTION IN SUPERCRITICAL FLUID

YASUSHI ENDO, KENSHIRO FUJIMOTO
Faculty of Agriculture, Tohoku University, Sendai 981, JAPAN
KUNIO ARAI
Faculty of Engineering, Tohoku University, Sendai 980, JAPAN

ABSTRACT

Enzymatic interesterification between tricaprylin and methyl oleate was carried out in supercritical carbon dioxide (SC-CO_2) using an immobilized lipase. A temperature-controlled reflux column was attached to the SC-CO_2 reaction system to preferentially extract methyl caprylate, a byproduct formed during enzymatic interesterification. The selective removal of methyl caprylate by SC-CO_2 extraction from the reaction mixture promoted the incorporation of oleic acid to triglycerides. The enzymatic interesterification in SC-CO_2 was affected by moisture and pH. The supplement of water or phosphate buffer in proper quantity activated the enzymatic interesterification. Chiral esters from secondary alcohols and short-chain fatty acids could be also produced by the SC-CO_2 system using a lipase.

INTRODUCTION

Recently the supercritical carbon dioxide (SC-CO_2) has been used for the extraction of oils and flavor compounds in biological materials, and the medium of enzymatic reactions instead of water or organic solvents. In fact, our group applied SC-CO_2 system to the removal of cholesterol from butter oil [1], the extraction of oil with low acid value from rice bran [2], and the defatting of fish meats [3]. On the other hand, the utilization of microbial lipases has been attempted as catalysts for producing useful triglycerides and esters. The purpose of this research was to produce the useful triglycerides and various esters efficiently by applying the SC-CO_2 system using immobilized lipases.

Enzymatic interesterification between tricaprylin and methyl oleate

A schematic diagram of the SC-CO_2 bioreactor system using an immobilized lipase is shown in Figure 1. A temperature-controlled reflux column was set in the SC-CO_2 reaction system to preferentially extract methyl caprylate produced during the enzymatic interesterification. A mixture of tricaprylin and methyl oleate (1:3, 30g) together with Lipozyme IM20 (Mucor miehei, Novo Ind. Japan) as an immobilized lipase at the concentration of 5 or 10%, were put in a reactor, and then incubated at 40°C with agitation in SC-CO_2 under the pressure of 10MPa for 12hr. The incubation was also attempted with CO_2 extraction at 5L/min.

Oleic acid levels in triglyceride after enzymatic inter-esterification by a closed or a flowing SC-CO_2 system are shown in Table 1. The level of oleic acid incorporated in triglycerides with flowing SC-CO_2 was lower than that with a batch type after interesterification although a SC-CO_2 extracted about 70% of methyl caprylate produced during the enzymatic reaction.

The water was supplied to a reactor in the range of 0-1.2mL/hr to confirm whether the water activity in an immobilized lipase was reduced by SC-CO_2 extraction. As shown in Table 1, the supplement of water (0.6mL/hr) increased the oleic acid level in triglycerides during lipase-catalyzed interesterification to the same level as that obtained by a batch type. This observation demonstrated that the removal of moisture in an immobilized lipase by SC-CO_2 impaired the enzyme activity. Phosphate buffer at pH6.98 and pH8.04 were supplemented to a reactor. Enzymatic interesterification was notably enhanced by the supplement of phosphate buffer (pH6.98) compared with water (Table 1). These results show that adjustment of pH besides supplement of water is necessary for enzymatic interesterification of oils when using a flowing SC-CO_2 system.

Enzymatical production of chiral esters from short-chain fatty acids and secondary alcohols in supercritical carbon dioxide

Short-chain fatty esters provide the specific flavors which are common in tropical fruits. Their flavor characteristics depend on the chiral structure. Our group attempted the enzymatical production of flavoring chiral esters from short-chain fatty acids (C6-10) and secondary alcohols (C6-8) in SC-CO_2 using immobilized lipases. Lipozyme (Mucor miehei, Novo Ind. Japan) and Lipase OF (Candida cylindracea, Meito Sangyo) catalyzed the production of esters from short-chain fatty acids and secondary alcohols in SC-CO_2 as well as in n-hexane. Lipozyme showed higher activity for octanoic and decanoic acid than hexanoic acid. Moreover, catalytic activity of the lipase depended on the chain length of secondary alcohols (2-octanol>2-heptanol>2-hexanol). Lipase

OF catalyzed to produce esters from both 2(R)- and 2(S)-hexanol, while Lipozyme showed activity only for 2(R)-hexanol and catalyzed to produce 2(R)-hexyl hexanoate.

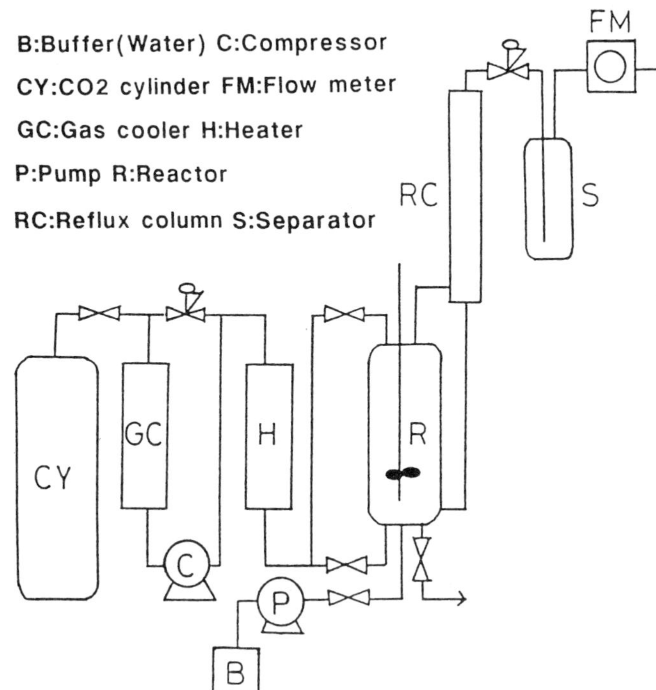

Figure 1. Supercritical carbon dioxide (SC-CO2) bioreactor system.

TABLE 1
Interesterification in SC-CO_2 using an immobilized lipase

	Incorporated oleic acid in triglycerides (%)
Batch	28
Flowing	17
Water	
0.36mL/hr	24
0.60mL/hr	30
1.20mL/hr	18
Buffer	
pH6.98	39
pH8.04	34

REFERENCES

1. Shishikura, A., Fujimoto, K., Kaneda, T., Arai, K. and Saito, S., Agric. Biol. Chem., 1986, **50**, 1209-1215.
2. Zho, W., Shishikura, A., Fujimoto, K., Arai, K. and Saito, S., Agric. Biol. Chem., 1987, **51**, 1773-1777.
3. Fujimoto, K., Endo, Y., Cho, S.-Y., Watabe, R., Suzuki, Y., Konno, M., Shoji, K., Arai, K. and Saito, S., J. Food Sci., 1989, **54**, 265-268.

SUPERCRITICAL CARBON DIOXIDE AS PROCESSING MEDIUM FOR ENZYMATIC INTERESTERIFICATION

B.MOSHAMMER, R.MARR, A.BILADT, F.FRÖSCHL, W.PREITSCHOPF
Institute of Thermal Process and Environmental Engineering,
Technical University Graz, Inffeldgasse 25, 8010 Graz, Austria

ABSTRACT

Supercritical carbon dioxide (SC-CO2) is used as processing medium for the enzymatic interesterification of D,L-Menthol with triacetin catalysed by the lipase from Candida cylindracea. Studies on the reaction and solubility-measurements were performed in a batch-mode reactor. The activity of the lipase was maintained to a high percentage after the exposure to SC-CO2. The kinetic parameters and the conversion-rate could be highly influenced by changing the pressure and the water-content of the system. A new continous process based on enzymatic reactions in SC-CO2 was developed and constructed for further studies.

INTRODUCTION

The industrial application of biocatalysis has attracted attention with the knowledge that enzymatic catalysis can take place in nearly anhydrous media (1,2). By using supercritical fluids, especially supercritical carbon dioxide (SC-CO2), as solvents for enzymatic reactions a new wave of research in the field of biocatalysis in unconventional-media has emerged (3,4) resulting in new aspects for the industrial biocatalysis as the demand for solvent-free as well as enantiomeric-pure products gets more and more distinct. The special properties of the enzymes as biocatalysts and the well-known advantages of SC-CO2 as solvent combined in one process lead to a continous production/product-recovery-process with a promising large application field in the food- and pharma-industry.

In this work SC-CO2 and the lipase of Candida cylindracea are used for the study of a new reaction-system in SC-CO2. The model-system is the transesterification and racemic resolution of D,L-Menthol with triacetin to L-Menthyl-acetate, Diacetin and D-Menthol.

The main objectives of the research-programme on enzymatic catalysis in SC-CO2 at TVT/UT-TU Graz consist of a study about the effects on the enzymatic reaction in SC-CO2 in a batch-reactor-system, the collecting of solubility-data and the comparison to organic solvents. The aim is the processing in a continous-mode with an integrated product-separation in a closed circuit based on the enzyme-catalysis in SC-CO2.

EXPERIMENTAL

The basic studies on the potential effects on the reaction and the solubility-measurements are realized in a Batch-reactor-system, which consists of a high-pressure-pump, an enzyme-reactor or an equilibrium-cell with a sapphire-window and a sampling-unit.

The Batch-system is a part of our High-Pressure-Enzyme-Reactor-System (HP-ERS) having been developed and constructed by our own (see figure 1). The special feature of the HP-ERS-plant, seen in figure 1, is on the one hand the online-SFC-unit, as an appropiate online-technique is favourable in the case of monitoring the reaction and the process. On the other hand a product/substrate-fractionation in the down-stream-phase, based on the different solubilities of the substances at different temperature/pressure-conditions, is provided in two steps.

The batch-reactions and the soluibility-measurements are carried out at constant pressure and temperature by magnetic-stirring. The production of L-Menthyl-acetate is followed by manual withdrawals of aliquots from the reaction-system using a HPLC-valve with a sampling loop of 100 µl instead of the online-SFC-unit. The samples are then analyzed offline gaschromatograhically. The online-SFC-technique is used and optimized by determining the solubility data of the substrates and products in CO2.

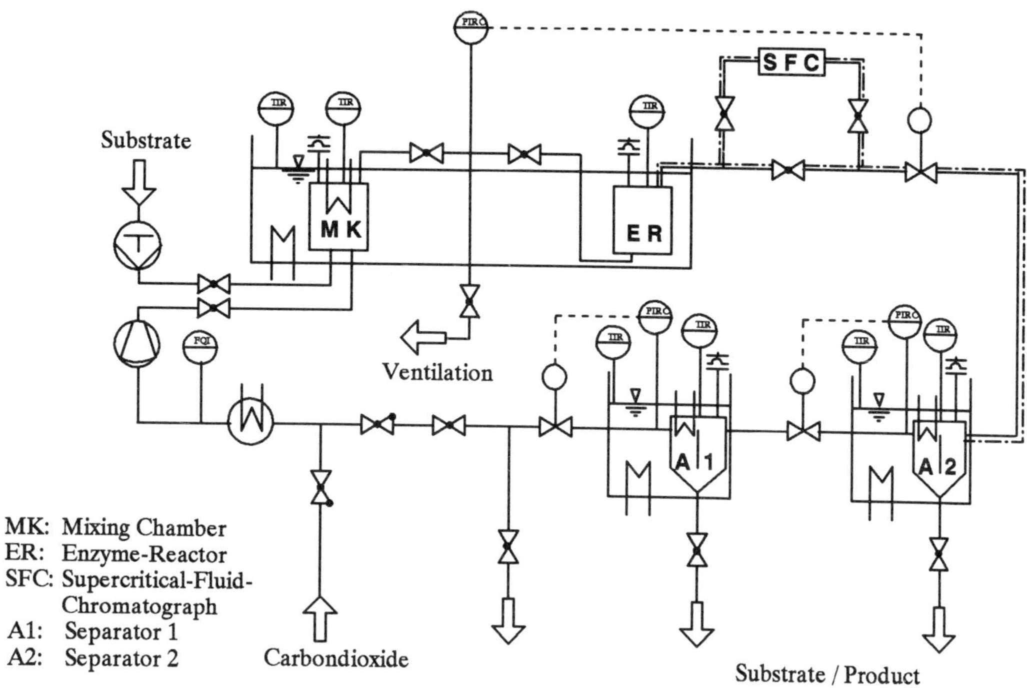

Figure 1. High-Pressure-Enzyme-Reactor-System at TVT/UT - TU Graz.

RESULTS AND DISCUSSION

The typical run of the studied racemic resolution of D,L-Menthol in SC-CO2 shows in the beginning of the reaction an increase of the conversion-rate, which reaches after a certain time a steady-state. The reaction-rate, the time to the equilibrium and the achievable conversion were found to be dependent of several external effects. The best studied parameters till now are the water content of the reaction-medium and the pressure-effect.

It is a fact that water for enzymatic activity in SC-CO2 is needed, but the amount of water differs from system to system and by using organic solvents (5,6). The water-content of the studied reaction influences distinctively the conversion- and reaction rate. The optimum value of water in the fluid is situated between 0.06 % and 0.1 % (w/w) water in the fluid. Increasing the pressure at constant temperature leads to a drastic reduction of the conversion rate whereas the reaction rate at different pressure-steps is comparable. In contrast to high pressures up to 200 bar, small pressure changes near the critical region of the CO2 lead to significant changes in the conversion rate corresponding to the properties of fluids in the critical region, where small changes in temperature and pressure result in great changes of density and diffusion directly effecting the mass transfer and the solubility parameters. As the study on the reaction demonstrates a substrate-inhibition of D,L-Menthol, it can be assumed additionally that by increasing pressure an increasing substrate-solubility leads to a less conversion-rate compared to reactions in the critical region.

The lipase of Candida cylindracea has been used for the first time in SC-CO2 for enzymatic conversions. The experiments show a good stability and activity of this lipase. After the exposure to SC-CO2 a deactivation, ranged about 20 %, is observed. A clear tendency can be deduced; the higher the pressure and the time of exposure to SC-CO2, the higher the deactivation of the lipase is found. Additionally the deactivation is dependent of the number of ventilation-steps and the kind of ventilation. The described reduced activity of the lipase after the CO2-exposure is confirmed by fluorescence-emmission-spectra of exposed enzymes. There is a shift in the height of the maximum of the tryptophane-fluorescence which is due to a partial unfolding of the proteins, according to (7), depending on the incubation time and used pressure.

The performed studies on the reaction result in a good and promising base for the continous processing in the new developed High-pressure-enzyme-reactor-plant.

ACKNOWLEDGEMENTS

The authors wish to express their cordial thanks to Amano Enzyme Europe Limited (Milton Keynes, U.K.) for their generous donation of lipase-samples. We thank J.Wolfgang carrying out the batch-experiments and Univ.Doz.Dr.A.Hermetter/TU-Graz for the experimental work on the fluorescence-emission-spectra. This work was supported by FWF (Vienna, Austria).

REFERENCES

1. Randolph, T.W., Blanch, H.W., Clark, D.S., Biocatalysis in Supercritical Fluids. In Biocatalysis for Industry, ed. J.S.Dordick, Plenum, London, 1991, pp 219.

2. Zaks, A. and Klibanov, A.M., Enzyme-catalyzed processes in organic solvents, Proc.Natl.Acad.Sci.USA, 1985, **82**, 3192-3196.

3. Aaltonen, O., Rantakylä, M., Biocatalysis in supercritical CO2, Chemtech, 1991, 240-48.

4. Nakamura, K., Biochemical reactions in supercritical fluids, TIBTECH, 1990, **8**, 288-92.

5. Dumont, T., Barth, D., Perrut, M., Enzymatic reaction kinetic comparison in organic solvent and in supercritical CO2, 2nd International Symposium on High Pressure Chemical Engineering, 1990, Erlangen.

6. Marty, A., Chulalaksananukul, W., Condoret, J.S. Willemot, R.M., Durand, G., Comparison of lipase-catallysed esterification in supercritical carbon dioxide and in n-Hexane, Biotechnol.Letters, 1990, **12**, 11-16.

7. Kasche, V., Schlothauer, R., Brunner, G., Enzyme denaturation in supercritical CO2: Stabilizing effect of S-S bonds during the depressurization step, Biotechnol.Letters, 1988, **18/8**, 569-74.

ENZYME INACTIVATION BY PRESSURIZED CARBON DIOXIDE

MURAT O. BALABAN, S. PEKYARDIMCI, C. S. CHEN, A. ARREOLA, M. R. MARSHALL
Food Science and Human Nutrition Dept. University of Florida
Gainesville, FL. 32611, USA.

ABSTRACT

Pectinesterase in orange juice, polyphenoloxidase in apple juice, and various polyphenoloxidases in model systems were used in the determination of the kinetics of inactivation of the enzymes by carbon dioxide treatment at different pressures, temperatures, pH values, and times. Controls were used for pressure, temperature and pH. Kinetics parameters are presented, effects of the treatment on the enzyme molecules examined, and implications to juice processing are discussed.

INTRODUCTION

Supercritical carbon dioxide (SC CO_2) is used in food processing in extraction, separation / fractionation of lipids, essential oils, and pigments [1]. Another application of pressurized CO_2 is the inactivation of enzymes in foods and beverages [2,3]. Polyphenoloxidase (PPO) causes browning in fruits, vegetables and juices. Different sources of PPO such as mushroom, potato and shrimp have different activities and different thermal resistances. Pectinesterase (PE) causes cloud loss in orange and pineapple juices. These enzymes can be thermally inactivated. However, this causes undesirable changes in their flavor and aroma. Therefore, non-thermal inactivation of these enzymes is of great interest. Our purpose was to determine the effect of pressurized CO_2 treatment conditions such as time, temperature, pH and pressure on the inactivation of PE in Valencia orange juice, PPO in apple juice, and PPO from various plant and animal sources in enzyme solutions in pure water. Inactivation kinetics of these two enzymes were also determined. Preliminary investigations of CO_2 treatment on secondary structure of PPO is presented.

MATERIALS AND METHODS

Mushroom tyrosinase was purchased from Sigma (St. Louis, MO). Fresh Florida spiny lobster (Panulirus argus) PPO was purified in our lab [4]. PPO from non-sulfited brown shrimp (Panaeus aztecus), and from Russet potato tubers were similarly prepared. Florida apples were crushed and pressed at 400 kPa at room temperature, and SC CO_2 treatments were performed immediately after juice extraction. Fresh squeezed Valencia orange juice (OJ) was frozen in cans by the FMC Corp. (Lakeland, FL). Before treatments, OJ was thawed with tap water while in the unopened can. PPO activity was determined by the pyrocatechin assay [5], and that of PE by the titration method [6].

Mushroom PPO in pure water was treated at a) atmospheric pressure by bubbling CO_2 through the solution at 33, 43 and 50°C, with T and pH controls; b) 2 MPa and 5.5 MPa by placing the solution in a pressure vessel immersed in a constant temperature (T) bath (33, 43 and 50°C) with pH controls; c) 33.7 MPa at 35, 45 and 55° with T controls. Periodic samples were taken and PPO activity determined. Spiny lobster PPO was treated at atmospheric pressure at 33°C, 38°C, or 43°C. T and pH controls were also run. Lobster, brown shrimp and potato PPO solutions were treated at 5.8 MPa and 43°C. T controls were run. Apple juice was treated at 33.7 MPa, 45°C for 4 hrs, with T controls. Orange juice was treated at pressures between 6.9 MPa and 34.4 MPa, temperatures between 35°C and 60°C, for times from 15 to 180 min.

RESULTS AND DISCUSSION

The 1st order inactivation rate constants (k) for mushroom PPO treated with CO_2 were calculated, and the energy of activation computed (Table 1).

Table 1.
Energy of activation for mushroom PPO inactivation at different CO_2 pressures.

	E (KJ/mol)		
Pressure	T control	pH control	CO_2 treatment
Atm. pressure	93.9	194.7	50.0
	117.6		46.8
2 MPa		26.1	44.8
5.5 MPa		19.1	37.4
33.7 MPa	110.3		43.1

Results show a drastic decrease in the E values of CO_2 treated samples compared to T controls. There is some variability in the results since different batches of the same commercial enzyme was used. There is also a general trend of decreasing E values of the treated PPO as pressure increases.

Spiny lobster PPO treated at atm. CO_2 gave E=42.9 KJ/mol, while T control gave E=26.2 KJ/mol. A single protein band at the same position was observed (255 kD) for the control and CO_2-treated PPO on the nondenatured PAGE gel, meaning that there was no alteration of the protein with CO_2. Secondary structure changes may be responsible for the loss of activity. Treatment at 5.81 MPa resulted in a very rapid inactivation of spiny lobster PPO : after 1 min there was 2% of original activity left. Spiny lobster PPO was more sensitive to CO_2 than all other enzymes used in this study.

Brown shrimp PPO treated at 5.8 MPa and 43°C yielded $k=0.79\ min^{-1}$, and T control gave $k=0.0082\ min^{-1}$. Potato PPO was the most resistant to CO_2 treatment at 5.8 MPa and 43°C ($k=0.061\ min^{-1}$). T control gave $k=0.0023\ min^{-1}$. Far UV circular dichroic spectra of shrimp, potato and lobster PPO showed a change between untreated, and CO_2 treated samples (Table 2). Lobster and brown shrimp PPO showed the most noticeable changes in composition of α-helix and random coil. Only minor changes in secondary structure occurred in potato PPO, which may explain its higher resistance to CO_2 treatment.

Table 2.
% Secondary Structure Estimates of Spiny Lobster, Brown Shrimp, and Potato PPO from Far UV Circular Dichroic Spectra.

PPO		α-helix	β-sheet	β-turn	random coil
Lobster	non-treated	24.4	26.2	21.4	29.9
	CO_2-treated	19.7	25.9	15.2	39.3
Brown shrimp	non-treated	20.1	22.3	15.2	42.4
	CO_2-treated	29.6	18.9	18.2	33.3
Potato	non-treated	14.8	34.6	28.4	22.2
	CO_2-treated	17.8	35.9	25.9	20.4

Treatment of apple juice at 33.7 MPa resulted in an increase in PPO activity after an hour, followed by a drastic decrease by 4 hrs (8 % of original activity). k was calculated as 0.0154 min^{-1}. T control resulted in $k = 0.0007\ min^{-1}$.

Analysis of PE inactivation kinetics in orange juice for T control samples resulted in E = 166.6 KJ/mol. Orange juice has 3 isoforms of PE. z values (increase of temperature necessary to decrease the decimal reduction value D by 90%) for these were 6.5°C for PE-I and PE-III, and 11°C for PE-II (7), or 6.5°C and 10.5°C (8). Our z value was 8.8°C, which is between reported values of heat stable and heat labile forms of PE, and represents a "lump-sum" value of z. Treatment at 31 MPa resulted in E = 97.4 KJ/mol.

REFERENCES

1. Rizvi, S. S. H.; Benado, A. L.; Zollweg, J. A.; Daniels, J. A., Supercritical fluid extraction: Fundamental principles and modeling methods. Food Technol. 1986, 40, 55-65.
2. Taniguchi, M.; Kamihira, M.; Kobayashi, T., Effect of treatment with SC CO_2 on enzymatic activity. Agric. Biol. Chem. 1987, 2, 593-594.
3. Haas, G. J.; Prescott, H. E.; Dudley, E.; Dik, R.; Hintlian, C.; Keane, L., Inactivation of microorganisms by CO_2 under pressure. J. Food Safety. 1989, 9, 253-265.
4. Chen, J. S., Balaban, M. O., Wei, C. I., Gleeson, R. A., and Marshall, M. R., Effect of CO_2 on the inactivation of Florida Spiny Lobster polyphenol oxidase. J. Sci. Food Agric. 1993, 61, 253-259.
5. Traverso-Rueda, S.; Singleton, V.L., Catecholase activity in grape juice and its implications in winemaking. Am. J. Enol. Viticult. 1973, 24, 103-106.
6. Rouse, A. H.; Atkins, C. D., Pectinesterase and pectin in commercial citrus juices as determined by methods used at the Citrus Experiment Station. University of Fla. Agric. Exp. Stn. Bulletin. 1955, 570.
7. Versteeg, C,; Rombouts, F.M.; Pilnik., Purification and some characteristics of two pectin esterase isoenzymes from orange. Lebensn. Wiss. U. Tech. 1978, 11, 267-274.
8. Wicker, L.; Temelli, F., Heat inactivation of pectinesterase in orange juice pulp. J. Food Sci. 1988, 53, 162-164.

PURIFICATION OF ORGANIC ACIDS BY GAS ANTI-SOLVENT CRYSTALLIZATION

AKIHIRO SHISHIKURA
Process Development and Engineering Center, Idemitsu Petrochemical Co., Ltd.,
1-1 Anesaki-kaigan, Ichihara, Chiba 299-01, Japan.

ABSTRACT

Citric acid (CA) has successfully been separated from fermentation broth by a novel and unique purification process, which is characterized by organic solvent extraction and precipitation using compressed carbon dioxide (CO_2) as an anti-solvent. Also, CA could be separated from other organic acids which are the by-products of CA fermentation.

INTRODUCTION

Citric acid (CA) is an important compound used as an acidulant in food and beverages. CA is generally produced by the fungal fermentation with mainly *Aspergillus nigar*, and is purified by a firmly established process known as the method of calcium salt precipitation. However, this process includes several batch treatments which require large amounts of chemical reagents. More calcium sulfate is formed as an industrial waste than the weight of CA. Though these negative factors have a significant influence on production costs, improvements in this process have not yet been practically accomplished.

We previously proposed a novel and unique CA purification process which was characterized by organic solvent extraction and precipitation using compressed carbon dioxide (CO_2) as an anti-solvent (1). This paper addresses the outline of the new CA purification process and fractionation of the organic acid mixture by gas anti-solvent crystallization.

MATERIALS AND METHODS

The broth of CA and the acetone solution of crude CA were prepared by the same previously reported methods (1).

The experimental apparatus for gas anti-solvent crystallization was equipped with three volumes (80, 250 and 2000 ml) of transparent glass-type level gauges (LG). The precipitation conditions of impurities and organic acids were measured by batch treatments. CO_2 was dissolved in acetone solutions from the bottom of the LG. In the case of continuous treatment, the acetone solution and CO_2 were previously mixed in the mixing line, and the LG was used as a settler.

Analysis for CA and other organic acids was carried out by mean of HPLC. Sugar, minerals and water contents were determined by the standard methods of the Japan Food Additives Association(2).

RESULTS AND DISCUSSION

Precipitation of Impurities from Acetone Solution of Crude CA.

The CA concentrations in the solute extracted with acetone from the condensed fermentation broth (44.9wt% CA, 20.1wt% Water) reached 91 wt% by removing the majority of impurities (mainly sugar). However, this value was not sufficient for the food additive grade of CA. Therefore, in order to further remove the residual impurities, we applied gas anti-solvent crystallization as the second step in our purification process.

Figure 1 shows the pressure vs. concentrations of CA (A) and impurities (B) in the supernatant at 30 ℃. CA started to deposit at 28 kg/cm^2. On the other hand, impurities started to deposit at 9 kg/cm^2G. Most of the impurities were separated as precipitates from the solution in the vicinity of 15 kg/cm^2 without any deposition of CA. These results suggest that the effective separation of impurities from CA solution is possible at a pressure below 28 kg/cm^2, because of the difference in the pressure needed to start the deposition of the CA and the impurities. The purity of CA dissolved in the supernatants was increased to 99.6 wt%. The results of continuous crystallization were similar to those of the batch treatment shown in Fig. 1.

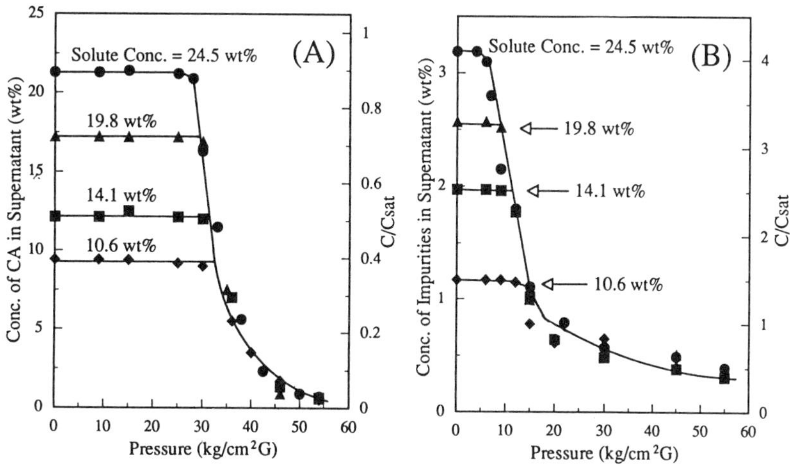

Figure 1. Effect of the pressure on the concentrations of CA (A) and impurities (B) in the supernatant at 30 ℃.

Selective Crystallization of Organic Acids by Gas Anti-Solvent Crystallization.

To separate and crystallize CA from by-products [oxalic (OA) and malic (MA) acids], we attempted gas anti-solvent crystallization.

Figure 2 shows the effect of the relative concentration on the pressure for the start of deposition for several organic acids at 30℃. The starting pressure for deposition decreased with increasing relative concentration in all acids. In the case of the crystallization of CA from the model acetone solution [CA 18.7wt% (C/Csat=0.89), OA 1.0 (0.03) and MA 0.4 (0.02)], the crystal grains of CA could be obtained with 99.9% purity and 99.6% recovery at 54 kg/cm^2 and 30℃. The particle size and the size distribution of the precipitated CA could be controlled by the rate of pressure rise (the introducing rate of CO_2) and the standing time. The crystallization of CA was finished within 1~3 min.

Novel CA Purification Process.

Figure 3 shows the flow diagram of a novel CA purification process. First, the fermentation broth of CA is filtered to remove microorganisms (F) and is dried to adjust its water content to about 10~20 wt% by multiple effect evaporation (C-1). CA is subsequently extracted with acetone from the condensed broth (SE). The residual impurities are then removed as precipitates from the acetone solution of CA using the anti-solvent effect of compressed CO_2 at 25 kg/cm^2 and 30℃ (PC-1). The deposited impurities are readily separated from the supernatant using a settler (SET-1). CA is crystallized from the acetone solution by anti-solvent crystallization with CO_2 at 50 kg/cm^2 (PC-2). Crystal grains of pure CA can be rapidly obtained by simple pressure regulation.

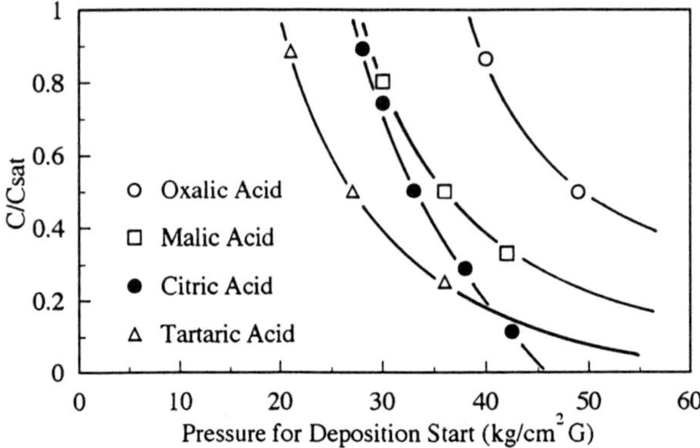

Figure 2. Effect of the relative concentration (the saturation ratio) on the pressure for the start of deposition for several organic acids at 30℃.

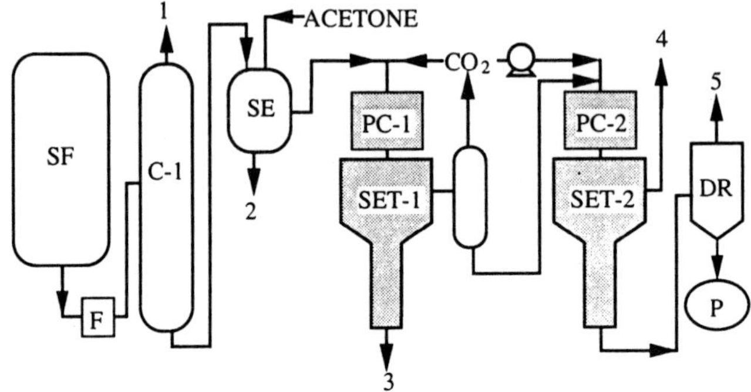

Figure 3. The flow diagram of novel CA purification process.
(SF) submerged fermentation tank, (F) filter, (C-1) multiple effect evaporator, (SE) solvent extractor, (PC-1,2) pre-crystallizer, (SET-1,2) settler, (DR) dryer, (1) water, (2) impurities, (3) residual impurities, (4) oxalic acid, (5) acetone and (P) product.

REFERENCES

(1) A. Shishikura, H. Takahashi, S. Hirohama and K. Arai, J. Supercri. Fluids, 1992, **5**, 303-312.
(2) "Food Additives Official Hand Book", p4-62, Japan Food Additives Association, 1986.

EFFECT OF HIGH PRESSURE ON ACTIVITY OF SOME OXIDIZING ENZYMES

YOSHIO AOYAMA, MASASHI ASAKA, and RITSUKO NAKANISHI
Biological Chemistry Division
Toyo Institute of Food Technology
23-2-4, Minami-hanayashiki, Kawanishi, Hyogo 666, JAPAN

ABSTRACT

The inactivation effects of high pressure treatment on some oxidizing enzymes were investigated, compared with thermal inactivation. Glucose oxidase and ascorbate oxidase were inactivated irreversiblly above 300 MPa and the inactivation was followed by first-order reaction. The activation volume of inactivation of these enzymes was determined from the pressure dependence of the rate constants. Tyrosinase and superoxide dismutase were stable against high pressure, but the former was thermolabile and the latter thermostable. Thus, these enzymes are divided into three types: thermostable and pressure-stable enzyme (superoxide dismutase), thermolabile and pressure-stable enzyme (tyrosinase), thermolabile and pressure-labile enzymes (glucose oxidase and ascorbate oxidase). Thermal treatment is more effective than high pressure treatment for irreversible enzyme inactivation.

INTRODUCTION

The residual enzyme activity in food processing seems to be one of the major problems. The effects of high pressure treatment on enzyme activity were different among enzymes (1)-(3). Although many studies have been performed about hydrolizing enzymes, there are only few studies about oxidizing enzymes related to food quality. So we studied effects of high pressure treatment on the activity of some oxidizing enzymes. The inactivation of the enzymes was investigated kinetically as compared with thermal inactivation.

MATERIALS AND METHODS

The enzymes were purchased from Wako Pure Chemical Industries, Ltd. and used without purification: glucose oxidase from *Aspergillus nigar,* ascorbate oxidase from cucumber, tyrosinase from mushroom, superoxide dismutase from bovine erythrocyte. High pressure treatments were performed with high pressure test machine MFP-7000 (Mitsubishi Heavy Industies, Ltd).

Enzyme activity was determined as follows (4)-(6):glucose oxidase, colorimetry with glucose and *o*-jianisijin and peroxidase; ascorbate oxidase, oxygen uptake by oxygen electrode; tyrosinase, colorimetry with tyrosine; superoxide dismutase, colorimetry with xanthine-xanthine oxidase/cytochrome c.

RESULTS AND DISCUSSION

The inactivation curves of glucose oxidase and ascorbate oxidase at different pressures are shown in Fig.1. The plot indicates that the inactivaton is first-order reaction. As pressure increases, the inactivation velocity of these enzymes increases. However, the extrapolation to 0 min does not show 100% of enzyme activity, showing very short time treatment of pressurization and depressurization could cause inactivation of the enzyme. The activation volumes for inactivation of these enzymes were determined from the pressure dependence of the rate constants (Table 1). Tyrosinase was stable against high pressure treatment(Fig. 2), but unstable against heat treatment (data not shown). Superoxide dismutase was stable against both pressure and heat treatment. These studies indicate that enzymes are divided into three types: 1) thermostable and pressure-stable, 2) thermolabile and pressure-stable, 3) thermolabile and pressure-labile.

TABLE 1

Effects of pressure and temperature on the inactivation of the oxidizing enzymes and activation volume ($\Delta V^{\neq}$) and activation energy (Ea)

	$P_{50/10m}$[a] (MPa)	$T_{50/10m}$[a] (℃)	$\Delta V^{\neq}$ (ml/mol)	Ea (kcal/mol)	P_{T10}[b] (MPa)
Glucose oxidase	440	56	-50	56	80
Ascorbate oxidase	410	56	-38	46	145

[a] The pressure or temperature which caused loss of half of the activity in 10 min.

[b] Pressure increase for the inactivation equivalent to temperature increase of 10℃.

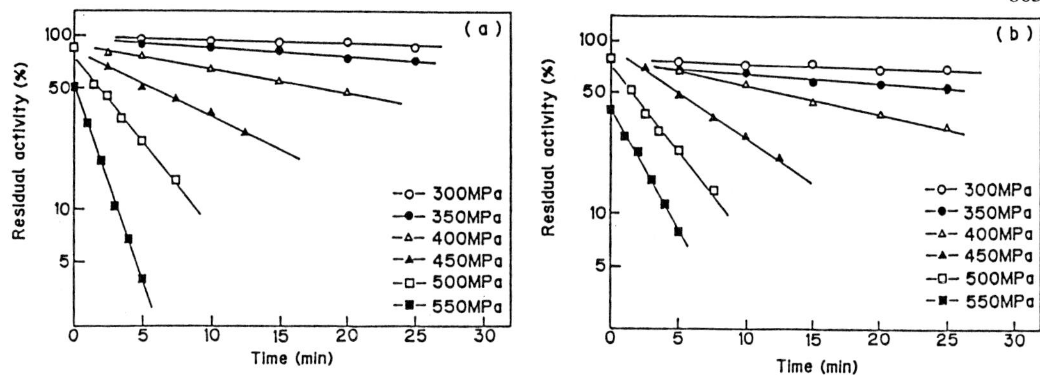

Figure 1. Inactivation of glucose oxidase(a) and ascorbate oxidase(b) by high pressure treatment at 20 ℃.

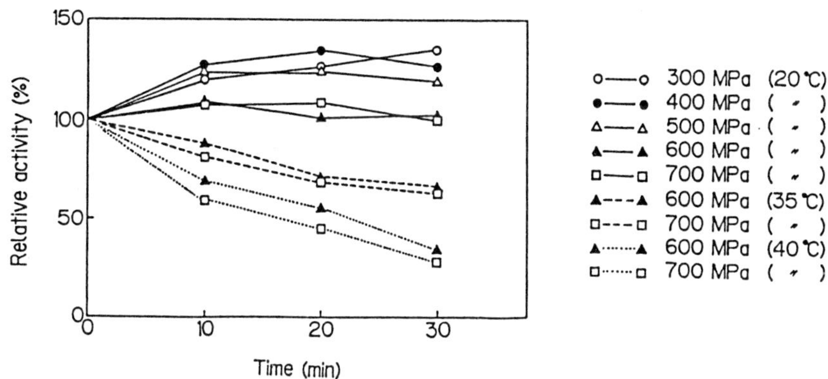

Figure 2. Effects of high pressure treatment on tyrosinase.

REFERENCES

1. Asaka, M. and Hayashi, R., Activation of polyphenoloxidase in pear fruits by high pressure treatment. Agric. Biol. Chem., 1991, **55**, 2439-2440.
2. Hara, A., Nagahama, G., Ohbayashi A. and Hayashi, R., Effects of high pressure on inactivation of enzymes and microorganisms in non-pasteurised rice wine (namazake). Nippon Nogeikagaku Kaishi, 1990, **64**, 1025-1030.
3. Ko, W. C., Tanaka, M., Nagashima, Y., Taguchi, T. and Amano, K., Effect of pressure treatment on actomyosin ATPase from flying fish and sardine muscle. J. Food Sci. 1991, **56**, 338-340.
4. Nakamura, T., Purification and properties of ascorbate oxidase from cucumber. J. Biochem., 1968, **64**, 189-196.
5. McCord, J. M. and Fridovich I, Superoxide dismutase ; an enzymatic function for erythrocuprein (hemocuprein). J. Biol. Chem., 1969, **244**, 6049-6055.
6. Swoboda, B. E. and Massey, V., Purification and properties of the glucose oxidase from *Aspergillus niger*. J. Biol. Chem., 1965, **240**, 2209-2215.

APPLICATION OF STERILIZATION TECHNIQUE BY HYDROSTATIC HIGH PRESSURE FOR GREEN TEA DRINK

TADAKAZU TAKEO, HITOSHI KINUGASA, *MASAMI ISHIHARA, *KENJI FUKUMOTO & *TETSU SHINNO

ITO-EN Co. Ltd., Central Research Institute
*Nippon Steel Co., R & A Laboratory I

ABSTRACT

Green tea extracted with water contained a few viable micro-organisms, which were sterilizad by pressurization at 400 MPa amd room temperature for 30 min.
Thermoduric spores of Bacillus species added to green tea infusion were not sterilized at 400 MPa and room temperature, but sterilized at 700 MPa and 80 C. degree. However, thermoduric spores survived at the pressure of 200–300 MPa were effectivly inactivated during storing tea drink at room temperature. It was revealed that thermoduric spores injured by hydrostatic pressure were inactivated by tea tannin (catechin)in tea infusion.
By hydrostatic pressure treatment at 700 MPa and 80 C. degree for 30 min, the induction of off-flavor and the changes of chemical compounds im green tea drink were effectivly depressed. These results showed that the pressurization technique at 300 MPa and below 100 C. degree might be possible to use as a new pasturization technique on making green tea drink.

INTRODUCTION

On the production of green tea drink, one of important problems is to make tea drink having fresh flavor. Usually, tea drink are sterilized by retort pasturization. By this treatment, tea flavor is deteriorated and strong retort smell is produced. Many trials were done to improve tea drink flavor, but sufficient results were not gotten yet. Recently, it has been reported that microbes in food and beverage were inactivated by a high pressure treatmewnt. On this time, low molecular compounds were not effected. Therefore, it was considered that the high pressure treatment might be a useful sterilization technique to make flavorous food or beverage.

EXPERIMENTALS

Material: Green tea infusion prepared with hot water was used.
Instrument: Performance of hydrostatic pressure equipment made by Nippon Steel. Capacity of pressuring chamber; 1000 ml. Max. pressure; 700 MPa. Max. temp.; 80 C. degree. Pressuring: Compressor type.
Microbial assay: Thermoduric spores belonging to Bcillus sp. were separated from vegetative cells by heating culture medium and inoculated in tea infusion. Microbes in tea infusion were assayed by counting the numbers of colony growing on agar plate after incuation.
Chemical analysis: Volatile compounds in tea drink were analyzed by GC and catechins and VC were determined by HPLC.

RESULTS

1) Microbes in tea infusion made with water were sterilized by the pressure treatment at 300 MPa under roon temp..
2) Thermoduric spores belonging to Bcillus sp. showed high resistibility for hydrostatic pressure. However, all these spores were inactivated perfectly under the pressure of 700 MPa at 80 C. degree.

3) In tea infusion, thermoduric spores were becoming sensible after the pressure treatment. As shown in fig. 1, spores survived after the pressure treatment were gradually inactivated and sterilized after several days in tea infusion. This effect was raised stronger by the combination of pressure and heating as shown in fig. 2.
4) It was revealed that the inactivation effect for spores found in tea infusion was due to the sterilization effecct for microbes of tea catechin dissolved in tea infusion (fig.3).
5) Aroma compounds and chemicals in tea drink did not suffer any changes by the pressure treatment and green tea drink treated by the high pressure kept fresh flavore.

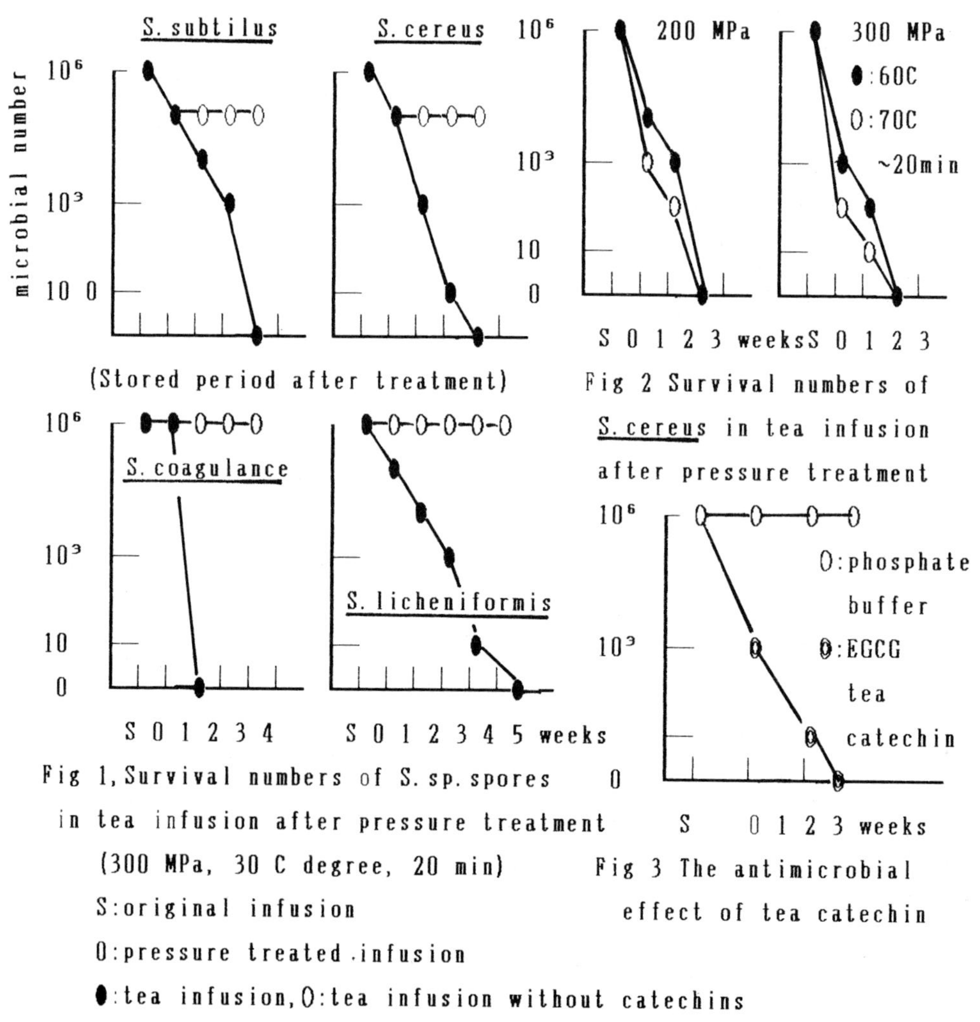

Fig 1, Survival numbers of S. sp. spores in tea infusion after pressure treatment (300 MPa, 30 C degree, 20 min)
S:original infusion
0:pressure treated infusion
●:tea infusion, O:tea infusion without catechins

Fig 2 Survival numbers of S. cereus in tea infusion after pressure treatment

Fig 3 The antimicrobial effect of tea catechin

STUDY OF HIGH PRESSURE EFFECT ON INACTIVATION OF *Bacillus stearothermophilus* SPORES

I. HAYAKAWA, T. KANNO AND Y. FUJIO
Department of Food Science and Technology, Faculty of Agriculture, Kyushu University
6-10-1 Higashi-ku, Fukuoka-shi 812, Japan

ABSTRACT

To establish a new sterilization method with minimal heating, the effect of elevated pressure on bacteriostasis was studied using a heat-resistant spore of *Bacillus stearothermophilus*. After exposure to 800 MPa for 60 min at 60 °C, the active spore count decreased from 10^6 to 10^2/m*l*. However, exposure to the same pressure at room temperature did not cause any significant change. The synergistic effect of high pressurization on the bacteriostatic action of sucrose fatty acid ester at low concentration (<10 ppm) was pronounced with sucrose palmitic acid ester but not with sucrose stearic acid ester. Repetitive pressurization was a more effective method for spore inactivation. Six times of 5-min pressurization with 400 MPa at 70 °C decreased the spore count from 10^6 to 10^2/m*l*, and with 600 MPa, complete inactivation was achieved.

INTRODUCTION

Spores are more heat resistant than living cells. D-values of *B. stearothermophilus* spores range from 0.1 to 14 min at 121°C. At present, sucrose fatty acid esters (SEs) are being used as additive in canned liquid coffee, because they inhibit the germination of bacterial spore contaminant, which would otherwise germinate at the lesser heat processing temperature employed in canning of liquid coffee and other foods. Some studies have already been done on the effects of SEs on bacteriostasis and at high concentration (100-1000 ppm) and on the germination of *B. cereus, B. licheniformis,* and *B. circulars* (Taki *et al.*, 1990). To assess the practical applications of high pressure technology to food processing and to clarify some critical conditions on high pressure sterilization, pressure resistant characteristic properties of the spores were studied using a strain of *B. stearothermophilus*. We also deemed it necessary to further clarify the integrated effects of pressurization and a method like repetitive pressurization without SEs, temperature, and low level addition of SEs (<10 ppm) on the sterilization of *B. stearothermophilus* spores under high pressure.

MATERIALS AND METHODS

Microorganism and culture method. Spores of *B. stearothermophilus* IFO 12550 were used. The spores were collected by centrifugation after 5 days' cultivation at 55°C in a soil medium, consisting of (g): bacto peptone (5), meat extract (3), bacto yeast extract (1), $MnSO_4$ (0.1), bacto agar (25), soil extract (250), H_2O (750), and pH adjusted to 7.2 by NaOH.

Sucrose palmitic acid ester supplements. Sucrose fatty acid esters(P-1695 and S-1695 (Mitsubishi Kasei Food Co., Ltd., Tokyo) were used as bacteriostatic agents. The spore suspension (1 m*l*) was treated with 0.1-10 ppm of SE, which was the concentration range that could not affect spore germination under normal conditions.

Pressure application. The spores were placed in a germ-free tube before pressurization. A pressure apparatus (P_{max}:1000 MPa, Yamamoto Suiatu Kogyosho Co., Ltd., Osaka) was used. Continuous and oscillatory pressurization were compared in this study. Survivors after pressurization were counted after 7 days incubation at 55 °C by plate culture in nutrient agar media (Eiken Chemical Co., Ltd., Tokyo). Heat treatment was used in tests of spore germination.

RESULTS

Temperature Dependence. Pressurization under 400-1000MPa for 20 and 60 min at 20°C gave no particular effect on spore germination. On the other hand, upon the pressurization under 800 MPa at 60 °C, the spore survivors slightly decreased from 10^6 to 10^{4-5}/m*l* after 20 min and about 10^2/m*l* after 60 min (Fig. 1).

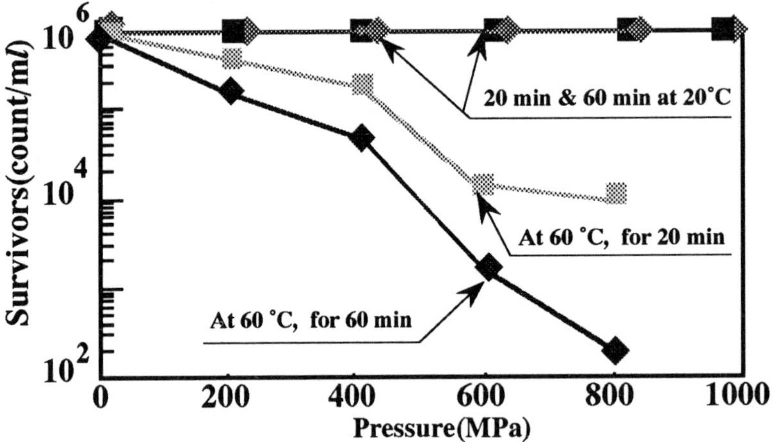

Fig. 1. Effect or pressure, temperature and treatment time on survivors of *B. stearothermophilus* spores.

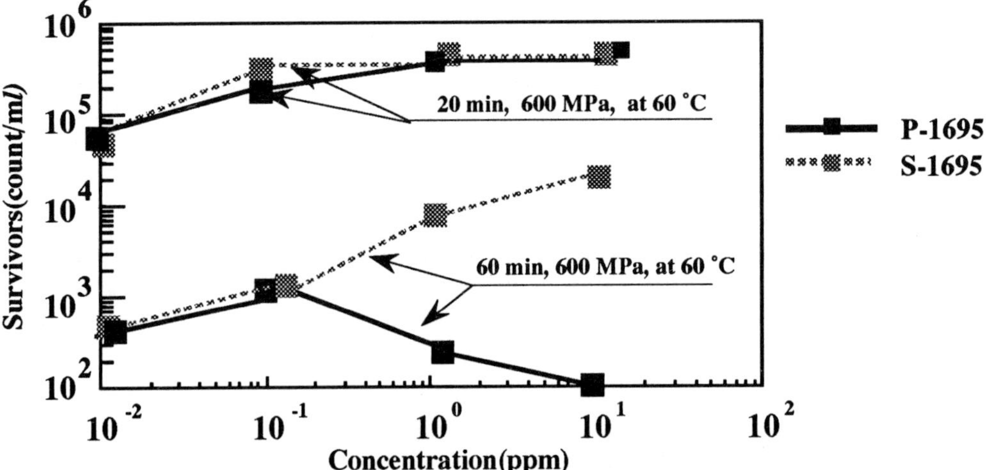

Fig. 2. Effect of concentrations of SEs on survivors of *B. stearothermophilus* spores at 600 MPa pressurization.

Effect of Sucrose Fatty Acid Ester. Pressurization for 60 min with increasing concentration of S-1695 enhanced survivability, but at such longer exposure P-1695 seemed unable to cause protective action against pressure sterilization(Fig. 2).

Repetitive Pressurization. Repetitive short pressurization(5 min internal 70 °C) was found very effective to inactivate the spore. The effect of continuous pressurization for 60 min under 600 MPa at 60 °C was similar to that of six times of oscillation under 400 MPa at 70 °C. Six times of the repetitive pressurization under 600 MPa at 70 °C resulted in complete inactivation (Fig. 3).

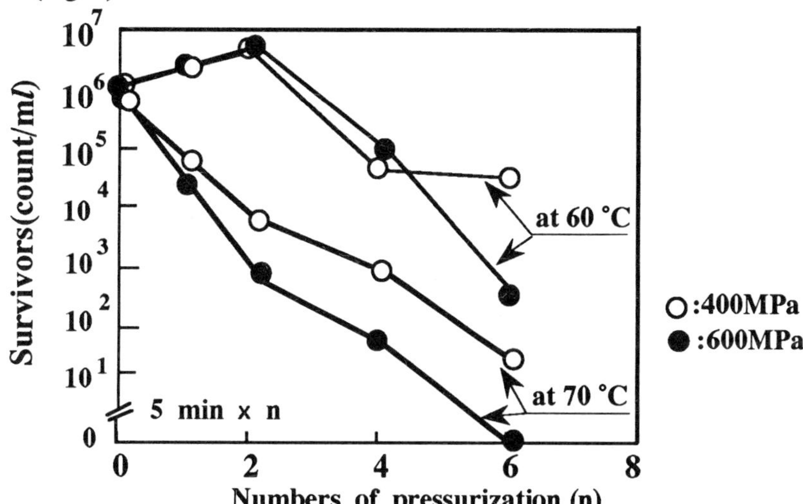

Fig. 3. Effect of pressurization times on survivors of *B. stearothermophilus* spores.

DISCUSSION

This research was undertaken in order to establish a new inactivation method for the spore by high pressure (400-1000 MPa). The *B. stearothermophilus* spore is one of the strongest end-spore of aerobic bacteria. If this spore can be sterilized by pressurization, food deterioration by microorganisms can be eliminated. Strong heat-resisting spores also showed strong pressure durability, as the spores were never affected by 1000 MPa for 60 min under room temperature. Even if, upon exposure for 60 min under 60 °C, some survivors (10^2/m*l*) were still found. Repetitive pressurization was more effective as the complete inactivation was achieved by 6 times of 5-min repetitive pressurization with 600 MPa at 70 °C. In this case, every spore was completely destroyed. This phenomenon may be due to the following two reasons: (1) the adiabatic explosion velocity of spore cell wall and that of high pressure water upon the release of highly pressurized, and (2) water permeability into the spore cell wall were promoted by the rise in temperature (70 °C). Of course, some physical changes could be given onto the spore cell wall by this significant temperature difference (from 20 °C to 70 °C).

ACKNOWLEDGMENT

The authors acknowledge Mr. K. Yoshiyama for his guidance on the spore observation by a scanning electron microscope. This work was supported by research grants from Mitsubishi-Kasei Foods Corporation, and Nestlé Science Promotion Committee is acknowledged with pleasure.

REFERENCES

Taki, Y., Awao. T., Asada, K. and Tsujimoto., S., Sterilization of *Bacillus* sp. Spores by Hydrostatic Pressure. In "pressure-Processed Food Research and Development". (Ed.), R. Hayashi. Sanei Publishing. Co. Ltd., Kyoto, 1990, pp. 143-155.

DEVELOPMENT OF FIBER-CONTAINING NON-EXPANDED PRODUCTS BY EXTRUSION COOKING PROCESS

WENCHANG CHIANG MING-JYH SHIH ANN-RONG YIAO JYH-HUAR WENG
Graduate Institute of Food Science and Technology
National Taiwan University, Taipei, Taiwan, R.O.C.

ABSTRACT

Wet soybean residue was directly used as a raw material to mix with starchy and proteinaceous substances to make cereal flakes and wet-type texturized products with a twin-screw extruder. The cereal flakes could be made from the pellets which were baked in a band infrared oven. While using a two-stage cooling system linked up with the extruder, a good fibrous texturized product containing dietary fiber could be manufactured.

INTRODUCTION

Wet soybean residue, a by-product of soymilk and tofu processing plant, contains high dietary fiber content (1). It is a good raw material for producing low calorie, high dietary fiber and cholesterol-free products. Our laboratory has used the mixture of rice flour and dried soybean residue to make directly expanded extrudates (2). Wet soybean residue contains about 80% moisture and is very hard to dry. Although if it is mixed with rice flour to reduce initial moisture content, the dehydration of the mixture becomes easier (3), we tried to use wet soybean residue directly to manufacture non-expanded extrudates in this study.

MATERIALS AND METHODS

Soybean residue was washed, soaked, ground and then filtrated to get wet soybean residue. Corn flour, rice flour, high gluten wheat flour, soy protein isolate (SPI), sucrose and sodium chloride were also used in this experiment.

A co-rotating and intermeshing twin-screw extruder (Clextral BC-45, France) was used with a barrel length of 100 cm. The screw profile was adopted with two reverse screws positioned in the middle part. The barrel was divided into four sections. The first barrel was cooled using tap water, and the barrel temperatures of other sections were controlled using an induction heater or tap water. There were two holes with a diameter of 0.4 cm on the die plate for cereal flakes processing. A special cooling system including two cooling stages and one central cooling pipe was linked up with the extruder.

Cereal flakes was milled to pass through a 60 mesh screen, and used to determine water solubility index (WSI) by the method of Anderson et al. (4). The supernatant, water-soluble fraction, was also used to analyze water soluble carbohydrate (WSC) by the phenol-sulfuric acid method (5), using the standard curve of glucose to convert the data. Hedonic scale was adopted to evaluate the acceptability of cereal flakes from 20 panels.

The hardness and chewiness of wet-type texturized products were calculated by the texture profile analysis using a rheometer (Fudoh NRM-2020J, Japan). A piano wire (adaptor No.30) of the rheometer was used to cut the sample from the parallel and vertical directions to measure tear strength separately. Dietary fiber content of the product was analyzed by the enzymatic method (6).

RESULTS AND DISCUSSION

Using the obtained extrusion conditions in the previous paper (7), the mixture of the basic recipe (corn flour : wet soybean residue : rice flour = 81.8 : 10 : 8.2 , on the total weight basis) and the additive of 5% SPI, 5% sucrose or 5% sodium chloride was extruded under feed moisture 33.5%, feed rate 26.4 kg/h, screw speed 150 rpm, and the 2nd, 3rd and 4th barrel temperatures being 90, 130 and 40℃ respectively to prepare non-expanded extrudates (pellets). After cooling, the pellet was flaked to the thickness of about 0.15 cm and pre-dried at 40℃ to reduce the moisture content around 25%. The pre-dried flake was then roasted at 140℃ for about 5 minutes in the infrared oven. Both WSI and WSC values were lower than 20%, and all the cereal flakes had acceptable appearance and mouth-feel quality according to the sensory evaluation.

The raw material (wet soybean residue : SPI : high gluten wheat flour = 60 : 30 : 10 , on the total weight basis) was extruded under feed moisture 60%, feed rate 25 kg/h, screw speed 160 rpm, the barrel temperatures in the 2nd, 3rd and 4th sections being 145, 185 and 165℃ respectively. As shown in Table 1, cooling water rate affected the rheological property and tear strength of the texturized product a lot. If controlling the 1st stage, 2nd stage and central pipe of cooling water rate at 1, 2 and 1 L/min respectively, a good fibrous texturized product could be obtained. Because the higher the Fv/Fp value is, the better the fibrous formability is. The soluble and insoluble dietary fiber content in the raw material was 3.6% and 12.3% respectively, but it was 8.5% and 13.8% in the texturized product. The increase of dietary fiber might be resulted from the polymer degradation and starch-protein interaction during extrusion.

ACKNOWLEDGMENT

Grateful acknowledgment is made for the financial support from the National Science Council of the Republic of China.

TABLE 1
Effect of cooling water rate on the rheological property and tear strength of the wet-type texturized product.

cooling water rate at each part, L/min			Rheological property		Tear strength	
1st	2nd	Central	Hardness (kg)	Chewiness (kg)	Fv (kg/cm^2)	Fv/Fp
0	2	0	0.97	0.75	1.08	1.10
		1	1.10	0.93	1.45	0.90
		2	1.01	0.89	1.40	0.95
1	1	0	0.82	0.53	1.20	0.77
		1	0.92	0.73	1.35	0.70
		2	0.97	0.79	1.46	1.10
1	2	0	0.75	0.60	1.40	0.70
		1	1.22	1.04	1.21	0.43
		2	0.99	0.80	1.16	1.10
2	2	0	0.95	0.53	1.14	0.74
		1	0.89	0.66	1.07	1.02
		2	0.98	0.78	1.12	1.23

REFERENCES

1. Chiang, W., Shih, M. J. and Chen, S., Effect of soaking and grinding conditions on quality of soymilk and physical properties of bean residue. J. Chinese Agric. Chem. Soc., 1988, **26**, 165-72.
2. Wu, T. P. and Chiang, W., Optimal extrusion conditions for manuacturing rice expanded products containing dietary fiber. J. Chinese Agric. Chem. Soc., 1991, **29**, 17-25.
3. Shih, M, J., Perng, M. Y. and Chiang, W., Drying of the mixture of soybean residue and rice flour. Food Sci., 1991, **18**, 286-94.
4. Anderson, R. A., Conway, H. F., Pfeifer, U. F. and Griffin, E. L., Gelatinization of corn grits by roll- and extrusion-cooking. Cereal Sci. Today, 1969, **14**, 4-7, 11-12.
5. Dubois, M., Gilles, K. A., Hamilton, J. K., Rebers, P. A. and Smith, F., Colorimeteric method for determination of sugar and related substances. Anal. Chem., 1956, **28**, 350-6.
6. Prosky, L., Asp, N., Schweizer, T. F., Devries, J. W. and Furda, I., Determination of insoluble, soluble and total dietary fiber in foods and food products : Interlaboratory study. J. Assoc. Off. Anal. Chem., 1988, **71**,1017-23.
7. Shih, M. J. and Chiang, W., Physicochemical changes of corn-based semi-product during extrusion processing. J. Chinese Agric. Chem. Soc., 1992, **30**, 454-61.

THE HEAT DENATURATION PROCESS OF SOYBEAN PROTEIN BY TWIN SCREW EXTRUDER

YOSHINOBU AKIYAMA
Research Laboratories,
Kyodo Milk Industry Co., Ltd.
180 Hirai, Hinode-cho, Nishitama-gun, Tokyo 190-01 JAPAN

ABSTRACT

Finely-fibered texturized tofu were prepared from tofu by use of twin screw extruder. The relations between the extrusion process parameters and the intensity of protein heat denaturations are mentioned. The screw revolution and material feed rate were controlled as variable process parameters in order to make extrudates have the different residence time. SDS-PAGE was applied to the protein denaturation analysis. Four major bands 7Sα, 7Sβ, 11S acidic and 11S basic were detected. They were decreased in proportion to residence time in the extruder. Cooking value(Cv) was defined as the index of denaturations, which indicates the integral of net heated time and temperature. The correlation coefficients between 7Sα, 7Sβ, 11S acidic, 11S basic and corresponding Cv were -0.98, -0.84, -0.82 and -0.81 respectively. It was suggested that Cv indicate the the degree of heat denaturation of proteins so accurate that it could be the rational process parameter in manufacturing of protein texturization by extruders.

INTRODUCTION

Textured vegetable products have been manufactured as meat substitutes, and it is expected that their consumption will increase. However, it seems that the qualities of textured vegetable products such as taste, flabour and texture are not enough to satisfy the consumers.

For successful texturization of vegetable proteins, the control of heat denaturaions in the extruder barrel and of flowing conditions of extrudate in the die section must be very important. The relations between residence time and protein denaturations were measured. The numerical model derived from the theory of heat sterilization was applied to estimate the amount of heat denaturation.

MATERIALS AND METHODS

Tofu were dehydrated partially and extruded by the twin screw extruder. Screw revolutions and material feed rates were controlled in order to make extrudates have the different residence time. To evaluate the heat denaturations, SDS-PAGE was applied to texturized tofu. Fine structures were examined by SEM.

RESULTS AND DISCUSSION

Table 1 shows the relations between the residence time and changes in four major subunits of soyprotein. All subunits decreased in proportion to residence time and 7Sα was most sensitive to heat.

TABLE 1
Residence time and changes in subunits

SPL No.	Residence time sec	7Sα'	7Sβ	11S acidic	11S basic
1	0	19.2	10.5	11.8	15.7
2	145	13.6	8.0	8.2	11.2
3	180	13.5	9.3	9.6	11.3
4	240	10.5	7.9	7.6	11.3
5	280	10.3	7.9	7.2	9.7
6	300	10.9	9.0	8.5	10.2
7	375	6.8	4.9	7.7	10.5
8	400	6.0	3.2	5.0	11.4
9	450	5.3	5.3	6.8	8.0

If the protein heat denaturation mechanism is regarded same as those of heat sterilization, it must be of the first order in reaction.

In Eq-1, θ_2 is maximum temp. of material, θ_1 is the lower critical temp. of denaturation and Z is constant. Amount of denaturation is given by S. Final amount of denaturation during extrusion cooking will be given by Eq-2, where t_1 is the beginning time of denaturation , and t_2 is terminating time. Cooking value Cv stands for, therefore, the degree of denaturation during extrusion.

Fig. 1 shows the relations between Cv and subunit 7Sα.
The correlation coefficient was -0.98. It is supposed that about 15 % of native 7Sα would be degraded or associated. Changes in denaturated subunit are in inverse proportion to Cv lineally.

$$S = 10^{\frac{\theta_2 - \theta_1}{Z}} \quad (1)$$

$$V = \int_{t1}^{t2} 10^{\frac{\theta_2 - \theta_1}{Z}} \, dt \quad (2)$$

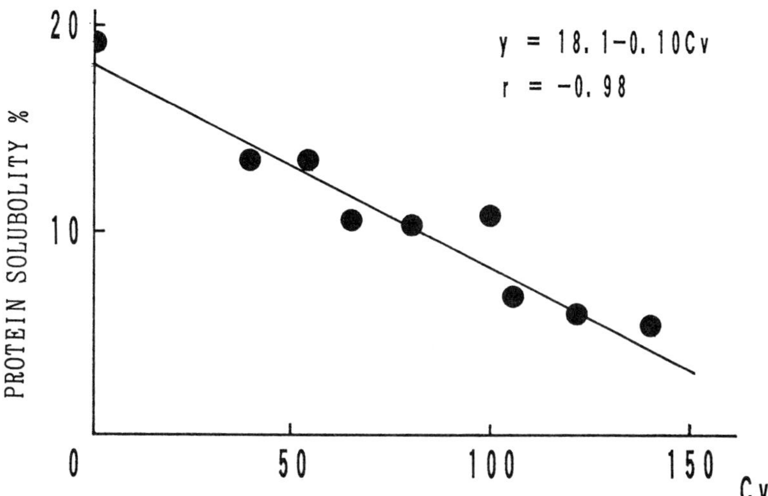

Figure 1. The relations between Cv and 7Sα.

CONCLUSIONS

It is concluded that subunit solubilities are closely connected with residence time and changes in amount of denaturated protein correlate with Cv. Cv stands for the degree of protein denaturation so accurate that it could be the rational index as extrusion parameter.

THE DEVELOPMENT AND CONTROL OF COLOUR IN EXTRUSION COOKED FOODS SIMULATED USING A MODEL REACTION CELL

LISA BATES, JENNIFER M. AMES AND DOUGLAS B. MACDOUGALL
Department of Food Science and Technology,
University of Reading, Whiteknights, P.O. Box 226, Reading RG6 2AP, UK

ABSTRACT

A reaction cell was designed to allow the rapid, small scale examination of colour development which occurred in an extrusion cooker. It simulated the temperature, time, rapid heating and pressure experienced during extrusion cooking, but not the mechanical shear. A starch-glucose-lysine mixture was extrusion cooked and heated in the reaction cell, and the colour of the samples was analysed. pH had the most significant effect on colour development; temperature, moisture and residence time were also significant. Reaction cell results were used to accurately predict L* and a* values of samples extruded at a predetermined pH.

INTRODUCTION

Extrusion cooking of expanded snacks is an intermediate moisture, high temperature and high mechanical shear operation. These conditions favour the Maillard reaction which reduces lysine availability and leads to colour development. Many previous studies have shown the effect on lysine availability, while only a few have examined the effect on colour [1]. This study uses a reaction cell to model colour development in an extruder over a range of conditions.

MATERIALS AND METHODS

A mixture of wheat starch type A, D(+)- glucose and L-lysine monohydrochloride (96:3:1, m:m:m) was cooked in either a twin screw corotating extruder or in the reaction cell. Tristimulus CIE L*a*b* values of ground and sieved samples were obtained and hue angle values, θ, $\tan^{-1}(b^*/a^*)$ were calculated. Statistical evaluation of the results was done by analysis of variance and regression analysis using SAS software. Further experimental details are available on request from the authors.

RESULTS AND DISCUSSION

Preliminary experiments were performed at 3 levels of pH, 3 levels of moisture, 3 temperatures and 2 target mean residence times. The most significant parameter affecting colour was pH, followed by temperature, pH / temperature interaction and moisture. Residence time had the least significant effect. Statistical analysis of the reaction cell data is summarised in Table 1.

TABLE 1
Preliminary Reaction Cell Study

VARIABLES & DEGREES OF FREEDOM		L* R-Square: 0.93		a* R-Square: 0.95		b* R-Square: 0.98		θ R-Square: 0.96	
		F Value	P[a]	F Value	P[a]	F Value	P[a]	F Value	P[a]
pH	2	107.49	***	191.71	***	419.93	***	241.68	***
Moisture	2	10.27	***	0.13	NS	10.00	***	7.41	**
Temperature	2	39.37	***	37.13	***	48.29	***	31.40	***
Time	1	3.60	NS	5.24	*	9.69	**	14.38	***
pH*Moisture	4	0.38	NS	1.62	NS	5.13	**	1.10	NS
pH*Temp	4	4.39	**	12.67	***	5.44	**	11.52	***
pH*Time	2	1.85	NS	1.31	NS	2.73	NS	2.97	NS
Moist*Temp	4	1.35	NS	1.19	NS	2.95	*	1.99	NS
Moist*Time	2	1.22	NS	1.80	NS	0.67	NS	0.80	NS
Temp*Time	2	0.05	NS	1.03	NS	0.07	NS	1.96	NS

P[a] (significance level) = *** at $p<0.001$, ** at $p<0.01$, * at $p<0.05$, NS not significant.

A detailed study of the most significant factor, pH, was then carried out, holding the other factors constant at 15% moisture, 140°C and 32s. Overall, more colour developed in the extruded samples than in the reaction cell samples. This may be due to differences in the mode of heat transfer, or in material fluidity, between the 2 systems. However, the increase in colour with increasing pH was very similar. Plots of L*, a* (see Figure 1) and hue angle against pH gave very similar gradients for extrusion and reaction cell samples. In contrast, the gradients were not similar for the b* plots, possibly due to the different stages of colour development in the 2 systems. Using linear regression of the L* and a* responses from the reaction cell, L* and a* were predicted for the mixture extrusion cooked at pH 6.3. The predicted values were 63.9

and 5.6 respectively. The ranges of extrusion experimental data were 60.5 - 70.2 (L*) and 5.2 - 7.3 (a*). Thus, the predicted responses at pH 6.3 were within the experimental ranges.

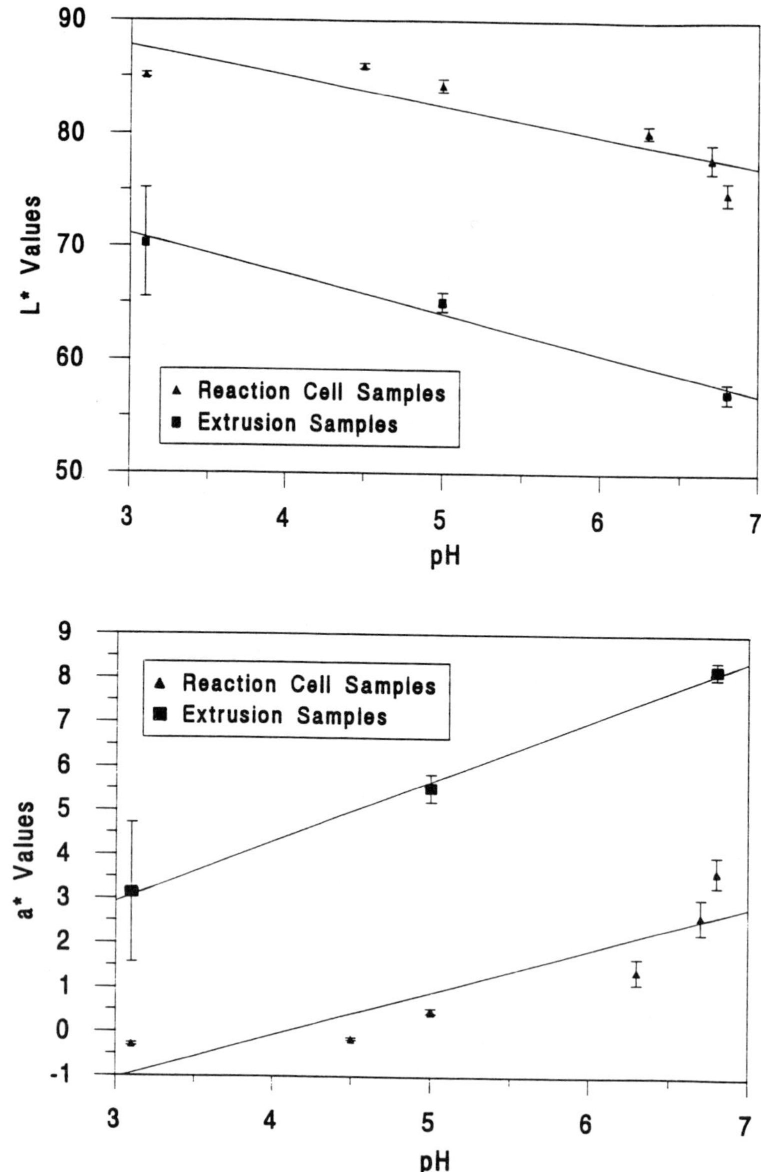

Figure 1. Changes in (a) L* and (b) a* values with increasing pH.

REFERENCE

1. Berset, C., Colour. In Extrusion Cooking, eds. C. Mercier, P. Linko and J.M. Harper, Am. Assoc. Cereal Chem., St Paul, MN, USA, 1989, pp. 371-85.

APPLICATION OF EXTRUSION-COOKING FOR FEED PREMIXES STABILIZATION

LESZEK MOŚCICKI & STANISŁAW MATYKA
Department of Food Process Engineering, Agricultural University
Doświadczalna 44, 20-236 Lublin, Poland

ABSTRACT

The premixes (microelements, vitamins and chemotherapeutics) are the most expensive components and play important role in feed production. Usually added in small amounts,thay can be incorrectly mixed with other components because of its bulk density and suppleness to segregation. Moreover, the water absorbtion and chemical activation taking place during the storage of microelements can reduce the quality of the feed.

In the paper,the results of premixes stabilization are reported using the extrusion-cooking technique.

The extrusion-cooking process conditions and its influence on physical and chemical products properties are discussed to show how the stabilization of feed premixes can be achived with minimum energy costs and maximum quality effects.

MATERIALS AND METHODS

The typical mineral premix, applied usually to the production of fodder mixture for farm animals, was used in this research. It includes microelements as: Fe, Cu, Zn and Mn. 50 kg samples of mineral premixmixed with a vegetable carrier [corn,I variant and the horse bean (vicia faba),II variant] in the different proportions of the mineral to the carrier (30/70; 35/65; 40/60; 45/55; 50/50 and 55/45) were prepared for the experiments. Each sample of the mixture was exposed to extrusion-cooking to gain homogenic extrudates. The extrusion-cooking was conducted on the single screw extrusion-cooker of TS-45 polish production, with the application of thermal treatment from 140 to 170°C, the o 6 mm die, the screw of the 1:3 c.r. and with a screw speed of 1,66 * 1/s.

The prepared mixtures were extruded without the addition of water. The obtained product was blended to the size of the bruised grain, the size of the most commonfraction of the typical components included in typical feed mixture.

Homogeneity of extrudates was estimated on the basis of the analysis of the Zn, Cu, and Co content in the probe taken in the initial, middle, and final phases of extruding of the 50

the analysis of the Zn, Cu, and Co content in the probe taken in the initial, middle, and final phases of extruding of the 50 kg propor tion of the previously mentioned mixtures. The elements mentioned above were present in the mixture in widely varying proportions. What is more,in order to estimate the homogeneity and resistance to selfsegregation, the corn-extrudates was crumbled then fractio nized by means of sieves into three fractions of the 1 mm; 0,5 mm and 0,1 mm grain.

The Zn, Cu and Co content was analyzed in each fraction by the technique of ASA Atomic Absorbtion. The obtained numerical data were described statistically, counting average values, standard deviations and coefficients of variability. Student's Test was applied to compare average values.

RESULTS

Table 1 presents the results of the Zn, Cu and Co designation,both in corn and horse bean extrudates.The data in the table indicate that the average content of the elements in the probes, taken in the initial, middle, and final phases of extrusion-cooking, did not differ significantly from the prescribed content ($p>0,05$) in the entire range of the examined relations of the mineral premixed with the vegetable carrier.

It needs to be emphasized that the variability of the content of the elements appeared to be extremely small and, in the case of Zn,it did not exceed 2%. The variability of the concentration of Cu did not exceed 7% in any case, and Co did not exceed 8%.

Bearing in mind that these results were also influenced by the degree of blending before the extrusion-cooking, the obtained coefficients of variability V testifies to the high repeatability of the composition of the product. This was also confirmed by the stability of the composition of the corn extrudate after blending and fractionating with the sieves.

In the 50/50; 45/55 and 40/60 mixtures, the relationship of the mineral fraction to the organic one, the coefficients of variability V did not exceed 2% for Zn, 5% for Cu, and 5,5% for Co.

According to the above discussion we may conclude that during extrusion-cooking, the mineral fraction dilutes regularly in the whole mass of the organic carrier, stabilizing the product in this way. The blended extrudates, which, even after the self-segregation of the grain during the transport, prevents its primary mineral content.

The examination of the physical features of the stabilized feed premixes showed that their bulk density approximated the value of the typical cereal feed components. Moreover, the capability of absorbtion of water or the tractability of dissolution decreased in relation to the qualities of non-extruded materials. Accordingly, the process of extrusion-cooking, even on that scale, appeared to be promising.

TABLE 1
Zn, Cu, Co concetration in corn and horse bean extrudates [g/kg]

Mineral/ Carrier	Components	Extrudates						Composition
		Corn			Horse bean			
		M	SD	V%	M	SD	V%	
30/70	Zn	39.100	0.245	0.62	38.310	0.802	2.09	39.420
	Cu	3.950	0.090	2.27	3.950	0.082	2.04	4.050
	Co	0.153	0.008	4.98	0.155	0.004	2.58	0.160
35/65	Zn	45.360	0.681	1.50	45.270	0.328	0.72	45.990
	Cu	4.670	0.999	2.14	4.630	0.311	6.71	4.720
	Co	0.177	0.011	6.29	0.184	0.006	3.48	0.187
40/60	Zn	51.630	1.118	2.16	52.410	0.336	0.64	52.560
	Cu	5.320	0.100	1.88	5.200	0.229	4.40	5.400
	Co	0.204	0.014	6.99	0.205	0.005	2.44	0.214
45/55	Zn	59.030	0.980	1.66	58.430	0.608	1.04	59.130
	Cu	6.030	0.100	1.65	5.920	0.142	2.39	6.070
	Co	0.232	0.005	2.21	0.229	0.018	7.96	0.240
50/55	Zn	65.490	0.499	0.76	65.180	0.724	1.11	65.700
	Cu	6.650	0.136	2.04	6.710	0.121	1.79	6.750
	Co	0.249	0.015	6.18	0.261	0.003	1.23	0.267
55/45	Zn	71.170	0.879	1.22	72.210	0.626	0.87	72.270
	Cu	7.220	0.204	2.83	7.310	0.373	5.10	7.420
	Co	0.280	0.013	4.72	0.293	0.006	2.07	0.294

TABLE 2
Zn, Cu, Co concentration in blended corn extrudates [g/kg]

Mineral/ Carrier	Components	Fraction			M	SD	V (%)	Composition
		A	B	C				
50/50	Zn	64.80	63.30	62.30	63.47	1.2297	1.94	65.700
	Cu	6.43	6.65	6.48	6.52	0.1153	1.77	6.750
	Co	0.23	0.25	0.23	0.24	0.0115	4.88	0.227
45/55	Zn	57.60	58.30	58.20	58.04	0.3763	0.65	59.130
	Cu	5.98	6.10	6.00	6.03	0.0643	1.07	6.070
	Co	0.22	0.23	0.23	0.22	0.0061	2.69	0.240
40/60	Zn	49.50	50.50	51.20	50.43	0.8934	1.77	52.560
	Cu	5.59	5.32	5.07	5.32	0.2600	4.88	5.400
	Co	0.18	0.20	0.19	0.19	0.0104	5.43	0.214

EXTRUSION OF Castanea sativa

F. SILVA, A. CHOUPINA, I. M.N. SOUSA and M. L. BEIRÃO da COSTA
*S.A.C.T.A./ Instituto Superior de Agronomia, Technical University of Lisbon
Tapada da Ajuda, 1399 Lisboa Codex. Portugal

ABSTRACT

In Portugal, chestnut (Castanea sativa) is mainly consumed unprocessed. Only a small amount of the exceeding production is used in the candy industry. The objective of this study was to find an alternative industrial utilisation for chestnuts, by extrusion-cooking.

The properties of chestnut starch were studied by the Brabender amylograph. The amylose/amylopectin ratio and the starch granules dimensions were obtained by light microscopy. The performance of chestnut flour in single screw extrusion cooking was studied setting a response surface method (R.S.M.). All the responses linearly decreased with initial moisture content and die temperature being independent of the screw speed.. The best conditions were low initial moisture and low die temperature for crispness, but the expansion ratio at this conditions was very poor. To overcome this, a new R.S.M. was set considering the incorporation of corn grits to enhance expansion.

The extruded starches showed an interesting preservation of the integrity of the granules with a huge swollen effect in the case of chestnut and some disrupted granules mainly among the corn starch.

INTRODUCTION

Chestnut is used in sweets and candy specialities. In Portugal its industrial use is still quite insufficient. Extrusion-cooking is known to be a process of high versatility for starchy products. Chestnut presents a high percentage of starch, consequently in this paper it was tried to assess the possibility of extruding chestnut flour. There are very few references on chestnut starch and one of our concerns was its characterisation.

Extrusion conditions for chestnut flour and this mixed with corn grits at different ratios were optimised using objective textural evaluation as dependent variables.

MATERIALS AND METHODS

Dry chestnuts and corn grits are commercially available. Both were milled and passed by a 0.71 mm sieve.

Extrudate Production and Evaluation

A single screw laboratorial 20 DN Brabender extruder was used, operating at 3:1 compression ratio, using a die of 3 mm diameter. Feed hopper speed was 40 r.p.m. and temperatures in the barrel respectively T_1=T_2= 423 K.

A response surface methodology (R.S.M.) with a central composite rotatable experimental design was used to optimise extrusion of chestnut and corn grits mixtures using as independent variables moisture (X1), ranging from 12 to 20%, die temperature (X2), from 423 to 473 K, and percentage of chestnut flour in the mixture of chestnut plus corn grits (X3), from 20 to 100%. The extrusion conditions of chestnut flour was optimised using the same methodology. The independent variables were moisture (X1), ranging from 10 to 18%, die temperature, T_3 (X2), from 423 to 473 K, and screw speed (X3) from 120 to 220 r.p.m.. The dependent variables used in both cases were expansion ratio measured with a micrometer gauge, rupture strength evaluated in a TAX-T2 Texturometer (Stable Microsystems) with a Warner-Bratzler cell, and consistency of the slurries of extrudates flour determined in a Brabender Amylograph.

Starch Characterisation

Starch in chestnut flour was quantified by the Lintner polarimetric method (AOAC, 1990). Starch size and shape granules were studied by light microscopy. Gelatinization properties were studied by Amylograph method using convenient dilution (8.5% starch in the suspension). Amylose/amylopectin ratio was determined by G.P.C. using Sepharose CL-2B eluted with NaN_3 (0.02% v/v) at a flow rate of 0.5ml.min^{-1}.

RESULTS AND DISCUSSION

Chestnut starch granules when studied by light microscopy showed an average size similar to corn starch. Nevertheless, after extrusion chestnut starch show a much higher swelling degree keeping simultaneously its integrity. It can also be seen that inside the granule, components are melted. Corn starch granules are much more susceptible to rupture.

When subjected to amylograph evaluation, chestnut starch began gelatinization earlier than corn starch (532 K and 343 K, respectively) showing higher values for maximum consistency (545 U.B.) than corn starch (230 U.B.). After extrusion, chestnut flour showed that some extruded starch had not yet been gelatinised what was evidenced by a new increase in amylograph consistency.

The results from the first set of experiments, with the chestnut/corn grits extrudates were fitted to a second order polynomial equation by a multiple regression analysis. When the response is expansion ratio (E.R.), the respective equation is:

$$Y = 1.683 - 0.1698\ X1 - 0.293\ X2 - 0.301\ X3 + 0.188\ X1\ X3 \quad (1)$$

with a R2 = 0.89. The equation for the rupture strength (R.S.) as the dependent variable is:

$$Y = 4.274 - 2.471\ X1 - 1.227\ X3^2 \quad (2)$$

with R2= 0.97. From the above equations one can see that both E.R. and R.S. of the extrudates decreased with moisture (X1), die temperature (X2) and chestnut flour incorporation (X3) or moisture and chestnut incorporation, respectively. Nevertheless, the experiments with 100% of chestnut gave extrudates with some expansion (1.67) and good organoleptic characteristics. Our next step was the optimisation of the extrusion of the chestnut flour de per si using this time as a third variable the screw speed. The respective equation for E.R. was:

$$Y = 1.683 - 0.173\ X1 - 0.166\ X2 \quad (3)$$

with $R^2 = 0.97$ and for R.S. :

$$Y = 2.166 - 0.849\ X1 - 0.891\ X2 \quad (4)$$

with $R^2 = 0.79$. From these equations it can be seen that the screw speed did not affect the E.R. or R.S. and the die temperature had a negative influence on both characteristics. These results are in agreement with previously reported work (1,2) on corn grits extrusion.

The higher E.R. value (1,9) was obtained for the extrudate produced with 10% moisture and 175° C die temperature, this extrudate also showed the higher rupture strength (5.18 N) that seems to be related with good crispness.

The consistency of chestnut / corn grits extrudates flour at 20°C, used as another response for the RSM, is an index of the degree of gelatinization undergone during extrusion and is expressed by equation:

$$Y = 292.94 + 102.71\ X1 - 191.80\ X3 + 58.91\ X1^2 + 68.63\ X3^2 \quad (5)$$

with an $R^2 = 0.90$

When extruded alone, the starch of the chestnut flour shows a consistency at 20°C as expressed by:

$$Y = 559.48 + 74.64X1 + 127.79\ X2$$

with an $R^2 = 0.79$. From the above equations one can see that the consistency of the extrudates slurries is linearly dependent on moisture in both cases. For the mixtures it also depends on chestnut flour incorporation but on the chestnut extrudates consistency is dependent on die temperature.

From GPC results the presence of amylose and amylopectin was determined to be 43 and 57% respectively in chestnut starch. It was also possible to assess that the molecular mass of the chestnut amylopectin is higher than the one of the corn starch (7413 kD and 3890 kD). The difference in the amylose molecular mass is not so big being higher for corn starch (346 kD, 490 kD).

CONCLUSIONS

It is possible to obtain expanded extrudates from chestnut flour or chestnut flour / corn grits mixtures with acceptable textural properties specially at low moisture and low die temperatures. Chestnut starch has a higher amylose / amylopectin ratio than corn starch, having simultaneously an interesting sweet taste and preservation of the integrity of the starch granules in the extruded products.

REFERENCES

1. Launay, B. and Lisch, J.M., Twin screw extrusion cocking of starches. J. Food Eng., 1983 **2** : 259.

2. Chinnaswamy, Y. and Hanna M.A.,. Optimum extrusion cooking conditions for maximum expansion of corn starch. J. Food Sci., 1988, **53** : 834.

EFFECTS OF PREPARATION PROCEDURES AND PACKAGING MATERIALS ON QUALITY RETENTION OF CUT CHINESE CABBAGE

EERO HURME, RAIJA AHVENAINEN, EIJA SKYTTÄ, MARGARETA HÄGG[*], MIRJAMI MATTILA[†], RAIJA-LIISA HEINIÖ AND ANNA-MAIJA SJÖBERG
Technical Research Centre of Finland (VTT), Food Research Laboratory, P.O.Box 203, FIN-02151 Espoo, Finland, [*]Agricultural Research Centre, FIN-31600 Jokioinen, Finland, [†]Department of Food Technology, P.O.Box 27, FIN-00014 Helsinki University, Finland

ABSTRACT

The effects of different washing procedures, packaging materials and long-term storage before processing on the quality retention of packed cut chinese cabbage were studied. A shelf life of at least 7 days at 5 °C could be achieved by two separate washings, the first of which could contain chlorine, and by highly permeable packaging films (O_2 perm. 5200 - 5800 cm^3/m^2 24 h 101.3 kPa, 23 °C, RH 0 %). Long-term storage before processing was detrimental to the quality of the packed samples.

INTRODUCTION

The shelf life of prepared cabbages can possibly be prolonged by using highly permeable films for preventing anaerobic respiration and by using chemical treatments for controlling microbial contamination. Several senescence reactions occur in cabbages after harvesting and therefore long-term storage before processing can affect the quality maintenance of prepared cabbages. The aim of this study was to determine the effects of different processing procedures on the shelf-life of packed cut chinese cabbage. In addition, the effects of long-term storage before processing were studied.

MATERIALS AND METHODS

All external parts of the cabbages (*Kingdom*) were first removed. The cabbages were then cut into 5 mm thick strips, washed (1 min/wash) in 5 °C water and/or in 0.01 % chlorine solution (as sodium hypochlorite), spin dried and packed. Five packaging types and two packaging sizes were used (1.0 kg cabbage/package, vol. 3000-3500 cm^3, film thickness 40 μm; PE-film impregnated with ceramic powder, PE-EVA-film and OPP-film, 1.5 kg cabbage/package, vol. 5000 cm^3; PE-paper-lid + PP-tray and PET-carton package). The O_2 permeabilities of the films were 5800, 5200, 1200 and 1500 cm^3/m^2 24 h 101.3 kPa, 23 °C, RH 0 %, respectively. The packages were stored at 5 °C for up to 7 days.

The percentages of CO_2 and O_2 in the package headspace were determined with Servomex PA 404 and Servomex 507A analysers, respectively. All analyses were carried out from four replicate packages.

Microbiological determinations using pour plate techniques were carried out on duplicate sample packs. Aerobic plate counts (APC) were obtained after incubation at 30 °C for 3 days (Plate Count Agar, Difco), the number of lactic acid bacteria (LAB) after incubation at 30 °C in 5 % CO_2 for 3 days (MRS agar, Oxoid), and coliforms after 24 h at 37 °C (Violet Red Bile Agar, Difco).

Sensory evaluations of the samples were performed by trained panel members. First, the odour and appearance of the samples were evaluated by 5 panelists, followed by flavour evaluations by 10 panelists (10-20 minutes after the first evaluation). All analyses were carried out on duplicate samples. Samples were scored on a scale of 0 to 5 (5 = excellent, 2 = unacceptable for retailing and 0 = unfit for consumption). The minimum level of acceptability for human consumption was 1.5.

β-carotene and ascorbic acid concentrations were analysed using HPLC (1,2). The residual free chlorine concentrations (samples ClW in Table 1) were analysed spectrofotometrically 16 hours after packaging (3). The detection level for free chlorine was 0.1 mg/kg.

The respiratory activity of the cut cabbages was measured by the flow-through system. Air was fed into an airtight chamber via one port and allowed to exit from another port. The O_2 and CO_2 concentrations of the outcoming gas were determined. Measurements were terminated when both the O_2 and the CO_2 concentrations had stabilized (four replicates).

RESULTS AND DISCUSSION

Washing in two separate washing waters improved the maintenance of sensory quality of the cut chinese cabbage (Table 1). The quality was maintained better when chlorine was used in the first washing water. The washing methods did not, however, significantly affect the microbial load, the CO_2 production and O_2 consumption, the ascorbic acid and β-carotene concentrations (12 mg/100 g fresh weight and 29 μg/100 g fresh weight, respectively, after 1 days of storage and 10 mg/100 g fresh weight and 17 μg/100 g fresh weight, respectively, after 7 days of storage) or the respiration rates (13 ml CO_2/kg h) of the samples. No residual free chlorine was found in chlorine-treated samples. Packaging in highly permeable films (PE and PE-EVA) improved the quality retention (Table 2). The rapid quality loss of the cut cabbage packed in OPP-film and PE-paper-lid + PP-tray was obviously due to the high CO_2 concentration in the packages. Drying was the main factor reducing shelf life of the cabbages packed in PET-carton packages.

Long-term storage before processing was detrimental to the quality keeping of the cut cabbage (washed once in water, packed in OPP-film). For example, the odours of the samples made from cabbages stored for 3 months were scored clearly better than those of samples made from cabbages stored for 5 months: after 7 days of storage the scores were 2.0 and 1.5 (season 1991) and 1.1 and 0.7 (season 1992), respectively. Obviously, this was due to respiration, enzymatic and microbiological activity of the cabbages. The β-carotene content of the uncut fresh cabbages decreased between 3 and 5 months of storage from 19 to 12 (season 1991) and from 12 to 9 (season 1992) μg/100 g fresh weight.

REFERENCES

1. Ollilainen, V.M., Heinonen, M., Linkola, E., Varo, P. and Koivistoinen, P., Carotenoids and retinoids in Finnish foods:meat and meat products. J. Food Comp. Anal., 1988, **1**, 178-88.
2. Speek, A., Schrijver, J. and Schreus, W.H.P., Fluorometric determination of total vitamin C and total isovitamin C in foodstuffs and beverages by high-performance liquid chromatography with precolumn derivatization. J. Agric. Food Chem., 1984, **2**, 352-5.
3. Test Method VTT-4426-91, VTT Food Research Laboratory, Espoo, Finland, 1991.

TABLE 1

The effects of washing method and storage time on the quality retention of cut chinese cabbage packed in OPP-film. W = washed in water, WW = washed twice in water, ClW = washed first in chlorinated water (0.01 % free chlorine) and then in pure water. The amount of washing water was 3 l/kg product. The cabbages were stored for 3 months before processing.

Quality attributes	Storage time/washing method								
	1 day			3 days			7 days		
	W	WW	ClW	W	WW	ClW	W	WW	ClW
CO_2 (%)	4 ±1	4 ±1	4 ±0	10 ±0	9 ±0	9 ±1	15 ±1	15 ±1	15 ±1
O_2 (%)	14 ±1	15 ±0	15 ±0	4 ±0	7 ±1	6 ±1	1 ±0	0 ±0	0 ±0
APC	5.7	nd	nd	nd	nd	nd	6.8	6.8	6.8
LAB	3.8	nd	nd	nd	nd	nd	5.9	5.9	5.7
Coliforms	1.5	nd	nd	nd	nd	nd	4.4	4.7	4.7
Odour	4.2 ±0.2	4.1 ±0.1	4.1 ±0.1	3.1 ±0.1	3.4 ±0.1	3.3 ±0.2	1.1 ±0.1	1.9 ±0.4	2.3 ±0.4
Appear.	4.4 ±0.5	4.5 ±0.4	4.4 ±0.4	4.0 ±0.4	4.0 ±0.4	4.0 ±0.4	3.4 ±0.9	3.4 ±0.9	3.4 ±0.9
Flavour	4.0 ±0.5	3.9 ±0.5	4.1 ±0.3	3.3 ±0.5	3.4 ±0.7	3.2 ±0.4	na	na	2.8 ±0.7

TABLE 2

The effects of packaging material and storage time on the quality retention of cut chinese cabbage. Cut cabbages were washed in water before packaging (W in Table 1). OPP = OPP-film, EVA = PE-EVA-film, PE = PE-film impregnated with ceramic powder, Pap = PE-paper-lid + PP-tray, Car = PET-carton package (not sealed). The cabbages were stored for 3 months before processing.

Quality attributes	Storage time/packaging material														
	1 day					3 days					7 days				
	OPP	EVA	PE	Pap	Car	OPP	EVA	PE	Pap	Car	OPP	EVA	PE	Pap	Car
CO_2 (%)	7 ±1	6 ±1	5 ±2	5 ±1	2 ±1	13 ±1	9 ±1	8 ±1	11 ±2	1 ±0	20 ±0	10 ±2	9 ±1	20 ±1	1 ±0
O_2 (%)	9 ±1	10 ±2	12 ±3	12 ±1	20 ±0	1 ±1	1 ±0	1 ±0	2 ±1	20 ±0	0 ±0	0 ±0	0 ±0	0 ±0	20 ±0
Odour	4.0 ±0.6	4.1 ±0.2	4.1 ±0.4	4.2 ±0.4	3.9 ±0.4	2.8 ±0.2	3.1 ±0.4	3.3 ±0.3	3.6 ±0.6	3.4 ±0.3	1.8 ±0.5	2.6 ±0.3	2.8 ±0.2	1.7 ±0.8	2.4 ±0.3
Appear.	4.3 ±0.3	4.2 ±0.3	4.2 ±0.3	4.2 ±0.2	4.3 ±0.3	3.7 ±0.5	3.6 ±0.5	3.6 ±0.4	3.7 ±0.5	3.8 ±0.3	3.0 ±0.4	3.3 ±0.3	3.3 ±0.3	3.3 ±0.3	1.1 ±0.3
Flavour	3.8 ±0.4	4.0 ±0.4	4.0 ±0.3	3.8 ±0.4	3.9 ±0.3	3.4 ±0.7	3.4 ±0.7	3.8 ±0.3	3.6 ±0.4	3.6 ±0.5	na -	3.3 ±0.8	2.9 ±1.0	na -	na -

nd = not determined, **na** = quality not acceptable for human consumption.
The O_2 and CO_2 concentrations (%) and the results of the sensory evaluations are presented as mean values ± standard deviations. The microbial counts are presented as mean values (log cfu/g).

EFFECTS OF PREPARATION PROCEDURES AND PACKAGING MATERIALS ON QUALITY RETENTION OF GRATED CARROTS

EERO HURME, RAIJA AHVENAINEN, EIJA SKYTTÄ, MIRJAMI MATTILA*, MARGARETA HÄGG†, RAIJA-LIISA HEINIÖ AND ANNA-MAIJA SJÖBERG
Technical Research Centre of Finland (VTT), Food Research Laboratory, P.O.Box 203, FIN-02151 Espoo, Finland, *Department of Food Technology, P.O.Box 27, FIN-00014 Helsinki University, Finland, †Agricultural Research Centre, FIN-31600 Jokioinen, Finland

ABSTRACT

The effects of different washing procedures, packaging materials and long-term storage before processing on the quality retention of packed grated carrots were studied. A shelf life of at least 8 days at 5 °C could be achieved by washing the peeled carrots in 0.01 % chlorine solution and in 0.5 % citric acid solution. The packaging materials tested did not affect the shelf life of grated carrots. Long-term storage before processing was detrimental to the quality of the packed samples.

INTRODUCTION

The shelf life of packed grated carrots can possibly be prolonged by using chlorine wash for reducing microbial contamination, by using citric acid wash for reducing enzyme activity and by using highly permeable films for preventing anaerobic respiration. Furthermore, the effectiviness of chlorine wash can be improved by high washing temperatures. The aim of this study was to determine the effects of different processing procedures on the shelf-life of packed grated carrots. In addition, the effects of long-term storage before processing were studied.

MATERIALS AND METHODS

The carrots (*Navarre*) were washed and all heavily contaminated parts were removed. The carrots were then peeled with carborundum, washed (Table 1), grated into 3 mm thick strips, gently sprayed with water, spin dried and packed. Five packaging types and two packaging sizes were used (1.0 kg carrot/package, vol. 3000 cm^3, film thickness 40 μm; PE-film impregnated with ceramic powder, PE-EVA-film and OPP-film, 1.6 kg carrot/package, vol. 5000 cm^3; PE-paper-lid + PP-tray and PET-carton package). The O_2 permeabilities of the films were 5800, 5200, 1200 and 1500 cm^3/m^2 24 h 101.3 kPa, 23 °C, RH 0 %, respectively. The packages were stored at 5 °C for up to 8 days.

The percentages of CO_2 and O_2 in the package headspace were determined with Servomex PA 404 and 507A analysers, respectively. All analyses were carried out from four replicate packages.

Microbiological determinations (pour plate techniques, duplicate sample packs) carried out were: 1) aerobic plate counts (APC) after incubation at 30 °C for 3 days (Plate Count Agar, Difco), 2) lactic acid bacteria (LAB) after incubation at 30 °C in 5 % CO_2 for 3 days (MRS agar, Oxoid), 3) yeasts and molds (Y & M) after 5 days at 25 °C (YGC Agar, Difco).

Sensory evaluations of the samples were performed by trained panel members. First, the odour and appearance of the samples were evaluated by 5 panelists, followed by flavour evaluations by 10 panelists (10-20 minutes after the first evaluation). All analyses were carried out on duplicate samples. Samples were scored on a scale of 0 to 5 (5 = excellent, 2 = unacceptable for retailing and 0 = unfit for consumption). The minimum level of acceptability for human consumption was 1.5.

α- and β-carotene acid concentrations were analysed using HPLC (1). The residual free chlorine concentrations (samples ClW in Table 1) were analysed spectrofotometrically 16 hours after packaging (2). The detection level for free chlorine was 0.1 mg/kg.

The respiratory activity of the grated carrots was measured by the flow-through system. Air was fed into an airtight chamber via one port and allowed to exit from another port. The O_2 and CO_2 concentrations of the outcoming gas were determined. Measurements were terminated when both the O_2 and the CO_2 concentrations had stabilized (four replicates).

The O_2 and CO_2 percentages and the results of the sensory evaluations are presented as mean values ± standard deviations and the microbial counts as mean values (log cfu/g) (Tables 1 and 2).

RESULTS AND DISCUSSION

Washing the peeled carrots in water containing 0.01 % chlorine or 0.5 % citric acid (5 °C) improved the maintenance of sensory quality of the packed grated carrots (Table 1). The same effect was obtained by washing the peeled carrots first in chlorinated water (30 °C) and then in pure water (5 °C). The washing methods did not, however, significantly affect the microbial load, the CO_2 production and O_2 consumption, the α- or β-carotene concentrations (2 mg/100 g fresh weight and 8 mg/100 g fresh weight, respectively, just after grating and 2 mg/100 g fresh weight and 6 mg/100 g fresh weight, respectively, after 8 days of storage) or the respiration rates (10 ml CO_2/kg h, measured from samples W, Cl and CA) of the samples. No residual free chlorine was found in chlorine-treated samples. Obviously, the chemical treatments lowered the enzymatic activity of the grated carrots. The packaging materials tested did not significantly affect the shelf life of grated carrots (Table 2). Drying slightly reduced the shelf life of the grated carrots packed in PET-carton packages. The shelf life retention of the samples OPP (Table 2) was clearly better than that of the similarly processed samples W (Table 1). This was possibly because of better manufacturing practices (e.g. hygiene) used in the processing of the samples OPP (the samples OPP were prepared in the laboratory and the samples W in an industrial processing line).

Long-term storage before processing was detrimental to the quality keeping of the grated carrot (not washed, packed in OPP-film). For example, the odours of the samples made from carrots stored for 2 months were scored better than those of samples made from carrots stored for 7 months: after 8 days of storage the scores were 1.4 and 0.7 (season 1991) and 1.4 and 1.2 (season 1992), respectively. Obviously, this was due to enzymatic activity of the carrots, because the long-term storage did not affect the α- or β-carotene contents (2-3 mg/100 g fresh weight and 7-8 mg/100 g fresh weight, respectively), the microbial load or the respiration rate (4 ml CO_2/kg h) of the peeled whole carrots.

REFERENCES

1. Ollilainen, V.M., Heinonen, M., Linkola, E., Varo, P. and Koivistoinen, P., Carotenoids and retinoids in Finnish foods:meat and meat products. J. Food Comp. Anal., 1988, **1**, 178-88.
2. Test method VTT-4426-91, VTT Food Research Laboratory, Espoo, Finland, 1991.

TABLE 1

The effects of washing method (3 l/kg carrot, 1 min/wash) of peeled carrots and of storage time on the quality retention of grated carrot packed in OPP-film. NW = not washed, W = washed in 5 °C water, Cl = washed in 5 °C 0.01 % free chlorine solution (as sodium hypochlorite), CA = washed in 5 °C 0.5 % citric acid solution, ClW = washed first in 30 °C chlorinated water (0.01 % free chlorine) and then in 5 °C pure water. The carrots were stored for 2 months before processing.

Quality attributes	Storage time/washing method: 2 days					4 days					8 days				
	NW	W	Cl	CA	ClW	NW	W	Cl	CA	ClW	NW	W	Cl	CA	ClW
CO_2 (%)	8	9	9	8	11	15	17	16	14	14	31	34	29	33	29
	±0	±1	±1	±1	±1	±2	±1	±2	±1	±1	±2	±3	±2	±5	±3
O_2 (%)	7	5	5	7	3	0	0	0	0	0	0	0	0	0	0
	±0	±2	±2	±1	±2	±0	±0	±0	±0	±0	±0	±0	±0	±0	±0
APC	5.1	5.0	4.5	6.1	4.9	nd	nd	nd	nd	nd	6.7	6.6	6.4	5.9	6.3
LAB	4.4	5.2	4.3	4.4	4.0	nd	nd	nd	nd	nd	7.7	7.8	7.6	7.7	7.7
Y & M	2.9	3.0	3.0	2.7	3.0	nd	nd	nd	nd	nd	3.8	5.0	3.7	3.5	3.5
Odour	4.0	4.1	4.0	4.2	4.1	2.8	3.3	3.5	3.5	3.8	1.4	1.3	2.9	2.9	2.6
	±0.2	±0.4	±0.2	±0.1	±0.2	±0.1	±0.5	±0.5	±0.5	±0.4	±0.8	±0.2	±0.5	±0.3	±0.6
Appear.	4.2	4.2	4.2	4.2	4.3	3.9	3.9	3.9	3.9	4.1	3.6	3.6	3.6	3.6	3.7
	±0.2	±0.2	±0.2	±0.2	±0.2	±0.4	±0.4	±0.4	±0.4	±0.1	±0.3	±0.3	±0.3	±0.3	±0.2
Flavour	3.5	3.5	3.6	3.8	3.7	3.2	3.3	3.5	3.3	3.5	na	na	2.9	3.0	3.1
	±0.5	±0.3	±0.4	±0.4	±0.3	±0.5	±0.5	±0.4	±0.5	±0.4	-	-	±0.5	±0.8	±0.6

TABLE 2

The effects of packaging material and storage time on the quality retention of grated carrots. Peeled carrots were washed in 5 °C water before grating and packaging. The carrots were stored for 2 months before processing. OPP = OPP-film, PE = PE-film impregnated with ceramic powder, EVA = PE-EVA-film, Pap = PE-paper-film + PP-tray, Car = PET-carton package (not sealed).

Quality attributes	Storage time/packaging material: 2 days					5 days					8 days				
	OPP	EVA	PE	Pap	Car	OPP	EVA	PE	Pap	Car	OPP	EVA	PE	Pap	Car
CO_2 (%)	10	6	7	7	1	19	10	13	17	2	25	15	15	27	2
	±2	±1	±1	±1	±0	±1	±0	±2	±2	±1	±3	±1	±4	±1	±1
O_2 (%)	2	4	1	9	20	0	0	1	2	20	0	0	1	0	20
	±3	±2	±0	±3	±0	±0	±0	±2	±3	±0	±0	±0	±2	±0	±0
Odour	3.9	3.7	3.0	4.1	3.7	2.9	3.0	3.1	3.5	2.9	3.0	2.8	2.8	2.9	3.1
	±0.3	±0.3	±0.3	±0.4	±0.4	±0.4	±0.4	±0.3	±0.5	±0.5	±0.4	±0.4	±0.6	±0.4	±0.8
Appear.	4.3	4.3	4.3	4.3	4.3	4.3	4.3	4.4	4.4	4.2	4.4	4.4	4.4	4.4	3.7
	±0.3	±0.3	±0.3	±0.3	±0.3	±0.7	±0.7	±0.6	±0.6	±0.5	±0.6	±0.6	±0.6	±0.6	±0.6
Flavour	3.6	3.5	3.7	3.9	3.8	3.3	3.2	3.3	3.0	3.4	3.1	3.4	3.1	3.0	3.1
	±0.6	±0.5	±0.5	±0.6	±0.6	±0.5	±0.4	±0.4	±0.6	±0.7	±0.7	±0.6	±0.9	±0.6	±0.6

nd = not determined, **na** = quality not acceptable for human consumption.

STORAGE TECHNOLOGY ON DEHYDRATED SWORD BEAN

ZHANG MIN KONG BAOHUA WANG CHENZHI YANG YONGLI
(Northeast Agricultural College, Harbin 150030,P.R. China)

ABSTRACT

In this paper, storage tests on dedydrated sword been were conducted, and the best storage technology was determined. Then, the sterilization mechanism of the methods before storage and the effects of temperature, moisture, and packing on storage are discussed.

INTRODUCTION

Every year, a great quantity of dehydrated vegetables absorb moisture because of improper storage which can lead to the loss of nutrition and pigment, even to the growth of molds and the presence insects which may cause the decay (C.Z. Wang,1990). So it is very importment to study the storage of dehydrated veget– ables for export.

The dehydrated sword bean, which is an important dehydrated product, was selected in this study to obtain the best storage technology.

MATERIALS AND METHODS

1.Materials

Dehydrated sword been with 8% moisture content was used as the sample for the storage tests. The handling methods before storage are mainly deoxidation, sterilization, and radiation .

2.Methods

Steps of the tests: Fresh sword bean was picked from the vegetable garden at our college, then pretreated (some of the sample was added with the bacte– ricide) and dehydrated.The 20g sample was put into the bottle with the corres– ponding saturated solution for the storage test. Meanwhile,the deoxidizer was put into some of the bottles, and the mouth of the bottle was sealed with vase–

line. Then some of the samples were irradiated. The storage tests were carried out at a temperature of 20–30℃. The contents of chlorophyll and vitamin C(Vc) were used as the test criteria, and the number of molds and bacteria were also considered.

RESULTS AND ANALYSIS

The primary test have shown that the main factors which affect the above four criteria are the method of pretreatment before storage, activity of water, and time of storage. The three levels of every factor were arranged in Table 1.

Table 1 The factors and levels of the orthogonal storing test for dehydrated sword bean

factors / levels	A pretreatment before storage	B water activity	C time of storage (months)
1	deoxygenation	0.85	0
2	sterilization	0.75	3
3	radiation	0.68	6

The results and analysis of the orthogonal test were given in Table 2(omitted). For comparison, normal technology test (10#) was arranged.The results presented in Table 2 show that the results of test 10# are poor in the four criteria,so the improvement of the technology is needed.

The results in Table 2 show that the order of importance of the factors and their best levels are A_1, C_1, B_2 for the contents of chlorophyll and V_c (K_Y,K_{VC}), and B_3, $A_3(C_2)$ for the number of molds and bacteria (K_M,K_X).

To judge the criteria synthetically,the proper powers for every criterion have been determined (Y.E.He,1986).The powers of the contents of chlorophyll and V_c, and the number of molds and bacteria are 0.3,0.2,and 0.25 respectively.The values with "*" in Table 2 are the comprehensive results. The order of importance of the factors and the best levels are B_3, C_1, A_1 after synthetic evaluation, that is,the greatest effect on the criteria is water activity, the second one is storage time and the least one is method of pretreatment before storage. The less water activity and storage time are, the better the comprehensive criteria. Water activities have little effect between 0.68–0.75, while storage times have little effect between 0–3 months. Deoxygenation before storage has the best effect on the criteria, while sterilization and radiation have little effect on the results.

All things considered, the best technology of storage is as follows: Controlling the moisture in dehydrated sword bean before storage,keeping the relative humidity less than 75%, deoxidizing the package, and keeping the storage time to less than 3 months.

DISCUSSION

The comparative analysis of the sterilization mechanism under three methods of pretreatment before storage.

It was observed that the main cause of the decay of dehydrated sword bean was that the number of molds and bacteria were beyond the National Standard. It has been reported (N.W.Desroier,1989) that bactericide,deoxygenation,and radiation can effectively inhibit the growth of molds and bacteria,but the results will vary with different material, bactericide, deoxydizer, and irradiating doses. Deoxygenation has the most significant effect under the test conditions. Sterilization and radiation have less effect, which may be because the concentrations of bactericide and dose of irradiation were too low. However,the primary tests have shown that chlorophyll and V_c will be destroyed heavily if the dose of irradiation is up to 8 kGy,and the recommended concentration of dehydrated sodium acetate (bactericide) in vegetable processing is 0.1%. Therefore, deoxygenation is the safest and most effective method of pretreatment.

CONCLUSION

1.The best technology of storing dehydrated sword bean is to control moisture before storage, keeping the relative humidity to less than 75%, deoxidizing the inside of the package, and keeping the storage to less than 3 months.

2.The sterilization mechanism of different pretreatments before storage has been analyzed comparatively.Deoxygenation is considered as the safest and the most effective method.

REFERENCES

1.C.Z.Wang,The developing prospect and countermeasure of food irradiation in China, Isotope (Chinese), 1990(2).

2.L.W.Zhang,The study on fresh egg's storage, the master paper of Northeast Agricultural College, 1988.

3. Y.E.He et al, The testing designing of the agricultural machinery, the Machinery Industry Press,1986.

EFFECTS OF SEVERAL TREATMENTS ON THE QUALITY OF COLD STORED 'CLEMENTINE MANDARINS'

F. PAZIR AND M. AZAK
Food Engineering Department of Engineering Faculty,
Ege University,
35101 Bornova-Izmir, Turkey

ABSTRACT

The effects of harvesting type (careful and rough harvesting), dipping into fungicide solution (TBZ + imazalil) at different temperatures (20, 40, 50, 60°C for 2 min.) and wax application on the quality of cold stored 'clementine' mandarins were investigated. Mandarins were stored at 6°C and 80–85% RH for 3 months. Juice content, titratable acidity, and ascorbic acid content decreased throughout the storage period. The inverted sugar content increased during the storage period although the changes of total sugar and soluble solids of mandarins were not significant. Surface deformation and off-flavour were detected in the ones which were dipped into fungicide solution at 60°C. Rough handling increased the amount of decayed fruits.

INTRODUCTION

One of the most important factors which lengthens the storing life of the cold stored citrus fruits is the way of harvesting. Most damages, during or after harvest are mainly due to some strikes, in other words composed by impact forces, which are produced by dropping from a height.[1]

Citrus fruits should be picked by trimming them off the branches, instead of plucking, since the risk of decay increases two times following plucking and carriage.[2]

Bruises due to the rough harvesting increase the rate of respiration of the fruit, with a great decay.[4]

Washing prior to storing and application of fungicide and waxing are effective practices in the prevention of the fruits decaying. Chilling injury is one of the main problems and such damage can be reduced considerably by hot-dip procedures, and the usage of TBZ increases the efficency of the procedure.[6]

MATERIALS AND METHODS

Clementine mandarins were harvested in December, 1991 and 1992. Fruits were then randomised into 12 treatment units of approximately 200 fruits each.

The mandarins were separated into two lots; careful and rough handling. In careful harvesting, the fruits were picked with clippers and placed in the picking box slowly. In rough harvesting, fruits were also picked by clippers but without considering their stem lengths and were allowed to drop from up to 50 cm heights. The treatment units are shown in Table 1.

Mandarins were dipped into fungicide solution for 2 min. All of the fungicide solutions and wax contained 2000 ppm TBZ–750 ppm imazalil.

Table 1 Treatment units which were applied in the research

A. Careful harvesting	*B. Rough harvesting*
a Control	a Control
b 20°C	b 20°C
c 40°C	c 40°C
d 50°C	d 50°C
e 60°C	e 60°C
f Wax.	f Wax.

Fruits were then held in wooden boxes and stacked in the cold room in the two replicate groups. All groups were stored for 12 weeks in a cold room with fan forced air circulation which operates at 6 ± 1°C. For all the groups, weight loss, juice content, titratable acidity, pH, ascorbic acid, total inverted sugar, color values (L,a,b) and sensory scores were measured every three weeks. Sensory analysis was made according to the diagram of Karlsruhe, but the scores have been changed to 1–9. The 60°C treatment units were not included in the research of 1992.

RESULTS AND DISCUSSION

At the end of three months storage in both years, the juice content (8.2–14.4%), titratable acidity (60–63%) and ascorbic acid (29.7–34.2%) had decreased. These decreases were highly significant ($p < 0.05$). The reduction in the titratable acidity tends to occur as a result of the use of citric acid in respiration and in the synthesis of amino acids.[3]

Hunterlab colour values of the peel were measured. L (lightness) value and a/b (orange colour) value decreased 3.3–4.5% and 6.8–7.5% respectively. L values of the 60°C processed samples were rapidly decreased during the storage. The reduction of L and a/b values were 20% and 33% respectively. This can be accounted for by browning as a result of casting surface oil off the peel and therefore oxidation.[5] In the 50°C processed samples orange color changed into yellow and a/b value decreased by 20%. The changes were found insignificant in all the samples except in those two samples. The sensory analyses of the treatment units were by appearance, color and taste. When examining the appearances of the mandarins of 1991, it was observed that during the process of storage the peel was removed from fruit segments, with sponginess occuring. In the samples of 1992, there was no considerable sponginess and other characteristics resembled those of the results in 1991. In the samples which were dipped at 60°C, some browning was detected due to surface deformation. After studying the samples in view of color it was observed that waxed samples preserved their color followed by the control, 20°C, and 40°C sample. It was found that 50°C treated orange color of the samples turned to yellow and 60°C processed ones were browned. In all of the samples loss of taste was observed because the acid–sugar balance was changed due to decrease in acidity. Off-flavour,

Table 2 The sum of the decay and weight loss at different temperatures and waxing according to rough and careful handling at the end of storage for two years. (%)

	1991						*1992*					
Treatment	*Careful harvest*			*Rough harvest*			*Careful harvest*			*Rough harvest*		
units	*1.M*	*2.M*	*3.M*	*1.M*	*2.M*	*3.M*	*1.M*	*2.M*	*3.M*	*1.M*	*2.M*	*3.M*
Cont.	1.8	9.5	53.2	12.1	28.0	64.6	2.6	4.0	17.0	3.6	6.9	21.6
20°C	1.7	7.2	14.5	6.7	13.8	29.9	2.8	4.3	8.2	2.4	3.8	8.5
40°C	1.4	4.9	36.1	1.3	3.4	32.9	3.3	4.8	8.9	3.9	5.1	9.5
50°C	2.9	7.4	51.4	2.7	5.5	39.7	3.0	4.1	8.0	2.8	4.1	9.9
60°C	2.6	11.3	69.3	2.6	10.5	74.5	—	—	—	—	—	—
Wax.	0.2	13.9	48.4	3.1	11.1	56.1	2.5	4.9	11.2	1.9	3.1	17.1

M: months.

especially was observed in 60°C treated samples. Weight loss and decaying were observed in all samples during the storage. The mandarins which were processed at 20°C showed the least weight loss and decay for each storage year. It was observed that waxing and fungicide application protected the fruit quality well until the late second month. Decaying suddenly increased in the fruits following this period when the effect of wax processing decreased. The rates of decaying were relatively less in all the treatment units in 1992, because of the falling rain and slight frost before the harvest of 1991. The decaying and the weight loss results are given in Table 2.

REFERENCES

1. Fluck, R. C. and Ahmed, E. M., 1985. Measurement by compression test of impact damage to citrus fruits. *J. Texture Studies* **4**, 4–500.
2. Grierson, W. and Ben-Yehoshua, S., 1987. Storage of citrus fruits. Chapter 20, in *Fresh Citrus Fruits* p. 484. Avi Publishing Company, Connecticut.
3. Karaçali, I., 1990. *Protection and marketing of orchard products*. Ege University Publishing, Izmir.
4. Munoz-Delgado, J. A., 1987. Problems in cold storage of citrus fruit. *Int. J. Refrigeration*, **10**, 229–233.
5. Usai, M., Arras, G., Fronteddu, F., 1992. Effect of cold storage on essential oils of peel of thompson navel oranges. *J. Agric. Food Chem.* **40**, 271–275.
6. Wild, B. L., 1990. Hot dip treatments reduce chilling injury during storage of citrus fruit at 1°C. *Food Research Quarterly*, **50**, 36–40.

POSTHARVEST TREATMENTS ON QUALITY OF MANGO FRUITS (*Mangifera indica* (L) var. ZILL) STORED AT LOW TEMPERATURE

MARIA AMALIA BRUNINI KANESIRO, JOSE FERNANDO DURIGAN,
RAUL ROBERTO DE SOUZA FALEIROS, DENISE MARIA ANDRIOLLI
Department of Technology,
College of Agricultural Science and Veterinary "Campus" Jaboticabal
San Paulo State University (UNESP)
(14870) Jaboticabal, SP, Brazil

ABSTRACT

The effect of postharvest treatments by using hot-water and benomyl solution and combination of both on the quality of mango fruits var. "Zill" during storage at 12°C, 95% RH and at normal atmosphere (24-27°C, 65-85% RH) were investigated. The parameters analyzed were weight, peel color, total soluble solid, starch content, titrable acidity, soluble carbohydrate and general appearance (presence of rottenness in the peel). Results allowed to conclude that the treatments did not interfere with the fruits' ripening process, which was showed by the increase in total soluble solid and decrease of titrable acidity and starch content; that neither heat or fungicide treatments were able to control the rottening process on fruits stored under low temperature; and that storage at 12°C was the most suitable to increase the shelf life these fruits.

INTRODUCTION

Mango fruits are becoming one of the most important commodity of Brazil, which recently has become the sixth producer-country of mango fruits[1]. With the increase of exportation and fresh consumption of mango fruits, many techniques of postharvest treatments and storage condition have been studied in order to extend the shelf life maintaining the quality and reducing the incidence of storage diseases. Among these techniques, hot water treatment and low temperature storage have been reported to be effective against latent infection of Collestotrichum gloeosporioides Penz, which is causative of anthracnose, and that hot water treatment is more effective when a fungicide is incorpored[2,3,4,5,]. Then, the objective of this study was to investigate the effect of postharvest treatment on the ripening quality and anthracnose development in mango fruits cv. Zill during storage.

MATERIAL AND METHODS

Mango fruits cv. "Zill" were harvested at unripe stage from the Agricultural Experimental Station of the College of Agricultural Science and Veterinary "Campus" of Jaboticabal/UNESP, Jaboticabal City, San Paulo State, Brazil and transported to the Laboratory of Technology, where they were randomly divided into four experimental groups of 80 fruits. After, experimental groups of 80 fruits were immersed in one of the following postharvest treatments: hot-water at 55°C for 10 minutes (HWT); 0.2% benomyl solution for 5 minutes (FT); and hot-water at 55°C for 10 minutes and subsequent 0.2% benomyl solution for 5 minutes (HWFT). One experimental group of 80 fruits was untreated and was used as control (Control). After, each experimental groups were separated in two lots of 40 fruits. One lot of 40 fruits of each experimental group was stored in refrigerated-room at 12°C, 95% RH and the second lot of 40 fruits of each experimental group was stored in room at normal atmosphere (24-27°C, 65-85% RH). For each stored lot of mango fruits, 20 fruits were separated and used to estimate the weight loss, peel color and presence of rotten in the peel.
The ripening qualities of the fruits were measured at harvesting time and during storage. Weight fruit, weight fruit pulp, peel color, general appearance, titrable acidity (TA), total soluble solid (TSS), soluble carbohydrate and starch content were measured as the component of ripening quality. Weight loss was measured by weighing of each fruit and expressed in %. The pulp weight was calculated from the difference between fruit weight and the weight of peel and seed, and expressed in %. The surface peel color was examined visually and assessed on a five index: 1= green color; 2= more green color than yellow color; 3= equal amounts of green color and yellow color; 4= more yellow color; and 5= yellow color with red color areas. The disease was also examined visually and assessed on a five index: 1= healthy ; 2= more healthy areas than rotten areas; 3= equal amount of healthy and rotten areas; 4= 2/3 of the surface peel with rotten; and 5= full rotten or more 2/3 of the surface peel with rotten. Titrable acidity was measured by titration method using 0.1N NaOH and the results was expressed in g of citric acid/100g of pulp. Total soluble solid was measured directly in the expressed juice by using a digital refractometer, and the results was expressed in °Brix. The soluble carbohydrate was determined by using the phenol-sulfuric method[6] and the values were expressed in g of glucose/ 100g of pulp. Starch content was hydrolyzed into reducing sugar, which was analyzed by using phenol-sulfuric method[6], and the starch content was calculated by multiplying 0.9 of reducing sugar content.

RESULTS

The results of this study gave evidence that storage temperature had effect on the ripening and storage life of treated and untreated "Zill" mango fruits. The storage at 12°C delayed the ripening and extended the storage life of the fruits until 26 days, while storage at normal atmosphere reduced the storage life of the fruits, accelerated the ripening and the senescence of the fruits (Table 1). The visual examination of treated and untreated- stored fruits showed that the fruits had greater peel color changes and disease development, and that the effect of HWFT on chlorophyll change was less detrimental than others treatments in fruits stored at 12°C (Table 1). It was observed that neither heat or fungicide treatments used in this experiment were able to control the incidence of anthracnose development, but these treatments associed with low temperature storage might retard the incidence of this disease.

Based on the pattern of changes that occurred in the chemical properties of stored fruits (Table1), it could be deduced that the postharvest treatments did not affect these properties. Also, it could be observed that there was no difference in the degree of ripening between treated and untreated fruits in both storage condition.

TABLE 1

Changes in "Zill" mango fruits quality at the end of storage

Quality index	Initial	Control NA*	mango 12°C	HWT NA	mango 12°C	FT NA	Mango 12°C	HWFT NA	mango 12°C
Storage life(days)	-	11	26	11	26	11	26	11	26
Weight loss(%)	100.0	86.5	91.2	86.7	89.9	87.7	92.2	84.9	90.1
Pulp weight(%)	67.1	75.1	81.5	77.8	75.8	79.8	82.6	78.1	80.7
Peel color index	1.6	5.0	4.8	5.0	4.0	5.0	4.0	5.0	3.8
Disease index	1.1	4.0	3.8	3.5	4.0	4.0	3.5	4.3	3.8
TA(g/100g)	1.3	0.2	0.6	0.2	0.8	0.2	0.9	0.2	0.8
TSS(°Brix)	7.8	15.5	16.0	15.0	14.0	15.0	16.5	18.5	15.5
Glucose(g/100g)	2.6	9.6	9.8	9.4	9.8	8.1	7.7	8.5	7.0
Starch(g/100g)	13.5	5.0	5.0	5.4	5.0	6.8	6.5	5.7	7.7

* NA= normal atmosphere storage

REFERENCES

1. FAO. Production yearbook. Roma, 1988, 42, pp. 218-219
2. Quimio, T.H., Mango anthracnose and low temperature storage, Philippine Agriculturist, 1974, 58, 192-199.
3. Dodd, J. C., Jeffries, P. and Jeger, M.J., Management strategies to control latent infection in tropical fruit, Aspect of Applied Biology, 1989, 20, 49-56. In Dodd, J.C., Bugante, R., Koomen, I., Jeffries, P. and Jeger, M.J., Pre-and post-harvest control of mango anthracnose in the Philippines. Plant Pathology, 1991, 40, 576-583.
4. Thompson, A.K., The development and adaptation of methods for control of anthracnose. In Dodd, J.C., Bugante, R., Koomen, I., Jeffries, P. and Jeger, M.J., Pre- and post-harvest control of mango anthracnose in the Philippines. Plant Pathology, 1991, 40, 576-583.
5. Dodd, J.C., Bugante, R., Koomen, I., Jeffries, P. and Jeger, M.J., Pre- and post-harvest control of mango anthracnose in the Philippines, Plant Pathology, 1991, 40, 576-583.
6. Dubois, M., Gilles, K.A., Hamilton, J.K., Rebers, P.A. and Smith, J.K., Colorimetric for determination of sugar and related substances. Analytical Chemistry, 1956, 28, 350.

FLAVOR CHANGE OF GRAPE JUICE DURING PROCESSING

HIDEAKI OHTA, YOICHI NOGATA and KOH-ICHI YOZA
Chugoku National Agricultural Research Institute
Ministry of Agriculture,Forestry and Fisheries
Fukuyama, Hiroshima 721, Japan

ABSTRACT

The flavor change of Concord (*Vitis labrusca* L.) grape juice during processing was examined. During manufacture of concentrated grape juice, 99.3% of the total peak area of headspace volatiles detected in fresh grapes had disappeared. The loss was greatest during the heating process (53%), followed by concentrating process by evaporator (27%) and clarifying process with enzymes (18%). On the other hand, both the heating process to extract color and clarifying process provided a larger amount of flavor extracted with solvent. Methyl anthranilate was abundant during the heating process to extract color, but it was reduced to half after the concentrating process.

INTRODUCTION

Little information concerning the flavor change of grape juice during processing has been reported, although flavor loss and change of fruit juices have been considered as one of the most important factors affecting the quality of fruit juice. Manufacture of concentrated grape juice involves two unique processes such as heat treatment for extracting color and filtration for removing argol. In this study, we clarified the flavor change of Concord grape juice during processing.

MATERIALS AND METHODS

The juices investigated were sampled during the commercial scale processing, including fresh grapes, heating to extract color (fruit blancher, 70°C, 15 min), extracting, pasteurizing, clarifying, concentrating and the final product. Analysis of headspace volatile components was performed by gas chromatography (GC) according to the Tenax trapping technique [1]. The flavor compounds from the sample juice recovered by continuous liquid extraction using with n-pentane and methylene chloride (2:1) [2]. The extract was dehydrated with sodium sulfate, then concentrated *in vacuo* (28°C, 10 mmHg) for GC analysis. 10 % benzyl acetate was used as internal standard.

Flavor compounds were analyzed quantitatively by GC using a Hitach 163 with a flame ionization detector. A WCOT glass capillary column (70 m x 0.50 mm I.D.) coated with PEG 20M was used. The column oven temperature was programmed from 60 to 190°C at 3°C/min. Nitrogen was used as the carrier gas at a flow rate of 2 ml/min with a splitting ratio of 10:1 for headspace analysis and 20:1 for solvent extract analysis, respectively. GC-mass spectrometry was performed as described previously [2].

RESULTS AND DISCUSSION

The presence of 27 peaks was demonstrated in the headspace of fresh grapes. Nineteen of these peaks were identified by comparing and matching the mass spectra and GC retention times. Figure 1 shows the relative changes in total peak area of headspace volatiles and peak area of ethyl crotonate during the processing of grape juice. During manufacture of concentrated grape juice, 99.3% of the total peak area of headspace volatiles detected in fresh grapes had disappeared. The loss was greatest during the heating process (53 %), followed by the concentrating process by evaporator (27%) and clarifying process with enzymes (18%). The quantity of ethyl crotonate, a key aroma component [3], decreased in a similar manner.

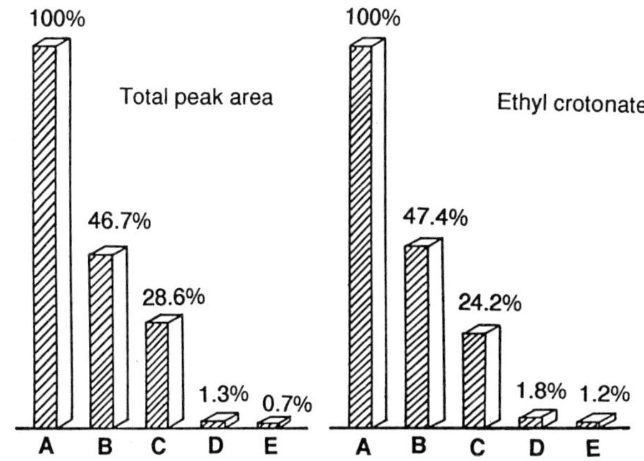

Figure 1. Retention of total peak area of headspace volatiles and ethyl crotonate from Concord grape juice during processing.

A: Fresh, B: Heating to extract color, C: Clarifying with enzymes, D: Concentrating, E: Final product.

The yield of flavor extract concentrate from the juice during the heating process to extract color was 3.22 mg kg^{-1}, followed by fresh grapes (2.56 mg kg^{-1}), clarifying with enzymes (2.37 mg kg^{-1}), concentrating process (0.52 mg kg^{-1}) and final product after filtration (0.50 mg kg^{-1}).

We detected 111 peaks in the solvent extract and identified 49 compounds. Figure 2 shows the change of four compounds, which contributed to the Concord grape flavor [3], during processing. As expected, 2-methyl-3-buten-2-ol and ethyl crotonate, the low-boiling compounds, decreased markedly during the heating, clarifying and concentrating processes, showing the similar behavior in the headspace analysis. On the other hand, methyl anthranilate, which is well-known as a foxy flavor [3] and high-boiling compound, was abundant during both the heating process to extract color and clarifying process, but it was half after the concentrating process.

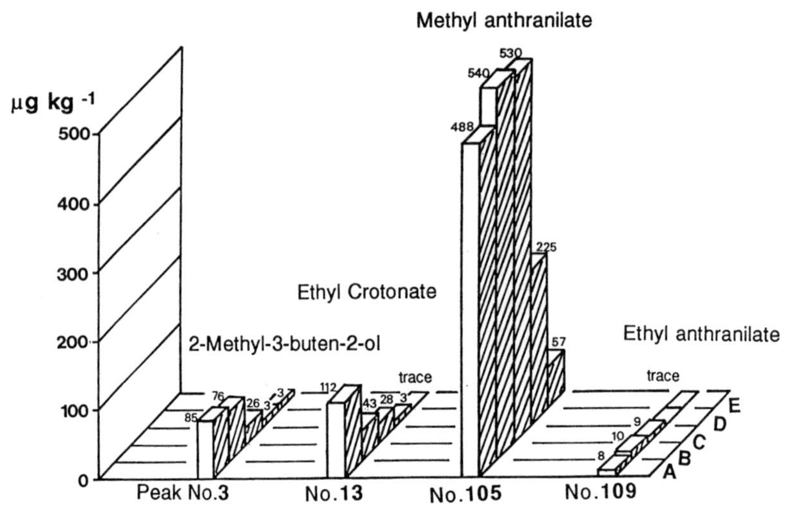

Figure 2. Retention of some flavor compounds from Concord grape juice during processing.
See Figure 1 for A, B, C, D and E.

These findings suggested that efforts should be made to recover the flavor from juices during both the heating process to extract color and concentrating process.

REFERENCES

1. Ohta, H., Tonohara, K., Watanabe, A., Iino, K. and Kimura, S., Flavor specificities of satsuma mandarin juice extracted by a new-type screw press extraction system. Agric. Biol. Chem., 1982, **46**, 1385-1386.

2. Ohta, H., Nogata, Y., Yoza, K. and Maeda, M., Glass capillary gas chromatographic identification of volatile components recovered from orange essence by continuous liquid extraction. J. Chromatogr., 1992, **607**, 154-158.

3. Stern, D.J., Lee, A. McFadden, W.H. and Stevens, K.L., Volatile from grapes: Identification of volatiles from Concord essence. J. Agric. Food Chem., 1967, **15**, 1100-1103.

PRODUCTION OF POWDERY PRODUCTS OUT OF CONCENTRATED FRUIT-JUICES AND OTHER FOOD AND BIO-PRODUCTS

T.Gamse, R. Marr, F.Fröschl, E.Moschitz

Insitute of Thermal Process and Environmental Engineering, TU Graz
Head of the institute: o.Univ.Prof.Dipl.-Ing.Dr.techn.R.MARR
Inffeldgasse 25, 8010 Graz, AUSTRIA

1.) Basic project idea:

At the Institute of Thermal Process and Environmental Engineering, TU Graz, experiments for the development of a new process were carried out. Because of the high percentage of water included in fruits there is the negativ effect on the storage and the water has to be transported. If it is possible to remove this water, there will be less transportation weight and the possibility for a longer dated storage. The new idea of this process is to freeze the fruit juice in fine drops and to remove the water in a subsequent vacuum dryer by sublimating it.

2.) State of the project:

For the realization of the basic idea a simple discontinous plant was built at the institute and in a feasibility study experiments with different raw materials were carried out.

The fruit-juice it is compressed in a storage tank (see figure 1). Leaving this vessel the fruit-juice is dispersed in fine drops into the crystallize with the help of a nozzle. The necessary performance for freezing will be reached by dropping the CO_2-pressure from 60 bar to atmospharic conditions. By this flash evaporation the CO_2

withdraws the crystallizer and the fruit juice the equal amount of heat and the fruit juice drops freeze immediately. It is necessary to coordinate the flow rate of the CO_2 and of the fruit juice. The intermediate product, which is collected at the filter cloth at the bottom of the crystallizer, is very fine crystalline and is put into the vacuum dryer.

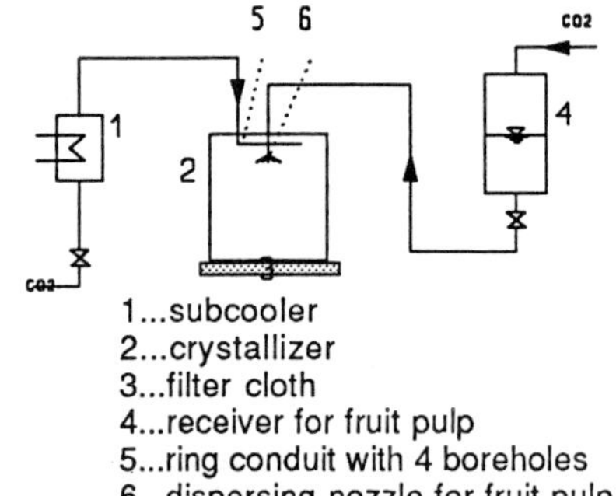

Figure 1: flow sheet of the dispersion

Figure 2 shows the vacuum dryer with the vacuum pump and the condenser. The pressure of the dryer has to be lower than the tripple point of water (5 mbar) because otherwise the water cannot sublimate and the intermediate product would melt. If that happens no powder can be produced. A very important factor in this part is the efficiency of the vacuum pump.

The condenser has to make sure that no water reaches the vacuum pump. If the seperation is insufficiently water reaches the oil of the pump, which has the effect of a decreasing pump-efficiency and an increasing pressure in the vacuum-dryer.

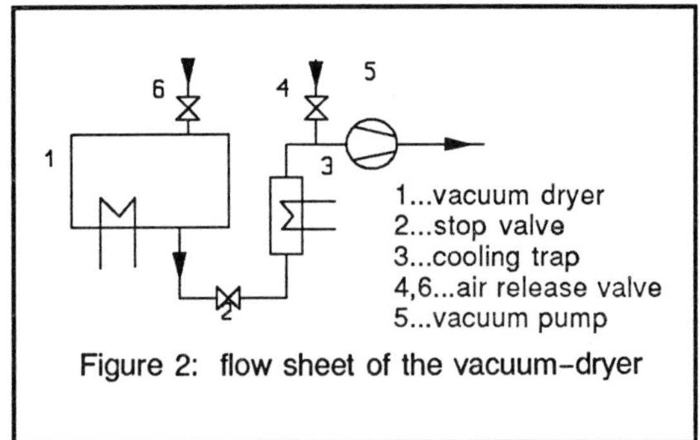

Figure 2: flow sheet of the vacuum–dryer

Another important point is the synchronization of the heating in the vacuum-dryer and the efficiency of the vacuum-pump and the condenser. If there is too much heating power the condenser and the vacuum pump cannot remove all the water and therefore the pressure will increase. On the other hand the time for drying decreases with increasing heating because more water sublimates from the frozen product.

Limiting factors for this process are the percentage of solid matter and of sugar in the raw material. With low percentage of solid matter a higher amount of water has to be removed in the vacuum-dryer and the drying time increases. But the dispersion of the fruit juice can only be achieved if there is a certain percentage of solid matter in the raw material. The other limiting factor is the percentage of sugar in the raw material. With increasing amount of sugar the problems with drying this product increase because of the bonding of water with sugar and it is difficult to remove the whole water.

Until now powdery products were produced out of sieved tomatoes, apple purre, orange pulp and liver. Commercial products (apple, orange and tomatoes) were used as raw material, which have the disadvantage of a high percentage of sugar and also preservatives are added. The produced powder was free-flowing and there were no problems to redissolve the powder in water. Further the resolved powder shows no loss of flavor in comparison to the raw material.

3.) Aim of the project

The next step of the project has to be the installation of a continous plant. Figure 3 shows a flow sheet of such a plant. The fruit pulp is pumped with a viscous material pump through a nozzle into the crystallizer. The necessary CO_2 for refrigerating capacity is compressed after the expansion with the help of a compressor to 60 bar. After passing the condenser it is liquid again and gets into the condensate tank. e the cycle for the CO_2 is closed. Further appliances for the inlet and the outlet to the vacuum dryer are necessary. The use of a heat pump is optimal, because the amount of heat for the dryer for sublimating the water is equal to the amount of coldness, which is needed in the condenser.

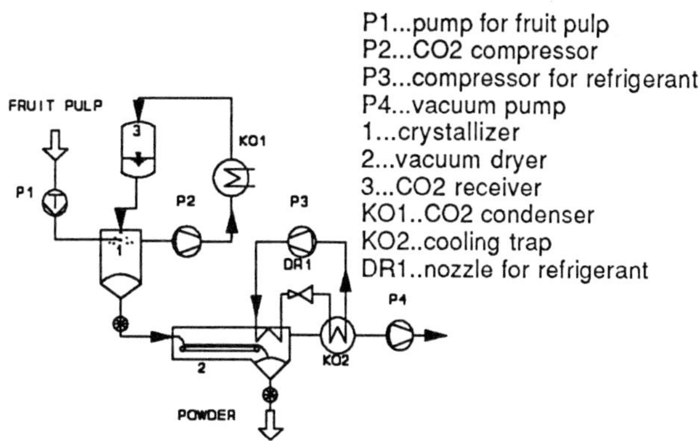

Figure 3: flow sheet for the continuous plant

The data of the discontinuously experiments cannot be transmitted directly for a continuous plant unit. With this plant it should be possible to vary different important parameters in dependence on the raw material and these data can then be used for a scale up.

PHYSIOLOGICAL ROLES OF MEMBRANE ALTERATION IN GAMMA IRRADIATED POTATO TUBER

S.TODORIKI and T.HAYASHI
National Food Research Institute, Ministry of Agriculture, Forestry & Fisheries,
Tsukuba, Ibaraki, 305 Japan

ABSTRACT

Gamma-irradiation of potato tuber brought about the increase of electrolyte leakage from the disks . This increase of membrane permeability began just after irradiation and continued for a few days after irradiation , and then decreased to the original levels. Lipids in irradiated potatoes were not significantly decreased by irradiation. However, for a few weeks storage after irradiation, phospholipid and linolenic acid content in galactolipid increased.

INTRODUCTION

Gamma-irradiation of potato tuber has been commercially utilized for the purpose of sprout inhibition. Not only sprouting inhibition, irradiation has been reported to bring about many other physiological changes related to membrane changes such as increase in respiration, accumulation of sugars etc [1]. However, very little was known about the effect of gamma -irradiation on the membrane lipid and property of potato tuber. This study was undertaken to clarify the effect of gamma-irradiation on lipid composition and membrane property of potato tuber in vivo.

MATERIALS AND METHODS

Irradiation of potato tuber

Potato tuber (*solanum tuberosum* cv. dejima) were obtained from a local market in Tsukuba and irradiated with a Gamma -cell 220 (2.1Tbq of Co-60, $4.6x10^3$Gy/h, AECL) at a dose of 0.5kGy, unless otherwise stated , and stored at 5°C in a dark room.

Extraction and fractionation of potato lipid

Lipid extraction was done according to the method of Bligh-Dyer after the boilingof tissues in 2-propanol for 5 minutes. Total lipid extract were fractionated into neutral, glyco-and phospholipid fractions on a silicic acid column and lipid component in glyco and phospholipid were separated by thin layer chromatography .

Lipid analysis

Fatty acid methyl esters was prepared from each lipid fraction in HCl-methanol and analyzed on a gas- chromatograph (GC 14A) equipped with capillary column (DB 225 J & W Inc Ltd 0.25mmx30m) and an FID. Pentadecanoic acid was used as an internal standard.

Quantifion of free fatty acid were carried out by the fluorescence measurement of ADAM derivative of free fatty acid after the separation by HPLC. The oxidative lipid degradation was evaluated by measuring TBA value of the extracted lipid.

Measurement of electrolyte leakage from potato disk

Ten disks(25mm diameter , 5mm thickness) were taken from one tuber, and rinsed with distilled water . The disks thus prepared were incubated at 27°C in 100ml of distilled water with gentle shaking, and the electrical conductivity of the water was measured with a Conductivity Meter ES12 (Horiba Ltd) at 15(C_{15}) and 75 min (C_{75}). The conductivity of the water incubated with frozen-thawed disks were measured to determine total electrolyte(C_t). The rate of electrolyte leakage was expressed by; $100x\ (C_{75} - C_{15})/C_t$.

RESULTS AND DISCUSSION

Membrane lipid composition of irradiated potato tuber

Total fatty acid content of potato tuber irradiated at 0.5kGy were slightly decreased 1 day after irradiation, but it recovered within a few days after irradiation (table 1). There was neither significant loss of poly unsaturated fatty acid nor accumulation of free fatty acid and lipid hydroperoxide up to the dose of 2 kGy. Long term storage after irradiation brought about the increase of phospholipid content rather than the reduction. Fatty acid composition of glycolipid , especially Monogalactocyl diglyceride (MGDG) and digalactosyl diglyceride(DGDG), were significantly changed by irradiation. The percentage of linoleic acid decreased accompanied by the increase of linolenic acid with increasing dose[2]

TABLE 1
Fatty acid content of ittadiated potato tuber

treatment	Fatty acid (μ mol/100g, f w)
unirradiated (0h)	138.02
unirradiated(10d)	139.96
0.5kGy(0h)	138.84
0.5kGy(6h)	138.94
0.5kGy(12h)	130.10
0.5kGy(24h)	114.16
0.5kGy(48h)	122.06
0.5kGy(72h)	136.24
0.5kGy(10d)	140.98

Effect of gamma -irradiation on membrane permeability of potato tuber

The increase in membrane permeability of irradiated potato tuber measured by increased rate of electrolyte leakage from potato disks(fig.1). Membrane permeability increased until 2 days after irradiation with increased radiation dose then decreased to the initial level.

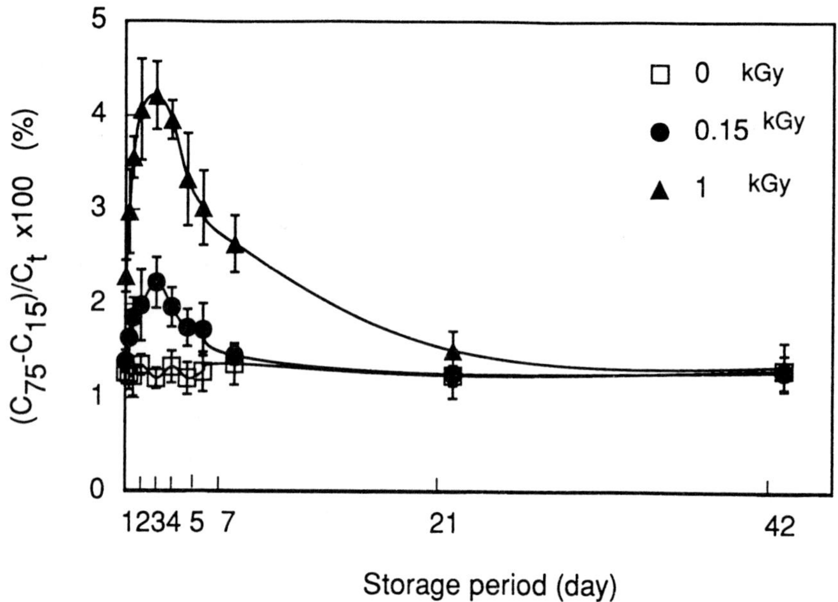

Figure 1. Electrolyte leakage of potato disks.

Conclusions

Gamma irradiation brought about the increase in membrane permeability of potato tuber immediately after irradiation. During the storage period after irradiation lipid composition of tuber altered , and the membrane permeability which once damaged was recovered.

References

1. Thomas, P. CRC Critical Reviews in Food Science and Nutrition,19,1984, pp. 343-349.

2. Hayashi, T. Todoriki, S. and Nagao, A. Effect of gamma-irradiation on membrane permeability and lipid composition of potato tubers. *Env. Exp. Bot.* , 1992, 32(3) pp. 265-271.

DEVELOPMENT OF A TIME-TEMPERATURE INDICATING DEVICE USING PHOSPHOLIPASE

S.H. YOON, C.H. LEE, D.Y. KIM, J.W. KIM, and K.H. PARK
Department of Food Science & Technology and Research Center for New Bio-Materials in Agriculture, Seoul National University, Suwon, 441-744, Korea

ABSTRACT

As an attempt to create a time-temperature indicator (TTI) which performs reproducible reaction with confidence in monitoring storage life of frozen foods, a TTI using phospholipase was developed and applicability of the TTI to frozen pork was investigated. Smaller standard deviations of the reaction rate constants for phospholipase system over the temperature range between 30°C and -18°C than those for lipase system indicated that the former is more reliable than the latter. Good correlation between the color change of the TTI and the TBA value of frozen pork at fluctuating storage temperature suggested the potentiality of the TTI for predicting the storage life of frozen foods.

INTRODUCTION

Various lipid hydrolyzing enzymes such as lipase could be used for a TTI, but stable substrate emulsion is highly required to obtain a reproducible reaction rate of the enzyme[1]. The emulsion of triglyceride, a substrate for lipase, has been known to be unstable. It is practically impossible to obtain a uniform droplet size in the emulsion and the emulsion cannot be stored over very long periods, especially at low temperatures. Therefore, the device using unstable emulsion system is not reliable for predicting enzymatic reaction to an acceptable degree of accuracy and reproducibility. On the other hand, phospholipid retains both hydrophilic and lipophilic portions in the molecule and is capable of maintaining a stable emulsion system. Commercial preparations of phospholipids derived from soy bean oil are used extensively in manufactured foods.

In present study, we report development of a new type of TTI using phospholipid-phospholipase system. The device was evaluated for the reproducibility of the enzyme reaction and the application of the system for predicting frozen pork quality change along the storage time and temperature fluctuation is also discussed.

MATERIALS AND METHODS

Preparation of TTIs.

Two types of TTI were prepared; one based on lipase-triglyceride system and the other phospholipase-phospholipid system. The TTIs were composed of an enzyme, substrate emulsion, salts, antifreezing agents, pH indicators, and emulsion stabilizer. The substrate emulsion for lipase was prepared by sonicating the mixture of 1.25 ml olive oil and 23.75 ml of 10% (w/v) gum arabic solution for 1.5 minutes (Fisher, Desmembrator Model 300). The substrate emulsion for phospholipase was prepared by sonicating 25 ml of 3% (w/v) soya lecithin for 3 minutes. In order to make the most appropriate pH indicator mixture for the TTI, bromothymol blue, neutral red, and methyl red were mixed at the ratio of 8 : 2 : 0.1 (v/v).

Reaction rate of lipid hydrolysis.

The hydrolysis reaction was initiated by adding a predetermined amount of each enzyme into the reaction solution of pH 8.0. The amount of free fatty acid liberated by enzymatic hydrolysis was determined by pH-stat method. One unit of enzyme activity was defined as the amount of enzyme which produces 1 μM of free fatty acid per minute. Production of free fatty acid by the enzymes followed the pattern of first-order reaction and the reaction could be expressed as the following equation:

$$-\frac{d[A]}{dt} = k_a [A] ,$$

where [A] is the concentration of substrate, k_a is the reaction rate constant, and t is the reaction time.

RESULTS AND DISCUSSION

The reaction rate constant and reproducibility of phospholipase reaction system.

The pH change in each TTI was monitored over the temperature between 30°C and -18°C. The pH of the reaction at temperatures between 0°C and the above dropped sharply during the first hour of the reaction and then slowed down significantly (data not shown). It could be due to the limiting amount of substrate in the reaction. The reaction rate constant, k_a, was calculated for each reaction temperature by assuming that it is a first order reaction. The reaction rate constant was largely dependent on the reaction temperature (Table 1) and were used to predict the time for color change of the TTI. Standard deviations of the initial reaction rate constant for lipase and phospholipase with 95% confidence were determined from the experiments that were repeated 6 times at each temperature by pH-stat test (Table 1).

From the result, we concluded that reproducibility of the reaction system of phospholipase was better than that of lipase, and it was more agreeable to employ phospholipase syste than to employ lipase system for monitoring food quality change during storage at very low temperature.

Table 1.
The reaction constants of lipid hydrolysis reactions by two lipases at various temperatures.

	Phospholipase		Lipase	
Temp(°C)	k (1/min)	SD	k (1/min)	SD
30	0.0012919	0.0001459	0.00195287	0.000228
20	0.0006749	7.942E-05	0.00134049	0.000282
10	0.0002048	3.875E-05	0.00079098	0.000210
0	3.706E-05	8.561E-06	0.00031010	0.000116
-5	1.498E-05	4.081E-06	0.00024044	5.79E-05
-10	2.416E-06	4.528E-07	2.3691E-05	6.58E-06
-15	8.772E-07	1.769E-07	7.6187E-07	1.56E-07
-18	5.524E-08	1.779E-08	9.5927E-08	1.2E-08

* Standard Deviation

Correlation between the TTI and quality change of frozen pork.
The degree of color change of the TTI was divided into three ranges; 1) between pH 8.0 and 7.5 usable, ii) between pH 7.5 and 7.0 uncertain, iii) between pH 7.0 and 6.5 unusable ranges. The TBA value of frozen pork during storage was also divided into three ranges; i) 0.0 and 0.5 usable, ii) 0.5 to 1.0 uncertain, and iii) 1.0 to 1.5 unusable ranges[2]. The amount of the enzyme in the TTI was adjusted so that the pH of the reaction mixture reaches 7.5, 7.0, and 6.5 when the TBA value of frozen pork became 0.5, 1.0, and 1.5, respectively, depending on the storage time and temperature. Thereby, the storage life of the frozen pork could be estimated from the color of the TTI. Both the changes of the TTI color and the TBA values of frozen pork were in good correlation with the predicted data.(data not shown). Therefore, we concluded that the TTI using phospholipid-phospholipase system is reliable for monitoring the quality deterioration of frozen pork.

CONCLUSIONS

A new type of TTI using phospholipid-phospholipase system was developed to monitor quality loss of frozen foods during storage. It was determined that the phospholipase system carried out more reliable reaction than the lipase system at all the temperatures tested. The pH of the TTI and the TBA value of frozen pork were diveded into three ranges to indicate usability of frozen pork. The color change of the TTI along the pH change correlated well with the change in TBA value of frozen pork. Therefore, it was concluded that the TTI developed in this study could be applied to other frozen foods.

REFERENCES

1. Taoukis, P. S., Fu, B., and Labuza T. P., Time-temperature indicators. Food Tech., 1991, **45**, 70-82.
2. Truner, E.W.,Paynter,W.D.,Montie, E.J.,Bessert, J.M., Struck, G.M., and Olson, F.C., Use of the 2-thiobarbituric acid reagent to measure rancidity in frozen pork. Food Tech., 1954, **8**, 326-330.

HEAT TREATMENT EXPERIMENTS ON HAY TO ELIMINATE POSSIBLE CONTAMINATION WITH HESSIAN FLY

SHAHAB SOKHANSANJ AND H.C. WOOD
College of Engineering, University of Saskatchewan
Saskatoon, Saskatchewan, S7N 0W0Canada

ABSTRACT

Alfalfa chops were exposed to hot air in a rotary machine to demonstrate that the heat treatment kills the insect Hessian fly *[Mayetiola destructor (Say)]*. The laboratory results showed that an exposure to 60 C for three minutes results in the total destruction of the pupae stage of the insect. Tests on a full scale rotary dryer showed that the chopped hay being dried in these machines reaches to the critical temperature of 60 C and remains at these temperatures for at least three minutes, and these temperature and time meet the kill conditions.

INTRODUCTION

Japan has permitted imports of heat treated chopped hay from Canada on the basis that the product temperature reaches 90 C for at least three minutes during artificial drying (Sokhansanj et al, 1990, Sokhansanj and Wood 1991). This high temperature was specified because experimental thermal kill data for Hessian fly *[Mayetiola destructor (Say)]* was not available at the time.

Experiments were conducted on a full number of insects but in a smaller rotary drum (Sokhansanj et al., 1993). About 48000 insects were heat treated in the drum and about 16000 insects were used as control. The results showed that a temperature of 58 C is critical for the survival of the insects. A theoretical analysis showed that the stem temperature of the infested plants during dehydration in a production dryer substantially surpasses the critical temperature of 58 C.

A series of tests were conducted on a production rotary drum to show that temperatures inside the drum can be measured experimentally; and the internal temperatures can be related to the exhaust temperature. This paper describes the dryer, method of testing and test results.

DESCRIPTION OF THE DRYER AND TEST PROCEDURES

The hay chop dryer was a rotary drum as shown schematically in Figure 1. The drum is 6.22 m long and 1.8 m in diameter and the body of the dryer is insulated. The dryer rotates at about 7 r/min. The source of heat is direct fired natural gas. Air flow through the dryer is about 500 m^3/min. The throughput of the dryer varies from 3 tonne per hour to 6 tonne per hour.

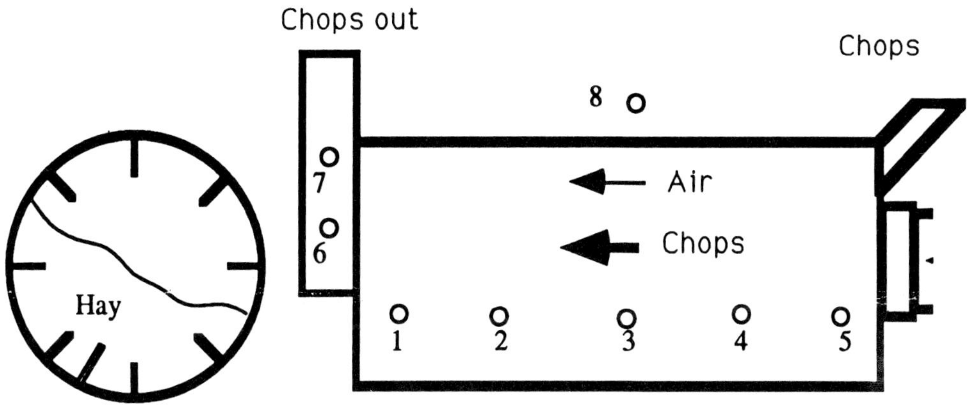

Figure 1. Schematic of the dryer showing the locations of the thermocouples

Instrumentation

Five thermocouples were inserted into the drum such that the tip of each thermocouple was inside the dryer, about 10 cm away from the wall. Hard Teflon tubing was used to hold each thermocouple in an upright position. Figure 1 shows the location of these thermocouples along the dryer. Two thermocouples were inserted into the inlet of the exhaust duct; one measured dry bulb temperature (thermocouple # 6) and the second one measured the wet bulb temperature (thermocouple #7). For wet bulb temperature, the junction of the thermocouple was wrapped in a wetted felt cloth. One thermocouple measured the outside ambient temperature close to the drum wall. Thermocouple wires were connected to a Campbell Scientific data logger which had been installed and secured on the body of the dryer drum. The data logger was pre-programmed to scan each channel every 10 seconds and store the acquired data on a data storage chip.

Procedure

Three tests were conducted on September 17, 1992. Prior to the tests, the dryer had been running normally. For preparation for the measurements, the dryer was stopped, thermocouples were installed, the data logger was initiated and the dryer was re-started to operate normally. The exhaust temperature was set to a desired value and this temperature was maintained automatically by gas flow rate to the burners. The exhaust temperature was set at 71 C for about 25 minutes (test A), followed by dropping the exhaust temperature to 61 C for 20 minutes (Test B), and finally increasing the temperature to 66 C for about 15 minutes (Test C).

During the tests, samples of hay chops from the inlet and from the end of the cooler drum were collected.

RESULTS

Temperatures - Figure 2 shows the plot of six dry bulb temperatures (thermocouples 1-6 shown on Figure 1). The three test durations are designated test A (exhaust temperature at 71 C) , test B(exhaust temperature at 61 C), and test C (exhaust temperature at 66 C). Thermocouple 6, marked by heavy line, shows the dry bulb temperature of the air at the exhaust.

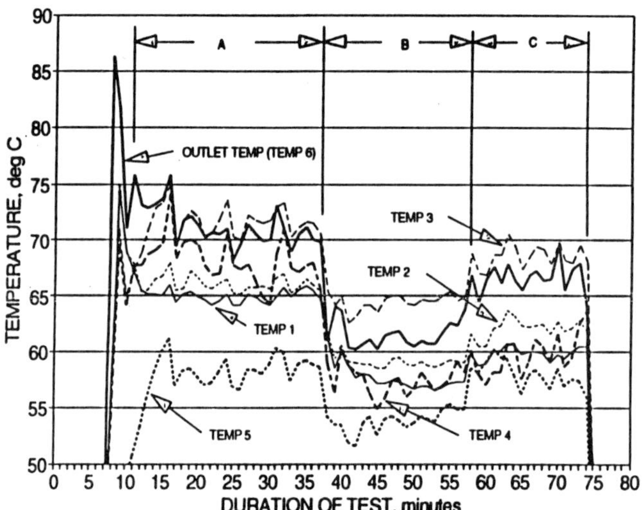

Figure 2. Temperatures in the drum during drying of chopped hay at three settings of the exhaust air temperature: duration A at 71 C, duration B at 61 C, and duration C at 66 C

The variation of the exhaust temperature in each setting, is about ±2 C. Thermocouples 4 and 5 registered low temperatures as these thermocouples sensed the temperature of the fresh chops entering the dryer. The largest temperature sensed was that by thermocouple 3 that was located at about the middle of the drum length. The average moisture content of the fresh chops during the test was 14.8% and the average dry chop moisture content was about 8.6%. The production throughput was about 5 tonnes per hour during the experiment. The relative humidity of air measured at the outlet of the dryer (6 and 7 on Figure 1) was 15% during test A (71 C), 45% during test B (61 C) and 25% during test C (66 C).

REFERENCES

Sokhansanj, S., H.C. Wood, J.W. Whistlecraft, and G.A. Koivisto. 1993. Thermal disinfestation of hay to eliminate possible contamination with Hessian fly. Postharvest Biology and Technology (Feb 10 1993).

Sokhansanj, S., V.S. Venkatesan, H.C. Wood, J.F. Doane, and D.T. Spurr. 1992. Thermal kill of wheat midge and Hessian fly (Diptera: Cecidomyiidae). Postharvest Biology and Technology 2(1992):65-71

Sokhansanj, S., and H.C. Wood. 1991. Simulation of thermal and disinfestation characteristics of a bale dryer. Drying Technology - an International Journal 9(3):643-656

BACTERIOCIN PRODUCING LACTIC ACID BACTERIA USED FOR BIOPRESERVATION OF FOOD

Birthe Jelle & Nicolai Peitersen*
Chr. Hansen's Laboratorium Danmark A/S, 10-12 Bøge Allé, P.O.Box 407, DK-2970 Hørsholm, Denmark
* Chr. Hansen's Laboratorium Danmark A/S, Siber Hegner Service Building Isogo, Nihon Siber Hegner K.K., IMD. 6-1 Shin Isogo-Cho, Isogo-Ku, Yokohama 235, Japan.

INTRODUCTION

Lactic acid bacteria are able to inhibit other microorganisms by producing organic acids hydrogen peroxide and sometimes other more specifically inhibitory substances like bacteriocins. Bacteriocins are proteins or protein complexes with bactericidal activity against organisms usually closely related to the producer organism.
Nisin, produced by *Lactococcus lactis*, is a polycyclic peptide, a so-called lantibiotic (Fig. 1). It is the best known and most extensively studied antimicrobial which tends to have a wider spectrum of activity than the majority of other bacteriocins produced by lactic acid bacteria. It inhibits the growth of several gram positive bacteria including the outgrowth of *Bacillus* and *Clostridium* spores.

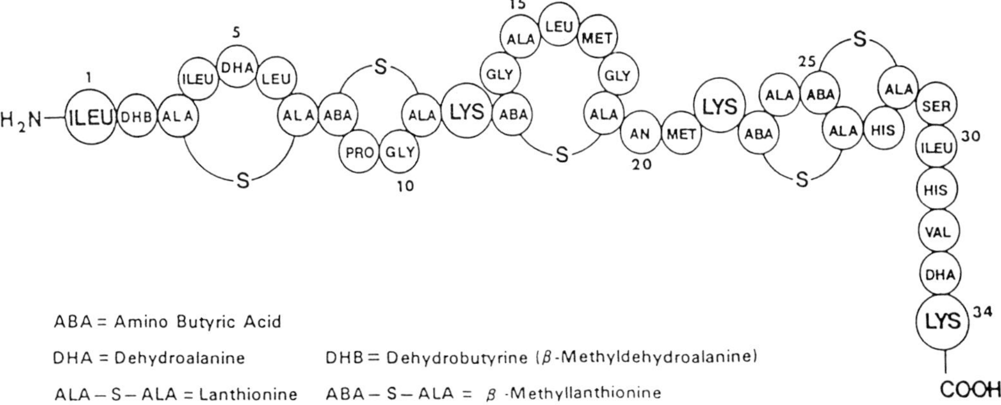

Figure 1. Structure of Nisin (from Gross and Morell, 1971 (1))

Numerous studies have been made on the mode of action of both susceptible vegetative cells and spores (2-5). Nisin is believed to affect the germination stage by inhibiting the swelling process (2,6). The effect against the spore formers is very important, especially for the dairy industry, where it has proved to be a most effective preservative in processed cheese spreads in preventing spoilage caused by *Clostridium tyrobutyricum*.

The use of nisin to prevent spore formers during production of hard and semi-hard cheese has been reviewed by Hurst (7). Experiments with Gouda cheese have indicated that approximately 40IU/g (1ppm) of nisin is sufficient to prevent the butyric acid fermentation, even at a relatively high number of clostridial spores (6/ml) in the cheese milk (8).

APPLICATION OF NISIN PRODUCING LACTOCOCCI

A number of nisin producing strains of *Lactococcus lactis* subsp. *lactis* has been screened at our laboratory for acidification and nisin production in reconstituted skim-milk. In general, they acidify very slowly due to limited proteolytic activity, which also limits the production of nisin. Addition of casein peptone (0.5%) to the milk results in a considerable increase in both the acidification rate and the nisin production (from 100 IU/ml to 500 IU/ml) during incubation for 18 hours at 30°C. These results are particularly of interest if the nisin producer can be applied as a bulk starter in the cheese manufacturing process.

Laboratory experiments combining a nisin producer and a thermophilic starter normally used in cheese production have been carried out in skim-milk using the procedure (time/temperature profile) for hard cheese processing, but without addition of rennet. Results from these experiments indicate that the nisin producer shall be applied as a bulk starter. When the nisin producer is added directly into the cheese milk (inoculum 10^7 CFU/ml) in the preripening step the growth is limited and nisin is not produced in detectable quantities during the process. The thermophilic starter added at 30°C (inoculum approx. 10^7 CFU/ml) causes a so fast acidification that the nisin producer is inhibited.

Application of the nisin producer as a bulk starter together with the thermophilic starter after the preripening step, gives an initial nisin concentration in milk of approx. 16IU/ml. In this experiment the thermophilic starter is inhibited by the nisin and nearly no acidification occurs during the process. It was therefore necessary to adapt the thermophilic starter to higher concentrations of nisin. This was done by repeated transfer into milk containing increasing amounts of nisin. Eventually, the adapted strain was able to grow in and acidify milk even in the presence of 150 IU/ml nisin. The experiment was repeated using the nisin producer together with the nisin resistent strain. Then the acidification rate increased and the end pH in the milk (after approx. 30 hrs) is nearly the same as for thermophilic starter applied alone, but the lag phase is extended by approx. 7 hrs. The growth of both organisms in the milk increased during the cooling process and further, nisin is produced which result in a final nisin concentration in the milk of 24 IU/ml.

A nisin concentration of 24 IU/ml is not sufficient to inhibit the germination of clostridial spores. However, all our experiments have been carried out in milk without the renneting process in which a 10 times concentration of the initial nisin content (16 IU/ml) in the milk can be expected. It is thus concluded that the strains developed may be applicable for making hard cheese.

ACKNOWLEDGEMENTS

The authors wish to thank Dorte Nørgaard and Karina Jensen, two dairy scientists, who made most of the practical work as part of their Dairy Engineering thesis.

REFERENCES

1. Gross, E. and Morell, J.L., The structure of nisin. J. Am. Chem. Soc., 1971, **93**, 4634-5.

2. Campbell, L.L. and Sniff, E.E., The effect of subtilin and nisin on the spores of *Bacillus coagulans*. J. Bacteriol., 1959, **77**, 766-70.

3. Ramseier, H.R., Die Wirkung von Nisin auf *Clostridium butyricum* prazm. Archiv für Mikrobiologie, 1960, **37**, 57-94.

4. Morris, S.L., Walsh, R.C. and Hansen, J.D., Identification and characterization of some bacterial membrane sulphydryl groups which are targets of bacteriostatic and antibiotic action. J. Biol. Chem., 1984, **259**, 13590-4.

5. Henning, S., Metz, R. and Hammes, W.P., Studies on mode of action of nisin. Int. J. Food Microbiol., 1986, **3**, 121-34.

6. Lipinska, E., Nisin and its applications. In: Antibiotics and antibiosis in agriculture. ed. M. Woodbrine. Butterworths, London. 1977, pp. 103-30.

7. Hurst, A., Nisin. Adv. Appl. Microbiol., 1981, **27**, 85-123.

8. Hugenholtz, J. and deVeer, G.J.C.M., Application of Nisin A and Nisin Z in Dairy Technology. In: Nisin and Novel antibiotics. Ed. G. Jung and H.-G. Sahl. ESCOM Science Publishers, The Netherlands, 1991, pp. 440-56.

STUDIES ON THE CONTROL OF THE GROWTH OF *SACCHAROMYCES CEREVISIAE* BY USING RESPONSE SURFACE METHODOLOGY TO ACHIEVE EFFECTIVE PRESERVATION AT HIGH WATER ACTIVITIES

AN-ERL KING
Dept. of Food Science, National Chung-Hsing Univ.
Taichung, Taiwan 40227, ROC

ABSTRACT

Water activity (0.93-0.97), pH (4-6), and sorbic acid concentration (0-50 ppm) were combined to study their effects on the growth of *Saccharomyces cerevisiae*. A three-variable and three-level design method, analyzed by response surface methodology (RSM), was used. The growth could be inhibited at water activity 0.94 and pH 4 with a sorbic acid concentration of 25 ppm. Predicted conditions of inhibition and experimental values were found to be consistent.

INTRODUCTION

Saccharomyces yeasts ferment sugars and often cause the spoilage of fruits and fruit products [1]. The water activity of fruit products might be lowered to avoid their spoilage [2]. However, flavors and characteristics are always affected appreciably by the reduction of water activity [3]. Preservation under higher water activities is therefore to be desired. Microbial inhibition might be effectively achieved at higher water activities using the "hurdle effect", in which several food preservation processes are applied simultaneously [4]. In this study, preservation techniques of water activity control and pH adjustment were combined with the addition of sorbic acid [5] to effectively inhibit the growth of *Saccharomyces cerevisiae* under higher water activities. Response surface methodology was used to study the effects of these three factors on the responses of the growth parameters, i.e. lag time, growth rate, and maximum population, and a three-variable and three-level design method was adopted.

MATERIALS AND METHODS

The culture strain obtained from the culture collection was maintained on Sabouraud agar slants as stock culture. Inoculated media were incubated in a rotary shaker at 27±1°C for the experiment. In the main experiment, media consisting of 1% peptone solution were used. Water activity, pH, and sorbic acid concentration were adjusted to the desired values according to the RSM experimental design. A three variables-three levels design was used [6]. Table 1 lists the real experimental values of various factors (X_i) adopted at different levels.

TABLE 1
Process variables and their levels in the three variables-three levels response surface design

Independent variables	Symbol		Levels	
	Coded*	Uncoded	Coded*	Uncoded
Water activity	X_1	a_w	1	0.97
			0	0.95
			-1	0.93
pH value	X_2	pH	1	6
			0	5
			-1	4
Sorbic acid (ppm)	X_3	HS	1	50
			0	25
			-1	0

$$\text{*coded variable} = \frac{\text{uncoded value} - 0.5\text{x(high value+low value)}}{0.5\text{x(high value-low value)}}$$

Lag time, growth rate in log phase, and maximum population in stationary phase of *S. cerevisiae* grown in various media were selected as responses. RSREG procedure in SAS software was used to conduct the RSM analysis. A second-order polynomial expression of three variables was fitted. Optimization was performed graphically.

RESULTS AND DISCUSSION

All the model regressions were found to be significant and lack-of-fits were not. This indicated that these fitted models could represent responses appropriately. Regression coefficients obtained from SAS analysis in the second-order polynomial were not all significant at a 5% level. Non-significant coefficients were deleted when graphing was conducted. On the other hand, since pH was less significant on the growth rate and maximum population compared to the other two factors, water activity and sorbic acid concentration were selected for plotting and pH was set to be constant at 4, 5, and 6, respectively. In the contour plot with pH=4, a region where the reciprocal of lag time fell below zero could be located. Meanwhile, the yeast could not grow at pH 4 with a water activity of 0.93 and a sorbic acid concentration of 25 ppm, and the growth rate was zero. When superimposing the contour plots of pH 4, the regions predicted for yeast inhibition were found not to overlap completely. This might be due to the errors caused in the experiments. For the reduction of errors, the overlapped region was adopted as the region for growth inhibition. Verification was conducted using a point in the region. Point of $X_1 = -0.8$, $X_2 = -1$, and $X_3 = 0$ was used for the test. The experimental result and the estimated value coincided and the model was proved to be adequate.

REFERENCES

1. Potter, N.N., Food Science, AVI Publishing Co., Westport, 1986, pp. 140-68.

2. Levi, A., Gagel, S. and Juven, B.J., Intermediate-moisture tropical fruit products for developing countries. II. Quality characteristics of papaya. J. Food Technol., 1985, **20**, 163-75.

3. Chirife, J., Ferro Fontan, C. and Benmergui, E.A., The prediction of water activity in aqueous solutions in connection with intermediate moisture foods. IV. A prediction in non-electrolyte solutions. J. Food Technol., 1980, **15**, 59-70.

4. Leistner, L., Rodel, W. and Krispeien, K., Microbiology of meat and meat product in high and intermediate moisture range. In Water Activity: Influence on Food Quality, ed. L.B. Rockland and G.F. Stewart, Academic Press, New York, 1981, pp. 855-916.

5. Cerrutti, P., Alzamora, S.M. and Chirife, J., A multi-parameter approach to control the growth of Saccharomyces cerevisiae in laboratory media. J. Food Sci., 1990, **55**, 873-40.

6. Box, G.E.P. and Behnken, D.W., Some new three level designs for the study of quantitative variables. Technometrics, 1960, **2**, 455-475.

THEORY AND APPLICATIONS OF A NEW VISCOMETER BASED ON ANNULUS LIQUID FLOW

KANICHI SUZUKI
Department of Food Science,
Faculty of Applied Biological Science, Hiroshima University,
1-4-4 Kagamiyama, Higashi-Hiroshima, 724 Japan

ABSTRACT

A new viscometer was developed by applying the flow theory of fluid in an annular channel. Pressure drop and shear stress were expressed as a function of flow rate, viscosity, and the dimensions of the annular channel. Both of direct measurement of wall shear force and pressure drop measurement were acceptable for evaluating viscosity with known flow rate. The viscometer measured wide range of viscosity with high accuracy. Non-Newtonian flow properties were also measurable. It was available for an on-line viscometer for liquid food processes.

INTRODUCTION

Liquid food processings need an on-line viscometer that measures wide range of flow properties accurately for quality and process controls. The viscometer also requires as simple structure as possible for sanitation. However, most conventional on-line viscometers have complicated structure, and require a bypass from the pipelines to introduce liquid foods for viscosity measurement. Thus, I have developed a new viscometer by applying the flow theory of fluid in an annular channel. This study presented the theory, structure and measured results of the new viscometer.

THEORY AND STRUCTURES

When a fluid flows through an annular channel between two coaxial circular cylinders of radii R and κR ($0 < \kappa < 1$)(Fig.1, (a)), the relationship between pressure depression $\triangle P$ for length L and flow rate Q in laminar flow region is expressed as follows[1),]

$$\Delta P = 8\,\mu\, LQ\,/\,\pi R^4[(1 - \kappa^4) - \{(1 - \kappa^2)^2\,/\,\ln(1/\kappa)\}] \qquad (1)$$

where μ is viscosity of the fluid. The shear stress acting on the surface of inner cylinder τ_i is also a function of $\triangle P$ as,

$$\tau_i = \Delta P R [\kappa - \{(1 - \kappa^2) / (2\kappa \ln(1/\kappa)\}] / (2L) \quad (2)$$

The shear force exerted by the fluid on the surface of inner cylinder F

$$F = - 2 \pi \kappa R L \tau_i \quad (3)$$

By combining equations (1), (2), and (3), F is expressed as a function of Q, μ, and dimensions of the annular channel. These relations suggest that two types of viscometer are possible if one measures ΔP or F for length L, and correlated with Q. Schematic illustrations of the two viscometers constructed in this study are shown in Fig.1 (b), (c).

1.Horizontal annular channel viscometer; a liquid flows through a horizontal cylindrical annulus. A differential pressure manometer measures pressure depression for a fixed cylinder length. Viscosity of the liquid can be evaluated from Eq.(1). This viscometer can be used as a perpendicular viscometer, too.

2.Perpendicular annular channel viscometer; a load sensor measures the shear force F on the surface of perpendicular inner cylinder of length L. The load sensor measures also line pressure of fluid and the difference of forces acting on the upper and lower surface of the inner cylinder. The former is measured by another load sensor, and subtracted. The latter value is theoretically correlated with F as follows,

$$\pi (\kappa R)^2 \Delta P = - F / [1 - \{(1 - \kappa^2) / (2\kappa^2 \ln(1/\kappa)\}] \quad (4)$$

The true value of F can be evaluated by correcting the measured force value using Eq.(4). The values of F, Q and the dimensions of the annular channel give the viscosity. For non-Newtonian fluids, the relationship between the shear stress and apparent shear rate calculated from the measured apparent viscosity and the shear stress gives the flow behavior.

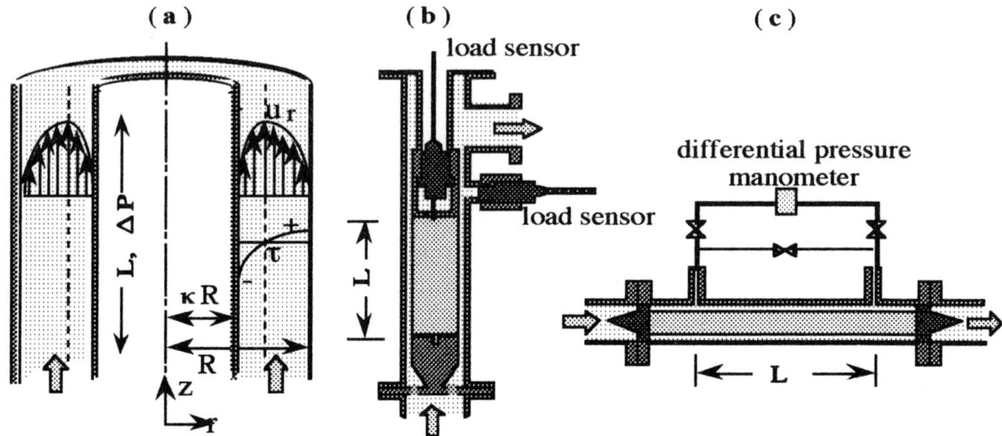

Fig. 1 Schematic illustration of fluid flow in an annular channel (a), and two types of viscometers examined((b), (c)).

RESULTS AND DISCUSSION

Both viscometers measured viscosities of low viscous liquids (e.g., water, sugar solutions) accurately, as well as of high viscous liquid foods (e.g., tomato puree, refinery molasses) by adjusting the radius of inner cylinder and/or sensitivity of load sensor or manometer (Fig.2). For non-Newtonian liquid foods, both viscometers measured apparent viscosities at different flow rates or shear rates. Simple structure and short length of these viscometers ensure easy installation on any process pipelines without large pressure drops, and real-time measurement of viscosity is possible. Computer monitoring of change in viscosity is also possible. The viscometers are suitable for laboratory use, too.

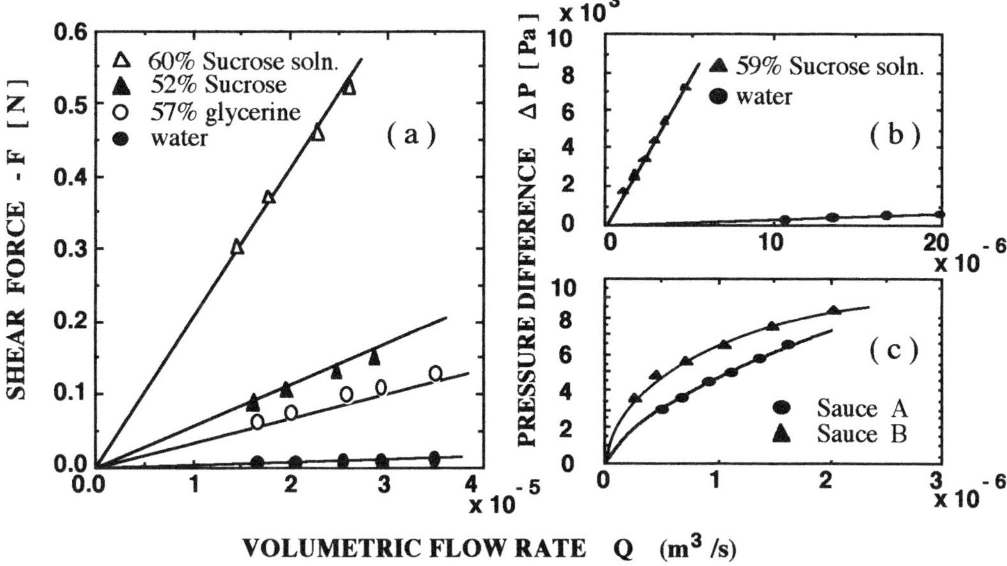

Fig. 2 Comparison of experimental results of flow rate Q vs. shear force F for the perpendicular viscometer (a), and Q vs. pressure difference $\triangle P$ for the horizontal one (b) with theoretical values (solid lines), and examples of non-Newtonian liquid foods (c).

CONCLUSIONS

Two new viscometers were developed by applying the fluid flow theory in an annular channel. The viscometers with simple structure measured wide range of viscosity with high accuracy.

ACKNOWLEDGMENT

The author wish to express his sincere thanks to Mr. S. Nakai, Technical Official, Hiroshima University, who contributed to make the experimental apparatus.

REFERENCE

1. R.B.Bird, W.E.Stewart, & E.N.Lightfoot : Transport Phenomena, pp.51-54, (John Wiley & Sons, Inc., New York, 1960).

AN EXTENDED HOT-WIRE METHOD FOR MONITORING FLUID VISCOSITY AND THE ONSET OF GEL FORMATION

TOMOSHIGE HORI, YASUHIKO SHIINOKI and KENSUKE ITOH
Technical Research Institute,
Snow Brand Milk Products Co. Ltd.
1-1-2 Minamidai, Kawagoe, Saitama 350-11, Japan

ABSTRACT

The practical feasibility of a hot-wire method for monitoring the change in viscosity and gel-setting point of fluids was critically examined on the basis of the theoretical relationship for free-convection heat transfer around a circular cylinder. By comparing the temperature of a hot-wire probe placed in a gelatin solution and in whole blood, this method is proved to be extremely effective for assessing coagulation.

INTRODUCTION

A new hot-wire technique that will detect the onset of gel formation accurately, in line and in real time as the change in fluid viscosity has been established since 1985 for true automation in the cheese industry.[1] The method uses the phenomenon of free-convection heat transfer to the surrounding fluid from a linear heat source. The rate of convective heat transfer from a solid to a fluid is inversely related to its viscosity, so that in the steady state, the heat source temperature will be lowest with fluids of the least viscosity. This is because, at equilibrium, a lower viscosity fluid possesses a thinner boundary layer at the interface with the solid heat source. During gel formation, the viscosity of a fluid changes, often very rapidly.

By using a hot-wire probe, we have been able to detect milk clotting during cheese production without disturbing the coagulum.[2] The probe placed in rennet-treated milk detected the clotting time, which is defined as the time at the inflection point of the time versus hot-wire temperature curve.

The objective of this study is to establish an extended hot-wire technique with strong potential for practical application in determining the viscosity change of a coagulating process fluid in both industry and clinical medicine, utilizing the practical characteristics of free convection heat-transfer around a circular cylinder placed in fluids with a wide range of both viscosity and thermal conductivity.

THEORETICAL BASIS

The following functional relationship in the equilibrium state of laminar free-convection heat transfer provides the basis for practical applications of the hot-wire method:

$$f(Nu, Gr, Pr) = 0 \tag{1}$$

where Nu, Gr and Pr are the Nusselt, Grashof and Prandtl numbers, respectively. If a fluid such as a gelatin solution consists principally of water, and if the solid content remains constant during the gelling process, the heat-transfer relationship can be obtained from Equation 1 as follows:

$$\nu = f(\Delta \theta_s) \tag{2}$$

where ν is the kinematic viscosity, and $\Delta \theta_s$ is the temperature difference between the surface of a linear heat source and the surrounding fluid.

The equation for conductive heat transfer in a cylindrical hot-wire probe with a built-in linear heat source results in a practical relationship between the surface temperature of the probe and the built-in hot-wire temperature.[3]

$$\Delta \theta_s = \Delta \theta_w - C Q \tag{3}$$

where $\Delta \theta_w$ is the hot-wire temperature referred to the fluid temperature, Q is the heat generated in a unit length of the heat source, and C is a numerical constant for the hot-wire probe. This C value is inversely proportional to the probe's thermal conductivity, which is practically constant with temperature.

Equations 2 and 3 assure the interchangeability of hot-wire probe and indicate conclusively that the temperature of the built-in hot wire can detect the onset of gel setting directly by the change in fluid viscosity.

In addition to Equation 1, the following relationships for a vertical linear heat source are of great use for extending the applications of the hot-wire method:[2]

$$Nu = (2\ell/d) / \{\ln(1 + 2\delta/d)\} \tag{4}$$

$$\delta = (d/2)\{\exp(2\lambda/\alpha d) - 1\} \tag{5}$$

where δ is the calculated thickness of an imaginary layer of stagnant fluid related to the temperature profile in the boundary layer, ℓ and d are respectively the length and diameter of the heat source, λ is the thermal conductivity of the fluids, and α is the heat transfer coefficient at the surface of the heat source.

Both λ and α in Equation 5 are measurable by the present hot-wire method, while the ℓ/δ value calculated from the experimental relationship for the Nusselt number of a vertical plate, Nu_P, such as

$$Nu_P = 0.638 Gr^{1/4} Pr^{1/2} (0.861 + Pr)^{-1/4} \tag{6}$$

[4] is useful for estimating the $\Delta \theta_w$ behavior in a potential process fluid.

MATERIALS AND METHODS

A hot-wire probe (ℓ =5cm, d=2mm, Q=5W/m) was placed in a 3% gelatin solution (G-2656 from Sigma Co., volume=200cm^3) to obtain the fluid temperature versus $\Delta\theta_w$ curve, the solution being cooled at 15℃/hr.

The activated partial prothrombin time of human whole blood diluted 10 times was determined at 37℃, monitoring the built-in hot-wire temperature of a probe designed for clinical use (ℓ =4mm, d=0.6mm, Q=3.8W/m) against time.

RESULTS AND DISCUSSION

Both the gel-setting temperature of the gelatin solution and the clotting time of whole blood were determined with inherent simplicity (Figure 1).

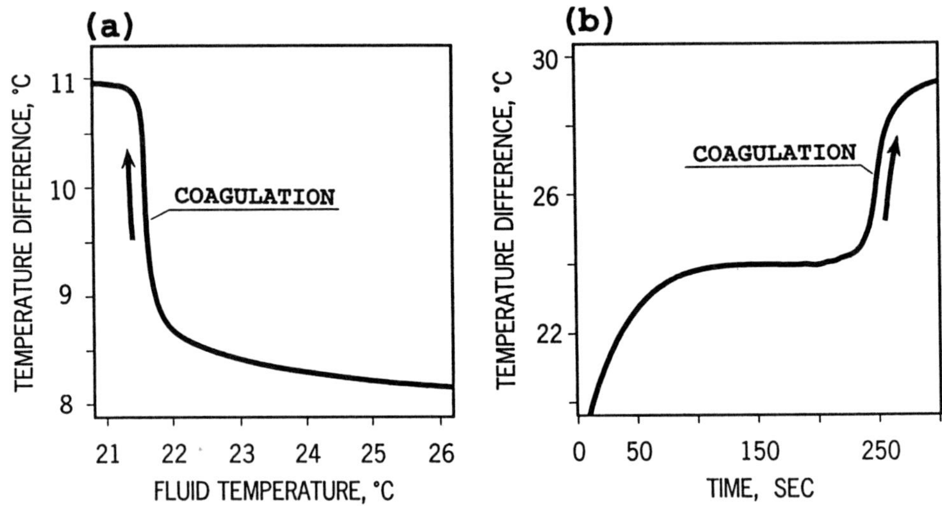

Figure 1. Measurement of the gel-setting point of (a) a gelatin solution and (b) human whole blood by the hot-wire technique.

CONCLUSIONS

The hot-wire method offers strong potential for practical application in determining gel-setting point by monitoring the change in fluid viscosity in both the food industry and clinical medicine.

REFERENCES

1. Hori, T., Objective measurement of the process of curd formation during rennet treatment of milks by the hot-wire method. J. Food Sci., 1985, 50(4), 911-917.
2. Hori, T., Advances in Food Engineering, ed. by Singh, P. and Wirakartakusmah, A., CRC Press, Florida, 1992, pp. 475-90.
3. Hori, T. and Itoh, K., Food Hydrocolloids, ed. by Nishinari, K., Plenum Pub. Co., New York, 1993, in print.
4. Ostrach, S., An analysis of laminar free-convection flow and heat transfer about a flat plate parallel to the direction of the generating body force. NACA Report, 1953, 1111, 1-17.

USE OF AN IN-LINE VISCOMETER IN THE MANUFACTURE OF SKIM MILK POWDER

COLM O'DONNELL*, NIALL HERLIHY**, BRIAN MCKENNA***
* Dept. of Agricultural and Food Engineering, Univ. College Dublin, Dublin 2, Ireland.
** Waterford Foods plc, Dungarvan, Co. Waterford, Ireland.
*** Dept. of Food Science, Univ. College Dublin, Dublin 4, Ireland.

ABSTRACT

Factors affecting skim milk concentrate viscosity were examined in an industrial milk powder plant using an in-line viscometer. It was concluded that using an in-line viscometer instead of a density meter to control the degree of skim milk concentration in an evaporator affords an opportunity to significantly reduce steam consumption and minimise evaporator fouling by ensuring skim milk concentrate of 100 cPs at 100/s viscosity is constantly produced.

INTRODUCTION

In the manufacture of skim milk powder (smp), falling film evaporators are used to concentrate skim milk prior to spray drying. The specific steam consumption of a modern multiple effect evaporator with thermal vapour recompression is 0.1 kg steam per kg water removed, while the specific steam consumption of a two stage spray drier is between 2.0 and 2.2 kg steam per kg water removed[1]. Thus considerable reductions in steam consumption can be achieved during smp production by maximising water removal in an evaporator prior to spray drying. However the degree of concentration in an evaporator is limited since the maximum recommended viscosity of skim milk concentrate (smc) in falling film evaporators to avoid pipeline blockages and excessive fouling levels is 100 cPs at 100/s shear rate[2].

In the dairy industry, the level of skim milk concentration in evaporators is generally determined with reference to smc density and hence a fixed total solids percentage (%TS). However problems may arise with pumping, storage or atomisation of smc due to smc with too high a viscosity being produced, particularly when the protein lactose ratio (PLR) is high or smp with a high heat specification is being produced.

The experimental work reported in this paper has been undertaken to investigate the main parameters influencing smc viscosity and to examine the feasibility of using an in-line viscometer to control the level of skim milk concentration in an evaporator.

MATERIALS AND METHODS

Research trials were carried out over a two year period in an industrial smp plant which consisted of a 6 effect falling film Laguilharre evaporator and a Niro two stage spray drier. A Brookfield TT100 in-line viscometer was installed after the evaporator to measure viscosity.

RESULTS

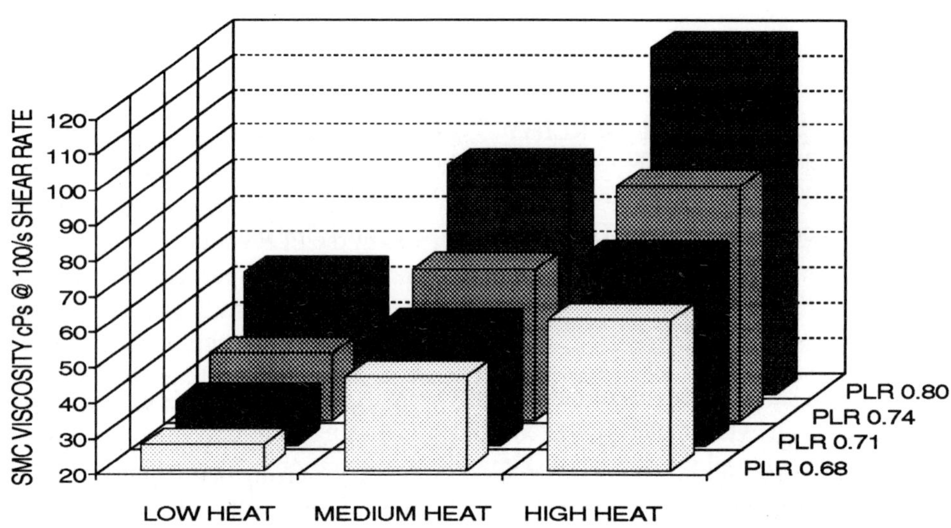

Figure 1. Effect of skim milk PLR and skim milk preheat treatment level on the viscosity (cPs) at 100/s shear rate of smc at 49% total solids +/-0.25%

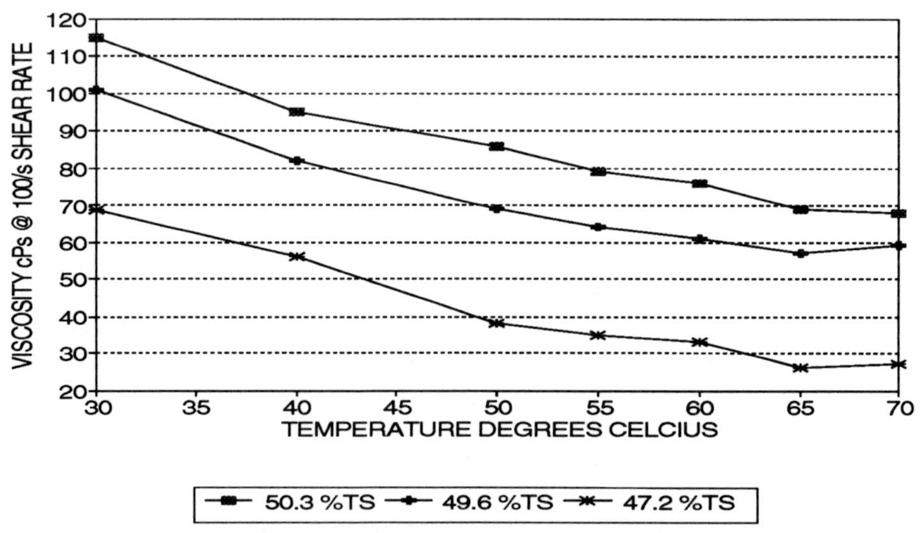

Figure 2. The effect of temperature and smc %TS on the viscosity of smc with a high heat specification and a PLR of 0.75

Table 1. The effect of temperature and agitation on age-thickening of skim milk concentrate.

	Smc viscosity cPs at 100/s shear rate			
Time (minutes)	50°C	55°C	60°C	60°C (agitated)
0	64	63	62.5	62.5
5	66.5	66	65.1	63
15	72.5	80.2	97.2	81
30	134	153	189	148

DISCUSSION

Figure 1 illustrates that if smc %TS and PLR are kept constant, smc viscosity may more than double with variation in preheat treatment. This rise in viscosity is caused by increased voluminosity of whey proteins due to heat denaturation. A similarly large increase in smc viscosity is shown to occur in figure 1 when variations in the PLR occur. This is due to the greater voluminosity of milk proteins over an equal mass of lactose particles. Large variations in the PLR from 0.62 to 0.91 are a characteristic of Irish milk production. Figure 1 clearly illustrates the difficulty in selecting a suitable smc density value to control the level of skim milk concentration in an evaporator. Variations in the skim milk preheat treatment or PLR are likely to cause the production of smc with a viscosity in excess of the maximum recommended value, or alternatively smc is produced which is below 100 cPs at 100/s shear rate and hence the maximum amount of water removal is not taking place, which significantly increases the steam consumption per tonne of smp produced.

Figure 2 shows that smc viscosity reduces with increasing temperature up to 65°C. This reduction is due to the decrease in the viscosity of the serum. An increase in the smc concentration level is also shown to sharply increase smc viscosity.

Table 1 shows that the age thickening rate of smc increases with temperature. It is therefore important to minimise the residence time for smc between the smc heat exchanger and the spray drier atomiser to prevent pipeline blockages occurring.

Frequently during the two year test period the in-line viscometer was checked for calibration against a laboratory viscometer especially near the end of long production runs. On no occasion was evidence of internal fouling or loss of accuracy detected.

CONCLUSIONS

It can be concluded that the installation of an in-line viscometer rather than a density meter in an evaporator to control the level of skim milk concentration affords an opportunity to significantly reduce steam consumption and prevent excessive evaporator fouling especially in situations where large variations in skim milk PLR or preheat treatments occur by ensuring smc of 100 cPs at 100/s shear rate is constantly produced.

REFERENCES

1. Fergusson P.H., Developments in the evaporation and drying of dairy products. J. of the Society of Dairy Technology, 1989, 42, pp. 94-101.

2. O'Donnell C.P., Ph.D. thesis , Univ. College Dublin, 1992.

DEVELOPMENT OF SHEAR STRESS BASED SENSOR TO MEASURE DRYING RATE AND ITS APPLICATION TO SNACK DRYING AUTOMATION

S.C. SHIN, I.J. YOO and J.K. CHUN
Department of Food Science and Technology, College of Agriculture and Life Sciences, Seoul National University, Suwon, Korea

ABSTRACT

A sensor based on shear strain was developed to measure the moisture content and volumetric changes of food chips in rotating drum dryer. The sensor was installed in pilot scale perforated drum dryer and studied for its characteristics in snack chips dehydration. The shear strains were correlated with the moisture content of chip and bulk volume of food. For the application of the sensor to the automation of drying process, transducing device and process controller were designed and fabricated.

INTRODUCTION

In snack industry rotary drier is widely used because of its advantage of the rapid and evenly drying capability. The dryer serves to adjust the final moisture content of the snack chip, which was produced in the first drying stage to moisture content around 20 % and tempered to equalize the moisture concentration and textual properties throughout chip.[1]
Humidity sensor is widely used to estimate moisture content of chip and in some food industry samples are withdrawn intermittently to analyze the moisture of food. In snack drying environment, crust and dust produced form a film layer on the surface of sensor and interfere the air circulation through the sensing element, and cause the inaccuracy and delay of the response of the measurement. Another difficulty faced is to transmit the data from the sensor installed inside of the rotating drum to outside. This study was investigated to measure the shear strain of the chip layer and correlate shear strain with the moisture content and volumetric changes of the material under drying operation, and aimed to apply the sensor technology to the automation of the snack drying process.

MATERIALS AND METHODS

Snack chip: Rectangular snack chip (3 x 3 x 30 mm) was prepared from flour sheet and then the wet chip of 40% moisture content was dried in first dryer to 19 - 20 % in a continuous belt drier. After being tempered in storage room, the dry product was used as the drying raw material in drum dryer.

Drum dryer: A pilot drum dryer of 1/ 90 scale downed in capacity is consisted of a cylindrical drum (70 cm diameter, 50 cm length). The perforated drum shell has 3 mm holes to pass air stream. Hot air of 75 C was blown into the bottom of the rotating drum(4 rpm) in cross-flow[1, 2].

Shear strain sensor: A dual beam spring element type strain sensor was fabricated with steel alloy sheet which was used in lumber saw. Fig. 1 shows the strain transducer mounted with strain gages(KFC-5-C1-II) to form full bridged circuit. The output signal was processed to meet 0 - 5 volt DC while load was applied in pilot dryer.

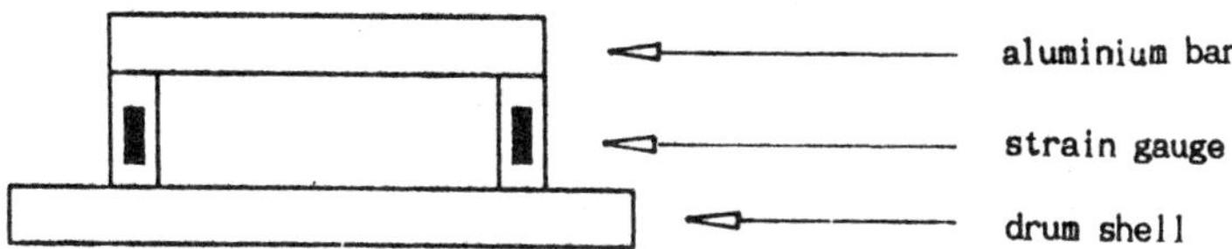

Figure 1. Schematic diagram of drum of snack dryer and strain sensor

Data processing and controller: Strain developed from moving bed of chip upon rotation of drum was measured with the strain sensor installed at inner surface of the drum shell. Data were sent to the process controller through a data transmitting device implemented at the end of the drum shaft. The device had 4 stationary carbon brush units to collect signals from the rotating copper rings which were soldered to lead-cable of the sensor. The information about the process variables were sent to the dryer controller, which was specially designed with Intel MCS8052AH as CPU to perform the necessary mathematical analysis and control[3].

RESULTS

The radial velocity profile of chips in opposite direction to the revolving shell of drum can produce shear stress gradient between the circular layers of chip.

For a given material recipe of snack pallet (flour, potato starch etc), strain may be described as;

$$\tau = \phi (M N S) \quad (1)$$

where M is moisture content of material, N is rpm of drum and S is surface characteristics of chip.

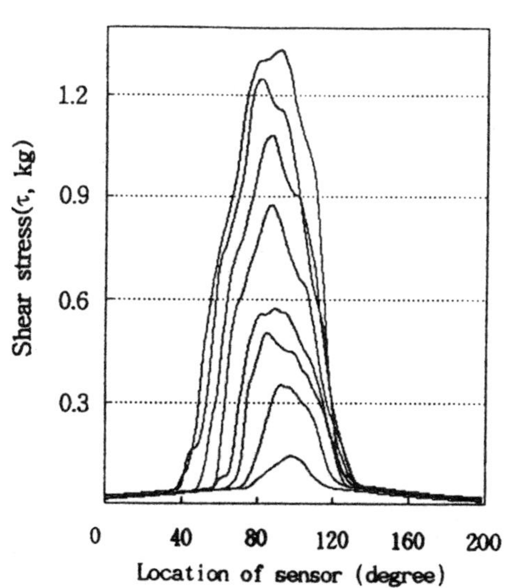

Figure 2. Strain curve obtained with snack chip loaded in drum.

The strain values obtained by the sensor were useful data to estimate the moisture content of drying material since surface properties closely relating to shear stress was affected by the moisture content. A sinusoidal strain curve as shown in Fig. 2 illustrates that the height and width of curve were influenced by the weight and volume of the material in the rotating drum.

From the measured strain of the chip having different moisture contents at a given working volumes, the relationship between moisture content (M %) and strain (τ_m, shear stress at peak of curve, kg) was described as the following linear straight line equation:

$$M = -42.2 + 0.79\,\tau_m \qquad (2)$$

And working volume (m^3) was also represented with the strain by the following second degree power equation(3):

$$V = 19.524 + 261\tau_m + 6\tau_m^2 \qquad (3)$$

or the volume also was estimated by arc length which was measured by the sensor. The coefficient of the above equations were experimentally determined for the given operation conditions.[4]

Drying curve: For the application of automation of snack drying, operation program was prepared to perform the measurement of the strain and convert into moisture content, and was resided in ROM of the controller. The process data such as operation time, temperature, moisture content and volume of material were processed in the controller for the local control of dryer or sent to the microcomputer.

Fig. 3 shows the snack drying curve constructed by the sensor and clearly illustrates the feasibility of the practical application in the drying automation system.

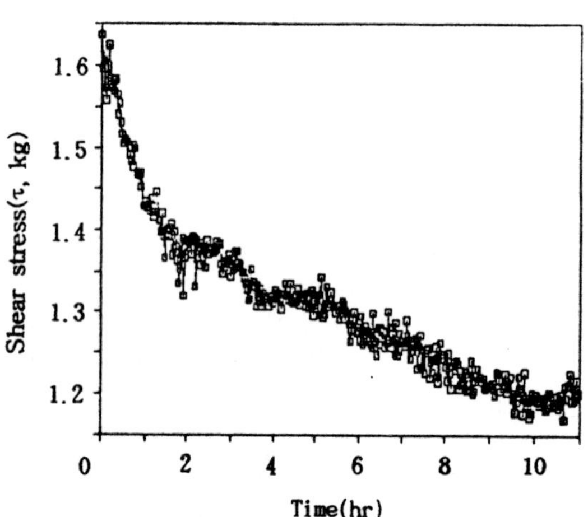

Figure 3. Snack drying curve obtained with the strain sensor and the drying controller

REFERENCES

1. Shin, S.C., *Development of shear stress sensor for the measurement of food drying rate and its application to snack drying process under the microcontroller based automation system*, MS thesis, Seoul National Univ., Korea(1989)
2. Yoo, I.J, *Automation element of snack drum drying process*, MS thesis, Seoul National Univ., Korea(1991)
3. Intel Co, BASIC-52 User's Manual(1983)
4. Shin, S.C., Yoo I.J. and Chun, J.K., Development of shear stress sensor for the measurement of food drying rate, submitted to *Korean J. Food Sci.Technol.(1993)*

MEASUREMENT OF FLUID THERMAL CONDUCTIVITY WITH A STEADY STATE HOT WIRE METHOD

YASUHIKO SHIINOKI, TOMOSHIGE HORI and KENSUKE ITOH
Technical Research Institute of Snow Brand Milk Products Co., Ltd.,
1-1-2 Minamidai, Kawagoe, Saitama 350, Japan

ABSTRACT

Thermal conductivity measurement apparatus was developed based on the steady state hot wire method. The measurement cell was made up of coaxial stainless cylinders and a stainless sheathed hot probe. The results of numerical analysis suggested that the effect of convection heat transfer was negligible small when the clearance between the outer cylinder and the hot probe was less than $0.75mm$. The average deviation of the measured thermal conductivity from known values was about 2 %. Thermal conductivity of skim milk, W/O and O/W emulsions were measured with the apparatus. The results indicated that this method is applicable to the measurement of each total solid content of skim milk, water or oil content of emulsions.

INTRODUCTION

Thermal conductivity of foods depends on their structure and chemical composition. Many measurement techniques have been discussed by Mohsenin [1], and others. The transient hot wire method has been recommended for most food applications because of its simplicity [2]. However, this method has several drawbacks in applying to a practical process. We have developed the thermal conductivity measurement apparatus, based on steady state hot wire method, for the practical process control. The principle of this method has already been established and the measurement apparatus has been developed [3], but that apparatus was not applicable to on-line measurement. The purpose of this research is to clarify the heat transfer in the measurement cell and to show that the apparatus presented in this paper can be used as a process sensor in food manufacture processes.

MATERIALS AND METHODS

Numerical analysis

The temperature and velocity profile in the measurement cell were calculated with

STREAM ver. 2.6 (Software Cradle Co., Ltd.) numerical analysis program which is based on finite volume method.

Thermal conductivity measurement

Figure 1 shows diagram of the measurement cell. Temperature of the heating resistor was directly calculated from the electrical resistance of itself. The length of the heating resistor was $30mm$ and the hot probe diameter was $2.5mm$. Temperature difference between the heating resistor and the thermostated fluid ($\Delta\theta$w) was measured using 16 liquid samples whose thermal conductivity is known. Heat amount of generated from hot probe was kept constant and the jacket was maintained at $25°C$ through the measurement.

Applications in food production process

Total solid content of skim milk was measured with the apparatus. Skim milk were prepared by dissolving four commercial skim milk powders. Water or oil content of emulsion was measured.

RESULTS AND DISCUSSION

Thermal conductivity measurement of the standard sample

As a result of numerical analysis, the calculated isotherms were parallel to the hot probe, and the velocity by natural convection was negligible small when the clearance between the outer cylinder and the hot probe was below $0.75mm$. The clearance between outer and inner cylinder was set to this value. $\Delta\theta$w was measured with the apparatus using liquid samples. The relationship between $\Delta\theta$w and the reciprocal thermal conductivity was confirmed to be linear as shown in Figure 2. The average deviation of estimated thermal conductivity was within 2 % from known values. The computed temperature difference $\Delta\theta$w was in good agreement with the experimental results.

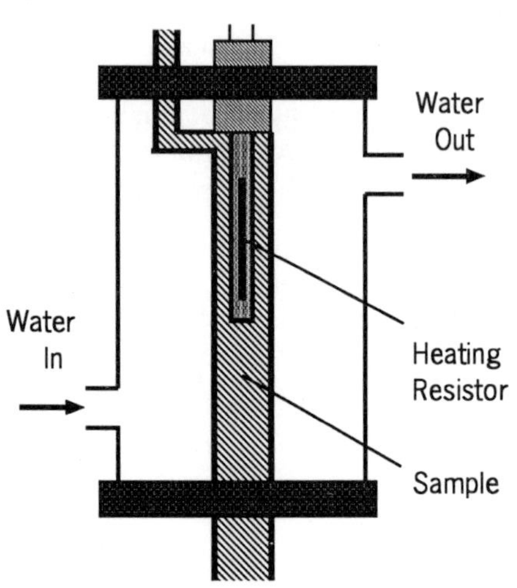

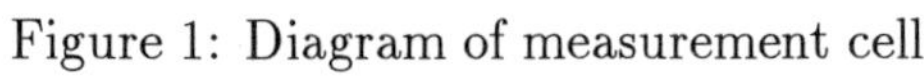

Figure 1: Diagram of measurement cell

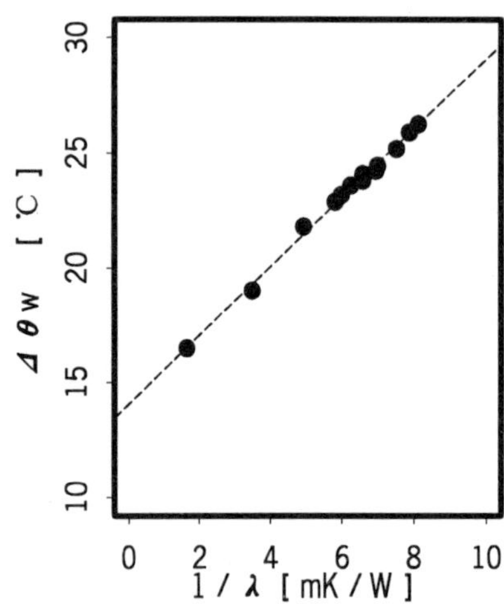

Figure 2: Typical relationship between $\Delta\theta$w and $\frac{1}{\lambda}$

Applications to food production process

Total solid content of skim milk: Figure 3 shows the relationship between the total solid content of skim milk and the calculated value from $\Delta\theta$w. The correlation coefficient was 0.998.

Water or oil content of emulsion: $\Delta\theta$w values on W/O and O/W emulsions were measured and the thermal conductivity was calculated. Figure 4 shows the dependence on ϕ_d (volume fraction of dispersed phase) of λ_d/λ_c (thermal conductivity ratio of dispersed phase and continuous phase). The thermal conductivity of emulsions was in good agreement with Maxwell Eucken's equation [4] (solid lines in figure 4).

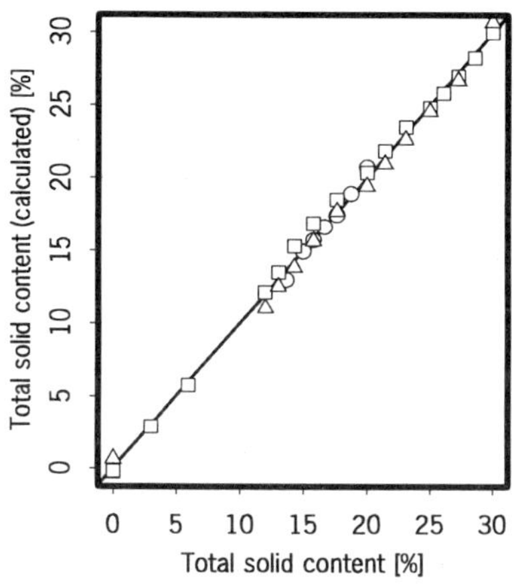

Figure 3: Relationship between total solid content of skim milk and calculated values

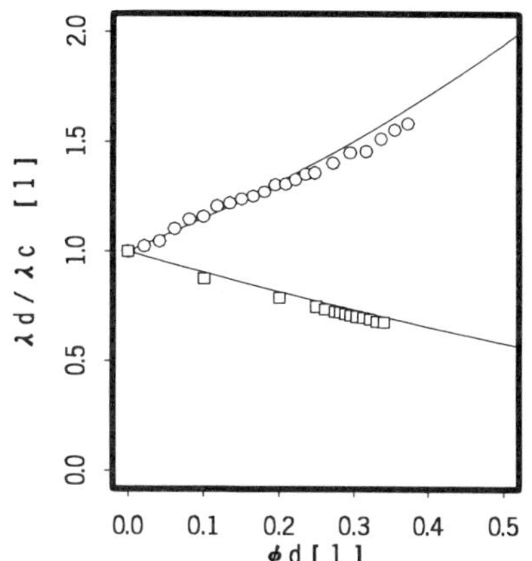

Figure 4: Dependence on ϕ_d of λ_d/λ_c (o :W/O, □ :O/W)

CONCLUSION

Thermal conductivity measurement apparatus based on the steady state hot wire method was developed. The results of the measurement of total solid content of skim milk, water or oil content of W/O and O/W emulsions indicated that the measurement method can provide high potential for practical applications.

REFERENCES

1. Mohsenin, N. N., Thermal Properties of Foods and Agricultural Materials, Gordon and Breach Science Publishers, Inc., New York, 1980.

2. Sweet, V. E. Experimental values of thermal conductivity of selected fruits and vegetables. J. Food Sci., 1974, **39**, 1080-1083.

3. Arakawa, K., Tanaka, Y., Kubota, H. and Makita, T., Thermal conductivity of freon mixtures. Proc. 5th Japan Symposium on Thermophysical Properties, 1984, 125-128.

4. Eucken, A., Allgemeine Gesetzmäßigkeiten für das Wärmeleitvermögen verschiedener Stoffanrten und Aggregatzustände. Forsch. Gebiete. Ingenieur., 1940, **11**, 6-20.

OPTICAL SENSORS (UV, VIS & NIR) FOR THE DETERMINATION OF CONNECTIVE TISSUE, LIPID AND PROTEIN FUNCTIONALITY IN MEAT

H.J. SWATLAND
Department of Food Science,
University of Guelph, Guelph, Ontario N1G 2W1, Canada

ABSTRACT

New developments in using fibre-optics (FO) to measure commercially important properties of meat are described. UV fluorescence is used to detect collagen and elastin in meat, which are important sources of toughness. Spectrophotometry has been enhanced by using multichannel FO to measure at different angles and positions relative to the point of illumination in the meat.

INTRODUCTION

Prediction of the composition of meat by rapid, non-destructive, optical methods is useful for the grading and sorting of fresh meat, and for controlling the functional properties of meat slurry in further processing using feed-back and feed-forward control. FO were used first to predict the water holding capacity of pork [1], but many other applications have now been developed involving ultraviolet (UV), visible (VIS) and near-infrared (NIR) light. Colorimetry and VIS and NIR spectrophotometry of food systems through FO are now well established and routine. The objective here is to high-light new developments of this technology.

UV FLUORESCENCE

Biochemical Types I and III collagen have different fluorescence emission spectra and optimum separation is obtained with excitation near 370 nm. Type I fibres emit a pre-quenching spectrum for longer than Type III fibres because they are larger in diameter and have a core that takes longer to quench. Thus, the emission spectrum contains information on collagen fibre diameters when duration and intensity of excitation are controlled. The principle has been used to develop a hand-held probe for detecting meat toughness, as shown in Figure 1.

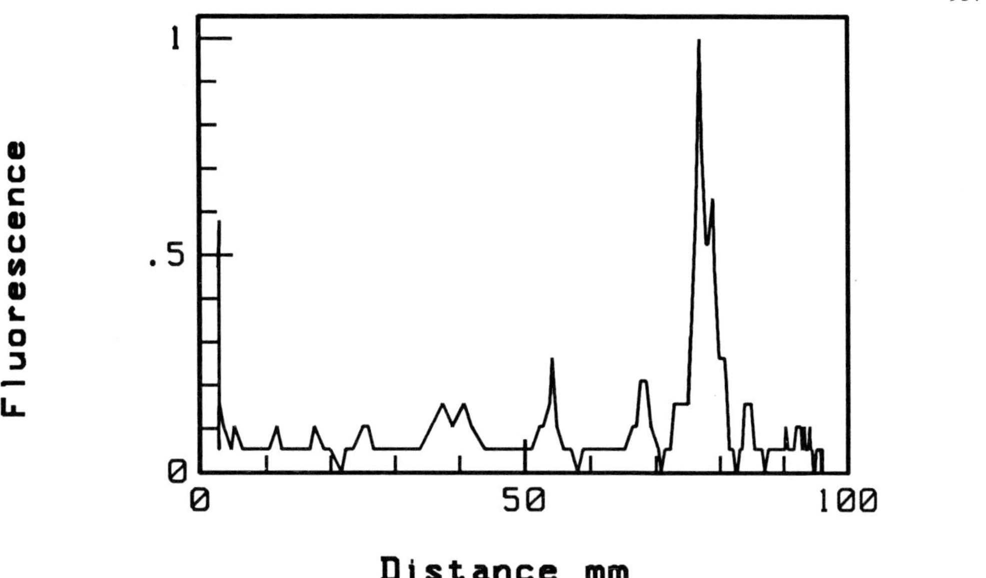

Figure 1. Signal from a beef carcass in a meat cooler using a hand-held FO probe to detect tough meat.

POLARIZED LIGHT

No applications have been attempted yet using polarized light, but basic research has shown that pH-related light scattering in meat, which has been known for years but never explained, originates from the scattering of light by the myofilament lattice of myofibrils.

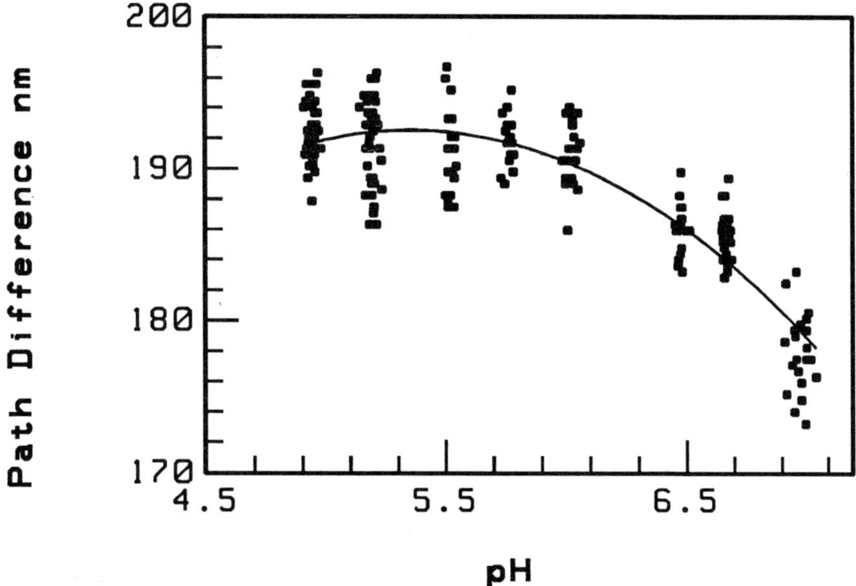

Figure 2. Effect of pH on meat birefringence.

In myofibrils, light splits into two components that travel at different velocities, the ordinary ray (O) and the extraordinary ray (E), with O $\perp$ E. Birefringence, which may be - or +, is given by $n_E - n_O$. Retardation, the decrease in velocity of light caused by interaction with the medium, may be detected as phase retardation, interference caused by path difference E $\neq$ O. The path difference of a depth of muscle (Γ_m) was measured by ellipsometry with a de Sénarmont compensator. Γ_m changed with pH (Figure 2), and thus is a major factor in pH-related light scattering in meat.

SPATIAL MEASUREMENTS AND GONIOSPECTROPHOTOMETRY

Light scattering in meat affects the pattern in which different wavelengths penetrate the meat, so that pH-dependent properties such as water-holding capacity may be predicted by both spectrophotometry and goniophotometry. When both are used together, the accuracy of prediction is improved [2]. Figure 3 shows the FO window pattern of a multi-channel probe used for the combined measurement of subcutaneous fat depth and the marbling, colour, water-holding capacity and connective tissue content of the meat.

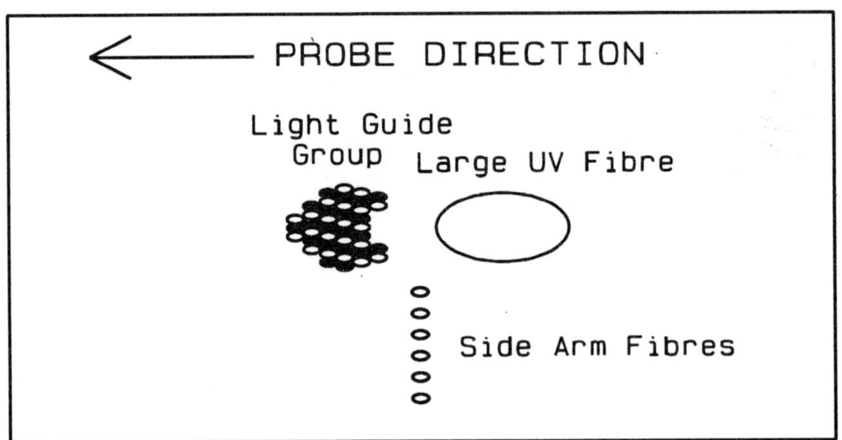

Figure 3. FO window pattern of multi-channel probe.

REFERENCES

1. Swatland, H.J., Fiber optic spectrophotometry and the wetness of meat. J. Food Sci., 1982, **47**, 1940-3.

2. Swatland, H.J. and Irie, M., Wavelength-position matrices in the fiber-optic analysis of meat. J. Comput. Assist. Microsc., 1992, **3**, 149-55.

QUICK DETERMINATION OF FAT CONTENT IN BEEF LONGISSIMUS BY NEAR-INFRARED SPECTROSCOPY WITH A FIBER OPTIC PROBE

MITSURU MITSUMOTO, SHINOBU OZAWA, TADAYOSHI MITSUHASHI, KIMIYUKI SHINOHARA*, KENICHI TATSUBAYASHI*
Chugoku National Agricultural Experiment Station, 60 Yoshinaga, Kawai-cho, Oda-shi, Shimane-ken 694, Japan
* Nireco Corporation, 2951-4 Ishikawa-cho, Hachioji-shi, Tokyo 192, Japan

ABSTRACT

Fat content in beef longissimus thoracis was compared with near-infrared (NIR) spectroscopy readings using a fiber optic probe. A partial least square regression was used to find the equation which would best fit the data. The correlation coefficient between optical densities and fat content was 0.975 (SE=2.1 %). A NIR spectroscope fitted with a fiber optic probe should enable quick determination of fat content in the longissimus thoracis at the quartering section during beef carcass grading.

INTRODUCTION

Marbling score of the longissimus thoracis at the site of quartering is the most important characteristic for carcass grading in Japan. Fat content in the muscle closely relates (r=0.93) with marbling score. Since conventional methods for determining fat content are time consuming, a more rapid and accurate technical tool is desired. Recently, near-infrared (NIR) spectroscopy has been developed for analyzing chemical compositions of foods. The objective of this work was to evaluate NIR spectroscopy using a fiber optic probe as a means of determining fat content of beef.

MATERIALS AND METHODS

Longissimus thoracis from forty–five Japanese Black steers were used. A 5.5 cm diameter x 6 cm deep sample was cut from the muscle using a template cutter and placed in a specially designed sample cup to prevent interference from outside light. Fiber optic spectra measurements (680 – 1235 nm) were performed by a Neotec Model 6250 Spectrophotometer. Scannings were performed on both sides of each sample to obtain the average value of individual beef cuts. A white ceramic disk was used as reference. Scanning reference was performed with a specially designed reference cup which had 2 mm of air distance between the fiber optic probe and the ceramic disk. Data were recorded at 2 nm intervals and 50 scans / 25 sec were averaged for every sample. Data obtained were saved as log $1/R$, where R is the reflectance energy, and then mathematically transformed to second derivatives to reduce effects of differences in particle size and sample composition.

Fat content was determined by ether extraction.

A partial least square regression was used to find the equation which would best fit the data.

RESULTS AND DISCUSSION

The correlation coefficient between optical densities and fat content was 0.975 (SE=2.1 %). Ben–Gera and Norris (1) reported a correlation coefficient of 0.974 between fat and ΔO.D. (1725 nm – 1650 nm) in 2 mm–thick emulsions of meat products using NIR absorbance. Iwamoto et al. (2) reported a multiple correlation coefficient of 0.996 for fat in ground pork using NIR reflectance. Kruggel et al. (3) reported that multiple correlation coefficients for fat were 0.92 in emulsified beef and 0.81 in ground lamb using NIR reflectance. Lanza (4) reported that multiple correlation coefficients were 0.998 for fat in emulsified pork and beef by NIR reflectance. Although our samples were not emulsified or ground, a high correlation coefficient for fat was obtained.

The conventional determination of fat content takes at least 3 days including time for grinding, freezing, freeze–drying, ether extraction and weighing. On the other hand, NIR determination of fat content can be done within ten seconds. A NIR spectroscope fitted with a fiber optic probe should enable quick determination of fat content in the longissimus thoracis at the quartering section during beef carcass grading.

CONCLUSIONS

Fat content of beef longissimus thoracis at the quartering section could be quickly determined by the near–infrared spectroscopy with a fiber optic probe. This would lead objective grading of beef carcasses.

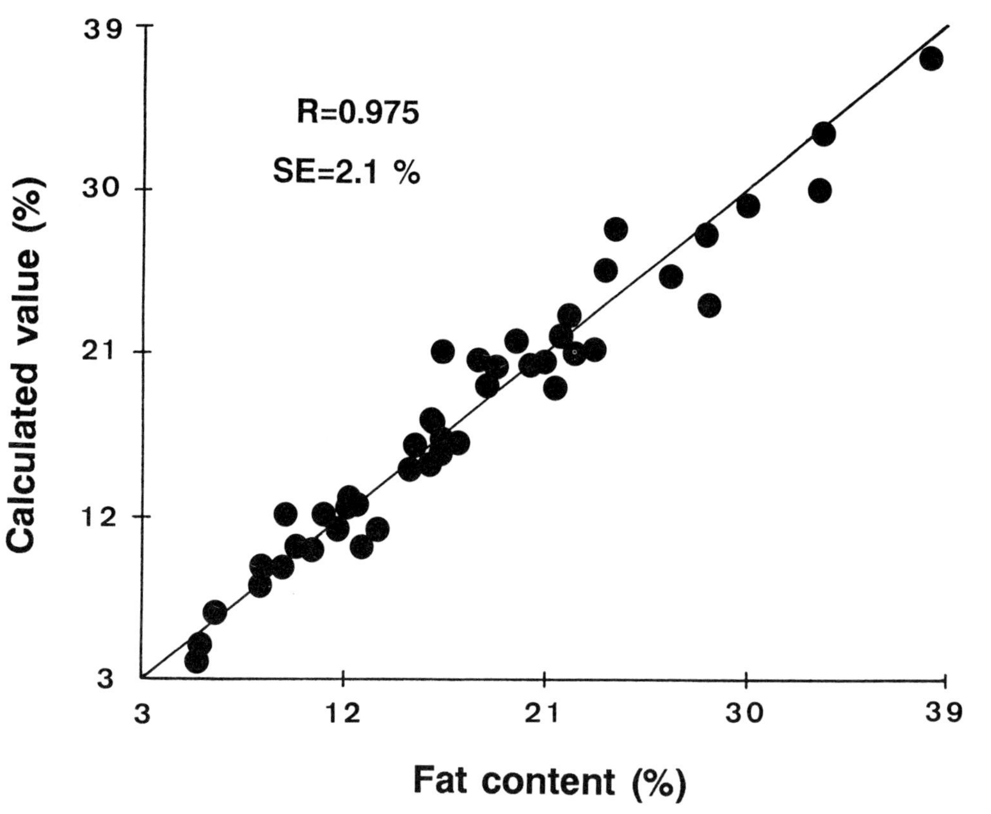

Figure 1. Relationship between fat content determined by ether extraction and values calculated by NIR spectroscopy (n=45).
R: correlation coefficient, SE: standard error

REFERENCES

1. Ben-Gera, I. and Norris, K.H., Direct spectrophotometric determination of fat and moisture in meat products. J. Food Sci., 1968, **33**, 64–67.

2. Iwamoto, M., Norris, K.H. and Kimura, S., Rapid prediction of chemical compositions for wheat, soybean, pork and fresh potatoes by near infrared spectrophotometric analysis. Nippon Shokuhin Kogyo Gakkaishi., 1981, **28**, 85–90.

3. Kruggel, W.G., Field, R.A., Riley, M.L., Radloff, H.D. and Horton, K.M., Near-infrared reflectance determination of fat, protein, and moisture in fresh meat. J. Assoc. Off. Anal. Chem., 1981, **64**, 692–696.

4. Lanza, E., Determination of moisture, protein, fat, and calories in raw pork and beef by near infrared spectroscopy. J. Food Sci., 1983, **48**, 471–474.

INDIVIDUAL SUGAR CONTENT CONTROL BY THE USE OF F.T.-I.R. SPECTROSCOPY COUPLED WITH AN A.T.R. ACCESSORY.

Véronique BELLON*, Céline VALLAT*, Olivier PAUWELS**
* CEMAGREF, BP 5095, 34033 MONTPELLIER Cedex 1, France.
** ARD, 27 rue Chateaubriand, 75 008 PARIS, France.

ABSTRACT

This paper describes the use of F.T.-I.R. spectroscopy coupled with an A.T.R. accessory to quantify the individual sugars (glucose, maltose, matodextrines...) in a mixture extracted from a process of starch hydrolysis. In a first step, binary and ternary model mixtures are prepared. Multivariate mathematical processing are applied to these spectra recorded from 1300 to 850 cm-1. The results show a very good discrimination of the different sugars: in the binary mixture, the standard error of prediction for glucose (ranking from 100 to 200 g/kg) and maltose (ranking from 250 to 290 g/kg) are respectively equal to 1.66 g/kg and 1.07 g/kg. In a second step, the experiment is successfully run on real samples.

INTRODUCTION.

Starch enzymatic hydrolysis can be a means to produce small polymeres or monomers of glucose: maltose (DP2), maltotriose (DP3), maltodextrine (DPn). The precision demanded by sugar industry customers for respectively glucose, maltose, maltotriose and DPn are typically 8g/kg, 10g/kg, 5g/kg, 5g/kg on samples containing over 800g/kg dry matter (in the so-called 60DE mixture). The concentrations are currently determined by H.P.L.C. which is a tedious and time consuming method. Our purpose is to develop an original method able to measure the concentrations of the different sugars fast enough and within the precision requirements cited before. Mid infrared spectroscopy (M.I.R.) could be specific enough for our application. Attenuated Total Reflection (ATR) [1] can be used to analyze visquous samples [2].

MATERIAL AND METHODS

Material

The F.T.-I.R. spectrometer is a BRUKER I.F.S. 25, equipped with a ZnSe Attenuated Total Reflectance accessory. The spectra are recorded from 1350 to 800 cm-1 with a 4cm-1 resolution and averaged on 100 scans.

Three experiments have been done with model and real solutions (60 Dextrose Equivalent or 60 DE) transformed by adding different quantities of sugar. The characteristics of the samples in the different experiments are given in Table 1.

TABLE 1
Characteristics of the samples used in the three experiments.

Exp	Binary	Ternary	Real Samples
Indiv. Sugars	Binary Model	Ternary Model	True samples
Glucose (g/Kg)	100-200	90-160	196-216
Maltose (g/Kg)	250-290	80-120	300-330
Maltotriose (g/Kg)	-	80-120	120-138
DPn (g/Kg)	-	-	177-197
Nr of samples	55	55	54

Mathematical processing

The spectra are related to the concentrations of individual sugars thanks to 3 kinds of mathematical processings: Mulilinear Stepwise Regression (M.L.R.), Principal Component Regression (P.C.R.) and Partial Least Squares (P.L.S.).

The initial set is always divided into 2 sub-sets: the calibration and the validation sets. The performances of the model are respectively given by the Standard Error of Calibration (SEC) and the Standard Error of Prediction (SEP).

RESULTS

As the individual sugars of interest are gluocse polymers (glucose, maltose and maltotriose and DPn i.e. sugars with a glucose polymerisation degree over 3), their spectra are very similar (Fig 1), what explains that the spectrum analysis will need powerful processing such as multivariate calibrations.

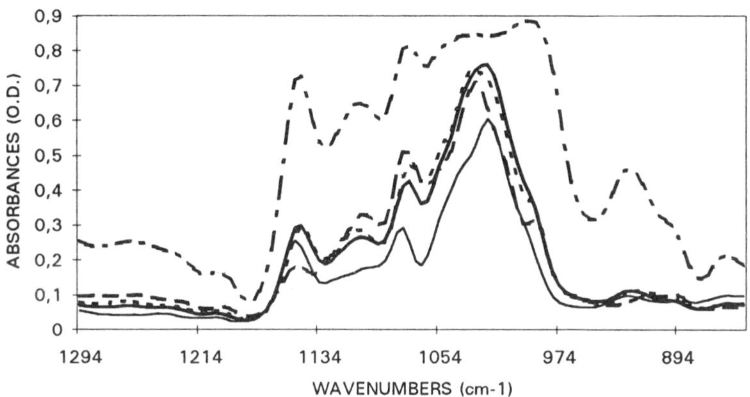

Figure 1. Infrared spectra of glucose and its polymers and of real sample: - - glucose, . . . maltose, __ (thick) maltotriose, __ (thin) DPn, _ . _ Real sample.

As shown in Table 2, in binary models, best results are got with P.L.S. mathematical processing in which 3 axis have been computed. In the case of multilinear stepwise regression, the wavelengths have to be selected according totheir correlation with sugar content. But as shown in Table 2, stepwise regression is not satisfying and will be then abandonned.

TABLE 2

Standard Error of Prediction (SEP) for glucose and maltose determination in binary solution.

S.E.P. (g/Kg)	PLS	PCR	Stepwise Regression
Glucose	1.66	2.07	9.09
Maltose	1.07	1.43	5.06

Table 3 shows the results of multivariate analysis on ternary models and on real samples.

TABLE 3

Performances of PLS2 and PCR for the quantification of individual sugars within the ternary model mixture and real samples.

Experiments	Ternary				Real samples	
Sugars	PLS		PCR		PLS	PCR
	SEC	SEP	SEC	SEP	SEP	SEP
Glucose	2.38	2.26	2.37	2.26	2.17	1.7
Maltose	4.12	5.18	4.89	5.27	4.71	5.04
Maltotriose	2.41	4.60	4.14	4.19	n.d.	n.d.
DPn	-	-	-	-	2.91	3.11
Total Sugars	3.16	2.30	3.25	2.14	-	-

These performances, although not so good as the ones of binary models, meet the requirements of the sugar industry in the ternary models. However, in the real samples, maltotriose cannot be detected within the limits of the industry..

CONCLUSION

Fourier transform A.T.R. spectroscopy processed by muktivariate methods shows very good results for the quantification of individual sugars in aqueous mixtures either in model or in real mixtures. Further experiments must be run in order to confirm the potential of this method for sugar analysis I.e. to test the influence of contaminants (proteins and salts), temperature...

BIBLIOGRAPHY

1 Belton, P.S., Saffa, A.M., Wilson, R.H., Use of Fourier transform infrared spectroscopy for quantitative analysis : a comparative study of different detection methods. Analyst, 1987, 112, 1117-1121.

2 Cadet, F., Bertrand, D., Robert, P., Maillot, J, Dieudonné, J., Rouch, C., Quantitative determination of sugar cane sucrose by multidimensionnal statistical analysis of their Mid-Infrared attenuated total reflectance spectra. Applied spectro., 1991, 45 (2) 166-172.

ELECTRICAL CONDUCTIVITY IN AVOCADO AS MATURITY INDEX.

Mar Montoya, V. López-Rodriguez, J.L. De La Plaza*.
Dept. Physics of Materials, Facultad de Ciencias, UNED, Madrid-SPAIN
*Dept. Fruit and Vegetables, Instituto del Frío, CSIC, Madrid-SPAIN.

ABSTRACT

Electrical conductivity, firmness, skin and pulp colour measurements were carried out in avocado (Persea americana, Mill) cv. 'Hass' during fruit ripening at +20°C. Conductivity measurements were taken by an impedance bridge and two needle electrodes inserted into the whole fruit. Respiration rate and ethylene production of the fruits were also determined.

Electrical conductivity increased slightly during the first five days. Afterwards, it grew rapidly and reached a maximum on the seventh day. A drop in electrical conductivity was observed when the fruits became overripe by about day 10. The onset of a steep rise in conductivity was followed by softening and colour change in the fruit and it reflected the beginning of the ripening process. On the other hand, the electrical conductivity peak arose at the same time as the ethylene peak. These results have allowed us to set the electrical conductivity as a maturity index simple and rapid to measure.

INTRODUCTION

The avocado fruit has a high rate of post-harvest respiration and limited shelf life [1]. It is important to be able to recognize the maturity stage of fruits. In order to find a physical test, simple and rapid to evaluate, some electrical properties in the low frequency range (100 Hz - 100 kHz) have been tested without success up to now because of the instability of registered data [2] and because the results taken at the lower frequency tested were not supported by those taken at the higher frequency [3].

Preliminary work [4] showed an enhanced technique for measuring electrical conductivity in the whole fruit which approaches the above-mentioned problems. Here, we measured this electrical parameter during avocado ripening as well as others physical properties. The time course of ripening was followed by the respiration rate and the ethylene production.

MATERIAL AND METHODS

Firm and unripe avocado fruits (Persea americana, Mill. cv. Hass) of uniform size were harvested and then kept at 20°C and 80% relative humidity for up 11 days.

Respiratory intensity and ethylene production were calculated from the flowing system derived from the Pratt and Mendoza method [5]. CO_2 and ethylene analysis were determined by gas chromatography of atmosphere samples of 1ml as described by Merodio and De la Plaza [6].

Electrical conductivity was determined by the method of M. Montoya et al [4]. This parameter was measured at 20 kHz over samples of five fruits. Four measurements were taken at the equatorial diameter of each fruit.

Skin and pulp colour were measured with a Hunter Lab Colorimetric, over an illuminated surface area of 13 mm. The expressions **10ab/L** and **(L*b)/100** were used as skin and pulp colour index respectively.

Firmness measurements were achieved with an "Instron" 1140 Universal Testing Meter. The penetrometer used was a double metallic plate designed by De La Plaza [7]. Five fruit were sampled and three measurements were taken at the equatorial diameter of fruit.

RESULTS AND DISCUSSION

Ethylene production by mature fruit remained below 0.5 μl/kg-h for 4 days after harvest. Ethylene production began to increase between days 4 and 5 and continued until the climacteric (Figure 1)

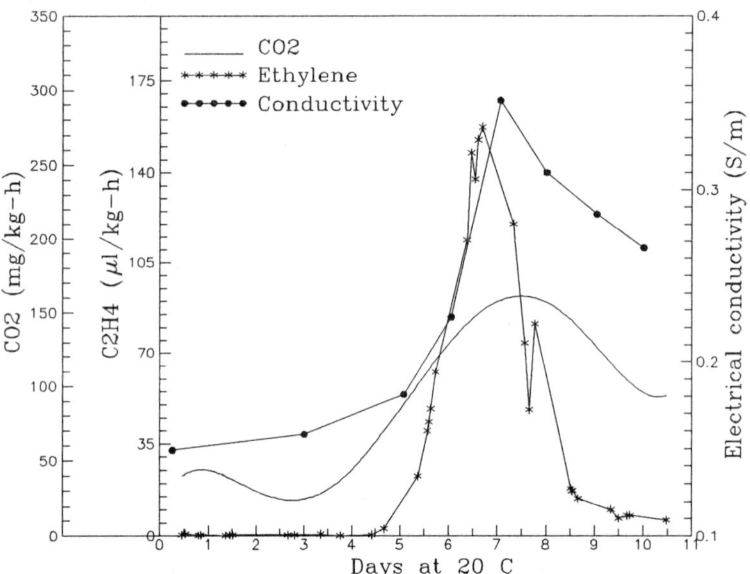

Fig.1 Respiration rate, ethylene production and electrical conductivity during avocado fruit storage at 20°C.

Accordingly, Figure 2 shows how physical properties tested, such as firmness, skin and pulp colour, changed suddenly starting from the fifth day. These parameters developed, either increasing (skin colour) or decreasing (pulp colour and firmness), until the fruit overshot the climacteric. Electrical conductivity, however, followed a pattern similar to the ethylene production one (Figure 1), that is, increased slightly during the first five days. Afterwards, conductivity grew rapidly and reached a maximum on the seventh day. A drop on electrical conductivity was observed when the fruit became overripe by about day 10.

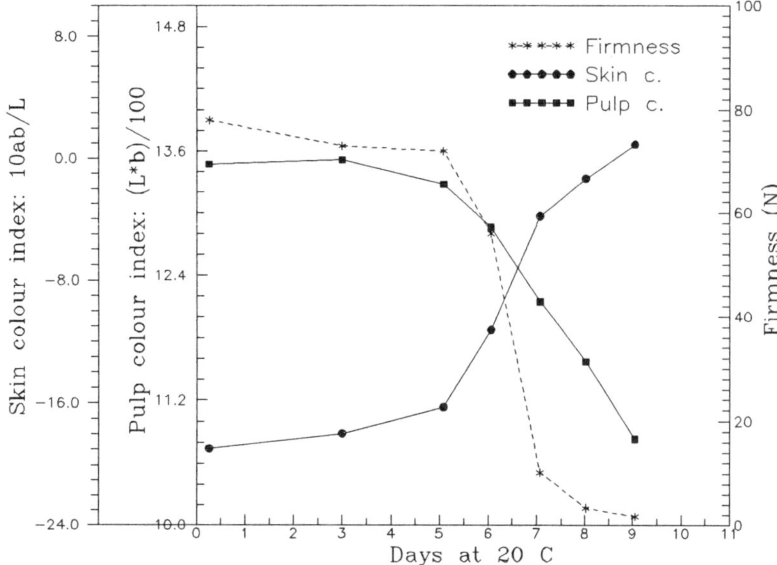

Fig.2 Evolution of some physical properties of 'Hass' avocado during storage at 20°C.

CONCLUSIONS

Avocado fruit ripening is characterized by a sharp increase in ethylene production which reaches a maximum at the same time as the climacteric peak in respiration. In practice, therefore, changes in ethylene production can be used to estimate the time course of ripening in avocado fruit. In the same way, our data show that electrical conductivity during ripening can be used successfully as a physical maturity index, the determination of which is simple and rapid and does not require expensive complicated equipment.

BIBLIOGRAPHY

1. Biale, J.B. and Young, R.E., The Avocado Pear. In The Biochemistry of Fruits and their Products, Vol II, Academic Press, London, (1971)

2. Lozano, Y.K., Rotovohery, J.V. and Gaydon E.M., Etúde de caractéristiques pomologiques et physico-chimiques de divers cultivars d'avocats produits en Corse, Fruits, **42** (5), 305-315, (1987).

3. Weaver, E.M. and Jackson, H.O., Electrical impedance, an objective index of Maturity in peach, Canad. J. Plant. Sci., **46**, 323-326, (1966).

4. Montoya, M., López-Rodriguez, V. and De La Plaza, J.L., An enhanced Technique for Measuring the Electrical Conductivity of Intact Fruits, Food Science and Technology, in press.

5. Pratt, H.K. and Mendoza, D.B., Colorimetric determination of carbon dioxide for respiration studies, HortScience, **14 (2)**, 175-176, (1979).

6. Merodio, C. and De La Plaza, J.L., Interaction between ethylene and carbon dioxide on controlled atmosphere storage of Blanca de Aranjuez pears, Acta Horticulturae, **258**, 81-88, (1989).

7. De la Plaza, J.L., Calvo, M.L., Iglesias, M.C. et Luechinger, R., La Lyophilisation des avocats en tranches, Proc. XIVth Int. Cong. of Refrigeration, Moscu, **3**, 679-689, (1975).

NEW ELECTRICAL METHOD FOR DENSITY SORTING OF SPHERICAL FRUITS

KORO KATO
Department of Agricultural Engineering, Faculty of Agriculture,
Kyoto University, Sakyo-ku, Kyoto 606-01,Japan

ABSTRACT

A new electrical dry method for density sorting of spherical fruits, which measures the volume by electrical capacitance and mass by electronic balance, was proposed. A exact relation between the capacitance of concentric double spheres and the volume of the inner sphere was used for measuring the volume of spherical fruit. The accuracy of measured volume was very high. The new automatic and continuous density sorting system for sorting hollow heart and over-mature watermelons based on this study was developed and tested. A new packing house for sorting hollow watermelons using this dry method was constructed and operated successfully.

INTRODUCTION

Density is one of the properties which can be used for non-destructive quality evaluation of fruits, vegetables and nuts. Density sorting has been employed by farms for some agricultural products since ancient times. However the conventional method is an inefficient wet method using brine solution which encounters complicated problems such as cleaning and drying the products and it is not practically used in automatic sorting system in the packing house. There is a change in density with maturity, cavity, soluble solid contents, damage, peel thickness, etc. If the dry method can be developed, density sorting will have a great potential for automatic sorting. Thus the dry method for density sorting of spherical fruits, which measures the volume by electrical capacitance and mass by electronic balance, was proposed.

Fig.1 Relation between density and cavity section of watermelons.

MATERIALS AND METHOD

Relationships between density and the internal quality of citrus fruits, Japanese pears and watermelons were investigated. It was apparent that the density of the watermelon was related to the cavity section (cavity volume) as shown in Fig.1. The relationships between density, cavity and ripeness were arranged and illustrated in Fig.2. The watermelon with a density of 0.94 to 0.96 was fully ripe, high in sugar content and had the best taste.

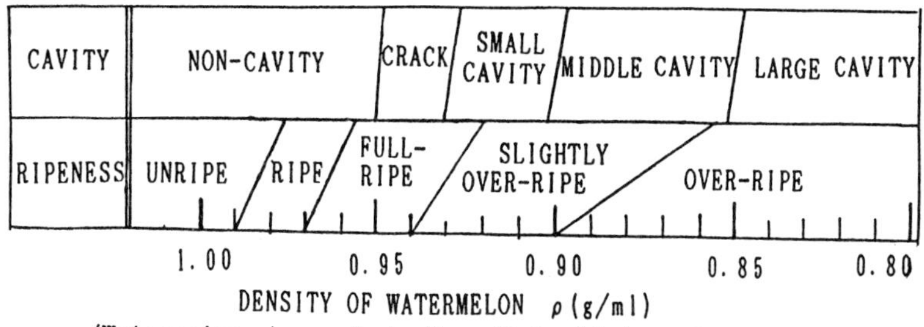

Fig.2 Relation between density, cavity and ripeness of watermelon.

It is very important yet difficult to measure the fruit volume precisely on the density sorting line.

There is an exact relationship between the electrical capacitance(Cs) of concentric double spheres and the radius (or volume) of the inner sphere as indicated in Fig 3.

This principle can be applied for measuring the volume of spherical and oval fruits having the ratios(k) of short to long axis in the range of 0.86-1.16, with small error within 2/1000 as shown in Fig.4, if the outer sphere diameter is much greater than that of the inner sphere.

The new automatic and continuous density sorting system for sorting hollow heart and over-mature watermelons based on this principle was developed as shown in Fig.5.

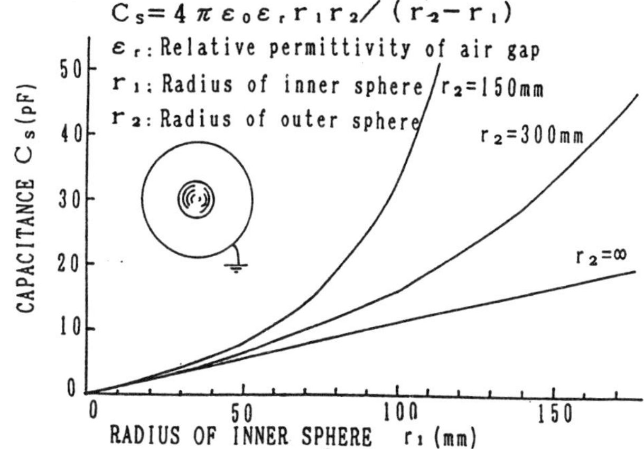

Fig.3 Capacitance of concentric double sphere.

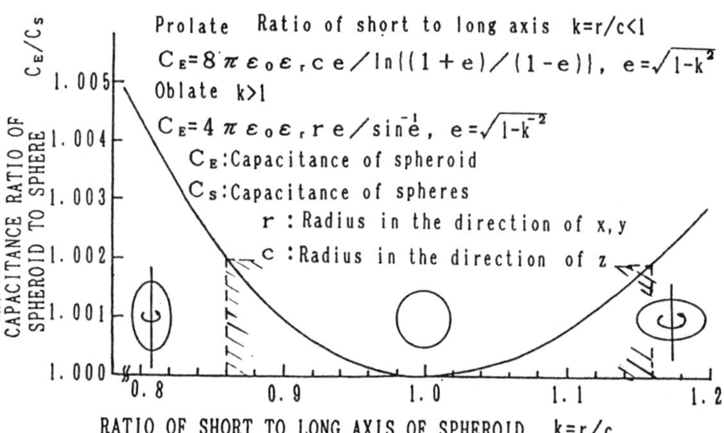

Fig.4 Capacitance ratio of spheroid to sphere.

Watermelon was put on a cup carrier with the fruit stalk on the horizontal plane pointing in forward direction(Fig.6). The volume of watermelon is measured while passing through the polygon external electrode tunnel, with conveying speed of 0.22m/s. The conductive rubber sucker electrode(Fig.6 & Fig.7), which is in contact with the fruit surface by suction, and the precision capacitance meter(Fig.7) were developed. The capacitance between the fruit and the external electrode was measured with a high frequency of over 1MHz to make good conduction through the fruit skin. The relation between the volume and the output of precision capacitance meter is shown in Fig.6. The density is calculated by computer from volume data and mass data which is measured by electronic balance before entering the tunnel while moving.

Fig.5 Experimental system for electrical density sorting of watermelon.

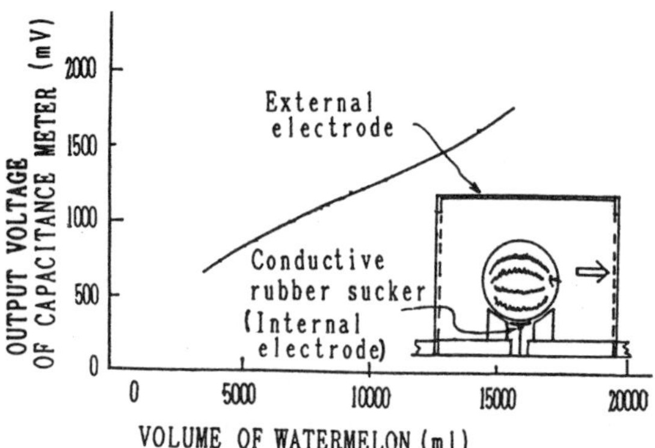

Fig.6 Relation between watermelon volume and capacitance meter output.

RESULTS AND DISCUSSION

The test results of the new automatic and continuous density sorting system for sorting hollow heart and over-mature watermelons was very satisfactory. The accuracy of measured volume was very high, approximately 0.4%, when the orientation of watermelon was maintained. The threshold value of density of watermelons can be set in multi steps, according to the cavity volume. It became clear that watermelon can be classified nondestructively according to the cavity volume ratio by this electrical dry method of density sorting.

A new packing house for sorting hollow heart watermelon using this dry method was constructed and operated. The system(Fig.8), which was controlled by a computer, consisted of two sorting lines that successfully performed at a capacity of 2800 watermelons /hr.

Fig.7 Conductive rubber sucker electrode and precision capacitance meter.

Fig.8 Watermelon volume measurement apparatus in the packing house.

HIGHLY SENSITIVE AND RAPID DETERMINATION OF DIACETYL IN LIQUID FOODS BY ELECTROCHEMICAL METHOD

N.Horikawa, K.Hayakawa, Y.Yamada, O.Miyawaki*
Research Institute, Kagome Co.,Ltd., 17, Nishitomiyama, Nishinasuno-cho, Nasu-gun, Tochigi 329-27, Japan, *Department of Agricultural Chemistry, Faculty of Agriculture, University of Tokyo, 1-1-1, Yayoi, Bunkyo-ku, Tokyo 113, Japan

ABSTRACT

Measurement of diacetyl(DA) in liquid foods is very important for its role as a flavor component with a low threshold. A sensitive electrochemical method for the determination of DA was developed. After sample was subjected to a pretreatment process such as filtering, centrifugal separation or distillation if necessary, DA was transformed to 2,3-dimethylquinoxaline (DMQ) by the condensation reaction with o-phenylenediamine(OPD) and electrochemical method was applied to determine DMQ with glassy carbon electrode as a working electrode. By use of differential pulse voltammetry(DPV), a detection limit of DA was 0.6μM(0.05ppm) with assay time of 15minutes. Optimum conditions for the assay were 4.5~5.0 for pH, 40~50°C for the temperature in the electrochemical measurement, and 8~10 minutes for the condensation time. DA in drinks could be determined by this method and the reliability was compared with the colorimetric method.

INTRODUCTION

DA seriously affects drinks and foods quality. Although DA is a flavor-providing component in fermented dairy products such as yogurt, it adversely affects the quality of processed drinks such as tomato juice and some fruit juices. In either situation it is important to measure the content of DA. Chromatography methods of measurement of DA have the advantage of providing high accuracy, but they require troublesome operations and take a long time for measurement. It is therefore the objective of our investigation to develop an accurate and rapid method of measuring DA in drinks and foods. This paper describes a DPV method for determination of DA. DPV method is compared with the colorimetric method based on the Voges-Prokauer reaction.

METHODS

<u>Procedure</u> : 5mL of sample solution was combined with 5mL of 1mM OPD solution in 100mM acetate buffer(pH5.0). The mixture was stirred for 10minutes, then DA was transformed to DMQ by the condensation reaction. The mixture was purged with nitrogen (50mL/min.) to remove dissolved oxygen for 10minutes and DPV measurement to determine DMQ by the electrochemical reaction was carried out. Yanaco polarographic

analyzer P-1100 was used for DPV measurement with glassy carbon electrode as a working electrode(i.d. 5mm), scanning between -0.2V and -0.9V vs Ag/AgCl.

Calculation method of peak current : In the calculation of peak current, normal peak current method and partial peak current method were used. In the normal peak current method, the start point and the end point of the reduction peak of DMQ were selected manually. In the partial peak current method, the data points of ±30mV of the peak top were fixed as the both ends of current peak. The reduction peak potential for DMQ was about -0.62V at pH5.0.

RESULTS AND DISCUSSION

Effect of condensation time : A standard solution containing 11μM DA mixed with 1mM OPD solution was subjected to DPV measurement with condensation time varied from 0 to 20minutes. In the presence of an excess of OPD the condensation reaction was rapid and finished in 6minutes at 30°C.

Effect of OPD concentration : A standard solution containing 11μM DA was mixed with OPD solution with concentration varied from 0.01mM to 5mM. After stirring for 10minutes, the reaction mixture was subjected to DPV measurement at 40°C. The condensation reaction with the OPD concentration higher than 0.5mM reached the equilibrium.

Effect of pH : The pH value of sample solutions seriously affected the electrochemical reaction rate. Dependence of the normal peak current of DA derivatives on pH was investigated. Britton-Robinson buffer were used for the adjustment of pH in the range 2~8. The highest peak current was observed at pH5.0. It was effective for the highly sensitive determination of DA to adjust the pH in the range 4.5~5.0.

Effect of temperature at DPV measurement : The peak current of DA derivatives increased with increase in temperature at DPV measurement with 10μM DA as a standard. Although increase in temperature was effective for high sensitivity, the experimental error might be increased at a high temperature like 70~80°C because boiling point of DA is 88°C. Thus the optimum temperature for the assay was estimated to be at 40~50°C.

Stability of peak current : A freshly polished electrode gives the highest sensitivity in electrochemical measurement, since the sensitivity decreases gradually by the pollution at the electrode surface. In the present case, however, the working electrode was stabilized by the immersion in a sample solution for 10minutes. The repeatability of the peak current both in the normal peak current method and the partial peak current method, was determined for tomato juice and for tomato juice filtrate with DA added at 5.5μM. Relative standard deviation for the normal peak current method was 2.9%-7.7% and that for the partial peak current method was 4.6%-13.0%.

Recovery of DA added in tomato juice : DA was added to distilled water and tomato juice and analyzed by DPV method. All the calibration curves were linear to DA concentration (Fig.1). DA recoveries in tomato juice were not much different from those in distilled water.

Comparison of DPV method with colorimetric method : DA concentration in off-flavored pineapple juices were determined by DPV method with and without distillation as a pretreatment. The normal peak current method and the partial peak current method were agreed well with the results by the colorimetric method. The colorimetric method was more reliable than HPLC and GC method. Correlation coefficients between DPV method with distillation and the colorimetric method were 0.984 and 0.996, for the normal peak current method and for the partial peak current method, respectively (Fig2). These correlation coefficients decreased to 0.948 and 0.945, respectively, when the distillation pretreatment was abbreviated. Thus, the distillation pretreatment was proved to be effective to increase in the accuracy in the measurement.

CONCLUSION

DA in liquid foods could be detected sensitively and rapidly by DPV method with glassy carbon as a working electrode. DA was transformed into DMQ by the condensation reaction with OPD before the measurement. Optimum conditions for the assay were determined to be 4.5~5.0 for pH, 40~50°C for the temperature in the electrochemical measurement, and 8~10minutes for the condensation time. DA in off-flavored pineapple juice could be determined by this method, the reliability of which was demonstrated by comparison with the colorimetric method.

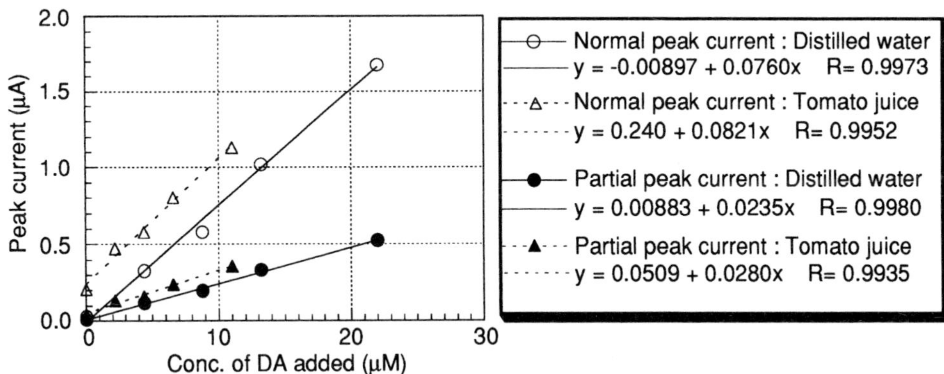

Fig.1 Relationship between concentration of DA added and peak current
Condensation reaction time : 10min.
Measurement temperature : 40°C

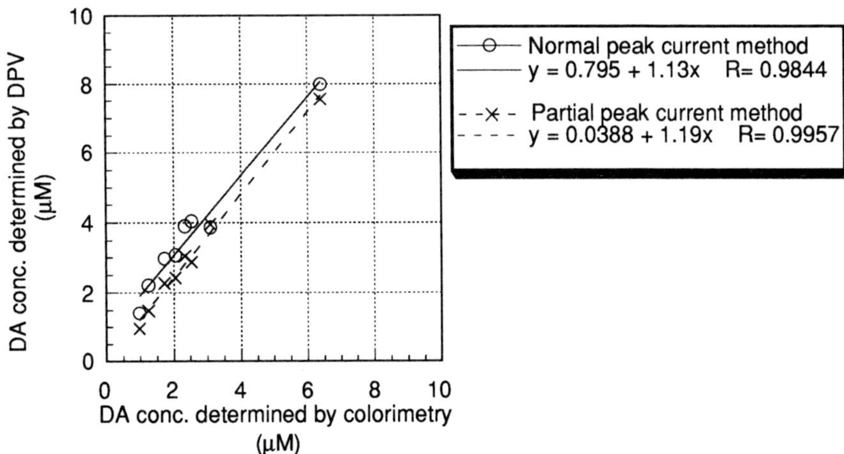

Fig.2 Relationship between diacetyl concentration determined by DPV method and that determined by colorimetric method in pineapple juice with distillation pretreatment

NONDESTRUCTIVE QUALITY EVALUATION OF MELONS BY ACOUSTIC TRANSMISSION CHARACTERISTICS

S.HAYASHI, J.SUGIYAMA*, K.OTOBE*.
Department of Agricultural Technology, College of Technology, Toyama Prefectural University, Kosugi-machi, Toyama 939-03, Japan, * National Food Research Institute, Ministry of Agricultural, Forestry and Fisheries, 2-1-2 Kannondai, Tukuba, Ibaraki 305, Japan

ABSTRACT

Acoustic impulse responses of muskmelons and watermelons were studied to evaluate their quality. We found that the impulse waveform of acoustic signals induced by impact was transmitted along the surface with uniform velocity in both fruits. The velocity varied drastically with the ripeness of muskmelons. The relationship between the velocity and hardness of flesh of muskmelons was examined. Two kinds of waves, however, were observed in watermelons: one was the wave transmitted along the surface with constant velocity, and the other was the wave going through the inside. The latter, however, was not found to occur in internally cracked watermelons.

INTRODUCTION

We obtained acoustic signals at points which divide the equator of the sample into 24 equal parts(see Figure 1). It became evident that the impulse waveform of acoustic signals induced by impact was transmitted along the surface of the equator with uniform velocity [1].

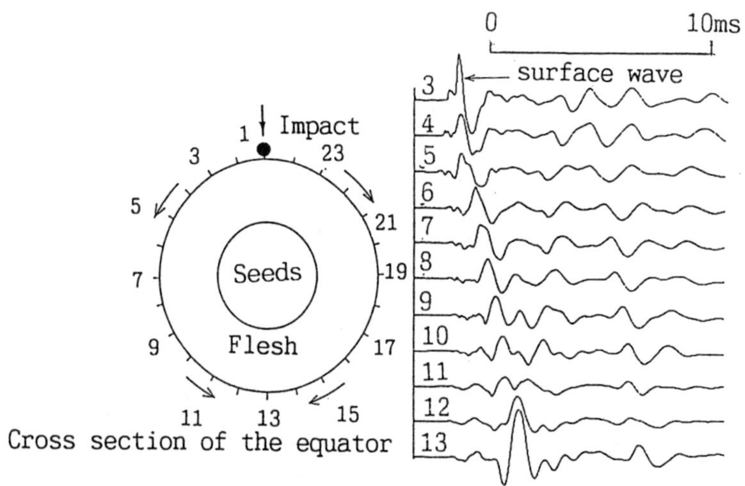

Figure 1. Transmission of wave forms along the equator of a muskmelon.

MATERIALS AND METHODS

MATERIALS

Thirty six muskmelons with various degrees of ripeness were used. Normal and internally cracked watermelons were also examined.

METHODS

We developed a system for determining the transmission velocity of acoustic signals. The system consists of 2 microphones, an A/D converter (100kHz sampling frequency) with a simultaneous sample-and-hold circuit and a computer(see Figure 2). The lag time, which is the difference between two microphone signals, was detected using cross-correlation techniques. Transmission velocity was determined by dividing the distance between the microphones by the lag time(see Figure 3).

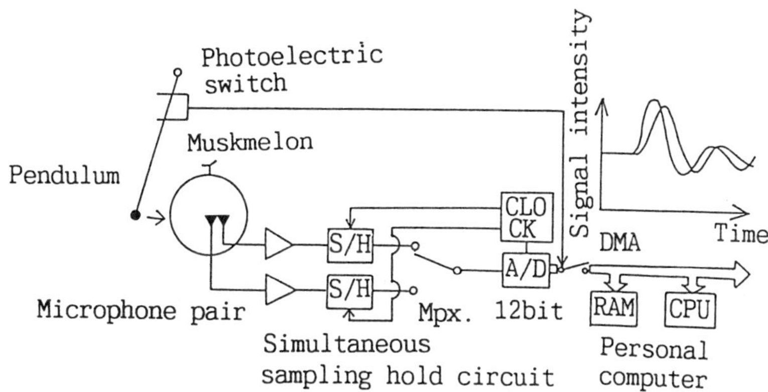

Figure 2. Measuring system of acoustic impulse response.

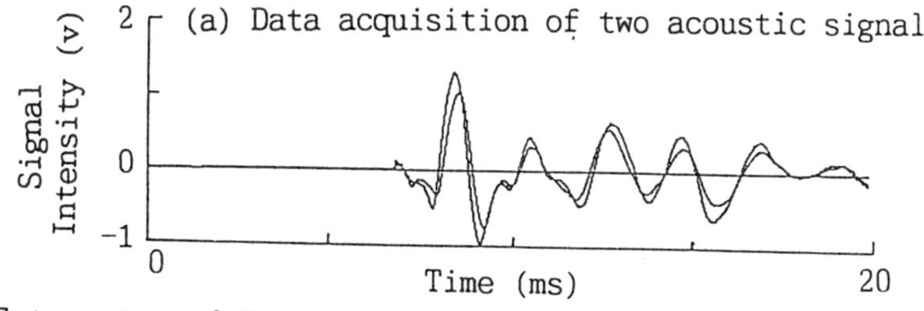

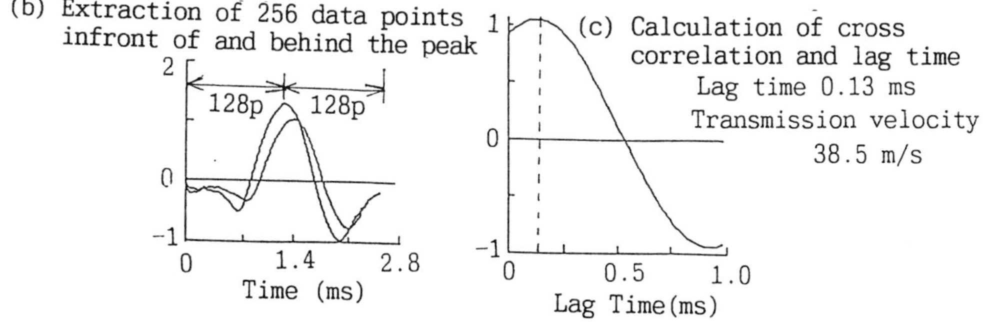

Figure 3. An example of the signal processing.

RESULTS

Experimental results in muskmelons were as follows : (1) The microphones should be located on the equater 45° - 90° from the impact point 0°. (2) Optimum microphone-to-sample distance was found to be 1 - 4 mm. (3) Five hundreds and twelve data points with 100 kHz sampling frequency were needed to calculate the accurate lag time by cross-correlation techniques.

(4) The velocity varied drastically with ripeness of fruits. The correlation coefficient between transmission velocities and fruit hardness was 0.832. The transmission velocity of edible muskmelons ranged from 37 to 50 m/s(see Figure 4).

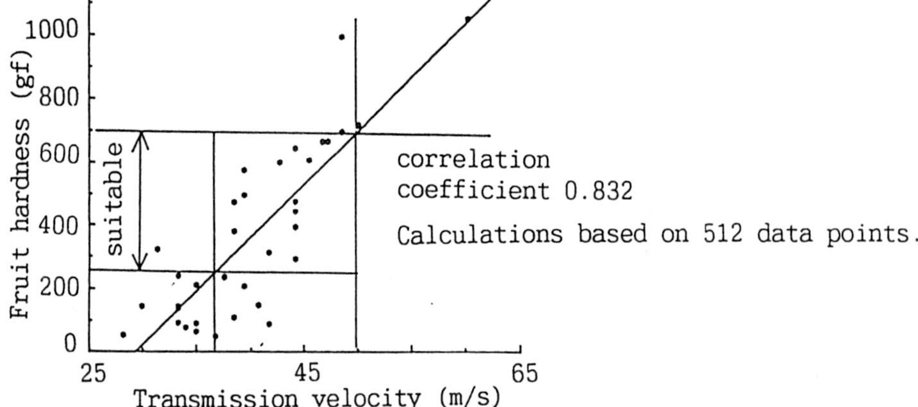

Figure 4. Relationship between transmission velocities and fruit hardness.

Two kinds of waves on the signal of a watermelon were observed: one was the wave transmitted along the surface with constant velocity, and the other was the wave going through the inside. The latter, however, was not found to occur in internally cracked watermelons [2].

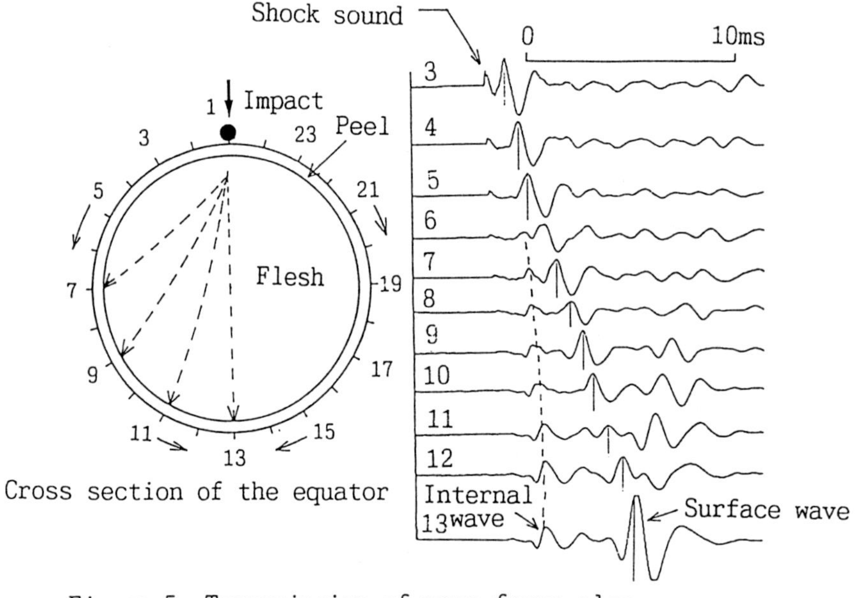

Figure 5. Transmission of wave forms along the equator of a normal watermelon.

REFERENCES

1. Sugiyama, J. and Usui, S., Nondestructive quality evaluation of muskmelons by acoustic impulse response. Trans.Soc.Instrument and Control Eng., 1990, **26**, 367-374.
2. Hayashi, S., Sugiyama, J., Otobe, K., Nondestructive internal crack detection of watermelons by acoustic signals. Bulletin of Toyama Prefectural University, 1992, **2**, 115-119.

NONDESTRUCTIVE INTERNAL DEFECT DETECTION OF AGRICULTURAL PRODUCTS USING SECONDARY ULTRASONIC WAVE

S.TANAKA, K.MORITA, S.TAHARAZAKO,
Chair of Agricultural Systems Engineering, Faculty of Agriculture,
Kagoshima University, Korimoto, 1 chome, Kagoshima 890, Japan

ABSTRACT

The purpose of the present work is to develop a method for nondestructive internal defect detection of agricultural products using secondary ultrasonic wave. The experimental apparatus consists of a function generator, a driver and detector, and a FFT Analyzer. The internal defects of several products and the ripeness of kiwi fruit were measured. Results obtained were as follows: (1)The amplitudes of waveform with respect to time for defective products were smaller than that for normal products. (2)There was a correlation between the amplitude of waveform and firmness of kiwi fruit.

INTRODUCTION

The requirement for nondestructive inspection of agricultural products is strict. Because it is necessary to detect internal defects of many agricultural products rapidly, simply, accurately, safely and with low cost. The purpose of the present work is to establish the fundamentals of a method for nondestructive detection of internal defect of various agricultural products using secondary ultrasonic wave instead of primary wave.

MATERIALS AND METHODS

The internal defects of watermelon, Japanese radish (longish common type and cv. Sakurajima), potato and apple, and the ripeness of kiwi fruit were measured. The experimental apparatus consists of a function generator, a driver and detector, sample holder and a FFT analyzer.

The products except for potato were held between the driver and the detector by a sample holder. Potato was held by hand. A pulse of sinusoidal secondary ultrasonic wave was input one side of a sample and the transmitted waveform was received on the other side. The waveform was analyzed with a FFT analyzer.

RESULTS AND DISCUSSION

1. On the detection of defective watermelon

Twenty normal fruits and forty-three defective fruits were inspected using the ultrasonic apparatus. After each fruit was divided into two parts, and the internal defect was studied.

Figure 1 shows waveforms with respect to time for watermelons, the left is for a normal fruit and the right is a defective one. There are obvious differences of amplitude and transmission time between normal and defective fruit. The amplitude of a defective fruit is smaller than that of a normal fruit, and therefore this technique should be able to detect a defective fruit.

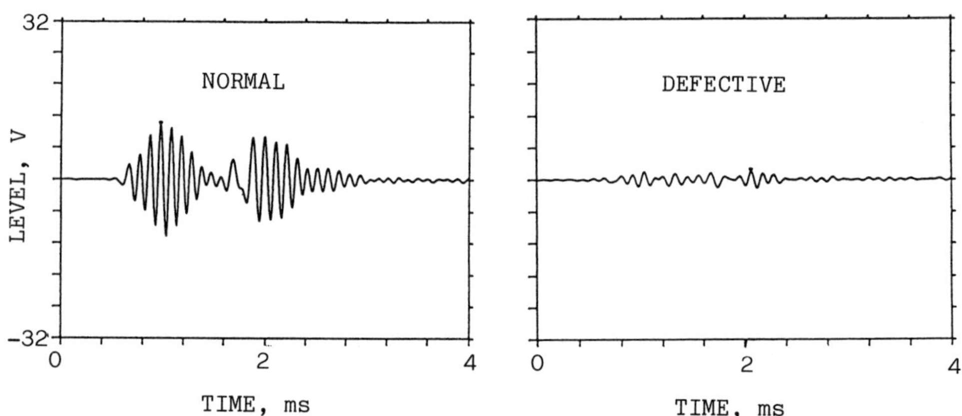

Figure 1. Waveforms with respect to time for watermelons

2. On the detection of bruised apple

Apples with artificial bruises greater than 1 inch diameter were inspected. Figure **2** shows the powerspectra of apples, the left is for a

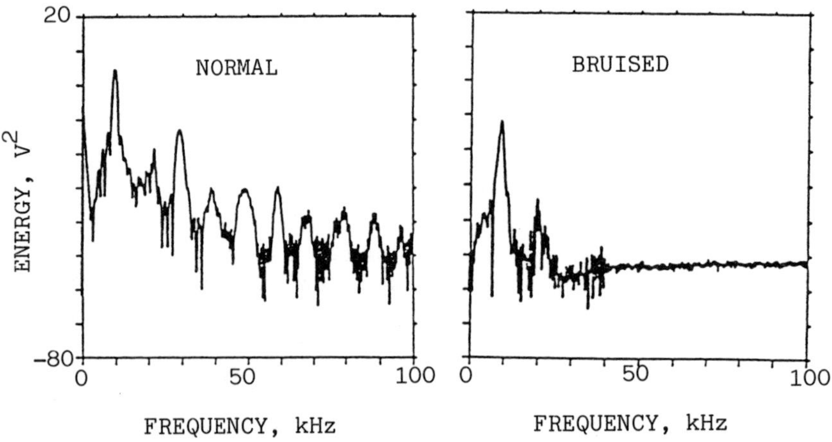

Figure 2. Powerspectra of apples

normal apple and the right is for a bruised fruit. There are differences in the output at the frequency band above 40--50 kHz, and therefore this technique was able to detect a bruised apple. However, in this case, the results were obtained by holding the driver and detector to the bruised part by hand. This is a problem to be solved in the future.

3. On the measurement of the ripeness of kiwi fruit

It can be considered that there is some correlation between the ripeness of a kiwi fruit with thin peel and its firmness. The authors compared the firmness of kiwi fruit measured with a pressure tester with a transmitted waveform of secondary ultrasonic wave. Forty kiwi fruits with almost the same firmness were left in a room at normal temperature, and five samples were tested per day for eight days.

Figure 3 shows waveforms with respect to time for kiwi fruits, the left is a sample of 570g for firmness and the right is a sample of 240g for firmness. From the fact that the amplitude for a sample which is less firm is small, it is assumed that a sample with small amplitude is riper.

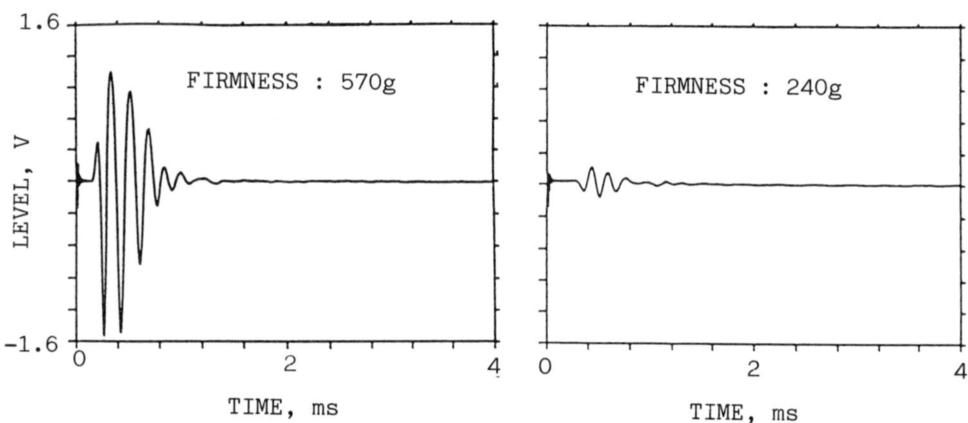

Figure 3. Waveforms with respect to time for kiwi fruits

CONCLUSIONS

The authors conclude from the experimental results described above that the method of using secondary ultrasonic wave is able to detect the internal defects of selected agricultural products and the ripeness of kiwi fruit.

The information provided by this method is less than that of X-ray CT, but for practical usage, this method offers some possibility.

WIRELESS IMAGE TRANSMIT UNIT AND HIGH SPEED INSPECTION SYSTEM FOR INNER SIDE OF CONTAINER

MASARU HOSHINO
Packaging Research Institute,
DAI NIPPON PRINTING Co.,Ltd.
591-10 Kamihirose,Sayama-City,Saitama,350-13 Japan

ABSTRACT

The Wireless Image Transmit(WIT) unit is the device to transmit plural video image data through closed pairs of special loop antennas shielded against electromagnetic wave.
It is designed for the rotary-carrier inspection system, and gives higher performance in speed and precision than conventional line-type or index-type inspection system.
Our inspection system with WIT unit for laminated tube enables inspection of inner side of shoulder part and nozzle part, also accomplishes high speed operation at 190 tubes per minute.

INTRODUCTION

As the demand for the product quality is getting higher and severer, manufacturers are aware of necessity to inspect all products. Automatic inspection system replaces traditional visual inspection in recent years for its convenience.
WIT unit improves inspection accuracy and processing capacity of image inspection device combined with rotary-carrier system. An inspection system for inner surface of shoulder part and nozzle part of laminated tube has been developped and the followings are the fundamental outline of the system.

HARDWARE COMPOSITION

Followings are the composition of 12 heads rotary-carrier tube inspection system.

Inspection head
Inspection head consists of coaxial optical fiber light and 250 mm long bore-scope directly connected to CCD camera.
The bore-scope axis is shifted from the tube axis and the bore-scope is designed to insert into the tube during inspection.

Figure 1.
Inspection system

Figure 2.
Inspection head

Figure 3. Antenna

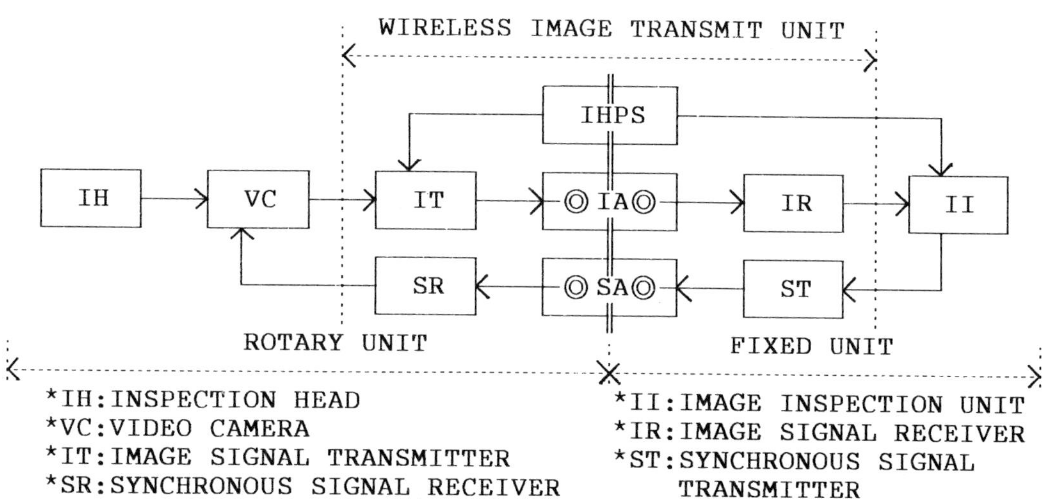

*IH:INSPECTION HEAD
*VC:VIDEO CAMERA
*IT:IMAGE SIGNAL TRANSMITTER
*SR:SYNCHRONOUS SIGNAL RECEIVER
*II:IMAGE INSPECTION UNIT
*IR:IMAGE SIGNAL RECEIVER
*ST:SYNCHRONOUS SIGNAL TRANSMITTER
* IA:IMAGE SIGNAL ANTENNA
* SA:SYNCHRONOUS SIGNAL ANTENNA
*IHPS:INSPECTION HEAD POSITION SENSOR

Figure 4. Block diagram of WIT unit.

WIT unit

Figure 4 shows block diagram of WIT unit.

WIT unit is generally divided into two systems, image data transmitter system and positioning data system.

Image data system consists of image signal system and synchronous signal system. Each system is composed of transmitter unit, receiver unit and antennas.

Transmitter unit consists of image-selector, modulator and mixer. It modulates four channels video signals into one RF signal.

Antennas are inpedance matched and shielded against electromagnetic noises from outside. Its loop shape enables constant transmission of signal during rotation. To perform mutual communication of image signal and synchronous signal,

two cocentric loop antennas separated electromagnetically are used.
Receiver unit consists of distributor and demodulator. It recovers RF siganl to video signals.

Carrying and handling unit
Tube packages with caps are carried by plastic tube holders in which the tube stands with the cap bottom.
During inspection, ring guide supports tube opening to fix the rotating axis.
The tubes revolve together with the inspection heads around the machine axis and they also repeat the motion of rotate themselves 90° and rest under the head.
Tube loading is performed with the screw-star wheel synchronized with the tube holder. At unloading unit, defectives are rejected by air suction and the rest return to line.

RESULTS AND DISCUSSION

Image transmission
Stable clear image is obtained without any noise interference or electric wave leak. As plural image signals are modulated differently, they are mixable and can be demodulated and separate without any interference.

Inspection accuracy
This system can inspect 0.2 diameter of black spots when a precise bore-scopes are used. As the bore-scope axis is shifted from tube axis, inner side of nozzle part can be inspected easily and efficiently.

Processing capability
The combination of WIT unit and rotary-carrier system performs high speed operation. This system is capable of inspection at the speed of up to 300 tubes per minute when appropriate carrier unit is combined.

CONCLUSION

It is confirmed that WIT unit performs positive improvement in tube inspection system at following points.

1. As the camera and the tube move together on inspection stage, their relative position can be set up freely. This enables effective inspection of complex shape product.
2. Less restriction to video camera type as input source expands the inspection field and enables more accurate inspection.
3. Commercial inspection device shows its potential efficiency by combining with WIT unit and it can achieve high speed inspection.

APPLICATION OF A BIOGAS BUBBLE COUNTER TO FERMENTATION PROCESSES

Y.J. LEE*, N.E. CHOI, D.H. WOO, K.M. KIM and J.K. CHUN
Department of Food Science and Technology, College of Agriculture and Life Sciences, Seoul National University, *Nong-Shim Co. Ltd, Korea

ABSTRACT

A biogas bubble counter was designed to count the number of bubble produced from a fermentation with photo-interrupter and IC chips. Gas production rate curve obtained with the sensor represented the growth associated microbial activities without disturbing the process environment. This system was applied successfully to monitor the Kimchi, alcohol and methane fermentation and found to be very practical to trace the progress of fermentation. To estimate gas production quantitatively, a measuring device of a individual bubble size was developed. A fermentation controller was built with one chip microcontroller to conduct the bubble counting and control the fermentation process.

INTRODUCTION

In most natural fermentations, the microbial flora and their activities are unpredictably varied depending on the raw materials and process variables. Once substrate is fed into the fermentation vessel and it is cured in a tightly packed state until the end of process. Therefore it is hard to estimate the state of the fermentation and the quality of the product. Conventionally broth is taken for the analysis of the process but it results in the disturbance of the fermentaion environment, which is undesirable and rather harmful.
Chun et al.[1] developed a biogas detector which could measure the number of gas bubbles produced during the curing process and they fabricated the biogas detectable sensor consisting of photo-interrupter and IC counter chip, and illustrated the possibility of its application in Kimchi, Tackjoo(alcohol brewing) and methane fermentation studies. [1, 2, 3, 4, 5,6]. This study is aimed to replace the conventional quality control method by a new monitoring and control system in the fermentation industry.

METHODS AND MATERIALS

Biogas bubble counter: Chun's bubble counter was used to count bubbles having mean volume of 0.02 cm3. Pulse signal obtained from the sensor was processed to convert to

Number of bubbles.[2, 4]

Fermentation materials: Kimchi, Tackjoo(alcohol) and methane fermentations were tested with the new sensor. (a)Kimchi: Chinese Cabbages of good quality and medium size were prepared by conventional method[2,3]. (b)Tackjoo: Korean rice-brewing was prepared with Koji and cooked rice in Korean traditional method[5]. (c)Methane: Rice straws-cut digested with various alkaline were used as substrate with C/N ratio adjusted.[6]

Fermentation controller: We used Motorola 68705 chip(EPROM) for the CPU in our controller and built the measurement module of bubble and temperature on the main PCB. The controller has necessary actuator control module for heater and agitator of fermentor. and it has a serial communication device for the monitoring system.

RESULTS

The gas production rate was analysed on the basis of the measured gas bubbles during the Kimchi fermentation process. Number of bubbles acquired with the monitoring system was plotted against the fermentation time and analysed their fermentation patterns of Kimchies. The curve patterns well reflected the influences of the fermentation variables and parameters such as material type, temperature and salt concentration.

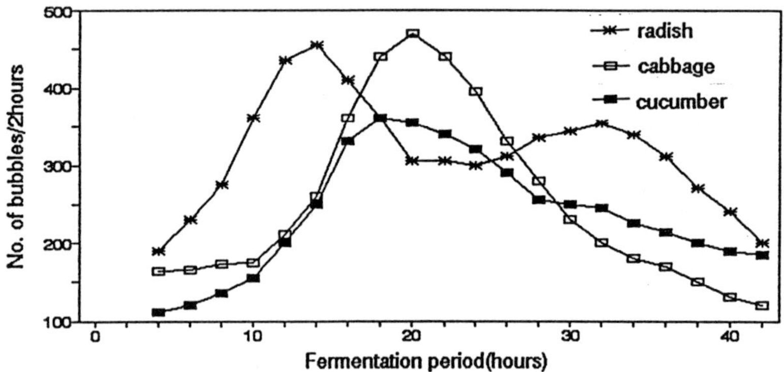

Figure 1. Kimchi fermentation curves constructed by the bubble sensor

When the sensor was applied to Tackjoo brewing in a open vessel fermentor, a special bubble collecting device was attached to the sensor. Ethanol content was in good agreement with the gas production measured with the sensor as shown in Fig.2.

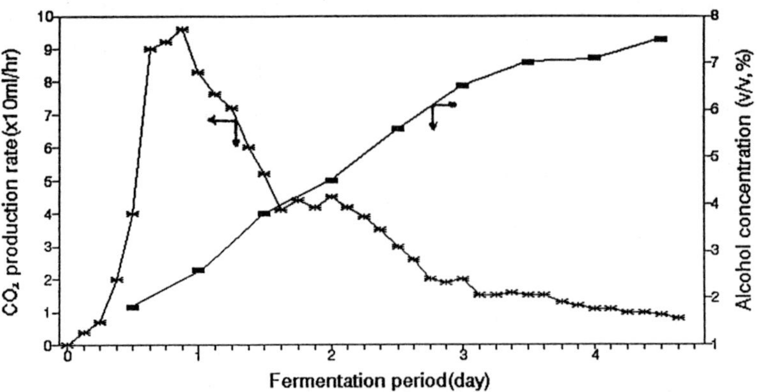

Figure 2. Gas and alcohol production curves in Tackjoo brewing

In methane fermentation the monitoring of gas production rate is more important than the analysis of gas composition, and the fermentation process is so sensitive to environmental changes that minor disturbances are undesirable. Therefore we tried to use the sensor in methane fermentor in an on-line manner to avoid the disturbance. Fig 3 shows the

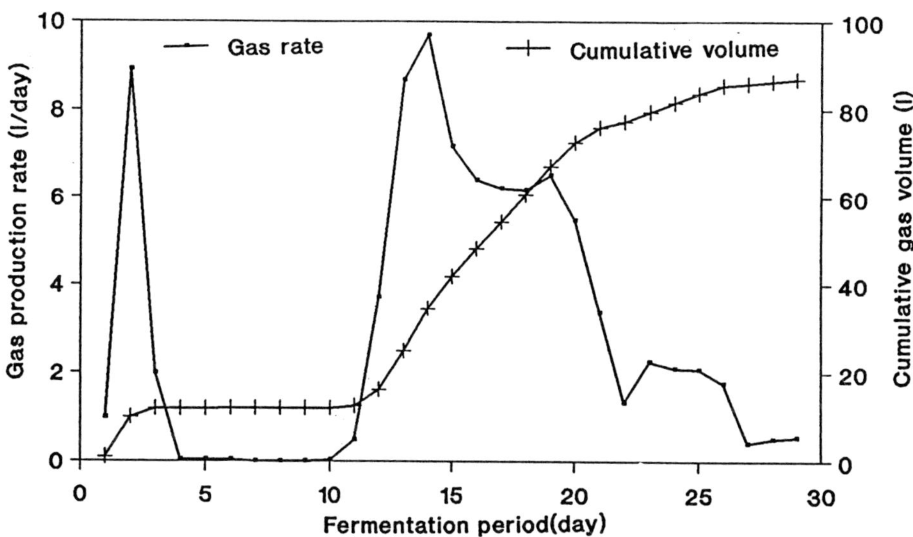

Figure 3. Methane gas production rate and cumulative curves

cumulative and gas production rate during the fermentation period under the enviromental control with the one chip controller. We also applied this sensor successfully to the multi-fermentor system, where the various alkaline treatment effects were investigated for several months.

REFERENCES

1. Lee, Y.J. and Chun, J.K. Development of gas production measurement system by bubble counting during fermentation, Korean J. Food Sci. Technol. 25(3)(in print,1993)
2. Lee, Y.J. and Chun, J.K., Development of pressure monitoring system and pressure changes during Kimchi fermentation, Korean J. Food Sci. Technol. 1990,22(6),686.
3. Choi, N, E., Application of Microcontroller to Plotting of Kimchi Fermentation Curve, MS thesis, Seoul National Univ., Korea,1989.
4. Woo,D.H., Development of the Bubble Size Measurement Sensor and its Application to the Monitoring of Kimchi Fermentation Process, MS thesis, Seoul National Univ., Korea,1989.
5. Kim, K.M., Development of Single Chip Microcomputer-based Auto-Measurement and Automatic Control Methods for Tackjoo Fermentation, MS thesis, Seoul National Univ., Korea,1989.
6. Hwang, C.H., The On-Line Monitoring System for Methane Fermentation Process, MS thesis, Seoul National Univ., Korea,1989.
7. Choi, N, E., Woo, D.H. and Chun, J.K., Development of the bubble size measurement sensor and its application to the monitoring of Kimchi fermentation, Korean J. Food

COMPUTER AIDS IN FLEXIBLE FOOD MANUFACTURING

CHRISTINA SKJÖLDEBRAND
SIK, Box 5401, S–402 29 Göteborg, Sweden

ABSTRACT

The future of food production plant is being intensely discussed in Sweden at present. The industry is facing tough international competition with the removal of trade barriers and deregulations. At the same time customers are demanding highest quality and greater varieties of food.

Tomorrow's plant will have to be more flexible, capable of producing products that may have only limited duration on the market. Many different products must be produced in the same plant – often with many shifts a day.

The consequences of these trends will be better production planning, production uniformity targeted towards predetermined quality short start-up and stoppage times, and minimum bottle necks.

Integrating computers will be one solution to achieve the needed flexibility. It must be possible for the food industry to automate further by enhancements without whole plants or production lines becoming obsolete.

This paper will present results from projects that have been carried out in Sweden at the moment. The results are – among other things – computer programmes that have been designed for production planning in a bakery, a meat factory, and a fine chemicals factory.

BACKGROUND

Quality assurance will be more and more important in the future food production plant. Together with the demand on increased productivity and better use of raw material the structure of production plant layout and planning will change.

Plant will have to be more flexible, capable of producing products that may have only limited duration on the market. Many different products must be produced in the same plant often with many shifts a day. The plant must have the ability to shift to another product, or accomplish a variation on the original recipe. The results will be reduced made-to-order inventory, lower costs to the manufacturer and better prices and quality for the consumer. Global marketing will be based on rapid response (Springer 1990, Le Maire 1990).

The consequence of these trends will be more effective production planning, production uniformity targeted towards predetermined quality, short start-up and stoppage times and minimum bottlenecks.

Integrating computers and automation will be one solution to achieve the needed flexibility. It must be possible for the food industry to automate further by making enhancements without whole plants or production lines becoming obsolete.

Integrating computers into the food manufacturing process is complex and requires more than just overbuying hardware and software as in the past. The operator in the plant has to be involved in the development.

THE DUP PROGRAMME

Since 1987 the Swedish Government through the Swedish National Board for Industrial and Technical Development (NUTEK) has funded a research and development action area programme where applications of new information technology are tested in the process industry. The programme is called Development of User Friendly Process Operation Systems (DUP).

The goal of the DUP R&D programme is to develop process control systems that cultivate and utilise the skill of the operator and facilitate a broadening of his or her role and at the same time contribute to a better and more even product quality by improved use of raw materials and energy and increased productivity.

Three different industrial areas are emphasised. These are the paper and pulp industry, the chemical industry and the food industry. The DUP programme is divided into two sub-programmes: basic studies and case studies. The basic studies are coordinated with the case studies and are intended to give support to their projects.

This programme is a six-year programme which means that it will end this summer (1993). The Government has, however, decided to extend the programme for 4 years more in order to give information about the results to representatives in the whole process industry. These information activities will be carried out in many ways.

The most important project within the DUP food industry programme concerns support and control systems to food industries with flexible production systems. A tool box will be developed. This box will contain different software for computer aids in a food production plant with flexible production.

FLEXIBLE MANUFACTURING SYSTEMS

Flexible manufacturing systems (FMS) as a concept have been around for many years. They are directly related to the computer era and the technologies that have evolved. FMS is a computer controlled array of semi-independent work stations and an integrated material handling system designed to produce a family of related products with medium variety and medium production volumes of each (Clayton 1987). The concept is often mentioned in connection with the engineering industry where it signifies a production plant system that has the ability to change production conditions. In the food industry, however, the development of FMS has been slow because managers are cost oriented rather than asset oriented and reluctance has left them with cheaper but unreliable equipment. The FMS will be very interesting in the food industry as, for example, batch processing will be the main type of production unit in this industry.

The traditional batch production process machine actual use has been observed to be only 6–8% of the available machine time. In contrast 45–55% machine utilisation is achieved in dedicated highly specialised transfer lines in which movement is in lockstep along a production line. For a typical machine product the cost using the most efficient mass production methods is less than, often much less than, 10% of the cost of batch production.

Why is there renewed interest in batch production? The increased demand on product quality and productivity of the production can be fulfilled by using, among other things, different computer systems. Flexibility is another keyword; the ability to use a manufacturing facility in such a way that one more piece of equipment with related auxiliary support devices for materials handling is integrated. The production organisation and working organisation are as important.

INCREASING FLEXIBILITY

As said before, tomorrow's plant will have to be more flexible, capable of producing products that may have only limited duration on the market. Many different products must be produced in the same plant. The customers are demanding higher quality and greater varieties of food.

This will demand a higher productivity and efficiency from the production plant. The most important issue for the food industry in order to increase flexibility is to develop equipment and control systems that optimize and control production lines and quality. These have to be user friendly. The operators have to be educated both in order to increase their knowledge and to take more responsibilities. It is important that they understand their role, their responsibilities and that they can influence their working conditions.

AUTOMATION OF CHEESE AND YOGURT MANUFACTURING PROCESSES

G. S. Mittal
School of Engineering, University of Guelph,
Guelph, Ontario, Canada N1G2W1

ABSTRACT

Milk coagulation and curd firmness in commercial operations is usually determined subjectively, either manually or visually. However, many factors that affect curd firmness do not remain constant. Curd firmness is influenced by acidity, heat treatment or cold storage of milk prior to cheese making, variation of calcium or inorganic salts in milk. Cheese is more uniform when the curd is cut at a constant firmness determined instrumentally. Line heat source probe, generally used to determine thermal conductivity, was used to determine coagulation time of curd and yogurt. This instrument can be used to automate cheese and yogurt manufacturing. The automation will maximize cheese and yogurt yield and provide optimum control of cheese moisture.

INTRODUCTION

Curd firmness in commercial operations is usually determined subjectively, either manually or visually. Some vats are cut automatically after a specified time. More objective instrumental determination of coagulation characteristics, curd firmness, and cutting time should refine cheese making, maximize cheese yield, and provide optimum control of cheese moisture. The overall objective was to develop techniques for on-line monitoring and control of the cheese and yogurt making processes.

METHODS AND MATERIAL

A line heat source probe was used to measure coagulation of milk for curd and yogurt production. The probe (Fig. 1) was inserted into the milk sample containing enzyme (rennet) for curd making or bacterial culture for yogurt manufacturing. A constant current of 200 mA was applied to the heater wire. The probe temperature was recorded at 2 s interval on a data logger. The thermal conductivity of the milk remains constant during curd/yogurt formation (1).

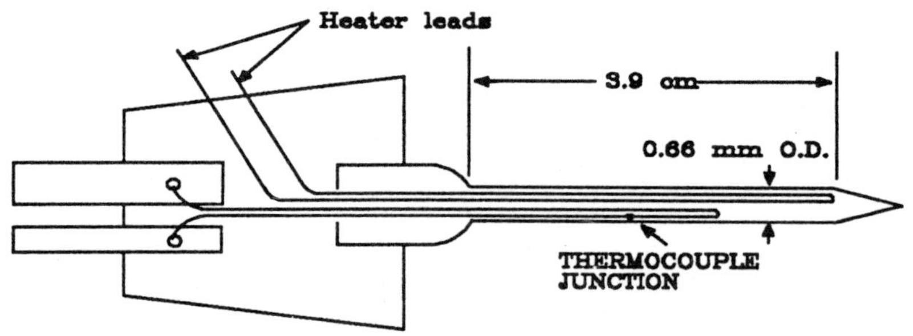

Figure 1. A line heat source probe

RESULTS AND DISCUSSION

Cheese/Curd

The milk is coagulated either by addition of rennet or acid or both. The coagulated milk (curd) is cut into cubes and excess water is expelled by continued action of the rennet, by acid development, by manual/mechanical stirring, and by heating at a desired temperature. The coagulation time (CT) is generally used to characterize changes in the physical state of coagulating milk. The CT of milk has been measured by using various instruments. Only a few equipments are appropriate for in-line use in a dairy plant. The syneresis in renneted milk is initiated by cutting the curd. Cutting curd at optimum curd firmness is important to minimise loss of milk solids and for proper drainage of whey (2).

The CT of milk was measured at 32°C and adding single strength rennet (Dairyland Food Lab., Waukesha, WI) at a rate of 0.2 mL/L of milk (3). Fig. 2 shows a typical plot between probe temperature minus initial milk temperature (ΔT) and time. A sharp increase in this temperature difference was noted at the time of milk coagulation. The tangent to the inflection point on the

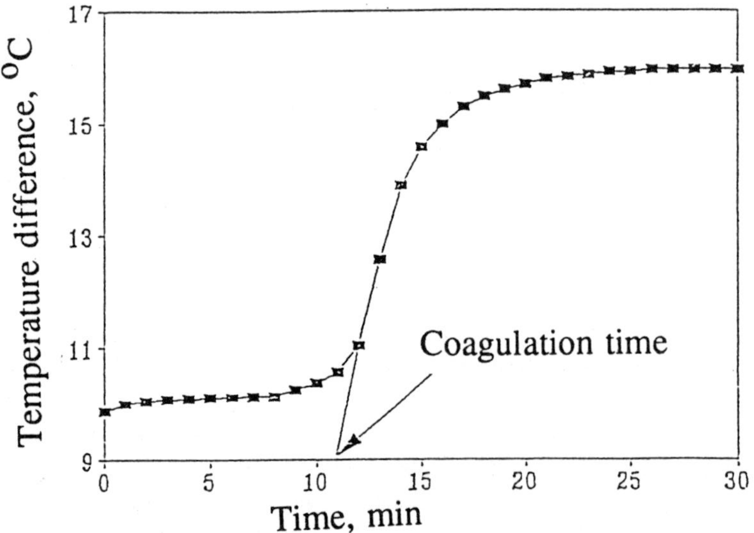

Figure 2. Measuring the coagulation time of the renneted milk

ΔT vs. time provided CT. Thus, the hot wire temperature could detect the onset of milk clotting in line and in real time without disturbing the milk coagulum as also observed (4).

The curd cutting time has not been measured by using this probe. With further work it may be feasible to determine the proper curd cutting time for various situations.

Yogurt

There are two predominant types of yogurt produced--set yogurt and stirred yogurt. After pasteurizing, the sterile milk is homogenized, and then cooled to about 41°C to 43°C before fermentation. It is inoculated with equal numbers of *Lactobacillus bulgaricus* and *Streptococcus thermophilus* at about 2% level. The automation of stirred yogurt is feasible by the sensor tested in this work. Fig. 3 shows a typical plot obtained for the probe temperature vs. time during yogurt manufacturing. The time at which probe temperature jumped to a high value provided coagulation time for the yogurt. This should also be evaluated to continuous process yogurt compared to batch operation.

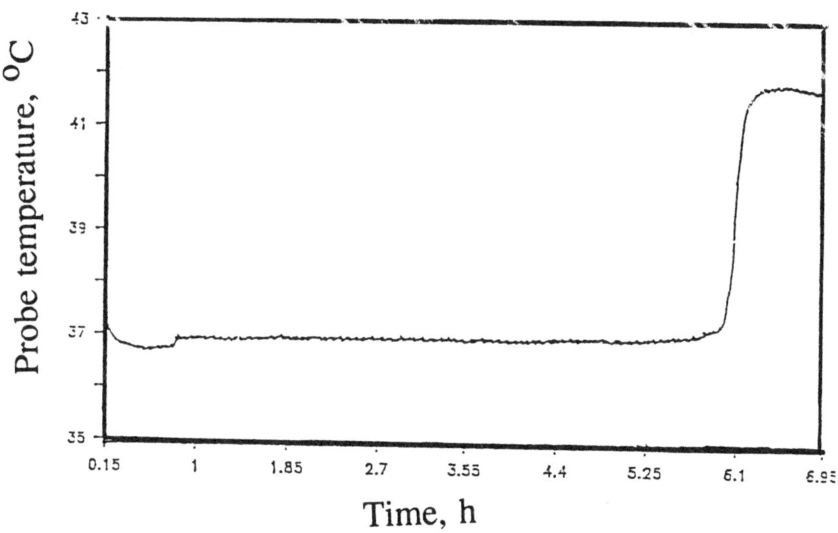

Figure 3. Measuring the coagulation time during yogurt manufacturing

REFERENCES

1. Hori, T., Effect of rennet treatment and water content on thermal conductivity of skim milk. J. Food Sci., 1983, 48, 1492-6.

2. Bynum, J.F. and Olson, N.F., Influence of curd firmness at cutting on cheddar cheese yield and recovery of milk constituents. J. Dairy Sci., 1982, 65, 2281-90.

3. Sharma, S.K. 1992. Kinetics of Enzymatic Coagulation and Aggregation of Ultrafiltered Milk. Ph.D. thesis, Univ. of Guelph, Canada.

4. Hori, T., Miyawaki, O. and Toshimasa, Y., In-line measurement of milk clotting by a hot wire method. In Engineering and Food, Vol. 1, ed. W.E.L. Spiess and H. Schubert, Elsevier Applied Sci. Publ., London, 1990, pp 743-51.

AUTOMATION OF SHRIMP QUALITY EVALUATION

MURAT BALABAN, SENCER YERALAN, YMIR BERGMANN, W. STEVEN OTWELL
Food Science and Human Nutrition Dept. University of Florida
Gainesville, FL. 32611, USA.

ABSTRACT

Current quality evaluation of shrimp involves subjective determination of melanosis, and manual measurements of count per unit weight, ratio of largest to smallest shrimp, and presence or absence of broken parts, foreign materials, etc. With the globalization of shrimp trade, objectivity, repeatability and standardization of quality evaluation is necessary. Computerized image analysis was used to evaluate count, uniformity ratio, and weight estimation. Weight of individual shrimp were predicted from surface area. Different correlations were tried to relate weight to surface area. Analysis of errors from automated evaluations is presented.

INTRODUCTION

Shrimp fisheries, imports and processing are important to Florida (92% of U.S. total value, 1990). About 70% of the processed shrimp is imported. Therefore, objective standards in trade will help in maintaining a high quality and safe shrimp supply. Access to rapid, objective and repeatable quality evaluations will help regulatory agencies in assuring high quality and safety that the consumer expects from a high-priced food. Current shrimp evaluation uses semi-quantitative sensory methods (visual, smell, texture). Development of an automated and objective quality evaluation device using computer vision, rapid ammonia determination, and texture measurement will improve standardization of shrimp quality, decision making in purchases from remote locations, and help in setting standards in marketing. With the accumulation of objective and repeatable quality data, prediction of quality from different locations, different species, different seasons etc. will also be possible. Machine vision has been tried in grading and orienting oysters [1, 2], in recognizing fish species [3, 4], and in shrimp processing [5-7]. Automation of count, uniformity ratio and weight estimation of shrimp is presented in this paper.

MATERIALS AND METHODS

A VCR camera was coupled to a ComputerEyes/RT color frame grabber (Digital Vision, Dedham, MA) installed in an IBM PC. White shrimp was purchased intact, frozen. A wide size range was selected. Headless tiger shrimp was purchased as a 2 kg frozen block. The package displayed a count of 21-25 per 453 g. A 5 mm x 5 mm black metallic piece served as an area reference square. Shrimp were placed one by one on a vision platform with the reference square. A computer program determined the surface area of shrimp relative to that of the reference square. Shrimp weight was determined by a balance. Sixty-three white shrimp were processed this way. Then, a) heads, b) shell except the tail and last segment, and c) all shell and tail were removed. Between each operation, shrimp were processed as before, resulting in four groups of view area-weight data for each shrimp. Tiger shrimp was separated into two batches of 50 shrimp each. These were processed as above, resulting in two independent sets of view area-weight data for headless, peeled, and tail off forms. The data were fitted the following equations : Linear, Y=A+BX, Power, Y=A exp(BX), Forced power (F. Power): $Y=A X^{1.5}$, where X=view area (mm^2), Y=experimental weight (g). The coefficients A, B, and R^2 were determined. The estimated weight of each shrimp, as well as the total weight estimation were calculated for each equation fitted. The count and uniformity ratio of shrimp were also determined for the actual data, and for the predictions of various equation fits. For the tiger shrimp, the A and B values for each equation fit for one set were used to estimate the weight of the other set, and vice versa.

RESULTS AND DISCUSSION

The A, B, and R^2 values are not given due to space restrictions. Experimental and estimated total weights, counts, and uniformity ratios based on various fits for white shrimp are shown in Table 1. For intact shrimp, the power and forced power equations are very close. For the linear fit there is a trend of decreasing intercepts and similar slopes as shrimp is deheaded, peeled and tail taken off. For the power fit, A's increase, while B's are around 1.5. For the forced power fit, exponent of the area is 1.5, and it increases with decreasing shrimp size (intact, peeled, tail off). Table 2 shows experimentally determined , and estimated total weight, count and uniformity ratios for batch 1 tiger shrimp. Batch 2 data is similar (not shown). In this table the view area of the batch 1 was used with the parameters of batch 2 ("By batch 2 parameters"). Since batches 1 and 2 came from the same box, their parameters should be interchangeable, and the parameters of one should accurately be used to predict the properties of the other. The only significant difference is in the peeled tail off form, where estimated weight of one batch by the parameters of the other is different by about 2%.

The weight, count, and uniformity ratio of white and tiger shrimp can be accurately evaluated by computer vision. We need to evaluate other product forms (such as butterflied), and other species to determine the maximum error associated with the "vision weighing" system.

REFERENCES

1. Diehl, K.C., Awa, T.W., Byler, R.K., van Gelder, M.F., Koslav, M. and Hackney, C.R. 1990. Geometric and physical properties of raw oyster meat as related to grading. Transactions of the ASAE. 33: 1270-1274.
2. Tojeiro, P. and Wheaton, F. 1991. Oyster orientation using computer vision. Transactions of the ASAE. 34:689-693.
3. Wagner, H., U. Schmidt, and J. H. Rudek. 1987. Distinction between species of sea fish. Lebensmittelindustrie 34: 20-23.
4. Pau, L.F. and Olafsson, R. 1991. Fish quality control by computer vision. Marcel Dekker, Inc. New York.

5. Marel, 1991. Model L-10 Vision Weigher for shrimp processing. Reykjavik, Iceland.
6. Ling, P.P., Searcy, S.W. and Grogan, J. 1988. Adaptive thresholding techniques for shrimp images. Presented during the 1988 international summer meeting of the ASAE. June 26-29, 1988. Rapid City, SD.
7. Ling, P.P., and Searcy, S.W. 1989. Feature extraction for a vision based shrimp deheader. Presented during the 1989 international winter meeting of the ASAE. December 12-15, 1989. New Orleans, LA.

TABLE 1
Experimentally determined, and estimated total weight, count and uniformity ratio values for different forms of white shrimp.

Form		Experimental	Linear	Power	F.Power
Intact	Total weight (g)	1762.6	1762.3	1756.9	1768.3
	Count (shrimp/453 g)	16.0	16.0	16.0	15.9
	Uniformity ratio	2.45	2.52	2.36	2.36
Headless	Total weight (g)	1217.7	1218.6	1217.5	1225.9
	Count	23.5	23.5	23.5	23.4
	Uniformity ratio	2.34	2.45	2.38	2.45
Peeled	Total weight (g)	1083.8	1084.3	1082.6	1091.8
	Count	26.4	26.4	26.5	26.2
	Uniformity ratio	2.36	2.42	2.36	2.41
Tail off	Total weight (g)	1008.2	1007.9	1005.2	1011.7
	Count	28.4	28.4	28.5	28.3
	Uniformity ratio	2.41	2.48	2.41	2.55

TABLE 2
Experimentally determined, and estimated total weight, count and uniformity ratio values for different forms of tiger shrimp, batch 1.

Form		Experimental	Linear	Power	F.Power
Headless	Total weight (g)	803.16	803.16	802.37	801.40
	By batch 2 parameters		807.91	807.08	806.90
	Count (shrimp/453 g)	28.3	28.3	28.3	28.4
	By batch 2 parameters		28.1	28.2	28.2
	Uniformity ratio	1.34	1.28	1.28	1.39
	By batch 2 parameters		1.25	1.25	1.39
Peeled	Total weight (g)	713.71	713.71	713.05	712.33
	By batch 2 parameters		714.53	713.94	712.93
	Count	31.8	31.8	31.9	31.9
	By batch 2 parameters		31.8	31.8	31.9
	Uniformity ratio	1.34	1.30	1.29	1.39
	By batch 2 parameters		1.29	1.29	1.39
Tail off	Total weight (g)	655.67	655.67	654.95	653.86
	By batch 2 parameters		666.96	666.53	669.56
	Count	34.7	34.7	34.7	34.8
	By batch 2 parameters		34.1	34.1	33.9
	Uniformity ratio	1.35	1.31	1.30	1.43
	By batch 2 parameters		1.32	1.32	1.43

AUTOMATIC CONTROL OF THE BISCUIT BAKING OVEN PROCESS

G. Trystram (1), M. Allache (2), F. Courtois (1)
(1) ENSIA, Food Process Engineering Department, France
(2) CTUC, Massy, France

ABSTRACT

The automatic control of a biscuit baking oven is studied from an experimental point of view. Sensors are adapted and identification is performed. A pole placement strategy is developped to control the colour of biscuits.

INTRODUCTION

The control of the baking oven process is important to improve process efficiency and the quality of baked cereal products. Few papers are dedicated to this problem. To minimize the energy losses and to maximize bread quality an approach with manual control is tested (2). A strategy for colour control based on air velocity control is discussed in (5). Some applications with feedforward and coupled control of moisture and colour are presented (4). The project presents a good opportunity to study sensors and associated control law of baking oven. First results are presented.

MATERIALS AND METHODS

The Baking oven process

A pilot plant is used to study the dynamic of the oven. It is a 15 meter long oven indirectly fired with multiple burners, heated with natural gas. 4 combustion chamber are equipped, in the roof and base, with 34 independently controlled burners. The band is 0.65 meter width. Six zones are available, and five exhausts permit the control of air velocity and of atmosphere composition inside the baking chamber. A Programmable Logic Controller interfaced with a computer is used for oven control and data handling. Classical control functions are implemented on the PLC (1, 6). This oven is used, in this research, for the baking of biscuits.

Sensors

In order to study baking oven control, sensors are adapted or developed specially for measurement quality of biscuits. Due to the drying phase occurring during baking, the final moisture content of the biscuit is an important parameter. A near infra-red analyser is used on line (Infrared Engineering, MM55G). This sensor is located at the exit (figure 1) of the oven.

Studies were performed in order to establish the best location and the accuracy of the measurement: 1%.

Moisture is not the only quality parameter for the biscuit. From a consumer point of view, other characteristics must be taken into account. The colour is one of these characteristics. Two measurement systems are studied. Firstly, image processing is compared with laboratory measurement (using Minolta CR200 colour measurement system). A good correlation is established between grey level analysis through image processing and luminosity (L): 0.987 is the correlation coefficient. Measurement is performed in a few seconds and is available for 5 biscuits at a time. Secondly, a real time measurement system is used. It is the Colourex (Infrared Engineering) which is able to measure the colour of the product in the CIE Lab coordinates. This sensor is located at the exit of the oven, near the moisture sensor. Distance between product and sensor is 0.03 m. Comparisons are made between laboratory measurement (Minolta and Gardner spectrometer) and Colorex. A very good correlation is obtained: 0.998.

Due to the cyclic derive of the band, strong and artificial disturbances are added to the real time measurements. As an example, Figure 1 presents a recording of colour with and without filtering.

Method

No publications are available for the description of the dynamics of moisture or colour in oven. The first part of this work is the systematic analysis of this dynamic behaviour. To do it, a point is chosen which is characterised by: baking time 8 min, air temperature (setpoint of the controllers): 200°C in each zone for roof and base, air exhaust profile: 20, 40, 80, 0, 0 (in % of control input). On the basis of this process tuning which is used to establish reproducibility of trials, step variations are performed for each control variable: 6 temperatures of air, 5 air exhausts and velocity of the band. Different step amplitudes are used. After filtering, the data are studied in order to establish the transfer functions between each control variable and moisture content and colour respectively. The controllers are designed on the basis of these transfer functions. Firstly the right control variable is chosen, in order to control moisture first and colour secondly. The control law is then established and parameters are calculated. Simulations of the performances are then performed.

RESULTS AND DISCUSSIONS

Figure 1 presents the comparison between measured colour variations versus time, due to air temperature step modification in zone 5, and predicted colour using dynamic model as identified from experimental measurements. The two curves are in good agreement. This numerical transfer function is a very simple model which is easy to determine from specific experimental results.

This method has been applied for all the measurable variables (water content, colour). All oven transfer functions are known. Air exhausts appear to be bad control variables. This is probably due to a bad design of the fan exhausts. The most significant variables are the air temperatures. Control variables are chosen from these results. For example for colour control, air temperature in zone 5 appears to be a good choice. The transfer function is used to determine the controller. A pole placement strategy is prefered and parameteres are calculated using PIM software (3). The oven responds quickly to this setpoint change. The effects of disturbances are well reduced. This illustrates the efficiency of the pole placement strategy. These results should be validated directly on the oven itself, but this study is an illustration of the capability of control science to provide progress in biscuit baking oven process.

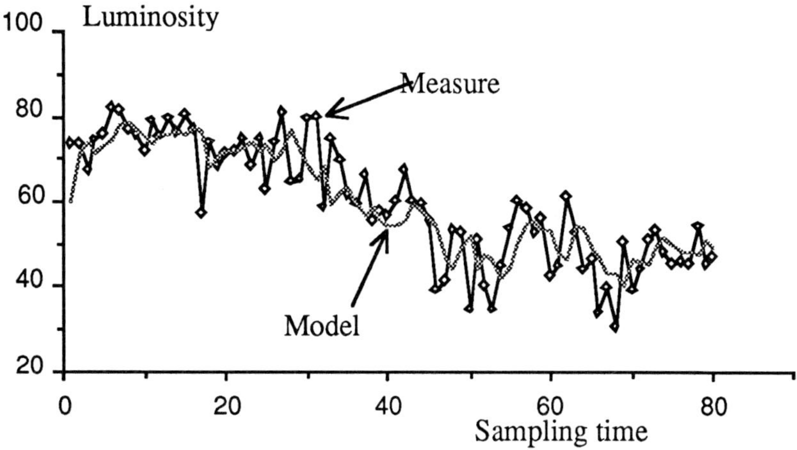

Figure 1: Recording of colour of biscuit at the end of the oven. Comparison between measured and modelled colour.

CONCLUSION

Even if a baking oven is a complicated process (distributed parameters system) it is established that progress with sensors is possible specially with on line moisture and colour sensors. Work is in progress for on line thickness measurements. On the basis of these measurements, the dynamics of moisture and biscuit colour is established. Pole placement controllers are tested and simulations are performed. In the near future pilot validation will be performed. This establishes that automatic control of biscuit qualities is possible.

REFERENCES

1. Emprun , C., Brunet, P. and Trystram, G., La cuisson des produits céréaliers dans un four à régulation automatique, Congrès innovation énergétiques et industries alimentaires, AFME,1990
2. German , H., Meuser , F., Z.F.L., 1987, 38, 25-30.
3. Landau I., Identification et commande, Hermes, paris, 1989.
4. MacFarlane, I. Automatic control of food manufacturing systems, 1988, Elsevier applied sciences, London.
5. Sato A., Sato, Y., Yamanoi, K., Method of controlling the baking of foods, United states patent, 1990, 4.963.375.
6. Brunet, P., Savoye, I., Trystram, G., Rapeau, F., Flexibilité d'un four de biscuiterie par régulation automatique, Congrès innovation énergétiques et industries alimentaires, AFME, 1990,

STUDY OF HANDLING TECHNIQUES FOR THE SOFT AND PLASTIC SUBSTANCE SUCH AS FOOD

SHINZO MAMMOTO
Mayekawa MFG.CO.,LTD.
Food process sect.
13-1 Botan 2, Koto-ku, Tokyo 135, Japan

ABSTRACT

For handling irregularly shaped,soft-textured materials such as food products, it is necessary to provide robots which have functions similar to human eyes and hands. The aim of this study is to construct a control system for grasping movement of mechanical gripper using visual and force sensors. Applicabilities of this system to typical Japanese food products such as tofu and kamaboko are confirmed.

INTRODUCTION

Sophisticated functions such as those provided by the human hands and eyes are required in order to achieve a clear grasp of irregularly shaped, soft-textured materails such as food products. For this reason, most of the dishing up processes, are performed manually.

In order to contribute the automation of food processing plants of the future, this study is aimed at construction of a system capable of grasping irregularly shaped, soft-textured materails such as food products.

The proposals contained herein make full use of modern computer and electronic technology and studies of visualization systems and grasping movement control systems.

PROCEDURE AND RESULTS

The construction of an automatic handling system to irregula rlyshaped, soft-textured materials such as food products, is obtained by means of the integration of the following basic technologies.

(1) Technique for measurement system of mechanical characteristics of the objective materials and development of the data base system.
(2) Technique for pattern recognition of irregularly shaped materials such as food products by visual sensor.

(3) Technique for reduction of the image processing time by means of hard ware and soft ware of computer.

(4) Technique for a real time control system for grasping movement of mechanical gripper with force sensor.

Table 1 shows the mechanical charactristics of typical food products in Japan. Food products have the peculiar mechanical characteristics because of viscoelasticity.

Figure 1 shows the composition of functional devices for visualization processing. The processing time for pattern recognition and visual information data was less than a few seconds.

TABLE 1

Grasping characteristic of typical food products

Item	Name of food products		
	Kamaboko	Processed Cheese	Tofu
Shape	Rectangular parallel-piped		
Shape of gripper	Flat plane		
Setting displacement	3 (mm)		
Rate of compression	20 (%)		
Grasping movement speed	Maximum grasping pressure (g)		
0.7 mm/s	966	774	258
1.1 mm/s	1092	804	276
3.3 mm/s	1054	876	234
Grasping movement speed	Stress relaxation rate (%)		
0.7 mm/s	15.5	32	26
1.1 mm/s	18.7	40	30
3.3 mm/s	22.0	51	38

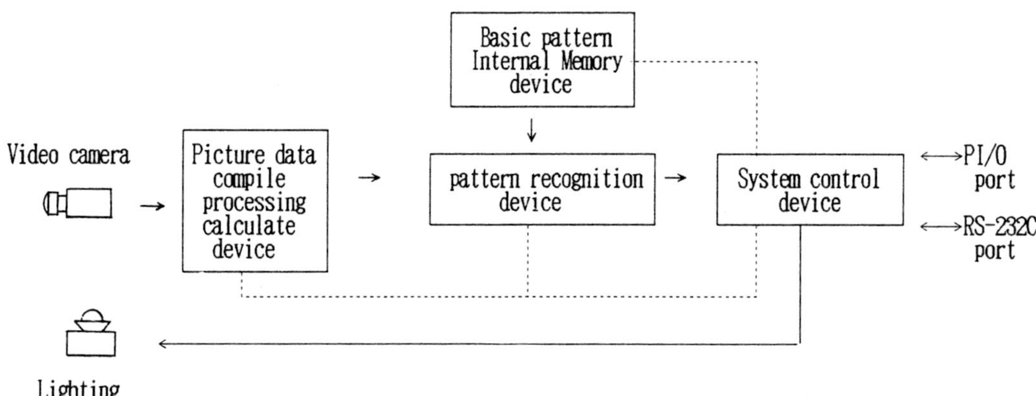

Figure 1. Composition of functional devices for visualization processing.

CONCLUSION

We have studied the necessary technologies for FMS (Flexible manufacturing system) of food processing. We then manufactured a test machine capable of grasping, transfer and positioning, and confirmed the usefulness of this system by typical food products such as tofu and kamaboko.
These technologies can be applied to most of the handling work at food production facilities which is performed manually.

ACKNOWLEDGMENTS

This study was done at research and development of Intellectualized Food Processing Technology Co.,LTD. which was invested by The Japan Key Technology Center.
We are deeply grateful to the members concerned.

REFERENCES

1. K.Tanie, T.Fukuda and N.Kitamura, Flexible handling by gripper with consideration of charactristics of objects. PROCEEDINGS OF THE IEEE INTERNATIONAL CONFERENCE ON ROBOTICS AND AUTOMATION, San Francisco, California, April 7-10, 1986

2. K.Hirota, Y.Arai and Y.Hachisu, Real time pattern recognition and its application to robot control, Hosei Univ. Techno. Report No.23,P 117

DYNAMIC MODELLING AND SIMULATION OF FOOD PROCESSES

J.J. BIMBENET, G. TRYSTRAM, A. DUQUENOY, F. COURTOIS
A. LEBERT *, M.L. LAMELOISE, F. GIROUX, M. DECLOUX
ENSIA, Food Process Engineering, Massy, France
* INRA, CRV, Theix, France

ABSTRACT

Based on the experience of the Food Process Engineering department at the ENSIA, several ways for building a dynamic model of food processes are presented and discussed. Examples of each approach are proposed.

INTRODUCTION

Food process engineering has three inherent objectives: to understand the phenomena which exist during processing, to design unit operations and to control them. Modelling offers a good opportunity to study and analyse the processes. Most of the models are steady state models, in which the time is not considered. This is not sufficient when qualities of food and heat and mass transfers are studied together. A dynamic model is often necessary because most of the phenomena are quickly performed and kinetics and transformation are carried out coupled with transport phenomena. The effects of time become important, and it is necessary to describe the phenomena versus time and operating conditions. For process control studies, dynamic modelling is also important. Chemical engineering takes dynamics into account and simulation packages were developed for dynamic simulation (Speed-up, Process, Batches, ...). In these libraries, some food processes are included, but the lack of models for physical parameters is a problem. Even if some parameters are available, variations with respect to water content, temperature and food components are not included. It is therefore necessary to develop original approaches. Different ways are available to build use dynamic models of a process. At the Food Process Engineering department, at the ENSIA, we have developed some studies in order to use modelling and simulation of food processes, specially for automatic control purposes. Several approaches were developed and this paper is a summary of these approaches illustrated with examples.

Three approaches are considered. First, the descending approach consists of the consideration of fundamentals laws. From assumptions about phenomena, the model structure is developed and validated with experiments. Different techniques are encountered. Often fundamental laws are not easy to model and then hybrid or analogue models are developed. They are simpler and easier to use with a computer. Secondly, knowledge is not always available. Then the ascending approach is possible. The model is built from data and techniques like Residence Time Distribution and identification of transfer function. Neural networks are also available for dynamic modelling. Thirdly, for process control purposes, an approach based upon the events that occur on the process is possible. Specific graphical methods like Petri nets or Sequential Functions Charts are available.

EXAMPLES

Knowledge based model; Dynamic adjustment of retort time

To establish the time-temperature requirements to achieve the sterilization of a canned product, one must run several pilot trials with variable retort times, using a trail-and-error procedure, until the desired sterilization value is obtained. The heating time to be applied at the industrial scale is deduced from the last trial. A computer programme as been developed by ENSIA and CTCPA (Technical Center for Caning of Agriculture Products) to follow the evolution of the temperature at the centre of cans placed inside a pilot retort. The interpretation of the evolution allows a real time evaluation of the heat transfer characteristics j and fh, and also heat transfer coefficients and apparent diffusivities of each can. Considering the can which appears to be the more difficult to heat up, the programme predicts its thermal behaviour during the forthcoming treatment: end of heating and also cooling. Based on a numerical simulation of the process, the prediction forecasts to know what could be the best time to stop the heating phase. Thus it is possible to proceed to trials that are very close to the optimal solution, saving time and giving a better knowledge of the variability of the sterilization value around the objective value. The programme is now being fitted into the hardware system for the temperature measurement and sterilization value evaluation of the French firm COMEUREG which will commercialize it under the name OPTIBAR.

Dynamic modelling and simulation of continuous chromatographic separations

Glucose and fructose separation from invert sugar may be achieved by adsorption chromatography on a strong cationic resin under Ca^{++} form. Since UOP established the principle of continuous chromatography in the 60s, glucose and fructose separation has been one of the most studied applications. However, optimisation of design and operating conditions, given quality and productivity requirements, as well as monitoring strategy are important questions which remain be answered. We have studied each of the phenomena responsible for this separation so as to build a predictive model of fixed-bed chomatography. Adsorption equilibrium isotherms determined by frontal analysis over a large range of concentration (0-400 g/l) were found linear and independent. Mass transfer resistance was shown to be essentially internal and to follow an homogeneous diffusion model, equivalent in the case of a linear system to a first order law with time constant td. Flow pattern in 50 p. 100 cm long columns, was well represented by the axial dispersed plug-flow model, with high peclet number. The chromatographic column transfer function could then be derived from the MCE model (Mixing Cell in series with mass Exchanges). Numerical resolution was achieved by the Fast Fourier Transform algorithm. Simulated curves were compared to experimental ones, in the case of pulse response experiments, and of elution curves; they showed satisfactory agreement. This linear fixed-bed chromatographic model has been used to simulate the performances and the dynamics of a continuous separator, the Simulated Moving bed adsorber. Both steady state and transient behaviour have been successfully investigated, for design and monitoring purposes.

Analog model; Production of yeast from whey by fermentation

Production of yeast from whey is performed on an large scale air lift fermenter. Physiological modelling of the culture is coupled with mass balance to simulate the dynamic behaviour of the fermenter. Three species are combined in the fermenter. The description of the physiological states of these species is much too difficult. Another approach is to consider an equivalent culture, which is a theoretical one, with an analogous physiological behaviour. This equivalent culture does not exist, the physiological states are functional states, and conditions of evolution from one state to another are described for the limiting factors of the culture; substrate, biomass and oxygen. From this description, a classical mass balance is developed to predict versus time the dynamic evolution of each variable of the culture. The model is based upon knowledge laws, but from a microbiological point of view, it is not a real model, because we create a species who is not a real one. The model is a description of functional states and not of real states, but this approach permits the use of fundamental transport laws for mass balances. Identification of unknown parameters is performed from literature and using optimization procedure from experimental results (error: 7%).

Compartmental model: corn dryer modelling
Corn is, in France, the second agricultural product after wheat. The objective of this work is to build up a simulation tool helpful for the design of dryers and control algorithms optimized with regards to energy, grain flow and quality. The influence of a thermal shock on wet-milling quality is modelled. The quality equation thus defined is used in a dynamic model of corn drying based on a compartmental method (two compartments) and developed from the thin layer to the industrial dryer. It permits the derivation of the characteristic equations for the dynamic behaviour of the grain under the influence of air temperature and moisture. The model, adjusted on drying kinetics under constant conditions, is used to predict the steady state of any dryer. It allows the modelling of any transient phenomena happening in industrial dryers : jumps of air temperature, drying breakdowns, cooling, condensation, air recycling . This model is also used to predict the dynamic behaviour of dryers when a disturbance is applied and thus to test the applicability of control algorithms. From the thin layer to the industrial dryer, simulations are compared to experimental results. A 5% mean error on the predicted moisture content of dried corn is obtained. The error on the wet-milling quality prediction is of the same order as the error due to the experimental procedure followed.

Identification: case of drum dryer
Automatic control science proposes different ways for dynamic modelling. A very simple structure is chosen and parameters are calculated from experiments. These experimental approaches are very useful, and many software packages exist which have identification algorithms. Such an approach is used for dynamic modelling of drum drying . Drum drying is a complicated process in which moisture content of dried product depends on steam pressure inside the drum, and drum velocity. Because the measurement of moisture content is established (using a near infrared on-line analyser) it becomes possible to perform identification. The model chosen is a recurrent one, moisture is modelled as a function of moisture at previous sampling times and drum velocity at present and previous time. It is the equivalent of a second order transfer function. This dynamic model is used to determine the design of a moisture controller. This technique is a quite simple one, the model is linear and is applicable in the experimental range used.

Neural networks applications
From 1985, artificial intelligence proposes new approaches for modelling. Use of neural networks is one such technique. Because of its structure, a neural network is able to take into account non linearities which are one of the characteristics of food processes. Several applications are performed in our research department for product formulation, drying or microfiltration. For example, for drying of solids with hot air a comparison between classical approaches and neural networks is made. Characteristic drying curves are obtained on an experimental basis, moisture evolution through time is observed depending on operating conditions like air velocity, moisture and temperature. For corn drying, previous work proposes different approaches. A neural network is built. Inputs are the operating variables and time. Output is only the moisture of dried corn. Two hidden layers are used with respectively 3 and 4 cells. Four experimental curves are used for learning in the neural network, and five curves for validation of the model. The network gives the same results as previous classical methods. Error is small; 5% maximum. Neural networks are a good opportunity when no previous knowledge about the structure of the model is available.

CONCLUSION

Differents ways are studied for dynamic modelling of food processes. Many different methods exist. The design of a dynamic model for food processes becomes, specially when real time measurement are available. Thus it is possible to study coupled phenomena in process control applications such as, for example, transport and reaction phenomena.

Neural Network control for extrusion cooking

K. Uemura, S. Isobe, A. Noguchi.
National Food Research Institute,
Tsukuba, Japan

ABSTRACT

Neural Network (N.N. for short) controller was designed and examined for extrusion cooking. The feed–forward and feed–back loops of N.N. were found to simulate the extrusion cooking. The control was realized by two kinds of N.N. that minimized the output deviation from the target value. The dynamic controlling can be achieved in the line of quick learning, feed–back–calculation of N.N. and renewed teaching data. N.N. will open the way for the AI control of extrusion cooking.

INTRODUCTION

The Extrusion cooking is one of the most versatile food process owing to its capability to convey, mix, homogenize, cook, gelatinize and denature of the food material, and to perform many other conventional food manufacturing operations and even biochemical and chemical processes. Nevertheless, the automation and control of extrusion cooker faces several difficulties because of the complex and unclear interaction between the operational parameters. Neural Networks(N.N. for short) are one of the Artificial Intelligences. The new control theory is applied all the controller s of home electrical equipments, because of the novel, the cost, the power, and the obscure. N.N. has strong points that are automatic designing and tuning. Then the switch over from the Fuzzy to N.N. is carried out. The practical aim of the present work was to apply N.N. to control of an extrusion cooking, and thus reduce the problems associated with the construction of fuzzy control systems.

The design of the Controller

Fig.1 represents an articulated extrusion cooking as a black–box which has three input(control) ports and four output(sensing) ports. The Figure illustrates the same

input/output set as N.N. with 10 neurons in the hidden layer. The example product investigated was flat bread produced with a Clextral BC 21 twin–screw extruder. The chosen input variables were mass moisture content, mass feed rate, and screw speed, and the output variables product expansion ratio, main motor current, pressure at the die outlet, and product bulk density. The neurons in the input layer transfer the input values within a range of [0, 1] to the appropriate neurons in the hidden layer. Teaching data for the N.N. consisted of 15 experimentally obtained data sets. Both the input and the output data were normalized between the range of values [0, 1].

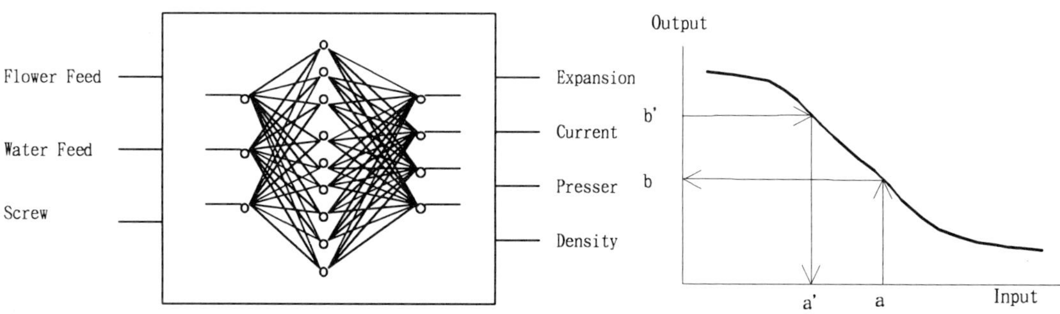

Fig.1 Block diagram of N.N. for Extruder Fig.2 Principle of the reverse calculation

Theoretical analysis of N.N. controller

we can calculate with the same routine. Fig.2 shows the principle of the reverse. If the state of present stay at the point(a), then the output is (b). Then if you want to increase the output (b) to (b'), following the response curve, you can get the renewal input data(a'). The result become as the renewal control factors.

Result

After 1000 times learning by B.P, the mean square difference decrease less than 1.0. The results of learning were illustrated by 3–dimensional contour graphs as shown in Fig. 5. In each graph the X–, Y– and Z–axes represented mass feed rate, feed moisture, and expansion ratio, respectively, with one of the input variables, in this case screw speed, kept constant at various levels. Whenever you set control condition to input ports of the N.N., after feed forward calculation, the result of the Extrusion cooking is appeared on output ports. In other words, the calculation realizes the simulation of the Extrusion cooking between the input and the output. The simulation has following tree merits. The first one is a clearing the operation. The second one is working well in the training for beginner operators. The last one is good for improvement of the controller. When the controller running and learning in line at the same time, we must decide to add ,change or ignore the new condition to the teachers sets and gradually improve the controller as follows.

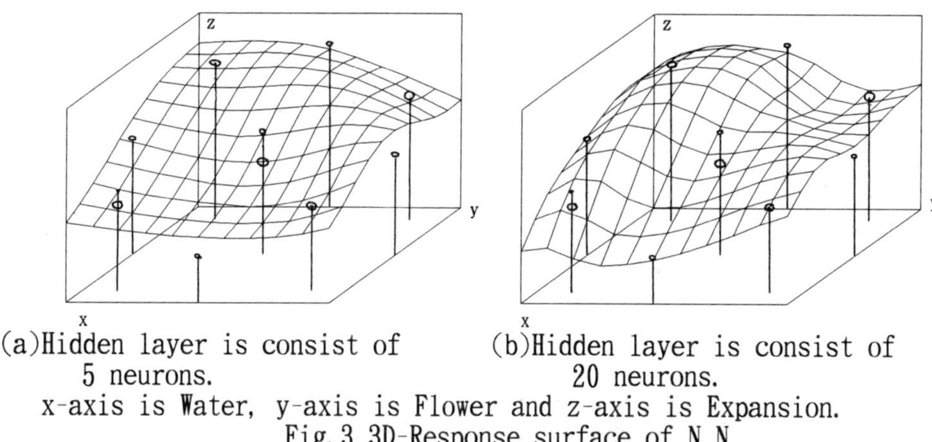

(a)Hidden layer is consist of 5 neurons. (b)Hidden layer is consist of 20 neurons.
x-axis is Water, y-axis is Flower and z-axis is Expansion.
Fig. 3 3D-Response surface of N. N.

Multi Reverse calculation

If each input values are set on the middle point of a preset range, the expansion ratio for example will be about 4.65 times. If a more expanded product would be desired, then the reverse calculation, the controller will fined out the revised input set. In this case, we can get the set of input parameters that are related to mark 4.99 times expansion. But, the expansion ratio was never global maximum point but local maximum point. We used complexed N.N. for avoidance the local maximum. Fig.3(a) and (b) shows the response surface that are build up by few–hidden layers node and large nodes respectively. The more the number increase, the more unevenness the response surface change. When the reverse calculation is executed with few nodes N.N. in the above example, the result is improved until the expansion become 5.52times. After that, next reverse calculation that uses large number nodes N.N. is started from the renewed point. After all, we get a set of input parameters that is related to 5.65times expansion.

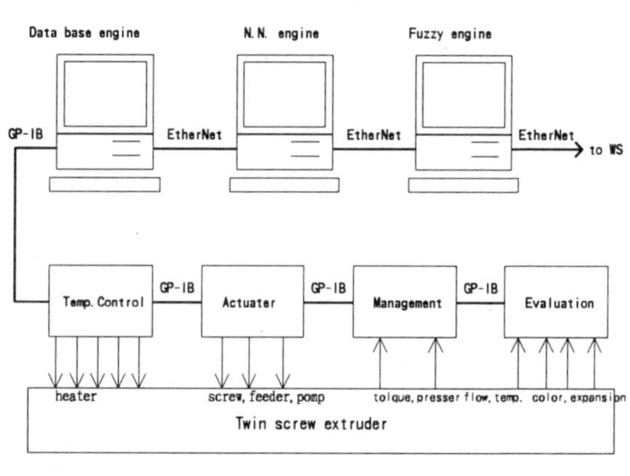

Fig. 4 Dispersed N. N. control system.

Conclusion

A neural network was constructed which could efficiently learn from prior extrusion cooking data sets. The complexed reverse calculation of the N.N. makes it possible to simulate and control the extruder simultaneously. We are currently working on the build up the dispersed AI controller system (show in Fig.4) that are consists of Database, more than one N.N. engine and Fuzzy engine. While they work independently, they work in cooperation with each other.

REFERENCES

1. Linko, P., Uemura, K., Y.–H. Zhu and Eerikainen, T., Application of network models in fuzzy extrusion control. Trans IChemE, vol70, PartC, Sept. 1992, pp. 131–137.

FUZZY TECHNIQUES FOR PROCESS STATE ESTIMATION

VALERIE J. DAVIDSON, RALPH B. BROWN and GORDON L. HAYWARD
School of Engineering, University of Guelph
Guelph, Ontario N1G 2W1 Canada

INTRODUCTION

For most food and biological processes, control information is available in three forms: continuous numerical data from sensors (e.g., temperatures, flowrates), discretely sampled data from the quality control laboratory or at-line analysis, and the operators' intermittent observations of process conditions and product characteristics. Operators usually express their observations in linguistic terms, frequently using words that are unique to the process, and which conventional automatic control systems cannot use as inputs. However, fuzzy set theory allows quantification of linguistic descriptors and makes it possible to integrate the operators' observations into the overall control strategy. In fact, a fuzzy logic/expert system controller allows all types of inputs to be processed. In this sense, working in the fuzzy domain is somewhat analogous to using the Laplace transform in conventional control. In the fuzzy domain all arithmetic and logical operations can be performed on mixed input types, and control rules that would make no sense in the traditional crisp numeric domain can be applied. For example, inputs like "temperature is high" and "exposure time is long" may be combined to yield the consequent "product is overheated". The resulting control decision, in this case "increase the product flowrate" must then be defuzzified (i.e., inverse transformation) to produce a numerical value for the setpoint change in flowrate.

Our objective is to develop computer-assisted control systems for processes that integrate human judgement values with electronic sensor inputs. Operators are an important link in the control strategy because their observations and interpretations are critical, and frequently cannot be replaced with on-line instrumentation. Computer-assisted control enhances the operator's decision-making process. In particular, the benefits are: recorded process history, improved forecasting, more consistent control actions among operators, and more flexibility for accommodating different products with the same process equipment. Two examples are used to demonstrate these concepts.

MOISTURE CONTROL SYSTEM FOR A CONTINUOUS DRYER

This system was developed to improve the current statistical process control (SPC) for a dryer on an industrial bakery line and to integrate the monitoring aspects of SPC with process knowledge that could assist the supervisor in making control decisions. The current quality control practice on this bakery line is to sample the product at the point of packaging, and to analyse the moisture content at-line. Moisture values are plotted on a control chart which shows the target moisture content (5.5 %) as well as the acceptable limits (4 to 7 %). Simple rules are stated on the control chart to guide the operator in making decisions about the process state, however some rules are vague and the operators are frequently so busy that they only notice results that are clearly unacceptable. The SPC protocol is simply a monitoring function. It is a tool to alert the operator to the need for control action but can not provide any recommendations as to what change is appropriate. Not surprisingly, control actions are not consistent among operators.

To be consistent with the SPC model, the numeric moisture values were transformed into three variables: an error value (difference between target and actual moisture), a variance estimate (the squared error for the last four observations) and an estimate of the derivative error (a linear regression through the last four observations). The control logic was expressed in the form of production rules. Two rule bases were compared. The first set of rules were based on proportional control action. The second rule set incorporated a derivative action. The manipulated variable was the setpoint for dryer air temperature and in the initial prototype equal changes are made across all three drying zones.

In the first rule base, the control action was proportional but it was moderated by the variance estimate. The measured variable (moisture content) was noisy. In our fuzzy controller, proportional control action was recommended only if the moving variance estimate became "medium" or "large". As the variance estimate increased, the gain of the controller also increased. This control strategy has been tested during actual production on the bakery line. Figure 1 shows the recommendations of the fuzzy production rules as well as the operators' actions during a ten hour operation. Early in the test, the operator made a setpoint change that was incorrect. The fuzzy rules catch this problem at the next observation and continue to recommend lower dryer temperature until the operator makes setpoint changes about half-way through the test. After these changes, the average moisture content moves much closer to the target value for the remaining 5 hours of the test.

A discrete proportional control algorithm was also defined based on the numeric value of error. The controller gain ($K_c = 3.0$) was the average of the fuzzy gains for medium variance (2.0) and large variance (4.0). A deadband of +/- 0.3 % error was also included to be consistent with the fuzzy value of zero error. As shown in Figure 1, this controller recommends frequent control actions, even during the last half of the test.

CONTROL SYSTEM FOR A BATCH-COOKING PROCESS

The control strategy for a batch process is quite different from that for continuous processes because the batch process is never at steady state. Key process variables are monitored to determine if the process trajectory is normal or abnormal. If it is abnormal, it is desirable to make compensating changes as early as possible so that the desired product characteristics are achieved at the end of the batch. If abnormal process dynamics can not be detected and corrected, there will be undesirable batch-to-batch variation and possibly batches of product that are completely unacceptable.

A control strategy was developed for a batch-cooking process in a pilot-scale smokehouse. The process was typical in that heat-transfer conditions were not uniform over the oven volume. This resulted in variable product heating rates with a range of normal

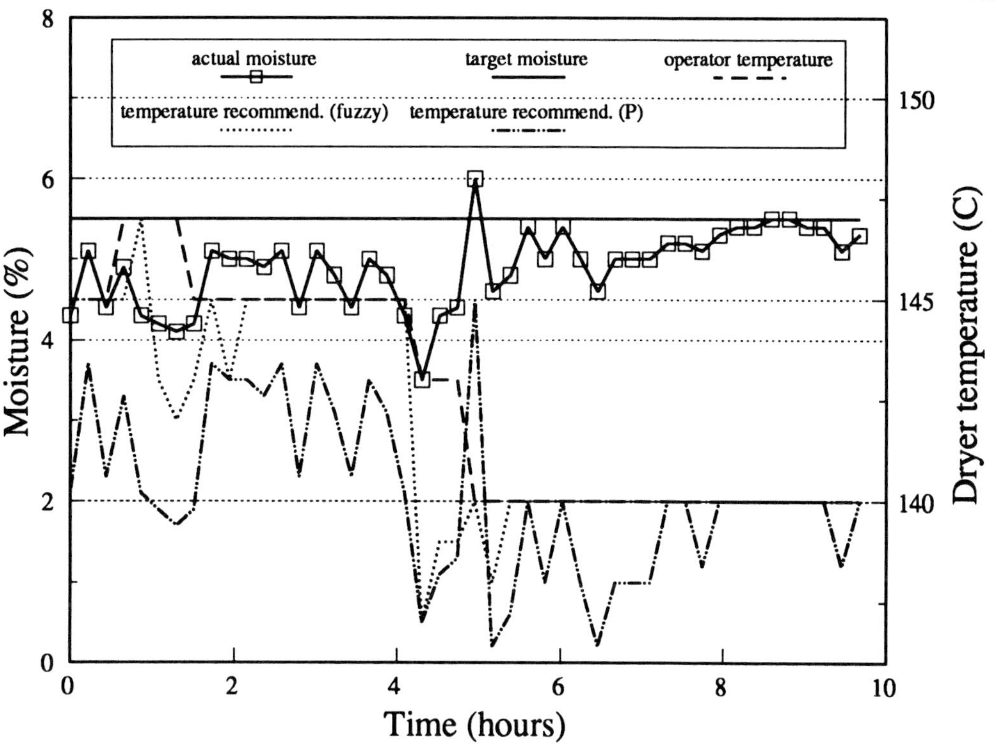

Figure 1: Continuous dryer production test

process trajectories instead of a single heating curve. The product heating rate (at the centre) was described by first-order transfer functions. Two time constants were defined at each fan speed: τ_{min} (fastest heating rate) and τ_{max} (slowest heating rate). Time constants were determined empirically over a range of typical operating conditions. During cooking, internal temperatures were monitored at several locations. The transfer function was used to estimate normal temperature limits (based on τ_{min} and τ_{max}) for the current process time. If the measured temperatures were within these limits, the fuzzy estimator predicted the remaining process time to ensure a minimum thermal exposure. A set of rules predicted the final product quality (moisture loss and texture) based on the estimated range of temperature-time trajectories. If the quality estimates were unacceptable, the control system suggested remedial actions (changes in air temperature and/or fan speed). The control system also recommended changes if the measured temperature(s) are outside the normal limits. The operator could accept or ignore the recommendations of the control system.

This control system integrates conventional control techniques with fuzzy mathematics. The process model is in a transfer-function form which explicitly defines the effects of air temperature and fan speed on heating rates, as well as the product temperature-time histories. This model is fuzzy because, at each operating condition, a range of time constants defines the possible internal temperatures. The second fuzzy component of the control system is the prediction of product quality. Fuzzy variables are convenient for describing food quality attributes in linguistic terms. Fuzzy rules are used to make inferences about the impact of process changes on product quality. In these inferences, a rule-based model is acceptable and more tractable than equation-based models.

JIT AND CIM CONCEPTS, APPLIED TO A LARGE SCALE CATERING PRODUCTION PLANT

TOON MARTENS
ALMA University Restaurants, Catholic University of Leuven
E. Van Evenstraat 2 C, B-3000 Leuven, BELGIUM

ABSTRACT

Japanese assembly industry has shown that quality, cost and market flexibility can be improved by applying the JIT-philosophy (material flow) and integration of computers, to speed up information processing from machine control to strategic planning. The Catholic University of Leuven has build a new production plant for "sous-vide" meal components, that tries to integrate the newest ideas in production management and food technology.
Although the general concepts of JIT and CIM were usefull, the implementation was very difficult. This was due to the biological variation of ingredients, the wide variety of products, the impossibility to automate certain tasks like tasting and lack of standardisation of hardware and software.
This experience was the basis to setup new research programs on computer aided design, computer aided quality control and intelligent equipment.

INTRODUCTION

Because of the high investment and operational cost of a new central kitchen that had to replace the classical warm kitchen, a study was done to identify the consumer needs and the technologies that can meet this requirements. We found that the product design criteria that were identified by different authors [1][2] were also true for the students and the personnel of our university. The major concern was the **variety**. More than 70% of the students want each day more than 5 menu items, about 20% would like to have more than 10 menus. The second concern was the **quality**. Sensory quality and freshness were the most important criteria for buying a product, especially the quality of vegetables was very important. Last but not least, the **price** should be as low as possible. We conclude that the new production should produce a wide variety of high quality products, free of additives, fresh and with a low cost. For quality reasons we became interested in the **"sous-vide"** technique, that has been applied by many French and Belgian chefs to improve quality in high class retaurants. We made some small scale experiments and start thinking how we could apply this technique on an industrial scale. We started discussions with researchers of the department of production management about layout, automation and logistics.

JUST IN TIME

Just In Time is not a well defined method or technique but a way of thinking. It is a set of ideas that are usually described with a number of zero's and one 1 : **"zero defects, zero setup-time, zero breakdown, zero inventory, zero lead time and a lot size of 1".**
In a JIT-approach all kinds of waste are eliminated and special attention is given to factors that influence market flexibility and/or productivity. Market flexibility is extreme in the catering business : diversity of products and fluctuations in numbers are very high in comparaison with other sectors. Another characteristic is that the catering business is very labor intensive and that productivity is underdevelopped due to the lack of planning and automation. For this reason JIT is very well suited for the catering industry. We found that the following points were very important in the implementation of JIT :

a) order and cleanliness
There is a direct relationship between the quality of the environment and the quality of the products. For this reason we decided to build a new plant instead of renovation. No intermediate stocks were allowed at different workplaces.
b) zero defects
In a JIT quality program the main attention is not focused on the output of the production but on product and process design. Quality problems should be avoided, this is certainly true for microbial problems. We stopped control of end products and started with HACCP-analysis and computer aided process design. Quality is in hands of the operator, who is also responsible for preventive maintenance and cleanliness.
c) uniform production load
JIT tries to synchronize the rhythms of production with the rhythm of consumers needs. The "sous-vide" system allows to level the peaks. JIT environment tries to make the most out of the available people and not of the machines. For this reason the personnel has to be multi-skilled and automation will help them :"computer aided cooking".
d) layout
The layout of the building has been studied with different computer layout-programs (CRAFT) to obtain a fast product flow, minimal space and minimal movements of materials. Material handling of the finished goods has been automated to guarantuee FIFO and minimize the work in cold storage.
e) reduction of setup times
For an operator it is impossible to remember all the settings of the machines for more than 1000 recipes. Also for safety reasons time-temperature of cooking is controlled by a process computer.
f) production control
To deliver just-in-time a good planning system is needed that is also very flexible because quantities are changing until the last minute. To optimize the usage of people and machines a scheduling software has been written to plan the batches in the right sequence.
g) network of suppliers
Suppliers have to be selected that can deliver just-in-time but also can guarantee the microbial and quality norms.

COMPUTER INTEGRATED MANUFACTURING

CIM integrates the information flow between the different departments of a factory : engineering and product design, marketing and sales, purchasing, production planning and control, maintenance [3].

a) computer aided process design
ALMA develops an expert system software for computer aided process design of minimally processed foods as a part of an EC-research program (FLAIR AGRF 0047) The effects of the

changes in product formulation, process parameters (time, temperature,...) the microbial safety and texture can be easily evaluated.

b) production planning and control

A Materials Requirement Planning software has been selected and adapted to our needs. The main problems were the variability in yield of the raw materials, frequent changes in the planning, substitution of ingredients depending on price.

A scheduling software has been written that tries to produce as much orders as possible within the shelf-life limits and capacity limits. The sequence of the batches is optimised in order to optimise the usage of the packaging machine, the bottleneck of the production.

c) computer aided manufacturing

To automate recipe handling a batch control software has been selected and adapted (FERRANTI PMS). The machine steering was implemented on a PLC (SIEMENS). Since no equipment was available that could be integrated in the batch control system, we had to program the PLC ourselves. The pyramid of information processing has 3 layers : PLC, real-time process computer and a UNIX-system at the highest level.

d) computer aided inventory control

Every batch has to be split in a different number of packs for each restaurant. Because of the high diversity, collection of all products for a restaurant is very time consuming. Also for safety reasons it is important that residence time in central storage can be controlled and FIFO is guaranteed.

CONCLUSIONS

An attempt was made to implement the concepts of JIT and CIM as far as possible. The installation is now running in full production for one academic year and we could achieve a substantial reduction of labour cost (ca. 50%). The changes in the recipes for the "sous-vide" system were very consuming. Also the lack of standardisation in the way cooks prepare a recipe caused many discussions. Software and hardware for recipe handling are not flexible and standardised. No equipment was available that could be interfaced very easily. We found that there is a lot of pep talk on JIT and CIM and that a lot of development is needed to make this JIT and CIM-concepts working in the food industry.

This study is in part supported by the European Commission, Food Linked Agro-Industrial Research Programme.

REFERENCES

[1] Fox, R., Plastic Packaging. The consumer preference of tomorrow. Food Technology 43 (12), 84, 1989

[2] Keuning, R., Food ingredients for the 90's. Chapter 8 in Food for the 90's edited by G.G. Birch, G. Campbell-Platt and M.G. Lindley, Elsevier Applied Sciences, London 1990.

[3] Scheer, A., CIM Computer Steered Industry, Springer Verlag, 1988

DEVELOPMENT OF NOVEL AGITATING DEVICE FOR BIOREACTORS

K.Maruyama,M.Ohmi,M.Imai,S.Urushiyama Department of Chemical Engneering, Division of Chemical & Biological Science, The Tokyo University of Agriculture and Technology. 2-24-16,Nagamachi, Koganei-C,Tokyo 184 ,Japan.

ABSTRACT

A novel agitating device called the "Cross Flow Agitator" was developed for bioreactors.The change of flow direction with flap angles were observed.By change of flap angles could successfully govern the flow direction even if the revolution direction of the cylinder is the same. Such flow direction inversion character is desirable to induce good mixing and circulation of multiphase fluid in bioreactors. The one dimensional energy spectra in the vessel were measured.The energy dissipation in the vessel was not localized compared with the case of turbine impeller.

INTRODUCTION

Development of a novel agitating device for bioreactors is important for the advancement of biochemical process systems. Agitation is effective for oxygen solubilization[1)] and homogenization of a culture medium , and it has been adopted in various bioreactors for culture of animal and plant cells and for enzyme reactions. Agitation, however, involves some practical issues such as damage of cultured cells due to fluid turbulence[2)] and formation of stagnant space in reactors. In this paper,we propose a novel agitating device "cross flow agitator" for bioreactors. Simple manner to change of circulation flow nearby the cross flow agitator and energy dissipation in the agitating vessel were investigated experimentally.

EXPERIMENTAL

Flow direction in the vessel

The schema of the experimental apparatus is shown in Figure 1. The agitating device was composed of the cylinder of cross flow agitator and two flaps. The cross flow agitator was composed of fifteen small wings. The cylinder of cross flow agitator was installed horizontally in the vessel. The diameter and axial length of the cross flow agitator were 100 mm and 296 mm respectively. The dimension of the agitating vessel was 224 mm(W), 230 mm(H) and 347 mm(L). The change of flow direction was seen using tracer particles which were photographed with a camera.

Measurement of one-dimensional energy spectra

A schematic diagram of the experimental apparatus is shown in Figure 2. The diameter and axial length of the cross flow agitator were 100 mm and 130 mm respectively in the vessel. When energy spectra was measured, the cross flow agitator was set up vertically, since accurate power consumption in the vessel could be measured simultaneously. The flow rate was detected by a probe electrode under the diffusion-controlling conditions. An autocorrelation function was calculated based on change of flow rate.The energy spectra were calculated with a Fourier transform of autocorrelation function. For comparison the case of a six-blade turbine impeller, the power consumption of the cross flow agitator was adjusted to be equal to that of the turbine impeller. The energy spectra were measured both in the axial and radial directions.

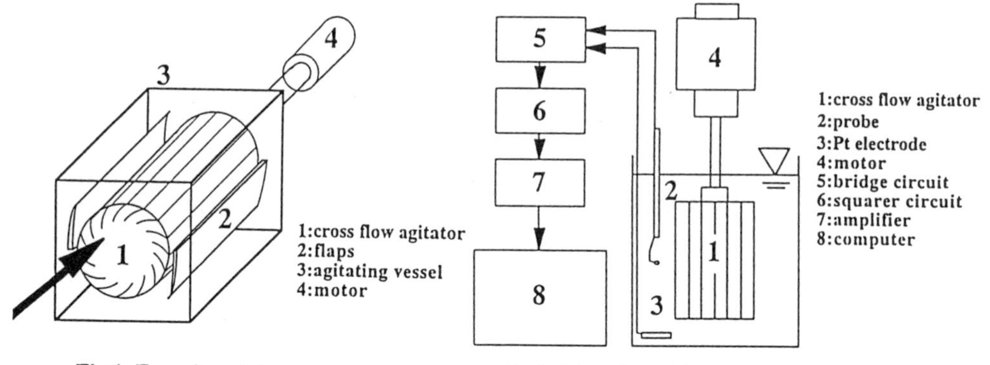

Fig.1 Experimental apparatus
The flow direction was observed
in the direction of the arrow

Fig.2 Experimental apparatus

RESULTS AND DISCUSSIONS

Flow direction in the vessel

Flow direction in the vessel using tracer particle are shown in Figure 3. The key (0°, 40°) express the flap angle of cross flow agitator as shown in this Figure 3.the tracer particles were introduced at the particular flank of the cylinder and exhausted out from the cylinder. According the loci of particles, the fluid was introduced into the cylinder along the small wings of cylinder. From this motion of particles, shear stresses during agitation will be reduced.The flow direction nearby the cross flow agitator was changed drastically with flap angles even if the direction of cylinder is the same. This change in flow direction

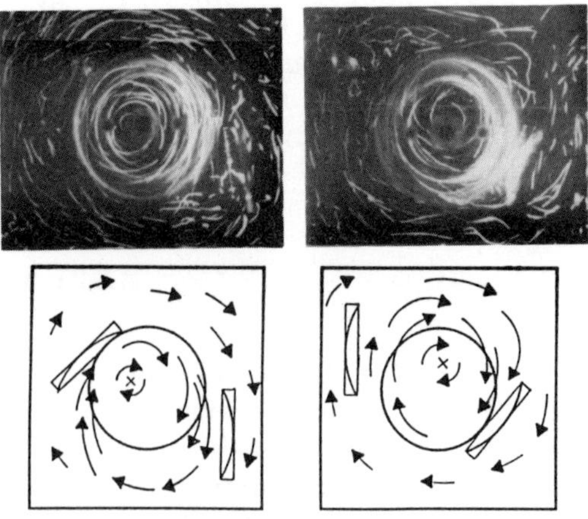

Fig.3 Loci of tracer particles in the vessel.
Flap angles (0°, 40°) , (40°, 0°)

is desirable to suppress the formation of stagnant space in the vessel.

Energy spectrum

The spectral distribution in the case of impeller used, presented in Figure 4, apparently changed in the axial direction. On the other hand, the distribution of the cross flow agitator slightly changed in the axial direction presented in Figure 5. In accordance with these results, the large energy dissipation in the local region did not occur in the agitating vessel , especially near the cross flow agitator.

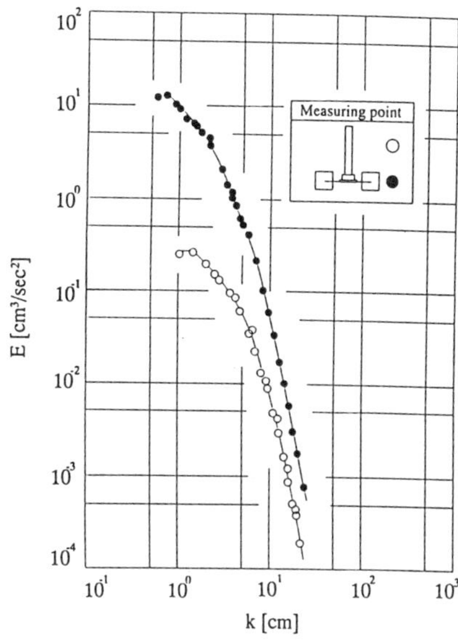

Fig.3 Energy spectra in turbine impeller

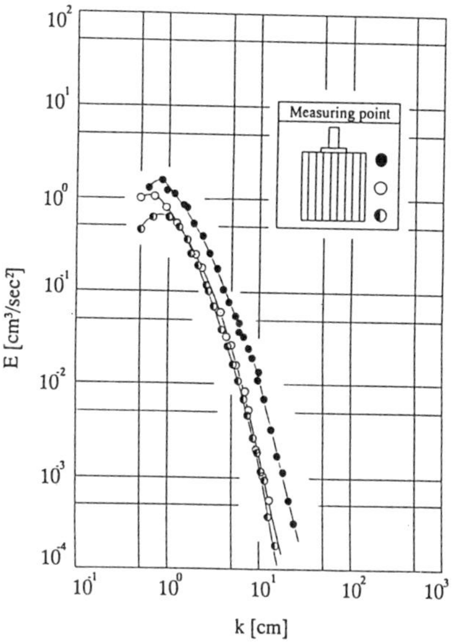

Fig.4 Energy spectra in cross flow agitator

CONCLUSIONS

A novel agitating device "cross flow agitator" was demonstrated. The shift of flow direction and the one-dimensional energy spectra in the agitating vessel were presented. The direction of flow near the cross flow agitator was changed by the flap angles, and this appears to suppress the formation of stagnant space. The energy dissipation in the vessel was not localized as with the turbine impeller. These characteristics are expected to reduce the biological damage due to shear stresses during agitation. The cross flow agitator shows the desirable characteristics for a practical bioprocess agitating device.

REFERENCES

1 Asai,T. and T.Kono ,Estimation of oxygen absorption coefficient and power consumption in a stirred tank fermenter. J.Ferment.Technol.,1982,60,265-268

2 Kawase,Y.,Design for bioreactors. Chem.Eng.(in Japanese),1987,7,646-650

TOTAL ENERGY MANAGEMENT SYSTEM (TEMS) AT FOODSTAFF FACTORY MODERATED ENVIRONMENTAL IMPACT PRODUCTION SYSTEM

SHIGERU SAKASHITA
Deputy Manager/Energy Management Block
Mayekawa Mfg.Co.,Ltd.
13-1,Botan 2, Koto-ku, Tokyo, Japan

ABSTRACT

In modern society, energy saving is an important theme in the operation of any production facility, as is the importance of limiting environmental pollution caused by release of CO_2, NOx and SOx into the atmosphere. Maximizing utilization of primary energy is a key solution to the above problems. Effective utilization of primary energy at a factory begins with the introduction energy-saving production engineering, for example,

- Saving input energy by recovering waste heat,
- Improving energy utilization efficiency by introducing co-generation
- Storing energy by some means such as thermal storage

All the above require instrumentation systems that assure that the hard ware used functions effectively. The above engineering is consistent not only with the energy saving goal but also the economic objective of a factory, that is ,to minimize the cost of energy used in production. The development of TEMS (Total Energy Management System)has been based on full consideration of the above and fully operational systems are actually in use at a beer brewery , a sake brewer and a ham production plant. The following can be mentioned as common characteristics of the above foodstuffs plant:

- Imbalance in daytime and nighttime energy load,
- Use of various energy sources such as electric power, steam, warm water and cold water(cold heat),
- High sensitivity to quality control aspects such as taste and sanitation.

The case of factories characterized by the above, the management of energy flow and storage using combinations of energy saving devices closely connected to the production process and the use of thermal storage units and computerized instrumentation systems greatly enhance efficiency.

What is TEMS

TEMS is the abbreviation for "Total Energy Management System. "As the name implies, it is a system aimed at achieving comprehensive and effective utilization of energy.
We call the system "comprehensive" because it covers the entire factory and also the management of various utilities such as electricity, steam, warm water and cold heat. As well, TEMS also involves the total flow of production, transport, storage, consumption and recovery of energy. TEMS save energy, recover waste heat and stock and or discharge utilities based on estimated energy consumption levels. Such management is achieved by of AI(artificial intelligence)and a heat accumulator.

Basic Policy of TEMS

Basically, TEMS consists of the following three factors:

1) Bringing a factory close to an a closed energy system by recovering and reusing waste heat.
2) Expending effective utilization of primary energy through co-generation.
3) Shifting and effectively utilizing energy and equipment through use of a heat accumulator and AI.

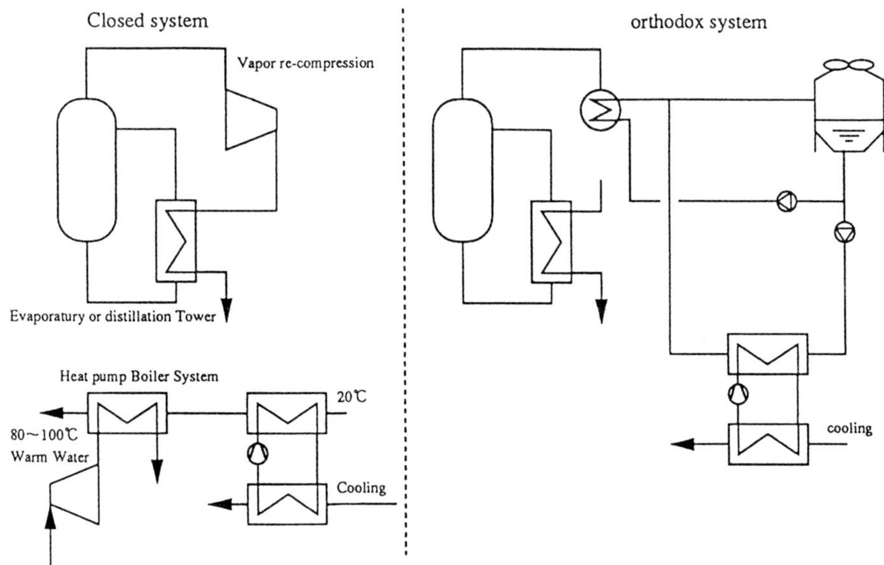

Figure 1. Closed energy System

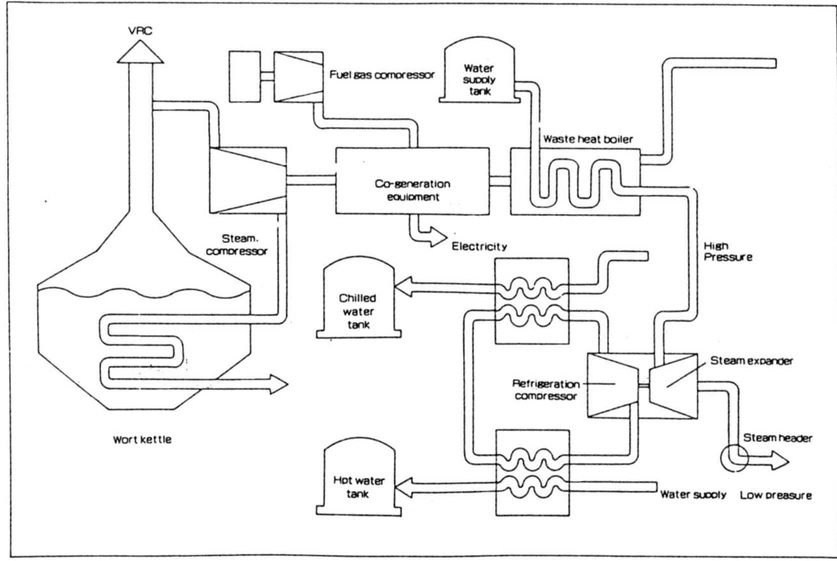

Figure 2. TEMS for Brewery

TEMS Instrumentation and Control System

Normally, TEMS uses the control system shown in Figure 3.
The Process control computer and the equipment sequence controller are independently mounted. Here, process control based on production management as well as sequence control and loop control based on feedback are carried out.
The system is basically a local control structure.
TEMS is controlled by a work station level computer. Optimum operation is determined upon receipt of production planning data and process information input into the process computer through signal communications. The value settings are then dispatched to the sequence process controller. The set values are renewed every minute to one hour.
The TEMS control system has the following four functions:

1) Utility load forecasting(energy demand forecast) ,
2) Optimum operation control,
3) Abnormality diagnosis,
4) Statistical analysis.

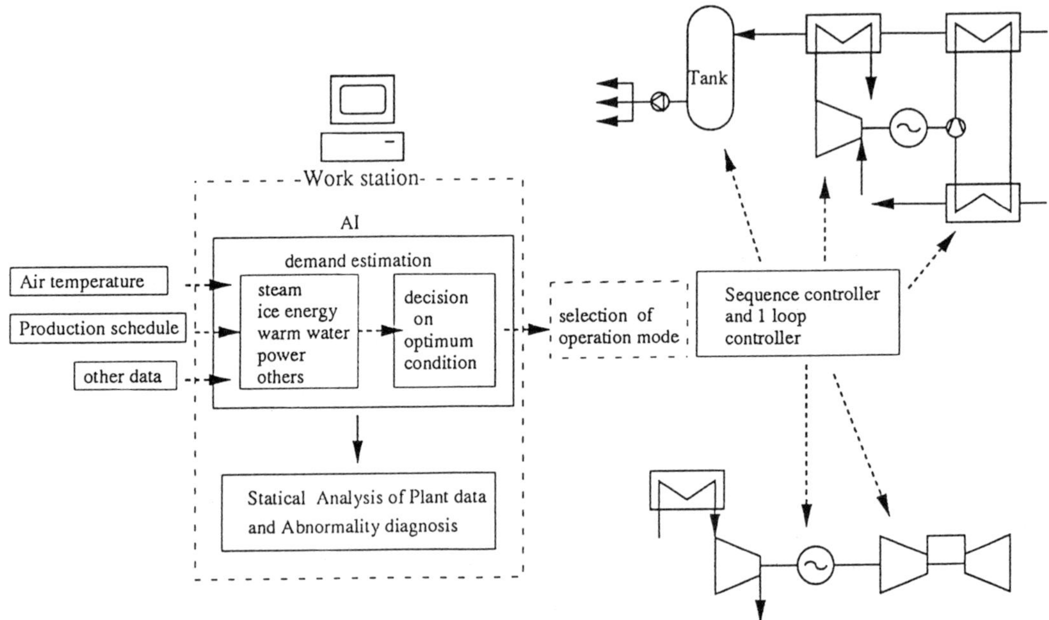

Figure 3. TEMS Control System

Conclusion

The total energy management system introduced here with first developed in 1987. Its high energy efficiency is unparalleled. The TEMS controlling computer system serves an important role in shifting energy. Problems do remain which must be surmounted, the most important being man/machine type problems.
Since the ability of the computer to communicate details of optimization and abnormality diagnosis to human beings is insufficient at present, the result is those high performance computer functions as a sort of 'black box.' The ironic result of this is that it may fall to attract the interest of human beings.
At present, the main development theme is improvement of the man/machine system to a capability level that takes advantage of the expressive power of the computer.
In closing may say that if this paper provides the reader with any useful advice, I would be most pleased.

A BLACKBOARD AND OBJECT ORIENTED FERMENTATION PLANT MODEL

VIRGINIE NIVIERE
Imeca, BP 94, 34800, Clermont-L'Hérault, France
PIERRE GRENIER, JEAN-MICHEL ROGER, FRANCIS SEVILA
Cemagref, BP 5095, 34033, Montpellier cedex 1, France
MORAD OUSSALAH
Eerie/Leri, parc scientifique Georges Besse, 30000, Nîmes, France

ABSTRACT

A model of simulation of wineries has been proposed. Each elementary set of knowledge used in the model has been defined as an entity gathering constant data, variable parameters, and expert calculation functions. The model has been structured hierarchically, and the interest of blackboards has been discussed. It has been implemented in object oriented programming and the language C++ has been chosen. This expert simulation software produces dynamic balances of resources.

RATIONALE

During the period of grapepicking, Enologists have to manage resources as varied as humam power, grapes, process equipments, and refrigeration power. Refrigerating grapes is a necessity in elaboration of most quality wines. A simple idea for optimising the use of these resources is the dynamic simulation of a winery operation. We have developed a knowledge based system with a hierarchical model of objects for simulation of wineries operation [1, 2].

RESULTS

Design of the model

Modeling: Wine-making follows one or several process lines, each one being a succession of operations such as, for instance, grapes reception, pressing, thermal transfer, settling, centrifugation, maceration, fermentation, or storage. Each operation gathers a set of units making the same kind of transformation on the materials. We have imagined a hierarchy between units, operations, and process lines, with respectively three levels of abstraction. At the first abstraction level, a set of knowledge has been defined with six objects, respectively Process-Unit, Input-Flux, Output-Flux, Material (processed by the unit), Phenomenon (taking place in the unit), and Equipment (container of the unit: a tank, an exchanger, ...). At the second level of abstraction, we have defined the objects Operation, Operation-Input-Flux and Operation-Output-Flux. At the third level of abstraction, we have the Process-line, containing a list of operations.

Dynamic simulation: The knowledge described at the first level of the hierarchical model is a discret event system. A continuous system may have the activities "On" and "Off". A discontinuous system may have the activities "Standby", "Refilling", "Storage", "Draining". The logics of state variables evolution of the objects of a system rely on its activity regulated by a set of functions.

Blackboard: for further extension of our knowledge based system, we have designed a blackboard architecture. One knowledge source, which contains the hierarchical model already implemented writes lists of tasks to achieve on an object called the Blackboard , and another knowledge source at present time under development, will work on these lists.

Validation

At Saint-Geniès-les-Mourgues (France), we have studied a process line for red winemaking where the alcoholic fermentation is split in two operations, maceration without temperature control, and fermentation in liquid phase with temperature control.

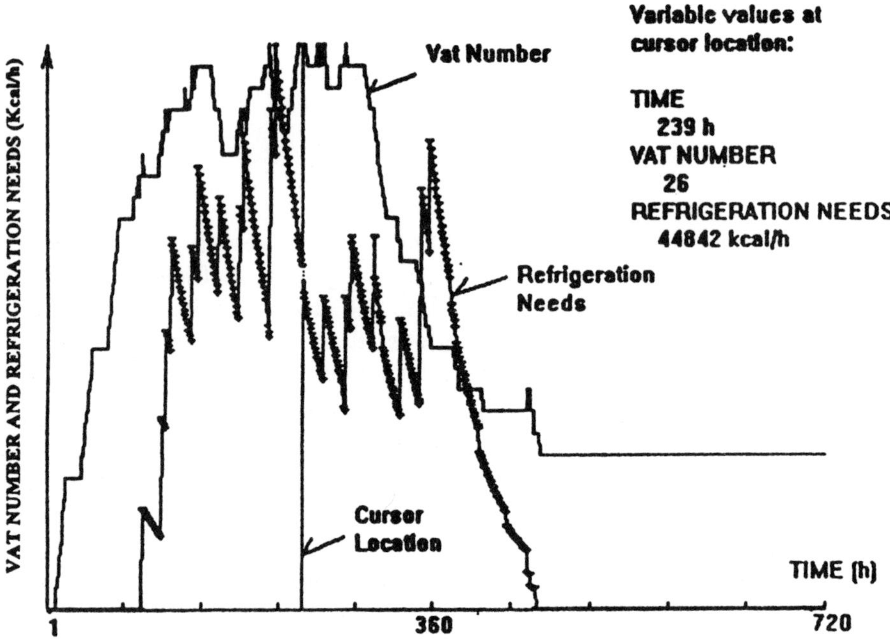

Figure 1. Example of computer output. Simulation of vats and refrigeration needs at the Cooperative winery of Saint-Geniès-les-Mourgues (France) during the 1991 harvest.

We have estimated the fluxes of grapes and musts through the various operations and units, and we have compared the computer simulation to these estimations. We could observe that the material fluxes were well simulated: the number of vats simulated was equal to the observed number by one unit, and this difference was due to the assumption that maceration was three days long when actually it was fluctuating. An example of computer output is presented on Figure 1.

PERSPECTIVES

We want to simulate more precisely the planification of the tasks. We can make the knowledge based simulation work by cycles of 24 hours. While the simulation is run a first time for a given day, the list of tasks to be achieved on this day is written and executed on a provisional basis. Then this list of tasks can be ordered, the simulation run for the second time, and the modified list of tasks executed. Tasks planification is a problem of Flexible Constraints Solving, and adequate theories are applied [3].

CONCLUSIONS

Dynamic simulation of equipments and refrigeration power requirements in wineries makes an excellent optimisation tool in the winery. Flexible Constraints Solving associated with fuzzy logics will permit in a near future to improve the simulation.

ACKNOWLEDGEMENTS

We thank Mr Orange, Director of the Winery of Saint-Geniès-les-Mourgues (France), for welcoming us during the 1991 grapepicking, and Dr Lopez, Professor at the University of Lerida (Spain), for his help in the model validation.

REFERENCES

1. Grenier, P., Feuilloley, P., Sablayrolles, J.-M., Development of an expert system for the optimization of the wine alcoholic fermentation. International Conference on Agricultural Engineering, March 2-5, 1988, Paris, European Society of Agricultural Enginneers, paper n°88.399.

2. Nivière, V., Grenier, P., Roger, J.-M., Sevila, F., Oussalah, M., Intelligent simulation of plant operation in the wine industry. Journal of Food Control. In revision, 1993.

3. Dubois, D. and Prade, H., La théorie des possibilités. Edition Masson, 1988.

TREATMENT OF SWEET-POTATO PROCESSING WASTEWATER USING UASB REACTORS. I – START-UP OF THE PROCESS

DE MARIA, M.O.; DEL BIANCHI, V.L. and MORAES, I.O.
Department of Food Engineering and Technology - UNESP
P.O. BOX 136, 15054-00, São José do Rio Preto, SP-Brazil

ABSTRACT

Three different UASB reactor designs were evaluated at ambient temperature, to determine the efficiency of COD removal, volatile acidity, total alkalinity and pH, with distinct initial sludge concentration and diverse pH and nutrient correction solutions. Each reactor was operated with anaerobic sludge originating from a UASB plant of an orange juice factory. At the end of each operation phase of the reactors, an average efficiency of COD removal greater than 80% was obtained.

INTRODUCTION

The UASB concept was developed by Lettinga and coworkers (1) in 1971, and showed the potential of anaerobic digestion systems using UASB reactors, leading to satisfactory results with industrial wastewaters from different sources. The fundamental property of a UASB reactor is the wastewater passage through a high concentration region of anaerobic microorganisms, called a sludge blanket, in upflow. This way, the wastewater suffers degradation, with the biodegradable components converted into biogas. This biogas is separated from the treated effluent for later elimination or utilization as energy source.

In this work, the start-up of three distinct reactors, with some design differences, was studied. Each reactor was used for the treatment of wastewater from the sweet-potato starch extraction laboratory process.

MATERIALS AND METHODS

The reactors used in this work were called UASB I, UASB II and UASB III. UASB I reactor was built of plastic material, with a useful volume of 2.5 liters, in cylindrical shape. UASB II and UASB III reactors were made from glass, with square section and volumes of 3.4 and 4.7 liters, respectively. They have some differences in the design of their phase-separator systems. To control the treatment process and to verify its efficiency, some periodic analyses were done, including COD, pH, total alkalinity and volatile acidity. These were done in the affluent and in the effluent of the reactors, according to the methods described in the literature (3).

RESULTS

Based on the medium characteristics of the brute wastewater (COD = 5880.0 mg/l; pH = 5.40; total nitrogen = 157.4 mg/l and total phosphorus = 153.2 mg/l) and using the optimum relation of 100/5/1 (COD/N/P) observed by Patza *et al.* (2), it was verified that correction of the nitrogen ratio and pH of the system affluent was needed. At first, this was done with NH_4HCO_3 and $Ca(OH)_2$ solutions, respectively.

1.2 liters of sludge were inoculated into the UASB I reactor (48% of its volume) and the operation began with a hydraulic retention time (HRT) of 24 hours, until the 16th day of operation, when it was reduced to 18 hours.

On the other hand, one liter of the same sludge was inoculated into the UASB II reactor (27% of its volume) and the operation began with an HRT of 24 hours, until 38 days, when the HRT was reduced again, now to 12 hours. In this phase, the affluent nitrogen ratio was corrected with NH_4HCO_3, and the pH was $Ca(OH)_2$/NaOH.

One liter of anaerobic sludge was inoculated into the UASB III reactor (21% of its volume). The nitrogen ratio and the pH were corrected at the start of the process, with NH_4HCO_3 and $Ca(OH)_2$. The operation began with an HRT of 24 hours until 18 days, when it was reduced to 12 hours, until 40 days. At the end of the 55th day of operation, it was observed that the total alkalinity of the system was falling, a fact which reflected the global efficiency of the system. Considering this, it was decided to correct the pH and nitrogen with a solution containing $NH_4HCO_3/Na_2CO_3/NaHCO_3$. This change caused a rise in the global efficiency of the system, in terms of COD removal, from 68.3% to 86.7%, three days after changing this parameter. The average results obtained for all reactors are presented in Table 1.

The initial sludge concentration in the reactor must also be considered. A high concentration results in a higher system efficiency, because of the presence of a higher number of active cells in the reactor. Table 2 shows these data.

TABLE 1.

Average COD affluent, COD effluent and COD removal data for each operation phase of UASB 1, UASB II and UASB III reactors.

Average value	UASB I	UASB II	UASB III	HRT
COD affluent (mg/l)	1248.9	1843.5	3754.8	
COD effluent (mg/l)	117.0	256.9	589.5	24h
COD removal(%)	90.3	85.4	82.1	
COD affluent (mg/l)	1277.5	1267.3		
COD effluent (mg/l)	237.5	241.3	–	18h
COD removal (%)	80.7	79.8		
COD affluent (mg/l)		1841.8	1616.0	
COD effluent (mg/l)		334.7	323.1	12h
COD removal (%)		82.0	79.9	

TABLE 2

Influence of initial sludge concentration on the average COD removal, for the three reactors, with an HRT of 24 hours.

Reactor	Initial sludge concentration	Average CD removal
UASBI	48%	90.3%
UASB II	27%	85.4%
UASB III	21%	82.1%

CONCLUSIONS

For all reactor start-ups, a fast adaptation was observed of the sludge bacteria to the new substrate used, and a rapid changing of process parameters, such as organic loading ratio, HRT and correction solutions, to pH and nitrogen ratios. This fact shows the versatility of the system, considering its adaptation capacity to new conditions and its application to different kinds of wastewaters.

REFERENCES

1. LETTINGA, G.; KLAPWIJK, A. and HOBMA, S.W. - Use of upflow sludge blanket (USB) reactor concept for biological wastewater treatment, especially anaerobic treatment. Biotechnology and Bioengineering, 22. pp 699-734, 1980.

2. PATZA, M.G.B.; PAWLOSKY, U. and GABARDO, M.T. - Tratamento anaeróbio de vinhoto de mandioca pelo reator de leito de lodo granulado. In: 12th Brazilian Congress of Environmental and Sanitary Engineering, 1983. 32p.

3. Standard Methods for the Examination for Water and Wastewater. American Public Health Association, 1965. 769p.

TWO STAGE CONTROLLED ANAEROBIC BIO-REACTOR FOR DIGESTING MIXED DAIRY WASTE

C.L. Hansen*, S.H. Hwang
Department of Nutrition and Food Sciences and Biological and Irrigation Engineering, Utah State University, Logan, Utah, 84322-8700, USA

ABSTRACT

As part of an effort to optimize each phase of a multi-phase anaerobic fermentation, a cheese processing waste mixture (chemical oxygen demand (COD) ≥ 25,000 mg/L) was digested in batch fermentations to optimize acid formation. Three different influent concentrations (5800, 10,000 and 17,000 mg soluble COD (SCOD)/L); n=10 for each concentration), were used in the acid forming fermentations. A mixture of ethanol (0.1 M) and acid solution including: acetate (0.7 M), propionate (0.1 M) and butyrate (0.2 M), (SCOD = 13,300 mg/L) was digested in the batch methane fermentation experiment (n=8). The acid forming fermenters produced the maximum amount of volatile organic acids (VOA) in about 30 h. The production of acids was directly dependent upon influent concentration. The maximum production of each acid in mg/L was: acetate, 3000, propionate, 2500 and butyrate, 3200. The methane fermenter removed about 70% of the influent SCOD in 20 d.

INTRODUCTION

Fermentation involves a series of catabolisms to produce methane which can provide fuel for many unit operations. First, one group of anaerobes, the acidigens produce volatile organic acids (VOA) from complex organic material. The VOA is utilized by a second group of bacteria, the methanogens to produce methane. Some research has been done on biological production of VOA from biomass (1, 2) and the results are encouraging, but this data is not sufficient to design an optimal fermenter. The purpose of this research was to learn more about production of VOA in the acid forming stage of fermentation of cheese processing waste.

MATERIALS AND METHODS

Acidogens: Cheese processing waste, with composition as given previously (3) was obtained from Gossner Foods, Logan, Utah in one batch during November, 1992 after which it was divided into smaller portions and frozen at ≤-25°C for use later on. This was thawed just prior to use and diluted with distilled water to three different concentrations (5,800, 10,000 and

17,000 mg Soluble COD/L) used in the non-mixed batch reactors. Thirty fermentations (125 mL serum bottles; T= 35°C) were run for the acid forming stage; ten each for each influent concentration. Two N NaOH was added when necessary to adjust the pH to 6.0 in the beginning of the experiment.

Methanogens: An acid solution and ethanol solution (0.7 mole acetate, 0.1 mole propionate, and 0.2 mole n-butyrate, 0.1 mole ethanol; SCOD = 13,300 mg/L) similar in composition to that used in separating the bacterial strain was tested for the methanogenic phase. Eight batch fermentations (125 mL serum bottles; T=35°C, pH adjusted to 7.0) were run. A concentrated nutrient and trace mineral solution was added to both phases to give a final COD:N:P ratio of 500:5:1 (4, 5). The mixture of substrate and microbial culture was put in the 125 mL serum bottles.

Sample Analysis: Soluble COD was measured by the ampule method (6). Solids concentration was determined according to Standard Methods (7).

A Varian gas chromatograph (series 2700) with Carbopack® column (Supelco Inc., 1-1825) and thermal conductivity detector was used to determine acetate, propionate and n-butyrate concentration.

RESULTS AND DISCUSSION

Figures 1-3 show the results of the acidogen experiments. Maximum concentration of VOA was achieved in about 30 h. After this same length of time, pH had stabilized to 4.8-5.0, volatile suspended solids production and soluble COD were also constant (data not shown). Maximum organic acid production for all acids was dependent on influent COD concentration. The experiment was designed to have the three COD concentrations set up in a ratio of 1:2:3: final COD values deviated slightly from these ratios due to difficultly of making exact COD concentrations with actual industrial waste. The amounts of acetate, propionate and butyrate produced followed about the same ratios in proportion to the influent SCOD. Experiments are continuing to determine the kinetics of the acid forming stage.

Soluble COD destruction and biomass production for the methanogenic reactor is shown in Figure 4. About 70% of the influent VOA and ethanol was removed in this less than optimum fermentation. Work is also continuing in production of methane from VOA in our laboratory; we have found that in a continuous fermentation of VOA to methane in a complete mix reactor, ≥95% of VOA can be removed (data not shown).

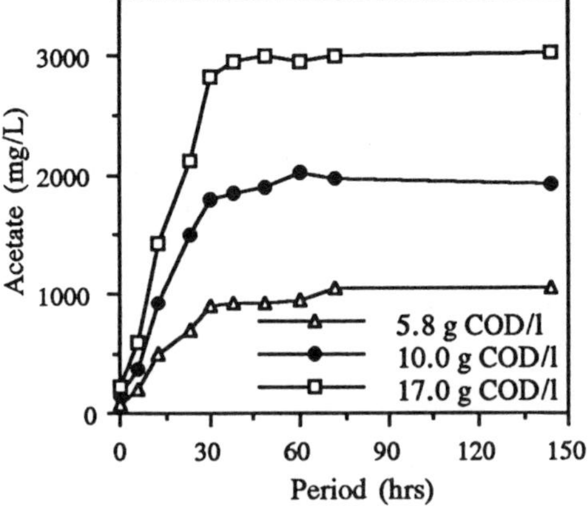

Figure 1. Acetate production versus time.

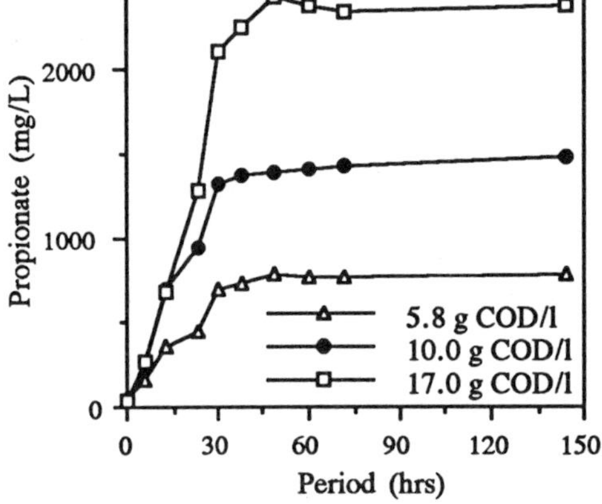

Figure 2. Propionate production versus time.

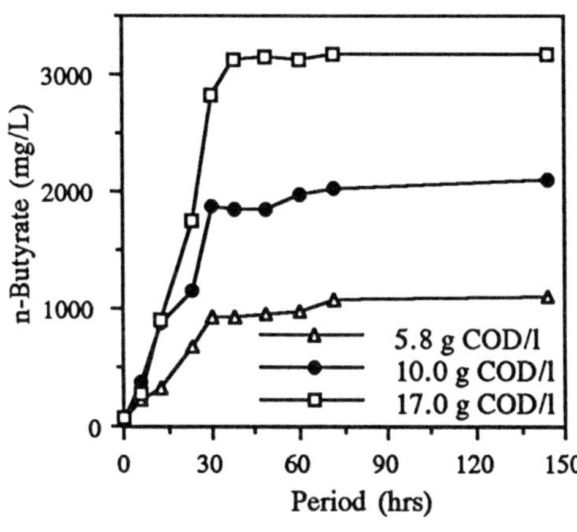

Figure 3. n-Butyrate prodution versus time.

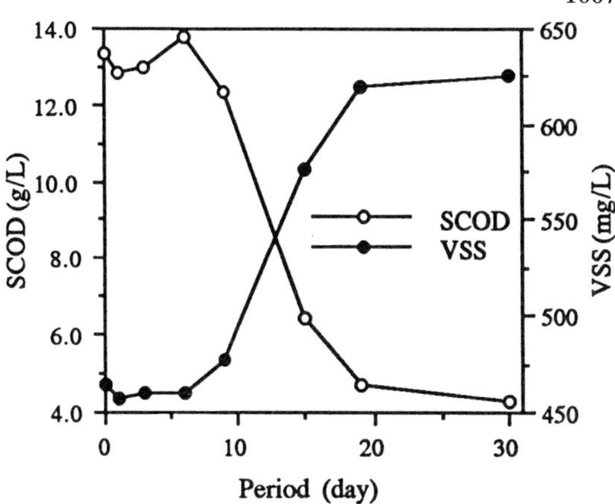

Figure 4. Soluble Chemical Oxygen Demand reduction and biomass production with time in the methane forming batch reactor

REFERENCES

1. Chiruvolu, C. and Engler, C.R. 1992. Biomass Conversion to Organic Acids by Rumen Microorganisms. American Society of Agricultural Engineers paper No. 92-6566, American Society of Agricultural Engineers, St. Joseph, MI.

2. Kisaalita, W.S., Pinder, K.L. and Lo, K.V. 1987. Acidogenic Fermentation of Lactose. Biotechnology and Bioengineering, 30: 88-95.

3. Hansen, C.L, S.H. Hwang. 1992. Two Stage Anaerobic Digestion of Cheese Processing Waste. Am. Soc. of Ag. Engr. Paper # 92-6605, presented at the ASAE Winter Meeting, December 15-18, Nashville, TN.

4. Zeikus, J. G. 1977. The biology of methanogenic bacteria. *Bacteriological Reviews.* 41(2):514-541.

5. Stronach, S. M., Rudd, T., and Lester, J. N. 1986. Anaerobic digestion processes in industrial wastewater treatment. Springer-Verlag press. Berlin: New York.

6. Adams, V. D., Cowan, P. A., Pitts, M. E., Porcella, D. B., and Seierstad, A. J. 1981. Analytical procedures for selected water quality parameters. Utah Water Research Laboratory, Utah State University, Logan, Utah.

7. APHA-AWWA-WPCF. 1989. Standard methods for the examination of water and wastewater. 17th Edition. American Public Health Association., Washington, D. C.

ACTIVATED SLUDGE TREATMENT OF MEAT PROCESSING WASTEWATER

A.P. ANNACHHATRE AND S.M.R. BHAMIDIMARRI
Department of Process and Environmental Technology
Massey University, Palmerston North, New Zealand

ABSTRACT

Feasibility of aerobic activated sludge process for treatment of meat industry wastewater has been demonstrated through the operation of a model activated sludge unit. High degree of nitrification and COD removal of more than 85% for loads upto 3.2 kg COD/(m^3.day) were achieved for reactor operated at SRT values of 3-13 days. Reactor operation at SRT of 6 days or more was highly stable and yielded near complete nitrification. Under these conditions aerobic treatment is likely to be more attractive than the conventional anaerobic lagoon treatment.

INTRODUCTION

New Zealand meat industry processes up to 40 million animals each year. A typical meat processing unit in New Zealand can produce wastewater upto 10,000 m^3/day with a pollution load equivalent to a city of around 200,000 inhabitants(1). A common practice in New Zealand for meat waste disposal is primary treatment followed by effluent irrigation (2). Many researchers have studied COD removal from meat waste through aerobic treatment (1,2) but no conclusive data on concomitant nitrification is available. Accordingly, this research demonstrates feasibility of aerobic activated sludge treatment for meat processing wastewater through the operation of a model activated sludge unit for COD removal and nitrification.

MATERIALS AND METHODS

Wastewater and sludge source:

Primary treated meat wastewater (Weddel Feilding, Feilding, New Zealand) was settled overnight to remove suspended solids and the supernatant was then used for further experimentation. Mixed liquor from an activated sludge treatment plant treating fellmongery effluent was acclimated to the slaughterhouse wastewater in a continuous reactor. Sludge from this reactor served as seed for all chemostat and activated sludge experiments.

Aerobic activated sludge process:

Experimental setup essentially consisted of fresh feed storage tank, 15 litres Brunswick aeration tank, sludge settling tank and pumps for fresh feed, sludge recirculation and sludge wastage. 1:1 diluted fresh feed with COD 600-750 mg/l was fed continuously to the aeration tank while the effluent from the reactor was withdrawn continuously and fed to the settling tank. The settled biomass was recirculated back into the reactor, while overflow from the settling tank was discarded. Sludge recirculation ratio of 1:1 was maintained throughout the investigation. Sludge was withdrawn continuously from the reactor and wasted in order to control Solids Residence Time(SRT) in the reactor. Investigations were performed with SRT of 13, 9, 6 and 3 hours at fixed HRT of 10 hours in the aeration tank. Finally, reactor performance was assessed at 5 hours HRT and 6 days SRT. Adequate air supply was maintained to the aeration tank so as to maintain DO concentration of about 6.5 mg/l. Fresh feed and reactor contents were analyzed daily for COD, NH_4-N, NO_2-N and NO_3-N. Besides TKN and fat content of the reactor feed and effluent were also analysed occasionsly.

RESULTS AND DISCUSSION

COD removal:

Kinetic parameters K_s, μ_m, Y and k_d were estimated as 176 mgCOD/l, 1.14 day^{-1}, 0.4 mg/mg and 0.043 day^{-1} respectively from continuous chemostat experiments (2). Figure 1 shows influent COD ranging from 650-750 mg/l while that of effluent to be always less than 100 mg/l, yielding COD destruction always over 85%. Figure 1 also shows COD removal rates achieved. Highest COD destruction rates in the range of 3-3.2 kg COD/(m^3.day) are recorded for HRT and SRT of 5 hours and 6 days respectively.

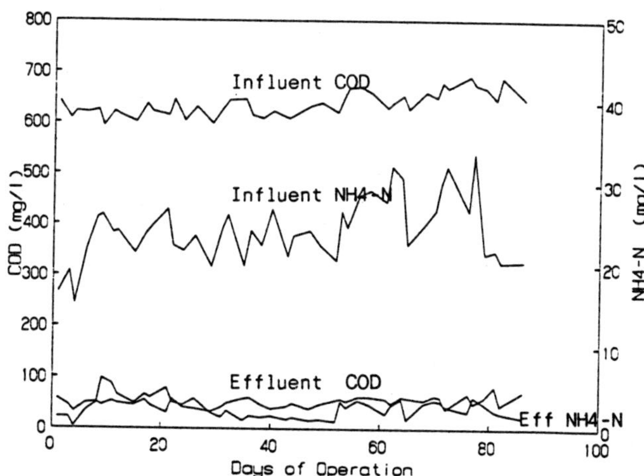

Figure 1 : COD and NH_4-N concentration

Nitrification:

Figure 1 shows the influent NH_4-N concentration ranges from 18-35 mg/l. The effluent NH_4-N concentration is remarkably steady at less than 6 mg/l. Likewise, NO_2-N buildups upto 10 mg/l and high effluent NO_3-N concentration in the range 15-35 mg/l (Figure 2) also confirm concomitant nitrification activity during activated sludge treatment. However, Figure 1 and Figure 2 also bring out that reactor performance with 3 days SRT is rather vulnerable to

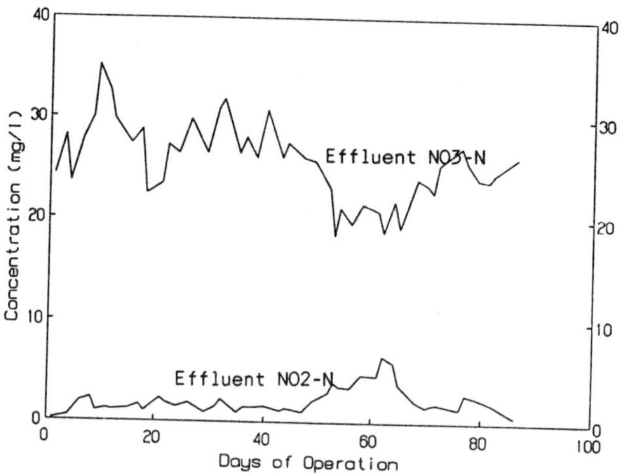

Figure 2 : Effluent NO_2-N and NO_3-N concentrations

drastic variations in influent NH_4-N concentration, is more prone to higher NO_2-N buildups (upto 5-7 mg/l) and requires more time to stabilize to lower NO_2-N levels thus confirming tendency for incomplete nitrification. In contrast reactor performance with 6, 9 and 13 days SRT is much more stable with low NO_2-N buildups (always less than 5 mg/l) and consistently higher NO_3-N concentration (22-35 mg/l) in comparison with SRT of 3 days (15-25 mg/l), thus signifying near complete nitrification.

These data suggest that although the reactor operation yielded satisfactory COD removal at 3 days SRT, its nitrification performance was vulnerable to variations in influent feed. In contrast, reactor operation at SRT of 6 days or more yielded high COD removal with steady and near complete nitrification. Figure 1 and 2 also show reactor performance at HRT of 5 hours and SRT of 6 days. For these operating parameters, both COD removal and nitrification are satisfactory. COD removal above 85% with loads upto 3.2 kg COD/(m^3.day) are achieved with low NH_4-N and NO_2-N concentrations (always well below 5 mg/l) and high NO_3-N concentration (upto 33 mg/l). TKN analysis of feed and effluent also confirmed more than 90% removal.

CONCLUSIONS

This study establishes that aerobic activated sludge treatment of meat industry wastewater is an effective way of COD removal and nitrification. Reactor operation at 6, 9 and 13 days SRT is much more stable and yields near complete nitrification as compared to reactor operated at 3 days SRT. Under theses conditions, aerobic treatment is likely to be attractive than conventional anaerobic lagoon treatment.

REFERENCES

1. Bhamidimarri, S.M.R., Appropriate industrial water management technologies : The New Zealand Meat Industry. Wat.Sci.Tech., 1991, **24(1)**, 89-95.

2. Annachhatre, A.P. and Bhamidimarri, S.M.R., Aerobic Treatment of Meat Industry Wastewater. In: Chemeca 93, 1993 (In press).

ANAEROBIC TREATMENT OF HIGH STRENGTH DAIRY PROCESSING WASTE IN AN UPFLOW ANAEROBIC SLUDGE BLANKET REACTOR

B GRAUER, S M R BHAMIDIMARRI and R L EARLE
Department of Process and Environmental Technology
Massey University
Palmerston North, New Zealand

ABSTRACT

An upflow anaerobic sludge blanket (UASB) reactor was used to study the anaerobic digestion of whey permeate. The 25L reactor was capable of reducing a soluble organic load of 15 kg COD/m^3.d by 95% while an increased load of 20 kg COD/m^3.d resulted in 90% COD reduction. The methane production rate increased linearly with the loading rate up to a loading rate of 15 kg COD/m^3.d beyond which there was no increase in methane yield. An analysis of the sludge granules revealed that the accumulation of inorganics in the granules was significant and the specific activity of the granules could be adversely affected by the concentration of calcium and magnesium salts, in particular the phosphate and carbonates. The concentration of methane in the biogas was approximately 52% for organic loads up to 15 kg COD/m^3.d but beyond this value the methane concentration dropped to a low 41%. The yield of methane was 4.1 m^3/m^3.d for organic loading rates of up to 15 kg COD/m^3.d.

INTRODUCTION

Large quantities of whey are generated as byproducts of the New Zealand cheese and casein industries. With the exception of a small quantity of sweet whey, it is considered a waste product, which is in general disposed of by spray irrigation. However, this strategy is no longer acceptable as its environmental impact is better understood. The anaerobic treatment of whey permeate from the ultrafiltration of lactic acid precipitated casein was successfully demonstrated with loads of 7.7 kg COD/m^3.d in a fluidized bed digester (1) and 4.3 kg COD/m^3.d in a sludge blanket digester (2). Although the performance of the fluidized bed digester was shown to be superior, operation of this system is complicated by the development and retention of the biofilm on the support surface in anaerobic reactors (3).

The sludge blanket reactor, on the other hand is a high rate digester which is simple to operate and more robust against fluctuations in operating conditions (4). Granulated sludge in UASB reactors has been shown to perform well even after extended periods of storage (4) and therefore the system is better suited to treat the waste streams from seasonal operations such as dairy processing.

The digestion performance of a UASB reactor treating lactic whey permeate is presented in this paper. The quality of biogas generated and the composition of the sludge granules are also discussed.

MATERIALS AND METHODS

Digester Operation

A 25L digester was constructed from QVF glass column sections and included a gas collection mechanism which minimised the sludge carry-over. The digester was charged with approximately 10 kg of granulated sludge which had been previously acclimated to lactic whey permeate. The average granule size was 2.2 mm in diameter. The whey permeate feed, with a COD of 48000 mg/L was supplemented with 1.0 g/L of agricultural grade urea and 50 μL/L of 10% (v/v) defoaming agent. The flow rate through the column was 1 L/min with recycle ratio ranging from 500-700.

Analytical Methods

Chemical Oxygen Demand (COD) was determined by digestion of samples with 0.035 M $k_2Cr_2O_7$ in concentrated sulphuric acid for two hours using a Hach COD apparatus (Hach, Ames, USA).

Biogas was measured using a Brand wet-test gas meter. Methane concentration in the biogas was determined using a Varian Aerograph 900 gas chromatograph with stainless steel column packed with 80/100 mesh Porapak S.

Volatile Solids (VS) were measured according to Standard Methods (5).

The elemental composition of granules was determined by Inductively Coupled Plasma Emission Spectroscopy whilst CO^-_3 was determined by the loss of mass on combustion at 1000°C.

RESULTS AND DISCUSSION

COD Removal

The digester was operated over a range of COD loading rates varying from 2.5 to 22 kg COD/m^3.d. The COD removal across the granular sludge bed was measured in terms of both the soluble and insoluble components. As shown in Figure 1 the performance of the UASB is significantly higher than that reported in the literature (2) with the soluble COD reduction greater than 95% up to a loading rate of 15 kg COD/m^3.d. Even at 20 kg COD/m^3.d loading rate, a 90% soluble COD removal was consistently achieved.

The total COD reduction was 90% up to 12 kg COD/m^3.d decreasing to 74% at a loading rate of 22 kg COD/m^3.d.

A COD mass balance over the entire period of digester operation of 237 days not accounting for COD incorporated in biomass and CO_2 revealed an 88.5% COD recovery through the process.

Methane Generation

The methane production rate as a function of organic loading rate is shown in Figure 1. The methane production rate of STP increased rapidly as the COD loading rate increased to a maximum of 4.1 m^3/m^3.d at a loading rate of 15 kg COD/m^3.d, but the methane yield decreased as the organic loading rate increased further. This reflects the incomplete digestion resulting in the accumulation of volatile fatty acids. Indeed, the concentrations of propionate and acetate were found to increase rapidly as the loading rate increased beyond 15 kg COD/m^3.d. The overall methane yield was found to be 0.3 m^3/kg COD removed.

Granule Composition

The granule elemental analysis revealed that there was an accumulation of inorganic matter is the granules. The major components were found to be Ca (0.36 kg/kg), P (0.17 kg/kg) and CO^-_3 (0.07 kg/kg) indicating calcium phosphate deposition. Interestingly, there was no significant potassium accumulation (0.001 kg/kg) although the potassium concentration in the permeate was high (1.37 g/L). These results imply that the specific activity of the granules could decline in UASB reactors operating over prolonged periods of time.

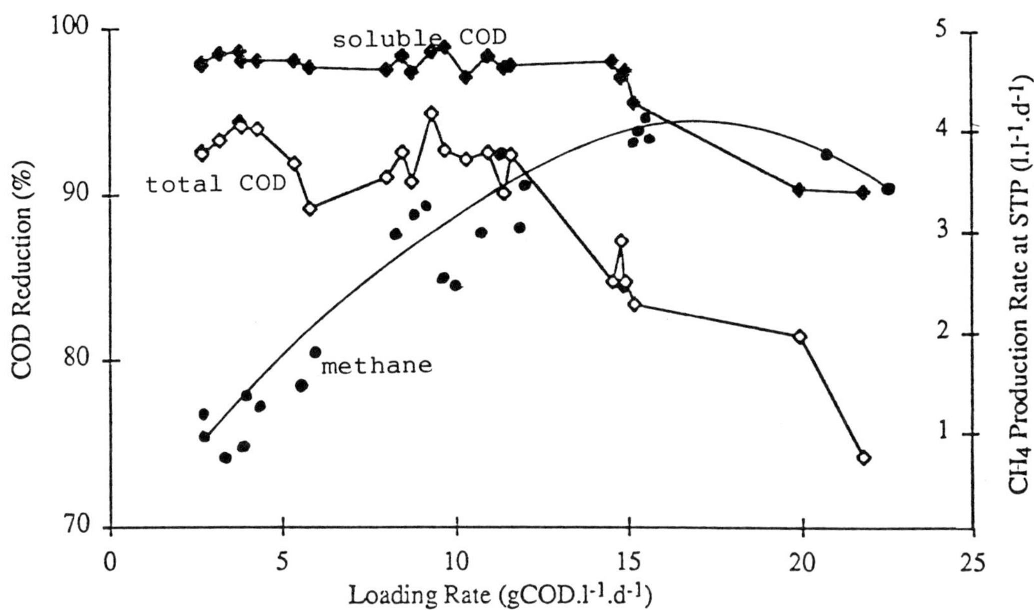

Figure1. Performance of UASB digester.

CONCLUSIONS

The high rate digestion of strong dairy processing waste is shown to be feasible. Organic loading rates of up to 15 kg CODm^3.d can be treated to achieve 95% soluble COD and 85% total COD removal. High methane yields of up to 4.1 m^3/m^{3}d are possible under optimal operating conditions. The results presented in this study confirm that a UASB reactor is simple to operate and is superior in performance to the biofilm systems. However, the specific activity of the granules is likely to decrease with time if the waste stream contains significant concentrations of inorganic salts.

REFERENCES

1. Boening, P.H. and Larsen, V.F. Anaerobic fluidized bed whey treatment. *Biotechnol. Bioeg.* 1982, **24**, 2539-2556.
2. Archer, R.H., Larsen, V.F. and McFarlane, P.N. Anaerobic digestion of whey permeate: A comparison of three reactor designs. Poster paper presented at the 3rd International Symposium on Anaerobic Digestion, Boston, August 14-19, 1983.
3. Henze, M. and Harremoes, P. Anaerobic treatment of wastewater in fixed film reactors - a literature review. *Wat. Sci. Tech.*, 1983, **15**, 1-101.
4. Lettinga, G., van Velsen, A.F.M., Hobma, S.W., de Zeeuw, W. and Klapwijk, A. Use of upflow sludge blanket reactor concept for biological wastewater treatment, especially for anaerobic treatment. *Biotechnol. Bioeng.*, 1980, **22**, 699-734.
5. American Public Health Association. *Standards methods for the examination of water and wastewater* (14th edition), Washington, D.C., (1975).

ANAEROBIC TREATMENT OF DISTILLED WASTE WATER OF WHEAT 'SHOCHU' USING TWO-PHASE FERMENTER

K. Morita, S. Taharazako, Y. Nishi
Chair of Agricultural Systems Engineering, Faculty of Agriculture, Kagoshima University, Korimoto, 1 chome, Kagoshima 890, Japan

ABSTRACT

Distilled waste water of wheat "shochu" which contained high concentration of organic pollutants was anaerobically fermented with practical scale two-phase fermenter.The purification of digestive waste water, the productivity and utility of biogas were also investigated. The performance of the fermenter was evaluated using COD removal rate, removal efficiency of COD and BOD. The generated biogas rate, the produced biogas ratio and the concentration of methane gas were measured, and the efficiency of biogas generation was evaluated.

INTRODUCTION

Waste from food processing operations has traditionally yielded severe pollution problems. Especially, it is difficult to purify the waste water containing high concentration of organic pollutants. In the field of food processing waste management, it is important to develop an improved and practical system to remove major pollutants such as COD, BOD, and utilize the produced biogas. This present study was undertaken to investigate the feasibility of anaerobic treatment using a Two-phase methane fermenter for distilled waste water of "shochu".

MATERIALS AND METHODS

Figure 1 shows a schematic diagram of the methane fermentation system. The two-phase fermenter(MBB) continuously loaded reactor consists of two cylindrical tanks (total volume 6.6m^3, phase1:phase2=1:9) equipped with separate biogas collector. The reactor was kept at about 38℃ and 7.4pH in this study. This experiment was conducted over a 3 month period. Hydraulic retention time(HRT) was about ten days and influent substrate concentration of COD from 10,000 to 14,000ppm, BOD from 20,000 to 37,000 ppm, SS fromm 6,000 to 8,000ppm were tested. The reactor performance was evaluated based on COD removal and biogas production. COD removal rate and biogas production rate were calculated with the following equations:

$$-dS/dt = K_1(S-S_f)$$

where, S : COD after anaerobic treatment(ppm)
S_f: undecomposable COD(ppm)
K_1: COD removal rate(1/d)
t : retention time(d)

$$dG/dt = K_2(G_t-G)$$

where, G : volume of actual generated biogas(m^3)
Gt: volume of theoretical generated biogas(m^3)
K_2: biogas production rate(1/d)

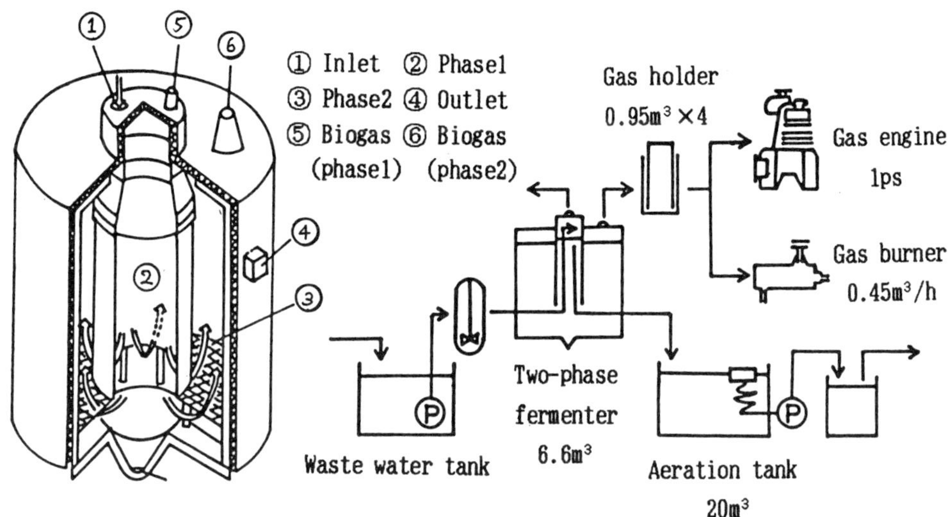

Fig.1 Block diagram of fermentation system and schematic diagram of two-phase fermenter

RESULTS AND DISCUSSION

Table 1 shows the change of COD concentration for anaerobic fermentation The process was started using 4.0m^3 of seed anaerobic sludge from a municipal waste treatment plant. The initial load consisted of sweet potato shochu waste water with low concentration of COD. Steady-state fermentation was obtained by a gradual increase of COD loading during five months. In these periods(A,B,C), COD removal efficiency were 66.3-79.0% at ten days of retention time and COD removal rate,K_1, varied from 0.112 to 0.166.

Table 2 shows the change of BOD concentration and effect of deterrent on anaerobic fermentation. The efficiency of BOD removal was low for high concentration of BOD. Dramatic changes were not observed in NH_4^+, VFA and T-N. SS removal efficiency greatly changed from 44.8 to 84.3%.

Table 3 shows the change of biogas production and methane concentration. In these periods, volume of biogas was more than 10m^3/d and volume ratio (phase2/total) was approximately 87%. Methane concentration maintained

Table 1 Change of COD concentration for anaerobic fermentation

period (d)	COD(ppm) charge	discharge	removal COD(%)	volume loading COD(kg/m^3)	HRT (d)	K_1
A(16d)	16,590	3,470	1,65	79.0	10.0	0.161
B(35d)	13,740	2,820	1,44	77.6	9.7	0.166
C(41d)	11,900	3,970	1,20	66.3	10.0	0.112

A: sweet potato and wheat shochu waste water
B: wheat shochu waste water
C: 〃 by low pressure distillation

Table 2 Change of BOD concentration and effect of deterrent on anaerobic fermentation

period (d)	Temp (℃)	pH	BOD(ppm) charge	removal BOD(%)	NH_4^+ (ppm)	VFA (ppm)	T-N (ppm)	SS(ppm) charge	SS(%) removal
A(16d)	38.5	7.35	25,600	46.1	1,450	-	2,000	18,200	84.3
B(35d)	38.2	7.41	25,500	57.9	1,990	-	1,900	8,500	68.1
C(41d)	38.6	7.49	35,200	51.6	1,970	5,325	3,000	6,100	44.8

Table 3 Change of biogas production and methane concentration

period (d)	biogas(m^3/d) phase1	phase2	ratio(%) phase2/total	CH_4(%) phase1	phase2	K_2	production ratio α (m^3/kgCOD)	β (m^3/m^3waste)
A(16d)	2.63	11.53	81.4	35.0	67.5	0.153	1.27	21.0
B(35d)	1.78	11.93	87.0	40.2	67.3	0.152	1.28	17.5
C(41d)	1.38	10.03	87.9	50.7	64.4	0.052	1.38	16.4

67% during phase 2. Biogas production rate, K_2, greatly decreased during period C, but biogas productivities α & β did not show any remarkable change.

CONCLUSIONS

This present study indicated that an anaerobic treatment system using a two-phase fermenter was practical to use for the disposal of high concentrated organic waste, and the produced biogas containing a high percentage of methane gas could be used as an alternate energy source.

PLEUROTUS PRODUCTION IN COFFEE BEAN WASH WATER

José E. Sánchez V.
Centro de Investigaciones Ecológicas del Sureste
Apdo. Postal 36, Tapachula, Chiapas 30700 México

Antonio M. Martin
Department of Biochemistry, Memorial University of Newfoundland
St. John's, Newfoundland, Canada A1B 3X9

ABSTRACT

In this work the growth of the fungus *Pleurotus ostreatus* in the form of pellets on coffee wash water is studied. By supplementing the medium with non-proteinaceous nitrogen, a biomass production of 3.05 g/L is obtained. The fungal growth causes a 25 % tannin content reduction and a 43 % BOD reduction.

INTRODUCTION

The run-off water from the processing of coffee beans is an environmental pollutant. Generally its COD value ranges between 5000 and 35,000 mg/L. It could be of interest to use this waste water as a culture medium for growing *Pleurotus* in the production of fungal pellets. *Pleurotus*, among other fungi, has been grown in submerged culture using various culture media [1,2,3]. In this work we have studied the technical viability of cultivating *Pleurotus* in the water used to depulp coffee.

METHODS

Half a kilogram of fresh, ripe coffee berries were washed and de-pulped manually in 3.5 L of water. The water thus obtained was separated from the pulp and the beans, centrifuged, filtered, and finally sterilized by filtration (0.22 µm Millipore filters). The INIREB-8 strain of *Pleurotus ostreatus* was inoculated into the sterile coffee wash water (CWW) and incubated for 10 days at room temperature, with constant stirring (50 rpm). The soluble sugar, total nitrogen (Kjeldahl) and tannins contents, as well as the pH, COD and BOD of the wash water were determined during fungal growth. At the end of fermentation, the pellet diameters and the dry material obtained were measured.

RESULTS AND DISCUSSION

When coffee wash water sterilized in an autoclave was used as medium, there was no growth, most likely due to degradation of carbohydrates and nitrogen, and to sulphur loss from the culture medium. Over 8 days of fermentation, the pH of the medium underwent an increase from a value of 5.5 to 7.9 due to the fungal metabolism, while the concentration of soluble sugars decreased from 1.20 to 0.45 %. Nevertheless, at the end of 8 days of fermentation, an important carbohydrate remnant remains, indicating incomplete fermentation. The nitrogen concentration in the medium at that time was 0.058 %.

At the end of 8 days of growth on CWW, the concentration of biomass produced, in the form of small pellets measuring 1.5 to 2 cm in diameter, reached 2.1 g/L. In terms of yield and productivity, this amount is small, since Martin and Bailey [3], working with *Agaricus campestris*, obtained between 3 and 4 g/L of biomass in 8 days of fermentation on hydrolysed peat. The lower result obtained here could be explained as a function of the nitrogen content of the waste water, since Sugihara and Humfeld [1] reported nitrogen values of around 1 g/L as optimal for the growth of *Agaricus* in submerged culture. In the present study, the initial nitrogen content was 0.089 %.

As seen in Table 1, when the medium was supplemented with 3 different sources of nitrogen (yeast extract, nitrogen without amino acids and nitrogen without sulphur-containing amino acids), an improvement in biomass production was observed in most cases at the end of 10 days of fermentation. In this time period, the carbon source was not entirely consumed (there was a 0.16 % remnant, on average). This could be due to a limiting factor in CWW, probably related to its low phosphorus and sulphur contents. *Pleurotus* utilizes tannins as a carbon source, and so the concentration of these compounds is reduced from 0.16 to 0.12 mg/L during fermentation (an average reduction of 25 %), implying that this fungus has anti-polluting activity.

In Table 2, the reductions of the BOD and COD in the CWW throughout the process of medium preparation and microbial growth are observed. At the start, due to the solid wastes that the water contains, the BOD and COD values reach levels on the order of 8,000 mg/L and 21,000 mg/L, respectively, giving a BOD/COD ratio of 0.38. This indicates a high content of non-biodegradable pollutants in the CWW. Centrifugation eliminated most of these waste products, and sterilizing filtration produced further elimination, leaving a BOD_5 value of 1,350 mg/L. Stirring and aerating, as well as the growth of *P. ostreatus* in the medium, brought about a reduction of this value on the order of 43 %, leaving the run-off water with a final BOD value of 770 mg/L, representing a more acceptable level in terms of pollution. The reduction of BOD and COD via fungal metabolism has previously been reported [4].

Coffee bean wash water contains high levels of pollutant materials, which for the most part are non-biodegradable (BOD/COD ratio 0.38). The treatment used leaves relatively low values of BOD_5 and COD in the run-off. To assure good growth it is necessary to supplement the medium with nitrogen, and probably phosphorus and sulphur, after centrifugation and sterilization.

TABLE 1
Pleurotus ostreatus growth on coffee bean wash water supplemented by nitrogen sources (10 days of fermentation).

Culture Medium Supplementation	Final pH	Biomass Produced (g/L of medium)	Soluble Sugars Residue (%)	N_2 (%)
Yeast Extract	8.27	1.99	0.17	0.070
N_2 without AA	6.09	3.05	0.17	0.055
N_2 without S-AA	5.60	2.87	0.13	0.075
No Additions	7.25	2.21	0.15	0.074

TABLE 2
BOD_5 and COD variation in coffee bean wash water during the different steps of the process.

Step	BOD_5 (mg/L)	COD (mg/L)
Initial sample	8,076	21,000
After centrifugation	2,285	6,000
After sterilization	1,350	3,000
After growth	770	800

ACKNOWLEDGEMENTS

The Rosario Izapa Experimental Station (INIFAP-SARH) is acknowledged for the cooperation provided to this study. The authors wish to express their thanks to M. del R. Hernández Ibarra, L.A. Calvo Bado and P. Bemister for their assistance.

REFERENCES

1. Sugihara, T.F. and Humfeld, H., Submerged culture of the mycelium of various species of mushroom. *Appl. Microbiol.*, 1954, **2**(3), 170-172.

2. Martin, A.M., A review of fundamental process aspects for the production of mushroom mycelium. *J. Food Proc. Eng.*, 1986, **8**, 81-96.

3. Martin, A.M. and Bailey, V.I., Growth of *Agaricus campestris* NRRL 2334 in the form of pellets. *Appl. Environ. Microbiol.*, 1985, **49**(6), 1502-1506.

4. Church, B.D., Erickson, E.E. and Widmer, C.M., Fungal digestion of food processing wastes. *Food Technol.*, 1976, **27**, 36.

BY PRODUCTS FROM FOOD INDUSTRIES: UTILIZATION FOR BIOINSECTICIDE PRODUCTION.

IRACEMA DE O. MORAES
Universidade Estadual Paulista -UNESP/IBILCE-DETA
CP 136 CEP 15054 000 S.José do Rio Preto SP Brazil

DEISE M. F. CAPALBO
Centro Nacional de Pesquisa de Defesa da Agricultura
CP 69 CEP 13820 000 Jaguariúna SP Brazil

REGINA DE O. MORAES
Universidade Estadual de Campinas-UNICAMP/FEAGRI
CP 6011 CEP 13081 970 Campinas SP Brazil

ABSTRACT

Agricultural residues and wastewaters from food, beverages and paper industries, are feasible substrates to produce microbial insecticides through *Bacillus thuringiensis* fermentations. These substrates are sources of carbon and nitrogen that are essential components to culture media composition. Since 1970 our research group are studying both the submerged (1) and semisolid fermentations (2) and two patents (3,4) of the processes were developed. Corn steep liquor and sugar cane molasses were determined as the principal components of the culture media. This paper deals with the process, advantages and problems and the use of low cost raw material availlable in Brazil. The group intends to develop local technology to produce *Bacillus thuringiensis*, just to use against agricultural pests that are responsible to 40% losses in field, harvested and stored products. These losses cause price increments in the final product and contribute to the maximize hungry and poverty problems, in Brazil.

INTRODUCTION

The practical use of entomopathogenic microorganisms to crop protection is possible when an industrial scale production of the organism is developed. That is the case of *Bacillus thuringiensis*, one of the most studied and commercially important entomopathogenic bacterium. Brazil is using imported products and our group is trying to develop local technology to massal production (5, 6), based in the use of low cost raw material, by products, residues or wastewaters from food, beverage or agroindustries to compose the culture media.

MATERIAL AND METHODS

Bacillus thuringiensis Berliner was maintened in nutrient agar slants to be used as inoculum, for the different media composition. Standard medium composition has sugar cane molasses and corn steep liquor, as stated in BR Patent (3).
Culture media were composed as follows:
MSG - wwater from monossodic glutamate industry + tap water;
SCG - monossodic glutamate industry ww + sugar cane molasse;
CWG - coconut water + ww from monossodic glutamate industry;
YBI - liquid residue from beer industry + tap water;
These media were adjusted to 8.3 g total carbohydrates.
MCF - cassava flour industry ww (75%) + tap water(25%).
pH was adjusted to 7.3, before sterilization, 15 min. 121°C.
The fermentation process were runned in shaker (flasks of 250 ml, with 50 ml culture medium sterilized), fermentors of 5 l and 10 l working volume. Temperature was controlled at 30 ° C, agitation at 150 and 300 rpm, respectivelly and 1 vvm aeration, for fermentors.
Optical density (600 nm), pH, viable spore count/colony forming units-CFU (7), were acomplished. Bioassay was performed against Anticarsia gemmatallis, to know about toxicity of the bioinsecticide obtained.

RESULTS

The fermentation processes, present the following results expressed on Table 1.

TABLE 1: Comparison of the fermentation processes development.

MEDIUM	Harvest time (h)	pH	Absorb. (OD)	Spores (CFU/ml)
MSG	0	6,5	4,8	$< 10^3$
	4	6,7	5,5	$< 10^3$
	7	6,8	6,4	$< 10^3$
	17	7,7	7,5	$1,8 \times 10^6$
	19	8,1	9,6	$5,5 \times 10^9$
	24	8,2	11,1	$> 10^{11}$
SCG	0	6,6	5,3	$< 10^3$
	5	5,7	6,6	$< 10^3$
	29	6,1	17,3	$1,8 \times 10^6$
	45	6,8	17,1	$7,7 \times 10^6$
	53	7,3	15,1	$2,8 \times 10^7$
CWG	0	6,1	5,3	$< 10^3$
	2	6,1	5,7	$< 10^3$
	21	5,5	16,0	$2,4 \times 10^4$
	45	6,8	21,9	$1,3 \times 10^7$
	66	7,3	19,3	$3,9 \times 10^8$

MEDIUM	Harvest time (h)	pH	Absorb. (OD)	Spores (CFU/ml)
YBI	0	7,0	3,7	$< 10^3$
	4	6,6	4,2	$< 10^3$
	22	7,8	7,1	$1,2 \times 10^{11}$
	25	8,0	7,7	$> 10^{11}$
	28	8,0	10,0	$> 10^{11}$
	41	8,4	8,7	$> 10^{11}$
MCF	0	6,5	10,6	$< 10^3$
	6	5,9	15,4	$< 10^3$
	46	6,1	21,8	$6,0 \times 10^5$
	70	6,9	19,5	$4,4 \times 10^6$
	76	7,3	20,5	$2,0 \times 10^9$

AKNOWLEDGEMENTS

The authors express their gratitude to **FUNDUNESP**, **CNPq**, **OEA** and **FAPESP**

REFERENCES

1. Moraes, I.O., Capalbo, D.M.F., Del Bianchi, V.L., Sifuentes, L.E., Technical aspects of the use of wastewater to get useful products through fermentative processes. IV Latinamerican Congress of heat and mass transfer, Chile, 1991, **1** , 93-96.

2. Capalbo, D.M.F. and Moraes, I.O., Production of proteic protoxin by Bacillus thuringiensis in semisolid process. 12th Simposio Anual da Academia de Ciências do Estado de S. Paulo. Campinas SP Brazil, 1988, **2** , 46-55.

3. Moraes, I.O., BR Patent 7608688, Brazil, 1976.

4. Moraes, I.O., BR Patent 8500663, Brazil, 1985.

5. Moraes, I.O. and Capalbo, D.M.F., Study of continuous fermentation for obtaining bacterial insecticide to control agricultural pests. IV Japan-Brazil Symposium on Science and Technology, Brazil, 1984, **II**, 248-255.

6. Moraes, I.O., Production and utilization of Bacillus thuringiensis for crop protection in Brazil. International workshop on Bt and its applications in developing countries. Egypt, 1991.

7. Thompson, P.J. and Stevenson, K.E. Mesophilic aerobic spore formers. In Compendium of methods for the microbiological examination of foods. Speck, M.L. 2. ed., Washington, Am. Public Health Assoc., 1984, **19**, 211-220.

NOx REMOVAL IN WATER BY BIOCHEMICAL REACTION

TADATAKE OKU, MITSUTAKA KONDO, HITOSHI SATO, TOSHIYUKI NISHIO, TEIICHIRO ITO, MIKIO HOSHINO* and HIROSHI SEKI*
Laboratory of Bio-organic Chemistry, College of Agriculture and Veterinary Medicine, Nihon University, 34-1, Shimouma 3-chome, Setagaya-ku, Tokyo 154, Japan,
*The Institute of Physical and Chemical Research, 2-1, Hirosawa, Wako, Saitama 351-01, Japan

ABSTRACT

NOx removal in water by the reduction of NO_2^- and NO to NH_4^+ was examined by using denatured cytochrome *c* (cyt *c*) by heating at 100°C for 30 min or by Co-60 gamma-ray irradiation with a dose of 10 kGy. Sodium dithionite and methyl viologen were, respectively, used as a reducing agent and an electron carrier in the reduction system. The conversion yield of NO_2^- to NH_4^+ were *ca.* 100% and *ca.* 50% in N_2 atmosphere and air, respectively. The ESR and optical studies have shown that nitrosyl complex of cyt *c* is produced as an intermediate for the reduction of NO_2^-. The conversion of NO_2^- to NH_4^+ is a net 6-electron and 8-proton reduction process in the same manner as nitrite reductases. Denatured cyt *c* is suggested to be used as a model of nitrite reductase for catalytic removal of NOx in water.

INTRODUCTION

Recently, removal of NO_2^- in water has been one of the important research area in environmental science because of the necessity of reducing water pollution. In nature, ferredoxin-nitrite reductase (EC 1.7.7.1) which reduces NO_2^- to NH_4^+ ($NO_2^- + 8H^+ + 6e^- \rightarrow NH_4^+ + 2H_2O$) has been found in higher plants[1)] and green algae[2)]. Purification method and some properties of nitric oxide oxidoreductase (EC 1.7.99.3) and nitric oxide reductase (EC 1.7.99.2) have been studied with denitrifying bacterium *Alcaligenes* S-6[3)] and *Alcaligenes faecalis* [4)], respectively. Although the reduction of NO_2^- to NH_4^+ by photoinduced enzyme (EC 1.9.6.1)[5)] or water soluble iron porphyrin[6)] as well as the reduction of NO_2^- to NO by NADH analog[7)] has been investigated, no nonenzymatic reduction of NO_2^- via NO to NH_4^+ by denatured protein has so far been reported. The present work shows that NO_2^- is reduced to NH_4^+ by a chemical reducing agent in the presence of denatured cyt *c* and methyl viologen.

MATERIALS AND METHODS

Cytochrome and its denaturation. Horse cardiac cyt *c* was obtained from Sigma Chemical Company and purified by gel filtration with Bio-gel P-10. Denaturation of cyt *c* in an aqueous solution was made by Co 60 gamma-ray irradiation with a dose of 10 kGy or by heating at 100°C for 30 min.

Reduction of nitrite by denatured cyt *c*. The reaction mixture which was composed of 1.5 ml of 0.2M buffer (0.75mmole, pH3-9), 2 ml of 0.01M sodium nitrite (20µmole), 2.5 ml of 0.003M methyl viologen (7.5µmole), 2.5 ml of 50µM cyt *c*, and 1.5 ml of 0.96M sodium dithionite (0.36µmole) dissolved in an aqueous solution of 0.29M sodium hydrogencarbonate, was incubated at 30°C. One ml of the reaction mixture was shaken vigorously by Vortex until the remaining methyl viologen was completely decolored.

Assay of nitrite and ammonium ions. NO_2^- and NH_4^+ concentrations in the reaction mixture were determined by using diazo-reaction method and HPLC, respectively.

Spectrometric measurement of iron-nitrosyl complex. Optical and ESR spectra of iron-nitrosyl complex were measured by a JOEL FE 3AX X-band spectrometer and a MILTON ROY 3000 spectrophotometer, respectively.

RESULTS AND DISCUSSION

Reduction of nitrite to ammonia in nitrogen atmosphere and air. In nitrogen atmosphere, a decrease of NO_2^- in the reaction mixture was observed with time; after 45 min the remaining of NO_2^- was 3% and all NO_2^- have disappeared after 60 min. The formation of NH_4^+ increased as the time proceeds; the yields were 92% after 45 min, 98% after 60 min and 100% after 75min.

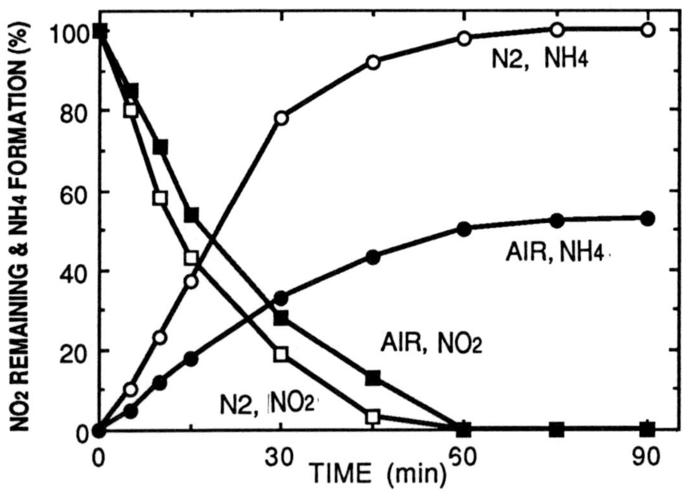

Figure 1. NO_2^- remaining and NH_4^+ formation when NO_2^- was reacted with denatured cyt *c* in the presence of sodium dithionite and methyl viologen in N_2 atmosphere and air at pH 7.0.

In the presence of air, the rate of NO_2^- reduction was slightly slow, compared to that in N_2 atmosphere. The formation of NH_4^+ was only 50-55% after 75 min. From these results, in the presence of air, the reaction other than the reduction of NO_2^- to NH_4^+ was expected as in the case of nitric oxide reductase[4].

pH Dependence of nitrite reduction. In the range of pH3-9, the rate of reduction of NO_2^- to NH_4^+ at lower pH was much faster than that at higher pH in both N_2 atmosphere and air.

Detection of iron-nitrosyl complex. A characteristic ESR of Fe^{III}-N=O complex of denatured cyt *c* was observed near 3,300 G. Optical absorption due to the formation of Fe^{III}-N=O was observed around 570 nm.

Reaction mechanism of nitrite reduction. From above results, the reaction mechanism of NO_2^- reduction by denatured cyt *c* in the presence of a reducing reagent and an electron carrier was proposed as follows.

$$NO_2^- \rightarrow NO \rightarrow NH_4^+$$

This reaction is a net 6-electron and 8-proton reduction process as in the case of feredoxin nitrite reductase[1,2]. These results suggest that the denatured cyt *c* can be used as a model of nitrite reductase for catalytic removal of NOx in water.

This work was partly supported under Nihon University Research Grant for 1992.

REFERENCES

1. Cardenas, B. J., Rivas, J. and Moreno, C. G., Purification and properties of nitrite reductase from spinach leaves. *FEBS Letters*, 1972, **23**, 131-132.

2. Zumft, W. G., Ferredoxin: Nitrite oxidoreductase from *Chlorella* purification and properties. *Biochim. Biophys. Acta*, 1972, **276**, 363-375.

3. Kakutani, T., Watanabe, H., Arima, K. and Beppu, T., Purification and properties of a copper-containing nitrite reductase from a denitrifying Bacterium, *Alcaligenes faecalis* strain S-6. *J. Biochem.*, 1981, **89,** 435-461, 463-472.

4. Miyata, M., Studies on denitrification XIV. The electron donating system in the reduction of nitric oxide and nitrate, *ibid.,* 1971, **70**, 205-213.

5. Willner, I., Lapidot, N. and Liklin, A., Photoinduced enzyme-catalyzed reduction of nitrate (NO_3^-) and nitrite (NO_2^-) to ammonia (NH_3). *J. Am. Chem. Soc.*, 1989, **111**, 1883-1884.

6. Barley, M. H., Takeuchi, K. J. and Meyer, T., Electrocatalytic reduction of nitrite to ammonia based on a water soluble iron porphyrin. *ibid.*, 1986, **108**, 5876-5885.

7. Fukusumi, S. and Yorisue,T., Acid-catalyzed one electron reduction of nitrite to nitric oxide by an NADH analog and 1,1'-dimethylferrocene in the absence and presence of dioxygen. *Chem. Letters,* 1990, 871-874.

RIPFADI: IBERO AMERICAN NETWORK ON PHYSICAL PROPERTIES OF FOODS

JOSE MIGUEL AGUILERA
RIPFADI Coordinator
Department of Chemical Engineering
Universidad Católica de Chile
P.O. Box 306, Santiago, Chile

ABSTRACT

This paper discusses the objectives and *modus operandi* of RIPFADI, an Ibero American Network of Physical Properties of Foods. Launched in 1992, RIPFADI already enrolls 88 groups from 11 countries and 263 researchers. Its objective is to promote scientific interactions between researchers of the region working on thermal, rheological, electrical, diffusion, thermodynamic and mechanical properties of foods and other important physical parameters.

INTRODUCTION

CYTED is an international cooperation program in Ibero America (Latin America plus Spain and Portugal) among research groups of universities, R&D centers and innovative companies. Its aim is to achieve scientific and technical results transferable to the production system and social development plans of the region to assist in improving the quality of life of people, technological modernization and ultimately, the economic development of Ibero American countries. Currently, twenty one countries participate in several of CYTED's sixteen subprograms, one of which is Food Technology and Preservation. A first successful project of this subprogram was finished in 1991 and dealt with intermediate moisture foods and combined methods technology [1].

Within CYTED a network is an association of research units of private or public organizations sharing scientific interests and activities in a common subject. Its objectives are to foster a continuous and stable scientific interaction, exchange technical and scientific information, assist in the formation of human resources and in the transfer of technology and helping in strengthening of research groups . Moreover, networks should be a nuclei for multinational projects and a means of coordinating research at the international level.

JUSTIFICATION AND OBJECTIVES OF RIPFADI

The agriculture and food sector plays a major role in the economy of countries of the region. The formidable double task of food technology is providing local consumers with wholesome and nutritious foods at a reasonable price while simultaneously exporting high-quality, value-added products that are in demand in industrialized countries. Technology and scientific knowledge in the latter case has to be state-of-the-art in order to compete in equal terms with foreign suppliers.

There are several reasons justifying the existence of a network like RIPFADI:

- Data on physical properties of foods are essential to design new processes, optimize unit operations and to develop local technologies.
- Needed information on local agricultural products and foods may not be available because they are not studied in industrialized countries.
- Local food companies are usually small and cannot afford research facilities to generate their own data.
- Research facilities in some countries may not be equipped to perform determinations on certain physical properties.
- Many products are of similar nature (e.g., tropical fruits) and duplication of research efforts should be avoided.
- Researchers in similar areas need to work in groups beyond critical mass.
- Synergistic effects are expected when research is coordinated and cooperation schemes are implemented.
- The almost common language of the network (all countries except Brazil and Portugal speak Spanish) facilitates communication, dissemination of results and fulfillment of exchange programs.

RIPFADI is expected to become an international forum for reporting research results and convey them to the productive system. It should also facilitate contacts between research groups and eventually provide access to an international data base on physical properties of foods.

A DIAGNOSIS OF RESEARCH IN PHYSICAL PROPERTIES OF FOODS

As a first step in the formation of RIPFADI a survey was conducted to assess the local capabilities of research groups willing to participate and their main areas of interest. In Table 1 a matrix of physical properties and products researched by groups is presented.

The fruits/vegetables group is the most cited, probably due to the indigenous nature of the products and their high perishability. Thermal, rheological and thermodynamic (water activity) are the most frequently mentioned properties under research. This initial survey also demonstrated the existence of some well staffed groups and the availability of modern equipment, which predicts good possibilities for high quality research and available training centers in the future.

TABLE 1
Matrix of physical property/product in research of institutions associated to RIPFADI (figures represent the number of groups working on a topic)

	Meat/Fish	Dairy	Fruits/Veg	Cereal Prod.	Other
Thermal	25	10	31	16	9
Rheological	14	22	24	18	16
Electrical	-	2	3	2	-
Diffusion	8	4	22	10	6
Microstructure	6	5	17	10	6
Thermodynamic	17	13	25	24	16
Mechanical	19	13	18	15	10

OPERATION AND FUTURE OF RIPFADI

As of today there are 88 research groups from 11 countries, representing 263 researchers, associated to RIPFADI. The CYTED program finances most of the coordination efforts, including publication of a newsletter, scientific exchange activities, technical courses and workshops, technical publications and the annual meeting of national coordinators. Participating countries through their national research councils, continue funding and organizing independently the research work based on scientific merit and provide financing for local coordination. Administration procedures for RIPFADI are minimum and quite simple, which is a plus for participants in the network. The system largely relies on the enthusiasm of those who collaborate that are also the direct beneficiaries.

Plans for 1993 include consolidation of the network, publication of an inventory of participating institutions and groups and printing of a book listing almost 300 abstracts of papers in international journals authored by local researchers. In order to become a truly collaborative experience some groups will be assisted in bringing up the level of their scientific background via courses and technical exchange visits. In turn, it is expected that beneficiaries of these actions will trickle down the experiences in their countries.

As RIPFADI promotes collaboration and creates contacts in the region it will become a nucleus for developing food engineering research, particularly in Latin America.

REFERENCES

1. Aguilera, J.M. and Parada, E., CYTED-D AHI: An Ibero American project on intermediate moisture foods and combined methods technology. Food Res. Intl., 1992, **25**, 159-165.

RELATIONSHIPS BETWEEN PHYSICAL PROPERTIES AND CHEMICAL COMPONENTS OF OKINAWAN BROWN SUGAR

TAKAYOSHI AKINAGA, YOSHIHIRO KOHDA
Associate professor/Professor/Department of Bioproduction,
College of Agriculture, University of the Ryukyus,
1 Senbaru, Nishihara, Okinawa 903-01, Japan

ABSTRACT

Some physical properties and chemical components of brown sugar produced in Okinawa were investigated to improve the quality standard and manufacturing methods. Hardness, viscosity, color, moisture content, water activity, and density of brown sugar were measured as the components of physical properties. Content of K, P, Ca, Mg, Na, N, and Cl in the brown sugar were measured. As the result of investigation, it became clear that the physical properties of brown sugars varied with the island where the sugar was produced; and P, N, and Ca significantly affected the quality of sugar.

INTRODUCTION

Brown sugar is one of the uncentrifuged sugars and a special product of Okinawa prefecture, and it have been manufactured only on small islands. Brown sugar manufacturing is the main industry that have been supporting the economy of those islands. However, it was reported that the quality of Okinawan brown sugars varied with the districts where it was produced and among layers in one package(1,2). Final products were inspected by the inspector under his experience and the traditional quality standard at each manufactory. A knowledge of quality standard is important and essential engineering data for quality control of brown sugar. Little literature on the quality of brown sugar was found because brown sugar is a local product. It is hoped to produce brown sugar which has constant quality. The subject of this study was to contribute to the improvement of the quality standard and manufacturing method of brown sugar.

MATERIALS AND METHODS

Brown sugars were collected at all manufactories in Okinawa prefecture and used for investigation of their qualities. TABLE 1 shows manufactories, islands, grades, and manufactured dates.

Physical properties

Hardness was defined as the maximum penetrating resistance and measured by a universal testing machine using 8 mm in diameter rigid plunger at 10 mm/min in penetrating velocity. Viscosity was measured by a Shimadzu model CFT-500 flowtester under preheated temperature of 75 °C, preheat time of 5 seconds, and pressure of 5884 KPa using a cylindrical specimen which recomposed from powdered brown sugar. Viscosity was calculated by Hagen-Poiseuille's equation. Surface color of brown sugar was measured by a Minolta model CR-200b chroma meter. Moisture content was measured by the vacuum oven method. Water activity was measured by a Rotronic model DT2-1 hygroscope of 25 °C. Density was measured by a Beckman model 930 air comparison pycnometer and a electronic balance.

TABLE 1
Manufactured data of brown sugars

Manufactory	Island	Grade	Date
Kohama Sugar Ltd	Kohama	Excellent	Jan. 28, 1988
Miyako Sugar Ltd	Tarama	Excellent	Feb. 12, 1989
Aguni Agri. Coop	Aguni	First	Jan. 17, 1989
Hateruma Sugar Ltd	Hateruma	First	Dec. 16, 1988
Iheya Agri. Coop	Iheya	First	Jan. 13, 1989
Iriomote Sugar Ltd	Iriomote	First	Jan. 14, 1989
Miyako Sugar Ltd	Tarama	First	Nov. 28, 1988
Yonaguni Agri. Coop	Yonaguni	First	Jan. 19, 1989

Chemical components

Brown sugars produced in Okinawa were composed of 10 to 11 layers. Samples were prepared from all layers and mixed after broken into pieces with a knife. K, P, Ca, Mg and Na were extracted by the wet-decompose method. K was measured by the flame photometry. P was measured by the Bernard-molybdic acid method and spectrophotometry; Ca and Mg were measured by the nuclear absorption photometry. N was measured by the semi-micro Kjeldahl method. Cl was measured by the silver nitrate titration method after ashed.

RESULTS AND DISCUSSION

TABLE 2 shows the average values of physical properties, and TABLE 3 shows average values of chemical components. TABLE 4 shows results of multiple regression analysis between physical properties and chemical components. From the multiple regression analysis, it became clear that P content affected the pH, viscosity and density, K content affected hardness, N and Na content affected the color, Ca content affected water activity, and Mg content affected the pH of brown sugar.

CONCLUSION

Content of inorganic composition such as P, K, N, Ca, Mg, and Na in brown sugar might be introduced in the quality standards of brown sugars.

TABLE 2
Average values of physical properties

Island	Grade.	M.C. (%WB)	Brix (°)	pH (-)	Hard. (MPa)	Visco. (KPaS)	Hue a(-)	W.A. (-)	Density (g/cm^3)
Kohama	E	5.33	97.8	5.51	2.39	8.50	5.99	0.686	1.606
Tarama	E	4.28	98.8	5.10	2.22	53.04	6.65	0.702	-
Aguni	F	4.58	97.1	6.45	1.85	89.62	4.57	0.684	1.572
Hateruma	F	4.35	98.8	5.97	1.16	12.22	5.78	-	-
Iheya	F	5.56	98.4	6.22	4.48	5.58	6.13	0.669	1.628
Iriomote	F	4.88	96.7	5.89	0.99	25.03	5.36	0.672	1.625
Tarama	F	4.59	98.6	5.88	5.41	15.59	6.52	-	-
Yonaguni	F	4.67	96.6	5.39	1.61	18.73	6.84	-	1.605

TABLE 3
Average values of chemical composition

Island	Grade	P_2O_5 (%)	K_2O (%)	N (%)	Ca (mg/g)	Mg (mg/g)	Na (mg/g)	Cl (mg/g)
Kohama	E	0.080	0.740	0.145	1.106	7.039	0.253	0.649
Tarama	E	0.053	1.499	0.213	1.519	7.226	0.534	0.668
Aguni	F	0.153	1.309	0.076	1.370	5.654	0.158	0.698
Hateruma	F	0.061	1.362	0.217	0.827	4.926	0.229	0.733
Iheya	F	0.056	0.826	0.148	0.687	4.908	0.157	0.800
Iriomote	F	0.046	1.558	0.172	0.991	4.902	0.147	0.676
Yonaguni	F	0.066	1.479	0.317	0.862	7.150	0.282	0.880

TABLE 4
Multiple correlation coefficients

	pH	Hard.	Visco.	W.Act.	Hue(a)	Chroma	Density
P	0.573	-	0.592	-	-	-	-0.929
K	-	-0.716	0.339	-	-	-0.506	-
N	-	-	-	-	0.739	0.982	-0.337
Ca	-	-	0.440	0.630	-	-	-0.260
Mg	-0.793	-	-	-	-	-	-
Cl	-	-	-	-0.231	-	-	-
Na	-	0.436	-	-	0.535	-	-
M.C.	0.187	-	-	-	0.319	-	-
Brix	-	-	-	0.403	-	0.364	-

REFERENCES

1. Okadome, H., Akinaga, T., Kohda, Y., Izumi, H. and Ueno, M., Fundamental studies on physical properties of brown sugars, Proccedings of 49th annual meetings of JSAM, 1990, 323-324.(in Japanese)
2. Akinaga, T. and Kohda, Y., Basic studies on the storage properties of Okinawan brown sugar -hardness, moisture content and pH-, Journal of Kyushu blanch of the JSAM, 1989, 38, 58-61. (in Japanese)

THE EFFECT OF PROCESSING VARIABLES ON THE PRODUCT QUALITY OF SOYA PLANTAIN BABY FOOD

P.O. OGAZI[1], H.O. OGUNDIPE[2], F.A. OYEWUSI[3], S.A.O. ADEYEMI[4]

1. RMRDC, P.M.B. 12873, Lagos, Nigeria
2. IITA, P.M.B. 5320, Ibadan, Nigeria
3. FIIRO, P.M.B. 21023, Ikeja, Lagos, Nigeria
4. NIHORT, P.M.B. 5432, Ibadan, Nigeria

ABSTRACT

Appropriate technologies were used to produce soya–plantain baby food enriched with minerals and vitamins. Plantain was peeled, and dried using cabinet dryer. Soybean seeds were dried, dehulled, milled and extruded. The extruded soybean and plantain flour were blended with sugar, multi–vitamin and calcium carbonate and subjected to proximate analysis and cooking tests. It was found that a mixture of 60% plantain flour, 32% soybean grit and 8% sugar produced a blend whose proximate analysis showed protein 15.8%, fat 8.0%, carbohydrate 70.8% and energy 457.4 kcal/100g. There was no significant difference between the soya–plantain foods produced using extruded and non–extruded soybean grits.

INTRODUCTION

Solid foods are introduced to children as from six months upwards. Most of these solid foods are made from our local crops like maize, rice, sorghum, millet, yam, cassava and plantain. But the problem of calorie–protein–mal–nutrition has always existed in rural and urban children between the age of four and eighteen months because the solid foods are not well balanced in the amount of protein, fat and carbohydrate content. In response of this obvious challenges of infact nutrition in Nigeria, a result–oriented research and development effort was initiated to produce a weaning formula that would meet local taste and siut Nigerian custom using plantain and soybean as the base raw materials.

MATERIALS AND METHODS

The main raw materials used for the experiment were green plantain, soybean, multi–vitamin and calcium carbonate. The stages involved in the development of soymusa are the production of plantain flour and soybean grit (extruded and non–extruded); formulation, laboratory production and preliminary sensory evaluation.

Plantain Flour Production: Green mature plantain fruits were hand peeled and sliced using knife or automatic dicing machine to an appropriate thickness of about 15mm. The slices were dried in a cabinet dryer. The drying temperature was 120°C for the first 2 hours and 80°C for the remaining 3 – 4 hours of drying. The drying was completed when the moisture content of the dried plantain was below 10%. The dried plantain was milled into flour using hammer mill.

Production of Non–Extruded Soybean Grit: 25 kg of raw soybean was weighed out and heated in a cabinet dryer at 110°C for one hour and cracked in a disc attrition mill for 40 minutes. The cracked beans were subjected to air aspiration for 15 minutes using wooden aspirator. The dehulled beans were washed and cooked for 4 hours. The cooked beans were filter pressed and dried in tunnel dryer at 150°C for 1½ hours.

Production of Extruded Soybean Grit: Dried soybean seeds were dehulled, winnowed, milled into fine powder, and extruded using the insta pro 600 Jr. model extruder. The feed intake was set to ensure a discharge of between 300 – 350 kg/hr at a temperature range of 140 – 150°C. The approximate residence time is about 30 seconds. All extruded soybean were dried into grit, hammer–milled, sieved, packaged.

Extruded and non–extruded soybean were blended with plantain flour and sugar respectively as follows:

	A	B	C	D	E
Plantain Flour	1500g	1458g	1393g	1313g	1222g
Soybean Grit	400g	500g	600g	700g	800g
Sugar	100g	125g	150g	175g	200g

The blended mixtures were milled in a roller mill to produce a homogenous mixture of uniform particle size. The milled samples were sieved using vibro sieving machine of aperture 315 microns. Multi–vitamin and calcium carbonate were added to the sieved blend and blended further for ten minutes using tumbling blender. The composition of the final blend is plantain flour 60%, soybean flour 32%, sugar 8%, multi–vitamin 0.15%, calcium carbonate 0.84%. Three desert spoon fulls of each blend were reconstituted with small quantity of cold water to obtain a homogenous liquid mixture. Boiling water was poured into the reconstituted mixture and cooked for about 3 minutes with continuous stirring to avoid any lumping until well cooked. The texture of the gel was consistent and smooth.

The moisture content of the blend was determined by drying in forced air oven at 130°C for one hour. Protein content, crude fat, crude fibre and ash contents were determined according to the AOAC Methods (1). Potassium, sodium, calcium and phosphorus were determined according to the method described by Pearson 1976 (2). Carbohydrate was determined by difference method.

The two samples produced using extruded and non–extruded soybean were submitted for sensory evaluation. The samples were compared for attributes of colour, flavour, consistency, mouthfeel and taste using 9 points hedonic type questionnaire (William 1982 (3)). The data collected were analysed by the student t–test to determine if the samples differed significantly.

RESULTS

The results of the proximate analysis are shown in Table 1. From the results, it was decided that plantain flour (60%), soybean flour (32%) and sugar (8%) will be maintained as the blend provides the required protein and energy level for the healthy growth of babies. The results from the sensory evaluation of the two samples showed that the extruded and non-extruded products did not differ significantly from any of the attributes for which they were evaluated.

TABLE 1
Proximate Analysis of Plantain/Soybean/Sugar Blends

		A	B	C	D	E
Moisture	%	5.5	6.0	5.3	6.2	6.3
Fat	%	3.6	4.5	5.3	8.0	8.6
Protein	%	8.7	10.5	12.0	15.8	16.5
Carbohydrate	%	82.5	76.8	73.51	70.8	70.4
Energy/Kcal/100g		417.7	420.7	419.74	457.4	458.7
Fibre	%	0.83	1.57	1.78	1.0	1.17
Ash	%	2.50	2.77	2.73	2.85	3.04
Potassium	%	0.1	0.2	0.1	0.3	0.2
Sodium	%	0.07	0.07	0.07	0.06	0.06
Calcium	%	0.30	0.40	0.35	0.35	0.35
Phosphorus	%	0.13	0.28	0.15	0.15	0.16

DISCUSSION

Plantain and soybean can be produced easily in the country. Plantain is a good source of carbohydrate, while soybean is superior to all other plant foods as source of protein and have good balance of the essential amino acids. The proximate analysis of plantain and soybean have shown that both contain all the necessary ingredients required for the formulation of a weaning food. The other required ingredients vitamins and minerals have been added as supplementary to the ones contained in the crops. The preliminary sensory evaluation has shown that the extruded and non-extruded products did not differ significantly from any of the attributes for which they were evaluated.

CONCLUSION

This study has shown that acceptable weaning food can be produced from plantain and soybean using appropriate technology. For a small-scale production, the non-extrusion method which is labour intensive with long process time can be adopted. For medium-large scale production the use of extrusion method which requires more capital for the equipment but minimum operational cost is recommended.

REFERENCES

1. A.O.A.C. Methods of analysis of the Association of Officials Analytical Chemists, Washington (1975).
2. Pearson, D., The Chemical Analysis of Foods. Published by Churchill Livingstone, Edinburgh, 7th ED. 1976, pp. 19-24.
3. Williams, A.A. Scoring methods used in the sensory analysis of foods and beverages at Long Ashton Research Station, J. Food Tech., 17, 1982.

PREDICTING TOFU PRODUCTIVITIES OF SOYBEAN VARIETIES BY TOFU GEL CENTRIFUGATION

S.-J. TSAI, T.L. HONG and S.C.S. TSOU*
Department of Food Science, National Chung Hsing University
250 Kuo Kuang Road, Taichung, Taiwan, R. O. C.
*Analytical Laboratory, Asian Vegetable Research and Development Center
(AVRDC) P. O. Box 42, Shanhua, Tainan, Taiwan, R. O. C.

ABSTRACT

Ten soybean varieties were selected to study their processing properties for tofu. Tofu yields varied from 70.83g to 98.70g per 30g of soybean. The yield of tofu was loosely correlated with the solid content of tofu ($r<0.5$). A tofu gel centrifugation method was developed as a substitution method for soybean quality evaluation in laboratory. A high regression coefficient was found on yield of tofu between the centrifugation method and conventional method ($R^2=0.84$). Texture of tofu gels prepared by the centrifugation method was also correlated with tofu made from varieties by the conventional method. Regression coefficients obtained for softness and jelly strength were 0.64 and 0.66, respectively. Experimental results suggested that the gel centrifugation method could be used to predict tofu productivity with small size of soybean in laboratory.

INTRODUCTION

The quality of tofu is characterized by the yield, color and texture. It is also known that the raw material of soybean can affect these characteristics. Several investigations have been conducted in identifying the soybean qualities that are related to its processing properties (1)(2)(3).

For laboratory operation, a centrifugation method as an evaluation technique for determining tofu making properties of soybean with small sample size was developed.

MATERIALS AND METHODS

Soybean samples
Ten soybean varieties grown in the experimental field of AVRDC were used.

Preparation of Conventional Tofu and Small-scaled Tofu (S-tofu)
250 g of soybean were washed and soaked. The soaked beans were removed from the water and ground for 5 min. by a Waring blender at the rate of 9:1 (water to bean). The filtrate was boiled for 5 min. and cooled down to 85°C before coagulated with calcium sulfate. After 20 min., it was poured into a wooden mold (15x15x6 cm) and covered with a piece of cheese cloth. A pressure of 62.2 g/cm^2 was put on the bean curd for 30 min. 30 g of soybean were used to make S-tofu. The procedures were the same as in the preparation of conventional tofu. Only the size of wooden mold was reduced to 7x7x4 cm.

Centrifugation Method
30 ml of soymilk was poured into a centrifuge tube and heated in boiling water for 5 min., and then poured into another centrifuge tube which contained 2 ml of water and calcium sulfate and incubated at 80°C for 20 min. The bean curd formed in the tube was centrifuged at various relative centrifugal forces (RCF) for 10 min. The supernatant was removed and the gel was weighed as the gel yield.

Evaluation of Tofu Texture
Texture of the tofu was evaluated by breaking test with a rheometer (Fudo NRM-2010J-CW) according to Tsai et al (4).

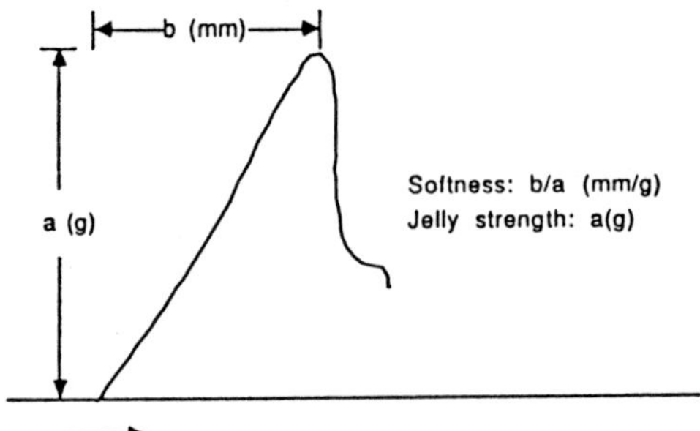

Figure 1. Rheogram of tofu texture

Statistical Analysis
All the experiments were performed in duplicate tests. By the SAS statistical package, simple regressions were used to obtain a regression model.

RESULTS AND DISCUSSION

Yields of Conventional Tofu and Small-scaled Tofu (S-tofu)
The results showed that there were a lot of differences in tofu yields (563.30 - 956.34g /250g soybean) were shown among the varieties (Table 1). The S-tofu yields of the ten samples ranged from 70.83g to 98.70g per 30g of soybean; whereas the highest and the lowest tofu yields were KS#9 and AGS302, respectively. A regression analysis was also made between yields of conventional tofu and S-tofu, and the regression coefficient (R^2) obtained was 0.86. There was a low correlation ($R^2<0.5$) between the yield and solid of conventional tofu and S-tofu. However, S-tofu making, as well as traditional tofu, was too time-consuming to be suitable for mass screening.

TABLE 1
Comparison of yields and solid contents among conventional tofu, S-tofu and gel.

Variety	Yield (g)			Solid (%)		
	Tofu*	S-tofu	Gel**	Tofu*	S-tofu	Gel**
AGS66	859.74	92.32	25.21	15.15	14.36	11.87
AGS129	756.04	89.99	22.14	14.29	11.01	11.67
FWCPT	637.51	83.01	17.72	13.88	11.35	10.58
PMT	644.46	82.84	17.70	12.12	11.98	9.55
CS#1	613.58	74.20	16.37	19.61	19.65	12.52
Clary	647.38	75.92	14.37	20.65	21.24	15.16
KS#5	735.94	87.10	17.80	15.42	16.50	12.28
AGS302	563.30	70.83	14.12	18.66	19.05	13.72
HKT	620.79	81.60	16.07	17.01	17.50	12.31
KS#9	956.34	98.70	26.70	17.19	17.84	12.45

* Conventional tofu. ** Gel was obtained by the centrifugation method.

Establishment of the Centrifugation Method

Irregular shaped gels were formed under the RCFs from 55 xg to 223 xg, and great variation of gel yields occurred at the range of 1397 - 2738 xg. The RCF of 894 xg resulted in the lowest C.V. % (1.40) and was selected to make gels for further experiments (Table 2).

TABLE 2
Gel yield and rotor speed by using the centrifugation method.

RCF* (xg)	Rotor speed (rpm)	Gel yield**	C.V. %***
55	1000	25.14±0.97	3.86
223	2000	17.17±0.43	2.54
503	3000	13.81±0.23	1.70
894	4000	11.61±0.16	1.40
1397	5000	10.02±1.09	10.81
2012	6000	8.24±1.03	12.50
2738	7000	8.67±1.32	15.22

* relative centrifugal force. ** mean±S.D.(standard deviation)
*** coefficient of variation.

Yields Comparison of S-tofu and the Gels Obtained by the Centrifugation Method

Varieties KS#9 and AGS302 respectively demonstrated the highest and the lowest yields of S-tofu as well as gel (Table 1). The regression coefficient (R^2) of yields between S-tofu and gel was 0.84, no matter how a low relationship in solid content they might be. Texture of tofu gels were also evaluated by rheological measurement as it was considered to be an another important quality indicator of tofu. The regression coefficient of jelly strength and softness between gel and S-tofu were 0.64 and 0.66, respectively. This result indicated that the centrifugation method could be a suitable way of evaluating the gel qualities.

CONCLUSIONS

The best way of evaluating soybean quality for tofu-making is to process tofu directly with the conventional method. Although small-scaled tofu method could be applied in soybean variety screening, it is time-consuming in the sample preparation. A simpler method, gel centrifugation, was therefore developed with a high regression coefficient in tofu yields among soybean varieties.

REFERENCES

1. Wang, H.L., Swain, E.W. and Kwolek, W.F. 1983. Effect of soybean varieties on the yield and quality of tofu. Cereal Chem. 60(3): 245-248.
2. Saio, K., Kamiya, M., and Watanabe, T. 1969. Food processing characteristics of soybean 11S and 7S proteins. Part I. Effect of difference of protein components among soybean varieties on formation of tofu-gel. Agr. Biol. Chem. 33(9): 1301-1308.
3. Skurray, G., Gunich, J., and Carter, O. 1980. The effect of different varieties of soybean and calcium ion concentration on the quality of tofu. Food Chem. 6:89.
4. Tsai, S.-J., Lan, C.Y., Kao, C.S., and Chen, S.C. 1981. Studies on the yield and quality characteristics of tofu. J. Food Sci. 46:1734.

USE OF THE MODERN TECHNOLOGY IN IMPROVEMENT OF CHINESE TRADITIONAL FOOD TOU-NAO PROCESSING

SHEN GOU-QI, CHENG CHUANG-JI, LI SHAO-JI*
Shanxi Agricultural University, Shanxi Taigu, 030801, P.R. China
*Taiyuan Industrial University, Shanxi Taiyuan, 030002, P.R. China

ABSTRACT

'Tou-nao', the Chinese traditional food was invented by the famous artist, scholar and Chinese doctor, Mr. Fu Shan, who lived during the Ming dynasty about 350 years ago, for his old mother as a morning drink in winter. It has continued through many families and restaurants. The ingredients of Tou-nao, which include wheat flour, essential mutton, Chinese yam and some other auxiliary materials, are satisfactory for old people's health from the modern nutritional point. Commercial formulae and procedures for Tou-nao are initially documented and the product is described from a traditional point of view. Then experiments to improve the pretreatment of the materials and the procedure in the laboratory are introduced.

MATERIALS AND METHODS

Tou-nao is a nutritious porridge. It is manually made and sold in restaurants for winter mornings. The formula and procedure used have been handed down from ancient times. It has been the habit for many old people to take a bowl of Tou-nao every morning in winter to keep good health. The process for making Tou-nao has received little systematic study despite its long history. The vast historical record and oral instruction about the magical effect of Tou-nao indicate that it is especially fit for weak old people and patients marked by deficiency of vital energy and lowering of body resistance. There are many records about the Tou-nao curing diabetes and maintaining a good status. The ingredients of the Tou-nao include wheat flour, essential mutton, Chinese yam and some other auxiliary materials. In the process for making Tou-nao, steam cooked wheat flour, pretreated Chinese yam chips and cooked shredded mutton are boiled together in a soup where some clean sheep bones have previously been boiled. Then some other auxiliary condiments are poured into the product to fit a personal taste. From a traditional viewpoint, an ideal Tou-nao has the following characteristics: brainlike shape and white colour cream without any muttony flavour. It has been verified by practice that the magical effect of the Tou-nao is mainly due to the correct prescription and adding of Chinese yam. In different areas of China, the Chinese yam is used with rice or seeds of Job's tears to make other nutritious porridges such as: Shen-xian (celestial) Zhou (porridge) and Zhu-yu (pearl and jade) Zhou. These porridges have significant effects on some patients suffering from dyspepsia, night sweats, listlessness, impotence and premature ejaculation. As a food material the Chinese yam is rich in protein (9.4%, dry) which is in a balanced amino acid pattern with a lysine content of 320 mg/100 g. Its effect is related to the chemical and amino acid compositions of its protein. From Table 1 it can be found there are balanced components and many kinds of trace elements, zinc, selenium and others.

Many studies have indicated that lysine is a limiting component of cereals resulting in a low biological value. The biological deficiencies of the protein quality do not really cause problems and

Table 1 The chemical and amino acid compositions of the Chinese yam (mg/100 g fresh)

Moisture	87.4%	Tryptophan	57	Lecine	120	Mg	20
Protein	2	Histidine	28	Cystine	25	Fe	0.3
Lysine	64	Arginine	178	Serine	121	Mn	0.12
Methionine	23	Tyrosine	47	K	213	Zu	0.27
Glutamic acid	307	Alanine	87	Na	18.6	Cu	0.24
Isoleucine	78	Aspartic	152	Ca	16		
Threonine	57	Valine	67	P	34		
Phenylalanine	57	Glycine	55	Cw	0.55		

they may be improved by: (1) supplementing with synthetic limiting amino acid(s) to upgrade the nutritive value, and (2) complementing one food material with another to increase protein quality as well as quantity. The Chinese yam is an ideal natural additive for this purpose. From the modern nutritious viewpoint there are gluelike substances, choline, amylase and polysaccharides in the Chinese yam. The Chinese yam can be used as a staple food material for the everyday diet of many people. In the last decade the output of Chinese yam has greatly increased in China (about 7500–10 000 kg/ha).

The Chinese yam is a stem-tuber crop. There are several species of it. The moisture of the fresh stem tuber is about 85–87%. The cuticle is very thin (less than 0.1 mm) in the fresh state and there are many fibrous roots scattered on the surface of the cuticle (about 4–6 sticks/cm^2 with 1–1.2 mm diameter). The fibrous roots extend into the pulp of the stem tuber about 1.2–1.5 mm deep. Under the nearest cuticle there is the precious gluelike substances which are very important for its nutrition and special effect in adjusting the physiology of the human body.

The traditional dehydration procedure for the Chinese yam is as follows: washing–peeling–shaping–sun drying, or washing–steam heating–peeling–shaping–sun drying. Another special pretreatment procedure is as follows: washing– peeling–sulphur smoking–air drying–kneading alternately (pass through much times). The product has a column-like shape and bright white colour with fine and close texture. It has a good mastication property after rehydration and is a traditional export to Southeast Asian countries.

Several problems exist in the pretreatment of the Chinese yam: (1) the loss of the precious substances is very high (about 17–20%), and (2) the efficiency of the traditional pretreatment is very low. Every year the loss of the Chinese yam is very high due to rough treatment.

According to the construction property and considering the difference between the cuticle and the fibrous roots we have treated it in a laboratory and induced a new procedure as follows: washing–hot air blowing–cutting the fibrous roots helped by vacuum apparatus–classing–laser melting the vestiges of the fibrous roots–flash heating with ultrasonic generator in a vessel filled hot water (about 100°C)–alkali treatment–scraping and cool water blowing–shaping (to cut into chips with 3–3.5 mm thickness)–drying in microwave oven–packaging. The cuticle and the vestige of the fibrous root have significantly different colours. The vestige of the fibrous root is deeper brown. The laser apparatus can easily hit the vestiges under the guidance system of optics according to the colour property and melt them out in very short time. Comparing this procedure with the traditional procedure, the loss of the precious substances diminishes from 17% to 5%. The mastication after rehydration is better than that of the procedure treated by much kneading and without sulphur vestige. The loss of other nutritious substances is also lower, especially the loss of the carotene and vitamin C.

CONCLUSION

(1) As the average age of the world population becomes greater and greater, it is important to study special food to fit old people and to find more green resources for the food material. (2) The Chinese food culture follows the principle: the food and the drug are from the same resources. Is it important to develop the world food culture to dig up these precious heritages? (3) There are many natural resources which have an abundance of nutritious substances. Combining these materials with

conventional food materials not only increases the food quality but also improves the food quality. (4) To extend the food resources and to diminish the loss of precious food materials, modern technology will play more important part in the treatment of the materials.

REFERENCE

1. Wang Guang-ya, *The Food Composition Table,* 1991, Renmin Sanitary Publishing Co.

FROZEN STABILIZED MINCE AS A SOURCE OF PACIFIC WHITING SURIMI

Simpson, R[1]., Kolbe, E[2]., MacDonald, G.A[3]., Lanier, T.C[4]., and Morrissey, M.T[1].
1 Dept. of Food Science, Oregon State University, Corvallis, OR, 97331. 2 Dept. of Bioresource Engineering, Oregon State University, Corvallis, OR, 97331. 3 CRI Crop and Food Research, PO Box 5114, Nelson, New Zealand. 4 Dept. of Food Science, North Carolina State University, Raleigh, NC, 27695.

ABSTRACT

Pacific whiting is available off the U.S. West Coast for about six months each year. A plan to extend the period of shore-based surimi production from whiting was investigated. Headed/gutted fish, and mince stabilized with cryoprotectants, were frozen, stored for up to six months, then processed into surimi. Measurements of texture and color were compared with those for a surimi control sample. Frozen mince stabilized with 6% sucrose and stored at -20 °C resulted in production of good quality surimi. Pilot scale yield and freezing rate studies indicated the potential feasibility of commercial-scale production.

INTRODUCTION

Pacific whiting (Merluccius productus) are in abundance off the U.S. West Coast for about six months per year; annual landings approach 200,000 tons. This fish is characterized by a soft texture and presence of a myxosporidian parasite linked to rapid and severe enzymatic softening when cooked.

Surimi is thus an attractive process because of the effectiveness of well-mixed enzyme inhibitors. In any case, acceptable quality of frozen fillets and surimi requires that the fish be rapidly chilled on board and processed within a day or two of capture.

Previous work has demonstrated the feasibility of making suitable surimi from a frozen mince of New Zealand hake stabilized with sucrose [1]. Such a process could lengthen the season for onshore surimi production on the U.S. West Coast.

The objectives of this study were to evaluate the potential of producting good quality surimi from frozen Pacific whiting, and to identify important process variables that might influence commercial feasibility.

MATERIALS AND METHODS

All samples were made from rapidly chilled whiting processed within 24 hours of harvest. Samples included headed and gutted (H&G), unstabilized mince (UM), mince stabilized with sucrose (6-12%) and 0.2% polyphosphates (SM), and surimi. Preparation of surimi with 8% cryoprotectants followed a standard process [2]. Samples were vacuum packed, frozen in a blast freezer, then stored at -20°C, -34°C, and -50°C until testing. At intervals from 0 - 6 months, samples were made into surimi and compared in quality with surimi stored at -34°C as a control.

Dimethylamine (DMA) content was determined by a modified copper dimethyldithiocarbamate colorimetric procedure. Color of surimi gels was measured using a Minolta Chroma Meter CR-300 which reports L^*, a^*, b^* color coordinates. Samples were compared for whiteness calculated as $100-((100-L^*)^2+a^{*2}+b^{*2})^{1/2}$.

To measure gel forming ability, frozen surimi was tempered and chopped with salt in a vacuum cutter, adjusting moisture content to 78%, salt content to 2%. The resulting paste was extruded into metals tubes, cooked at 90°C for 15 min then cut into samples and tested to failure in the torsion mode, as described by Simpson et al. [2].

Yield and freezing rate effects were tested over a 3 month period under conditions described by Simpson et al. [3].

RESULTS

Good quality surimi was made from mince stabilized with 6-12% sucrose (SM) and stored at -20 °C for up to six months. Failure strain remained above 2.2 (Fig. 1). Values of 1.9 and above indicate surimi that will pass the double fold test.

Surimi made from headed and gutted fish (H&G) and from unstabilzed mince (UM) was of poor quality (Fig.1). DMA formation in UM was also excessive (Fig.2).

Whiteness of surimi made from mince was less than that of control (77 vs. 81.3) but still considered commercially acceptable.

Surimi yield made from pilot-scale lots of 15 kg of mince was comparable to that of the control. Minimal differences were noted in surimi gels made from SM when frozen over a range of freezing times, to 2 days.

PROCESS IMPLICATIONS

Initial results suggest the potential to: land whiting quickly at nearby coastal processing plants where it can be minced, mixed, and frozen; partially process and freeze mince on mid-size factory trawlers which have no surimi processing ability; regulate and lengthen shore-based surimi process schedules; utilize continuous drum freezers or batch vertical plate freezers; manufacture new products from frozen mince, using reduced, or alternative stabilizers.

Future efforts will determine suitable stabilizer alternatives and levels, frozen storage stability of different product forms, and economics of production.

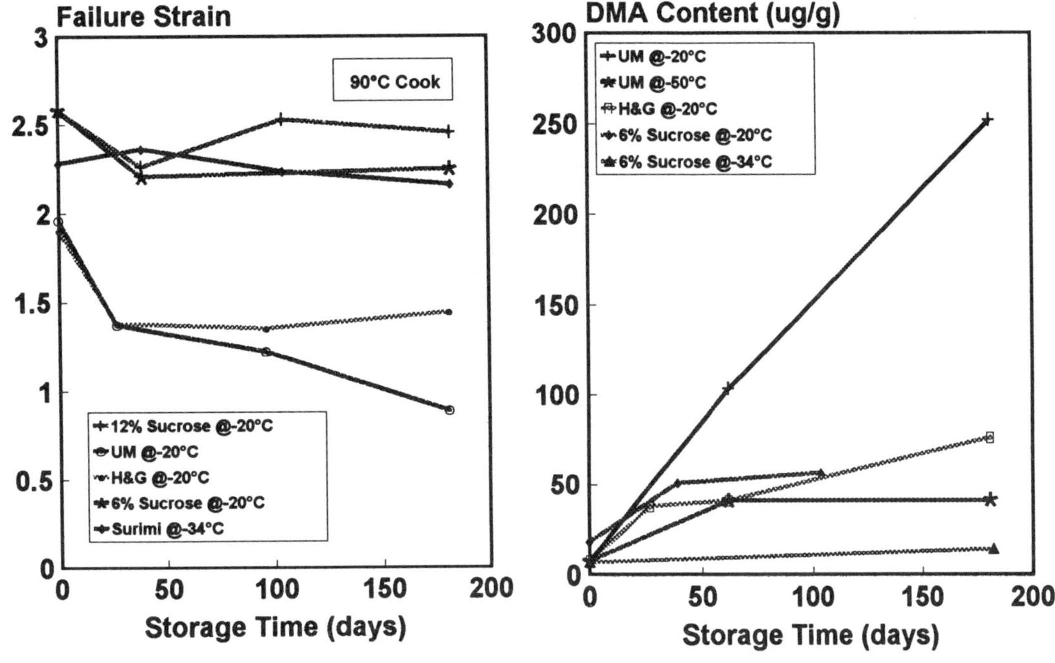

Figure 1. Failure Strain of surimi vs. storage time

Figure 2. DMA levels during frozen storage

REFERENCES

1. MacDonald, G.A., Wilson, N.D., and Lanier, T.C. 1990. Stabilized mince: an alternative to the traditional surimi process. IN Proc. of the IIR Conf. on Chilling and Freezing of New Fish Products. Sept 18-20. Torry Research Station, Aberdeen, Scotland.

2. Simpson, R., Kolbe, E., MacDonald, G.A., Lanier, T.C., Morrissey, M.T. 1993a. The feasibility of producing surimi from partially processed and frozen Pacific whiting (Merluccius productus). In Review, J. of Food Sci.

3. Simpson, R., Kolbe, E., MacDonald, G.A., Lanier, T.C., Morrissey, M.T. 1993b. Process variables affecting surimi made from frozen stabilized mince of Pacific whiting (Merluccius productus). In Review, J. of Aquatic Food Prod. Tech.

ACKNOWLEDNEMENTS

This publication is the result in part of research sponsored by Oregon Sea Grant with funds from the National Oceanic and Atmospheric Administration, Office of Sea Grant, Department of Commerce, under grant no. NA89AA-D-SG108 (project no. R/PD-57) and from appropriations made by the Oregon State Legislature. The U.S. Government is authorized to produce and distribute reprints for governmental purposes, notwithstanding any copyright notation that may appear hereon.

PRODUCTION OF BRASILIAN IMITATION CHEESE WITH A PARTIALLY HIDROGENATED VEGETABLE FAT

MIRNA L. GIGANTE*; SALVADOR M. ROIG**
*DETA - UNESP - Cx.P. 136, 15054-000 São José do Rio Preto - SP - BRASIL
**FEA - UNICAMP - Cx.P. 6121, 13081-970 Campinas, SP - BRASIL

ABSTRACT

Brasilian fresh cheese (minas type) were prepared from skimmed milk to which partially hydrogenated vegetable fat or milk fat was added.

The cheeses were compared as to the transfer level of milk componentes yield and sensory evaluation.

Results showed that there was a significant difference ($P<0,05$) between the fat and total solids transfer level, but there was not a significant difference between the transfer level of proteins and ashes.

Cheese with vegetable fat showed 2,86% increase in yield as compared to those made with milk fat. In the sensory evaluation a difference was detected between the cheeses, but there was no preference for neither one.

INTRODUCTION

Imitation cheese is a product in which milk fat is substituted by vegetable fat, on a protein system from milk origin or being a vegetable fat used in a product produced with caseinates, vegetable proteins, stabilisers, emulsifiers, flavour and colour (5).

Presently different imitation cheeses are being produced as: mussarela, cheddar, swiss, colby, gouda, etc., and are commercialized in different countries as England, United States, Sweden and Australia.

"Minas Frescal" cheese is a widely produced cheese in Brazil, and for this reason was choosen for this feasability study.

MATERIAL AND METHODS

Four set of cheeses were prepared (4) with skim milk and milk fat added (3,5% - control cheese - CC) or partially hidrogenated vegetable fat added (3,5% - imitation cheese - IC). The transfer level values were calculated with reference to milk and cheese whey weight and composition and yield was expressed as cheese kg per 100 milk kg employed for fixed cheese moisture. It was used "t" Student's test to verify the existence os significative difference between transfer level values and yield. Milk, cheeses and cheese whey (CC and IC) were evaluated with respect to total solids (TS), total proteins (TP) ash (A) (1) fat (F) (2) content and lactose (L) by difference. A 20 persons panel was used and run a Triangular and Paread - comparisons - preference tests, (3) for difference and preference respectively.

RESULTS AND DISCUSSION

The milk employed had normal composition with 11,94 TS, 3,55% F, 2,95% TP, 0,7% A and 4,71% L.

The transfer level value for fat was in average 89% for control and 94% for imitation cheese. There was a significative difference at 5% level, which resulted in a transfer level value for TS also statiscally diferent at 5% level for the imitation cheese, as can be seen on figure 1.

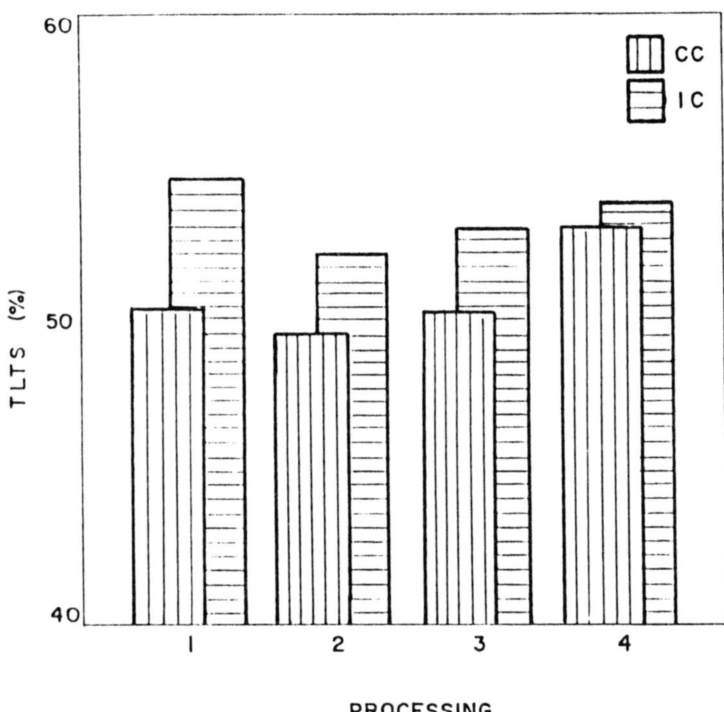

Figure 1. Transfer level of total solids (TLTS-%) for control (CC) and imitation (IC) cheese

There was no significative difference on the transfer level value for TP and A either for control cheese and imitation cheese.

With control and imitation cheese were obtained yields respectively of 14,69 and 15,11 kg cheese/100 kg milk signifying an yield increase of 2,86%.

On the sensorial evaluation, 11 from 20 panelist indicated the different sample, 8 and 3 respectively prefered control and imitation cheese. As result it was verifyed to exist a significative difference between treatments, but do not exist statistical significative preference at 5% level among them.

CONCLUSIONS

1. It was shown the feasability of "Minas Frescal" imitation cheese production.

2. Cheese whey of imitation cheese presented lower fat content as compared to control cheese whey.

3. Imitation cheese presented higher transfer level value for fat and total solids.

4. There was a significative difference ($P<0,05$) for transfer level value total solids for imitation and control cheeses.

5. There was not a significative difference for transfer level value for imitation and control cheeses.

6. Imitation cheeses presented yield increase of 2,86 with respect to control cheeses.

7. Sensorial analysis indicated to exist significative difference ($P<0,05$) for imitation and control cheese, however was not detected preference for neither one.

REFERENCES

1. A.O.A.C. Official methods of analysis of the Association of Official Analytical Chemists, 14 ed., Arlington, 1984, p. 1141.

2. Atherton, H.V. and Newlander, J.A., Chemistry and testing of dairy products. 4 ed., Westport, Avi, 1981, p. 386.

3. Moraes, M.A.C., Métodos para avaliação sensorial dos alimentos. 5 ed., Campinas Ed., UNICAMP, 1985, p. 85.

4. Oliveira, J.S., Queijos: fundamentos tecnológicos. São Paulo, Ed., UNICAMP, 1986, p. 146.

5. Walder, D., Imitation dairy produts - the identity problem. British Food Journal 90 (3), 1988, pp. 117-119.

Authors thanks to Drª. Maria Helena Damasio and to Refinadora de Óleos Brasil S/A.

IMPROVEMENT OF BREAD FLAVOR BY A DAIRY SUBSTRATE TREATED WITH ENZYMES AND LACTIC ACID BACTERIA

MASASHI YAMAMOTO, REIKO WATANABE,
TSUTOMU KANEKO AND HIDEKI SUZUKI
R&D Department, Central Research Institute, Meiji Milk Products Co., Ltd.
1-21-3 Sakae-cho, Higashimurayama-shi, Tokyo 189 Japan.

ABSTRACT

Iso-amylalcohol (iso-AmOH), iso-butylalcohol (iso-BuOH) and ethylcaprylate (EtOcapry) were the main aroma components in bread, and the content of these components with a dairy substrate treated with several enzymes and fermented with lactic acid bacteria (FCT) increased from 1.2 to 1.6 times those of control bread. The increase in iso-AmOH and iso-BuOH content in bread with FCT appeared due to increased leucine and valine content in dough by adding FCT. The increase in EtOcapry content may have been caused by the caprylic acid found in FCT. The addition of the hydrolysate of milk protein and milk fat to dough is thus shown to greatly improve bread flavor.

INTRODUCTION

In Japan, bread is widely produced by the sponge dough method, or short-time fermentation method, of straight dough. Generally, aroma components such as iso-AmOH and iso-BuOH in bread are present in small amounts. Thus, a method was sought for improving bread flavor. This was found possible by a dairy substrate treated with protease and lipase and fermented with two strains of *Lactococcus lactis* (FCT). The mechanism for improvement was studied.

MATERIALS AND METHODS

Preparation of bread

FCT (2.8% milk protein and 38% milk fat) and skim milk powder (Meiji Milk Products Co., Ltd. Tokyo Japan) were used. FCT was prepared by treating with commercial protease from *Aspergillus*, lipase from kid and lamb and strains two strains of *L. lactis* subsp. *lactis*. Agitation was carried out at 100 rpm during incubation at 35 ℃ for 48 hr. FCT was then heated at 90 ℃. The flour, yeast, yeastfood, salt, sugar, and shortening were used to prepare the bread by the sponge dough method.

Determination of flavor components in bread crumbs
This parameter was determined by head space gas chromatography (HS-GC) (1) with modification. Extracts of bread flavor were prepared by distillation and extraction methods and analyzed by gas chromatography-mass spectrometry (GC-MS). n-Butylalcohol was used as the internal standard. Individual flavor components were identified by relative retention times with standards and GC-MS analysis.

RESULTS

Analysis of bread flavor
Acetaldehyde, ethylalcohol, n-propylalcohol, iso-BuOH, and iso-AmOH were the main components. Fusel oils such as iso-BuOH and iso-AmOH were the most important of these components.

Improvement of bread flavor by FCT
Table 1 shows the effects of FCT on aroma component production. Iso-AmOH and iso-BuOH were the main aroma components in bread. The addition of FCT raised their relative amounts by 1.6 and 1.2 times that of control bread. EtOcapry and ethylcaprate (EtOcapra) content in bread crumb with FCT was 1.6 and 1.3 times that of control bread.

TABLE 1
Effects of FCT on the production of aroma components in bread crumb

Component	None	5% FCT
iso-AmOH	21.7 μg/g	34.4 μg/g
iso-BuOH	18.0 〃	22.3 〃
EtOcapry	45.6 μg/150g	59.6 μg/150g
EtOcapra	27.5 〃	33.9 〃

Changes in iso-AmOH and leucine content in dough during fermentation
Baker's yeasts produce alcohols such as iso-AmOH and iso-BuOH from amino acids through the Ehrlich mechanism. Changes in iso-AmOH and leucine concentrations in dough sponge during fermentation (Fig. 1) were investigated. Leucine in dough reacted with baker's yeast during sponge dough fermentation. However, its content in mixing dough with FCT exceeded that without FCT. At the final proofing step, this reaction occurred rapidly with the production of iso-AmOH.

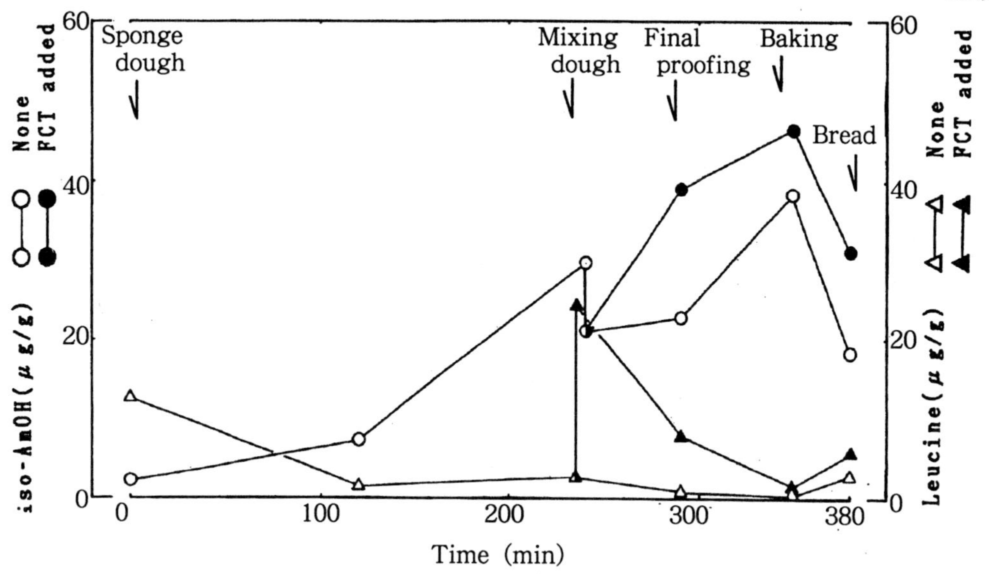

Figure 1. Changes in iso-AmOH and leucine concentrations in dough sponge during fermentation

DISCUSSION

Iso-AmOH, iso-BuOH, EtOcapry, and EtOcapra were the main aroma components in bread. Their amounts increased from 1.2 to 1.6 times those of control bread. Amino acids such as leucine and valine, which are present in FCT, are precursors of iso-AmOH and iso-BuOH, and they reacted with baker's yeast during dough fermentation to increase iso-AmOH and iso-BuOH content in bread. Baker's yeast contains an enzyme which converts caproic acid to EtOcapra. Thus possibly, increase in EtOcapry contents may have resulted from the capric acid found in FCT. Lactic acid bacteria also contribute to the hydrolysis of the substrate and produce lactate. It is evident from the above that the addition of the hydrolysate of milk protein and milk fat to dough greatly improves bread flavor.

REFERENCES

1. Kaneko, T., Suzuki, H. and Takahashi, T., Diacetyl Formation and Degradation by Streptococcus lactis subsp. diacetylactis 3022. Agr. Biol. Chem., 1986, 50(10), 2639-2641.

INTERACTION BETWEEN INHIBITORS OF CALCIUM PHOSPHATE FORMATION AND CALCIUM ION

KAZUTAKA YAMAMOTO, HITOSHI KUMAGAI, TAKAHARU SAKIYAMA, HIROYUKI OGAWA, AND TOSHIMASA YANO

Department of Agricultural Chemistry, The University of Tokyo, Bunkyo-ku, Tokyo 113, Japan

ABSTRACT

The interaction between the inhibitors of calcium phosphate formation and calcium ion was investigated. First, the induction time in base titration experiments was measured to evaluate the inhibitory activity against calcium phosphate formation. Inhibitors such as citrate, alginate, and poly-L-glutamate delayed the induction time. It was suggested that not only the monomer concentration but also the molecular weight of the inhibitor was the important factor for the inhibitory activity. Second, calcium ion activity was measured with a calcium ion selective electrode. The calcium ion activity gradually decreased in the same concentration range as tested in the base titration experiments.

INTRODUCTION

Calcium phosphate precipitation leads to clogging of the membrane pores, resulting in a lower flux during ultrafiltration of milk and whey. Moreover, the formation of calcium phosphate is also important from a physiological point of view, e.g. biological calcification of bones and teeth. Some inhibitors of calcium phosphate formation affect the biological calcification. However, it is not clearly understood how the inhibitor behaves in the inhibition of calcium phosphate formation. In this study, the inhibitory activity against calcium phosphate formation was evaluated by the induction time in the base titration experiments, and the calcium ion activity was measured to evaluate the interaction between the inhibitors and calcium ion. In addition, the effect of molecular weight of the inhibitor on the inhibitory activity was investigated using oligo- and poly-L-glutamate.

MATERIALS AND METHODS

Materials

Poly-L-glutamate (PLG) (Mw = 1×10^4 and 6×10^4 ; Sigma Chemical Co.), alginate (G1-L type ; Kimitsu Chem. Ind., Tokyo), and citrate (Kanto Chemicals Co. Ltd., Tokyo) were used. Oligo-L-glutamate (n = 10, purity 48%) was synthesized by the solid phase method using Fmoc protecting group.[1]

Inhibitory activity against calcium phosphate formation

The induction time, at which the transition of amorphous calcium phosphate to crystalline is known to occur,[2] was measured in base titration experiments as an index of the inhibitory activity[3]. Details of experimental procedure are described in our previous paper.[4]

Calcium ion activity

Calcium ion activity was measured by the ion meter connected to calcium ion selective electrode and to reference electrode (double junction type).

RESULTS AND DISCUSSION

Figure 1 shows the effect of concentration on the induction time. As shown in Fig.1 (a), the plots for polyelectrolytes were sigmoidal, while the plot being concave upwards in the case of citrate. These results suggest that the inhibitory mechanism of polyelectrolyte inhibitor is different from that of a low molecular weight inhibitor. In Fig.1 (b), the inhibitory activity of PLG was highest, and that of oligo-L-glutamate was lower than that of PLG. However, L-glutamate, the monomer of PLG, had almost no inhibitory activity. This result indicates that not only the monomer concentration but also the molecular weight was the important factor for the inhibitory activity of oligo- and poly-L-glutamate.

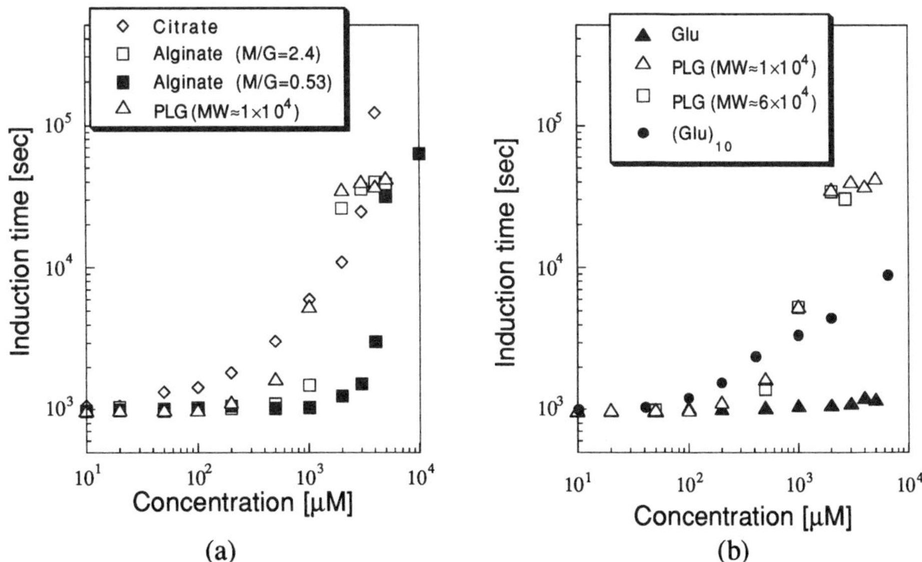

Figure 1 **Effect of concentration on the induction time**. Polyelectrolyte concentrations were expressed by the monomer concentrations.

Figure 2 shows the effect of concentrations on the calcium ion activity. The calcium ion activity gradually decreased in the presence of inhibitors, indicating that the inhibitors interacted with calcium ion.

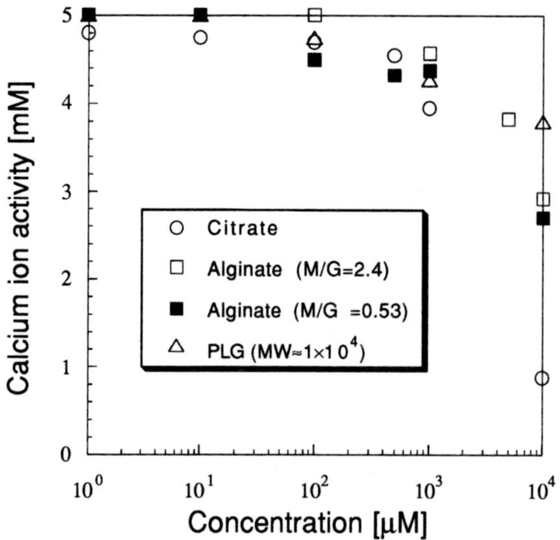

Figure 2 **Effect of concentration on the calcium ion activity.** Total calcium concentration was 5mM in 0.2M Tris-HCl (pH 7.4)

As shown in Figs. 1 and 2, the calcium ion activity decreased in the same concentration range tested in the evaluation of the inhibitory activity. This result indicates that the interaction between the inhibitors and calcium ion plays an important role on the inhibition of calcium phosphate formation.

REFERENCES

1. G. B. Fields and R. L. Noble, Solid phase synthesis utilizing 9-fluorenylmethoxycarbonyl amono acids, Int. J. Peptide Protein Res., 35, 161-214 (1990).

2 A. L. Boskey and A. S. Posner, Conversion of amorphous calcium phosphate to microcrystalline hydroxyapatite. A pH-dependent, solution mediated, solid-solid conversion, J. Phys. Chem. 77: 2313 (1973).

3. J. L. Meyer and E. D. Eanes, A thermodynamic analysis of the amorphous to crystalline calcium phosphate transformation, Calcif. Tissue Res. 25: 59 (1978).

4. K.Yamamoto, H. Kumagai, T. Sakiyama, C.-M. Song, and T. Yano, Inhibitory activity of alginates against the formation of calcium phosphate,.Biosci., Biotech., Biochem., 56, 90-93 (1992).

STUDIES ON THE OPTIMUM CONDITIONS TO UTILIZE BIOLOGICALLY ACTIVE PEPTIDES DERIVED FROM FOOD PROTEINS

MASAAKI YOSHIKAWA and HIROYUKI FUJITA*
Department of Food Science and Technology, Kyoto University, Kyoto 606, and
*The Nippon Synthetic Chemical Industry Co. Ltd., Ibaraki-shi, Osaka 567, Japan

ABSTRACT

As a model study to effectively utilize biologically active peptides derived from food proteins, the optimum conditions for the release of inhibitory peptides for the angiotensin-converting enzyme (ACE) were investigated. Peptides which showed apparent inhibition of the enzyme were classified into three groups: true inhibitors, substrates and pro-drug type peptides which are converted from the substrates to the true inhibitors *in vivo.* After intravenous administration, all of these peptides suppressed the blood pressure elevation caused by the angiotensin I injected. After oral administration in spontaneously hypertensive rats (SHR), however, only the true inhibitors and pro-drug type peptides lowered blood pressure. The thermolysin digest of "Katuo-bushi", dried bonito which contained many kinds of true inhibitors and pro-drug type peptides showed long-lasting anti-hypertensive effects. On the other hand, those obtained by proteases from digestive tracts were ineffective in lowering the blood pressure of SHR after oral administration. Thus, because of its good taste and anti-hypertensive effect, the thermolysin digest of dried bonito is an ideal food stuff for a physiologically functional food.

INTRODUCTION

Many kinds of biologically active peptides are released from food proteins during their enzymatic digestion [1,2]. The angiotensin-converting enzyme (ACE) is a dipeptidyl carboxypeptidase which catalyses the conversion of angiotensin I to angiotensin II, a strong pressor. Therefore, inhibitors for ACE show anti-hypertensive effect [3]. Many inhibitory peptides for ACE have been isolated from the enzymatic digests of food proteins [4-7]. However, we found that some of them were ineffective in lowering blood pressure of spontaneously hypertensive rats (SHR) after their oral administration. In order to obtain orally effective peptides, factors affecting the anti-hypertensive activity were studied and conditions for the enzymatic hydrolysis of proteins were optimized.

MATERIALS AND METHODS

The inhibitory activity for ACE was measured by the method of Cushman and Cheung and expressed by IC_{50} [8]. The inhibitory activity was also measured after preincubation of the peptides with ACE for 3 hrs. The anti-hypertensive activities of peptides against injected angiotensin I (100 ng/Kg) were measured with a pressure transducer which was cannulated into the carotid artery of normotensive rat. The anti-hypertensive effects of peptides were also measured by the tail cuff method using a UR-5000 (Ueda Seisakusho) after oral administration of the peptides. Stability of the inhibitory peptides for preincubation with ACE was tested by HPLC equipped with an ODS column.

RESULTS AND DISCUSSION

Peptides showing inhibitory activities for ACE were obtained from the thermolysin digests of dried bonito and chicken muscle and the peptic digest of ovalbumin (Table 1). These peptides are classified into three groups; 1) The true inhibitors of which IC_{50} values are not altered by preincubation with ACE, 2) Substrates for ACE, of which IC_{50} values are increased by the preincubation with ACE, and 3) Pro-drug type peptides which are converted from substrates to true inhibitors for ACE by digestion with ACE itself or other proteases *in vivo.* After intravenous administration, all of the peptides suppressed the hypertensive effect of angiotensin I. On the other hand, only the true inhibitors and the pro-drug type peptides lowered the blood pressure of SHR after oral administration. The pro-drug type peptides required a longer time to achieve their maximal effect than their activated forms. This 2-hr delay may be explained by the time required for the absorption of slightly larger peptides and that required for the enzymatic conversion into true substrates *in vivo.* The thermolysin digest of dried bonito, which contained many kinds of true inhibitors and pro-drug type peptides, showed long-lasting anti-hypertensive effects after oral administration. On the other hand, the digests obtained by gastrointestinal

Table 1

Relationships between IC_{50} for ACE and Anti-hypertensive Effects

Peptides	Origin	IC_{50} (μM)		Stability for ACE by HPLC (%)	Anti-An-I by i.v.	Blood Pressure by p.o. (max. ΔmmHg)	
		-Preinc.	+Preinc.				
Inhibitor type							
IY	Bonito	2.1	1.9	100	+	-19	2hrs
LW	OVA	6.8	6.6	100	+	-22	2hrs
IKW	Chicken	0.21	0.18	>95	+	-17	4hrs
IKY	Bonito	2.2	2.4	>95	+	-17	4hrs
IKP	Bonito	6.9	6.7	>95	+	-19	4hrs
<u>LKP</u>	Bonito	0.32	0.3	>95	+	-18	4hrs
<u>IWH</u>	Bonito	3.5	3.5	100	+	-30	4hrs
<u>IVGRPR</u>	Bonito	>300	>300	100	+	-17	6hrs
Pro-drug type							
IWHHT*	Bonito	5.8	3.5	0	+	-26	6hrs
LKPNM*	Bonito	2.4	0.76	80	+	-23	6hrs
IVGRPRHQG**	Bonito	2.4	23	0	+	-14	8hrs
Substrate Type							
FFGRCVSP	OVA	0.4	4.6	0	+	0	
FKGRYYP	Chicken	5.8	34	0	+	0	

*: Activated by ACE ; **:Activated by Trypsin.
Underlined peptides are activated forms of pro-drug type peptides.

proteases are ineffective in lowering the blood pressure of SHR after oral administration (data not shown). This means that we cannot expect such an anti-hypertensive effect by merely eating the fish. In this context, digestion by microbial proteases during the food processing step has an important meaning for the expression of potential anti-hypertensive activity.

By controlling the hydrolysis with thermolysin at 80°C, we can decrease the amount of enzyme and prevent the growth of bacteria. Furthermore, thermolysin digest has no bitter taste while other digests did taste bitter. Fishy flavor has been another problem in utilizing fish protein hydrolysates as a food constituent. This problem could be overcome by using Katuo-bushi, a Japanese traditional seaoning material made of dried bonito as a starting material. By using residues of dried bonito, which are obtained in large quantity after the industrial flavor extraction, we can obtain a hydrolysate with the anti-hypertensive effect and no specific taste.

Thermolysin was also effective in releasing inhibitory peptides for ACE from zein [6] and opioid peptides from gluten [9]. However, thermolysin digest of ovalbumin showed very weak inhibitory activity for ACE (data not shown). This means that the most suitable protease to be used is variable depending on substrate proteins and biological activities to be aimed.

Another factor to affect oral avairability of peptides is a physical state of the peptides. Suetuna and Osajima observed improvement in the oral availability of ACE inhibitors by emulsification in egg yolk [5]. However, we could not observe the same effect in our inhibitors. This may be probably caused because our peptides are small enough to be absorbed without emulsification. In fact, we observed an improvement in the oral availability by emulsification of a vaso-relaxing octapeptide, ovokinin, which we isolated from the peptic digest of ovalbumin (unpublished data).

REFERENCES

1. Brantl, V., Teschemacher, H., Henschen, A. and Lottspeich, F., Novel opioid peptides derived from casein (β-Casomorphins) I. Isolation from bovine casein peptone. Hoppe-Seyler's Z. Physiol. Chem, 1979, **360**, 1211-6
2. Yoshikawa, M. and Chiba, H., Biologically active peptides derived from food and blood proteins. In Frotiers and New Horizons in Amino Acid Research, ed. K. Takai, Elsevier Science Publishers, Amsterdam, 1992, pp. 403-9
3. Cushman, D.W. and Ondetti, M.A., Inhibitor of angiotensin-converting enzyme. In Progress in Medicinal Chemistry, ed. G. P. Ellis and G. B. West., Elsevier / North-Holland Biochemical Press, Amsterdam, 1980, **17**, pp. 42-104
4. Maruyama, S. and Suzuki, H., A peptide inhibitor for angiotensin I converting enzyme in the tryptic hydrolysate of casein. Agric. Biol. Chem., 1982, **46**, 1393-4
5. Suetuna, K. and Osajima, K., Blood pressure reduction and vasodilatory effect *in vivo* of peptides originating from sardine muscle. Nippon Eiyo Shokuryo Gakkaishi, 1989, **52**, 47-54
6. Miyoshi, S., Ishikawa, H., Kaneko, T., Fukui, F., Tanaka, H. and Maruyama, S., Structure and activity of angiotensin-converting enzyme inhibitors in an α-zein hydrolysate. Agric. Biol. Chem., 1991, **55**, 1313-8
7. Yokoyama, K., Chiba, H. and M. Yoshikawa, Peptide inhibitors for angiotensin I converting enzyme from thermolysin digest of dried bonito. Biosci. Biotec. Biochem., 1992, **56**, 1541-5
8. Cushman, D.W. and Cheung, H.S., Spectrophotometric assay and properties of angiotensin-converting enzyme of rabbit lung. Biochem. Pharmacol., 1971, **20**, 1637-48
9. Fukudome, S. and Yoshikawa, M., Opioid peptides derived from gluten: their isolation and characterization. FEBS Lett., 1992, **296**, 107-11

ANTI-PLATELET PRINCIPLE FOUND IN THE ESSENTIAL OIL OF GARLIC (*Allium sativum L.*) AND ITS INHIBITION MECHANISM

TOYOHIKO ARIGA, TAIICHIRO SEKI, KIYOSI ANDO, SACHIYUKI TERAMOTO AND HIROYUKI NISHIMURA*
Department of Nutrition and Physiology, Nihon University
School of Agriculture and Veterinary Medicine,
3-34-1 Shimouma, Setagaya, Tokyo 154, Japan
*Department of Bioscience and Technology, Hokkaido Tokai University,
Sapporo 005, Japan

ABSTRACT

Garlic oil and its component methyl allyl trisulfide (MATS) inhibited platelet aggregation induced by arachidonic acid (AA); however, the aggregation induced by AA metabolites was not inhibited by these *Allium* components. The *Allium* components inhibited production of thromboxane B_2 (TXB_2), 12-hydroxyheptadecatrienoic acid (HHT), and prostaglandin E_2 (PGE_2), as well as 12-hydroxyeicosatetraenoic acid (12-HETE), although the AA release reaction was accelerated to some extent, indicating that the inhibition would be caused by a metabolic impairment from AA to its metabolites. The direct interaction between MATS and cyclooxygenase was not revealed. MATS was absorbed by platelets, and some of it reached to the platelet organella.

INTRODUCTION

The plants that belong to *Allium* family have been widely used as nutritious and physiologically functional vegetables. Among the functions, antithrombotic activity afforded by essential oil of garlic is the most reputed [1]. The authors initially found MATS as an active principle for platelet aggregation inhibition in garlic oil, and suggested that the AA cascade in platelet is one of the possible sites of being blocked by MATS [2, 3]. In this paper, the interaction of MATS with human and animal platelets is extensively studied.

MATERIALS AND METHODS

The essential oil of garlic was prepared by using a Linkens-Nickerson's apparatus, and its component MATS was isolated by a preparative gas chromatography. Platelet aggregation was measured by an aggregation meter (DP-247E, Sienco Inc., U.S.A.). For the assaying of antiaggregatory activity, the essential oil or MATS was added to the platelet-rich plasma (PRP) from human or rabbits prior to the addition of an inducer. In radiolabeling, rabbit platelets were labeled with ^{14}C-AA (NEN Research Products) according to Bills *et al.* [4]. ^{14}C-AA-derived radioactive metabolites were separated by an HPLC system (Hitachi 655A, Tokyo) equipped with a column of Inertsil ODS (GL Science Co., Tokyo). For identification of the radioactive eluates, authentic materials obtained from Dupont/NEN Research Products were analyzed by the same system. ^{35}S-MATS was synthesized chemically by using sodium sulfide, ^{35}S (Amersham, England), allyl bromide and methyl bromide.

RESULTS AND DISCUSSION

Effects of Garlic oil and MATS on the AA-Releasing Mechanism.
^{14}C-AA-labeled rabbit platelets were stimulated with thrombin, and the amounts of the label released from various phospholipid moieties of the platelets were measured. In the presence of garlic oil, platelet aggregation was suppressed, whereas the total amount of the label released, which would be free AA and/or its metabolites, was markedly increased. The most prominent phospholipid liberating the label was phosphatidylcholine (PC). MATS also enhanced AA liberation from PC.

Inhibitory Effect of Garlic Oil and MATS on the AA Metabolism in Platelets.
We performed quantitative analysis of the ^{14}C-AA metabolites generated in washed human platelets treated or not with garlic oil or MATS. The platelets prepared could be aggregated fully in response to 50 μM AA containing ^{14}C-AA, and could be inhibited by either garlic oil (2 μg/ml) or MATS (2.6x10^{-6}M) by 50% of the maximum aggregation of the control platelets. MATS or garlic oil seemed to inhibit both cyclooxygenase and lipoxygenase, since they suppressed production of the metabolites TXB_2 and PGE_2, as well as 12-HETE. Cyclooxygenase appeared to be more susceptible to MATS than lipoxygenase, since a MATS concentration sub-inhibitory for 12-HETE production (10^{-6}M) caused appreciable inhibition of TXB_2 and PGE_2 formations.

Garlic Oil and MATS Do Not Inhibit TXA_2-Induced Aggregation.
Platelet aggregation induced by either TXA_2 or its analogues was not inhibited by garlic oil or MATS. LASS, a labile aggregation-stimulating substance prepared from human platelets [4], U-46619 (a PGH_2 mimic), and STA_2 (a TXA_2 analogue) were able to induce platelet aggregation even in the presence of garlic oil or MATS. These results strongly suggest that the *Allium* components specifically inhibit the metabolic pathways from AA to its aggregatory metabolites, PGH_2 or TXA_2.

Effect of MATS on the enzymatic reaction of cyclooxygenase.
In *ex vivo* experiments, garlic oil or MATS was found to interfere the pathway from AA to TXA_2. However, there is no evidence that such foreign chemicals interact with enzyme molecules, and cause inactivation. Then, we studied to know how MATS affects cyclooxygenase activity in *in vitro* reaction system. As shown in Figure 1, the cyclooxygenase was inhibited very weakly by MATS. In this experiment, the substrate ^{14}C-AA remained after reaction was measured (n=3; $\pm$SD). The inhibition would not be specific for the enzyme, since the complete inhibition was not obtained even at the concentration higher than 10^{-5}M MATS, by which platelet aggregation was fully suppressed [2]. Demonstration of real target(s) of MATS within platelets should await further study.

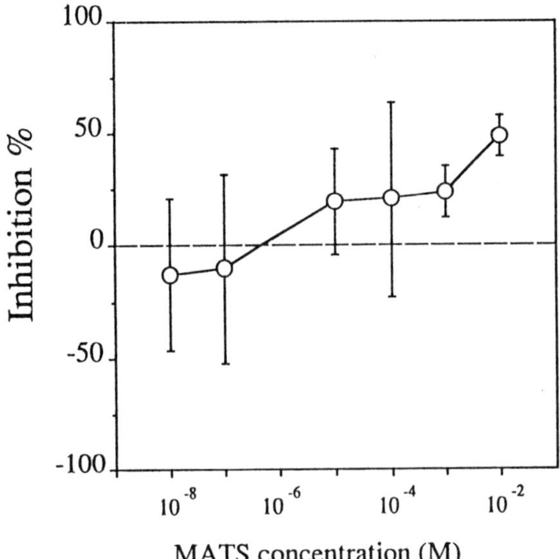

Figure 1. Inhibitory effect of MATS on cyclooxygenase activity.

MATS-Uptake by Platelets.
Ingestion of garlic or its essential oil or even an active principle, MATS, gives rise to decreased platelet aggregability. However, no evidence has been obtained to indicate that platelets absorb such garlic components. We studied the *in vitro* uptake of MATS using synthesized ^{35}S-MATS and washed human platelets. The majority of ^{35}S-MATS was absorbed by platelets. Distribution to the membrane reached to 92% of the total label absorbed, and only 8% was shared by mitochondria, microsomes and cytosol. This suggests that the antiaggregatory principle in garlic may positively be absorbed by the platelets *in vivo*.

ACKNOWLEDGEMENT

This work was supported by grant from the Ministry of Education, Science and Culture (Grant-in-Aid for Scientific Research (C), 03806015), Japan.

REFERENCES

1. Fenwick, G.R. and Hanley, A.B., The genus *Allium*. In CRC Critical Reviews in Food Science and Nutrition, CRC Press, Inc., London, Vol. 23, Issue 1, pp. 1-73.

2. Ariga, T., Oshiba, S. and Tamada, T., Platelet aggregation inhibitor in garlic. Lancet, 1981, i, 150-1.

3. Bills, T.K., Smith, J.B. and Silver, M.J., Metabolism of [^{14}C]arachidonic acid by human platelets. Biochim. Biophys. Acta, 1976, 424, 303-14.

4. Willis, A.L., Vane, F.M., Kuhn, D.C., Scott, C.G. and Petrin, M., An endoperoxide aggregator (LASS), formed in platelets in response to thrombotic stimuli. Prostaglandins, 1974, 8, 453-507.

LIPID PEROXIDE DECREASING ACTIVITY OF MICROBIAL CELLS

MASANORI ITO and KAZUOKI ISHIHARA
Institute for Intestinal and Environmental Microbiology
Advance Co.Ltd.,1-35-2 Shimoishihara,
Chofu-shi,Tokyo 182

ABSTRACT

Microbial cells decreased linoleic acid hydroperoxide (LPO) in phosphate buffer, by binding LPO and reducing hydroperoxide group of LPO. The LPO-decreasing activity of microbial cells was not so affected by digestive enzyme or bile.

INTRODUCTION

Peroxidatants in foods can cause damage to artery and other tissues because they are potential source of free radicals. It is important to decrease the level of peroxidants in foods and intestinal contents.

In this paper, we describe decreasing activity of microbial cells to linoleic acid hydroperoxide (LPO) in vitro.

MATERIALS AND METHODS

Preparation and determination of LPO

LPO was prepared as described by Terao *et al* (1). Concentration of LPO was determined by TBA method (2) and absorbance at 233nm due to conjugated diene of LPO (3).

LPO decreasing activity of various microbes

Living cells or heat-treated (115°C,10min.) dry cells of microbes were added to 2ml of 40mM phosphate buffer (pH6.8) containing LPO (250-500 µg/ml), and stood for 1 hour at room temperature. After centrifugation, concentration of LPO in the supernatant was measured.

Action of the cells of *S.serevisiae* on LPO

The heat-treated cells of *S.cerevisiae* was washed with water 3 times and freeze-dried. The dried materials(500mg) was added to 10ml of LPO solution (500 µg/ml). The mixture was centrifuged and amounts of LPO in the supernatant and acetone extract from the precipitate were measured. The acetone extract, LPO and reduced substances of LPO with $NaBH_4$ or $Na_2S_2O_3$ were developed on silica gel TLC plates with the developing sol. ethyl ether/ acetic acid, 100:1 (vol/vol) and spots were detected by conc. H_2SO_4 with heating.

Proteolytic digestion of heat-treated cells of *S.cerevisiae*
The heat-treated cells of *S.cerevisiae* were treated with pepsin, trypsin and chymotrypsin at 37°C for 2.5h, respectively, and the LPO decreasing activity of indigestible parts of cells were measured.

RESULTS AND DISCUSSION

LPO decreasing activity of various microbes and Scatchard analysis
The LPO decreasing activity was observed among all strains of microbes tested (TABLE 1). There is little difference in LPO decreasing activity between living cells and heat-treated cells. It can be considered that the LPO decreasing activity of microbial cells is physico-chemical action. According to Scatchard plots it was indicated that the adsorption of LPO to microbial cells was nonspecific binding.

Action of the cells of *S.serevisiae* on LPO
Percentage of LPO remaining in the supernatant calculated by using Abs. at 233nm was about the same as that calculated by using TBA method (TABLE 2). TBA value and Abs. at 233nm of the acetone extract from the precipitate were 16% and 82% of initial value, respectively. These data suggest that the heat-treated cells of *S.cerevisiae* have both hydrophobic binding ability and chemical-reducing ability which can reduce the hydroperoxide group but can not affect on conjugated diene structures of LPO. Reduction of LPO by heat-treated cells of *S.cerevisiae* was confirmed by TLC analysis (TABLE 3).

TABLE 1
LPO Decdreasing Activity of Various Microbes

Microbes	LPO Decreasing Activity (μg/mg-dry cells)	Scatchard analysis
E. faecalis AD1001	28	-*
E. faesium AD1060	30	nonspecific
L. acidophilus AD0006	36	-
L. casei AD0013	32	-
L. reuteri AD0002	27	-
L. plantarum AD0010	23	-
B. bifidum AD0060	29	-
B. breve AD0058	31	-
B. fragilis ATCC2528	19	-
S. cerevisiae 167-12	43	nonspecific

*Not done

Influence of proteolytic digestion and presence of bile on LPO decreasing activity

The LPO decreasing activity of the *S.cerevisiae* cells was not so affected by treatment of enzymatic digestion or by the presence of 0.1% bile.

These findings suggest that microbial cells are able to lower the level of LPO in foods and intestinal systems.

TABLE 2
Action of the Cells of *S.cerevisiae* on LPO

Method	% of LPO		
	Extracts from cells	Supernatant	Reduced
Abs. at 233nm	82	18	0
TBA Method	16	17	67

TABLE 3
TLC Analysis

	Rf Value of Main spot
LPO	0.82
Acetone extracts*	0.78
Reduced of LPO with $Na_2S_2O_3$	0.78
Reduced of LPO with $NaBH_4$	0.78

* Reduced substances of LPO with heat-treated cells of *S.cerevisiae* .

REFFERENCES

1.Terao,J.and Matsushita,S.,The isomeric compositions of hydroperoxides produced by oxidation of arachidonic acid with singlet oxigen. Agric.Biol.Chem.,1981,**45**(3),587-93.

2.Ohkawa,H.,Ohishi,N.and Yagi K.,Reaction of linoleic acid hydroperoxide with thiobarbituric acid. J.Lipid Res.,1978,**19**, 1053-7.

3.Terao,J.and Matsushita,S.,Products formed by Photosensitized oxidation of unsaturated fatty acid esters. J.Am.Oil Chem.Soc. 1977,**54**.234-9.

EFFECT OF WATER AND ALCOHOL ON THE FORMATION OF INCLUSION COMPLEX BETWEEN CYCLODEXTRINS AND *d*-LIMONENE BY TWIN SCREW KNEADER

HIDEFUMI YOSHII[1], TAKESHI FURUTA[2], TAKASHI KOBAYASHI[3], TOSHIMI NISHITARUMIZU[3], HIROSHI HIRANO[4] AND AKIRA YASUNISHI[2]
[1]Department of Biochemical Engineering, Toyama National College of Technology, Toyama 939, Japan.
[2]Department of Biotechnology, Tottori University, Tottori 680, Japan.
[3]Kurimoto Co. Ltd., Osaka 559, Japan
[4]Food Engineering Laboratory, Toyama Food Research Institute, Toyama 939, Japan

ABSTRACT

To enhance the storage stability of essential oils such as d-limonene, the molecular inclusion complex powder of d-limonene in α- or β-cyclodextrin was made by using a twin screw kneader. The influence of water and alcohol content on the formation of the inclusion complex was studied in comparison with the inclusion complex by the micro-aqueous method. There were many differences in the quantity of the inclusion complex between the two methods, particularly at a low water content. For α-cyclodextrin, the addition of alcohol inhibited the formation of the inclusion complexes.

INTRODUCTION

Powdery encapsulation of a hydrophobic liquid flavor is quite important to improve its storage stability, as well as to expand its application field. A popular method for the encapsulation is preparation in solution of cyclodextrin. In this method, the specimen is dissolved into an aqueous solution of cyclodextrin to form the inclusion complex in crystalline form [1, 2]. The crystalline complex is separated and dried by an adequate method. However, with so called kneading method, the specimen is mixed directly in a mixer to be transformed into the inclusion complex during kneading. This method is superior to the former, because no separation process is required and less energy will be needed in the dehydration process. In this study, the molecular inclusion of d-limonene (as a model of orange oil) in β-cyclodextrin was accomplished by using a twin screw kneader at a low water content. We have reported that the hydrophilic linear alcohol had increased the formation of the inclusion complex of d-limonene with β-cyclodextrin [3]. We examined the effect of water and alcohol content on the formation of inclusion complex between d-limonene and cyclodextrins.

MATERIALS AND METHODS

Materials

α- and β-cyclodextrin (α-CD and β-CD) was from Ensuiko Sugar Chemical Co. Chloroform stabilized with amylene was from Kanto Chemical Co., Inc. Ethanol of special grade was from Nakalai Tesque.

Preparation of Inclusion Complex Powder by Twin Screw Kneader

Twenty grams of the mixed powder of β-CD was moistened by distilled water, followed by the addition of d-limonene. After being mixed gently in a glass beaker, it was supplied into a twin screw kneader (KRC-S1, Krimoto Ltd.). The kneaded wet samples were taken at 30 and 60 min after kneading, followed by being dried *in vacuo* at 70° C for 15 hours. The content of d-limonene in the powder was analyzed by a gas chromatograph, using a solvent extraction procedure reported previously [3]. The inclusion fraction of d-limonene was defined as a molar ratio of d-limonene in the complex powder to CD.

RESULTS AND DISCUSSION

β-CD/*d*-Limonene System

Figure 1 shows the molar ratio of the inclusion complex between β-CD and d-limonene (inclusion fraction) formed by kneading for 60 min under various initial moisture contents. The solid line represents the experimental results of the inclusion fraction by the so-called micro-aqueous method reported previously. There are marked differences in the inclusion fraction between them, particularly below the water content of 10.

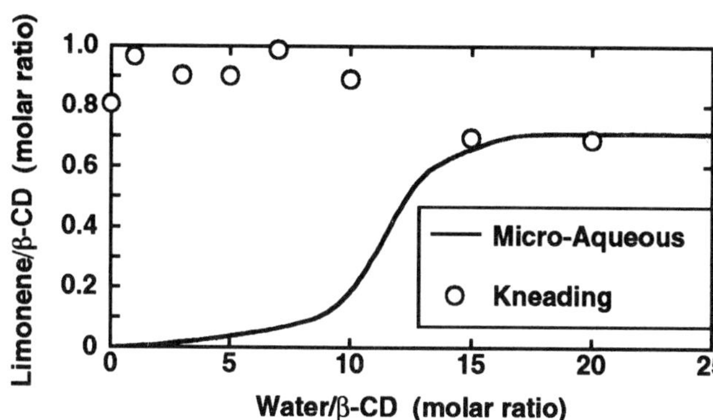

FIGURE 1 Formation of Inclusion Complex of β-CD by Kneading (Comparison with Micro-Aqueous Method)

Figure 2 shows a typical example of the influence of addition of linear alcohol in the case of methanol and pentanol. The amount of each alcohol added is 1 or 5 molar times as much as that of β-CD. For the addition of methanol, the inclusion fraction does not depend on the moisture content, and shows a much higher value than that by the micro-aqueous method shown by the solid line. However, at a higher methanol content, the difference between the kneading and the micro-aqueous method becomes less and tends to coincide with each other. For pentanol, on the other hand, the inclusion fraction depends markedly on the amount of pentanol added at a water content of less than 10, though the inclusion fraction shows no dependencies on the quantity of pentanol added in the case of the micro-aqueous method. Moreover, in the case of pentanol/β-CD = 5, the inclusion fraction by the two methods are nearly equal. This means that with the increase in the alkyl chain

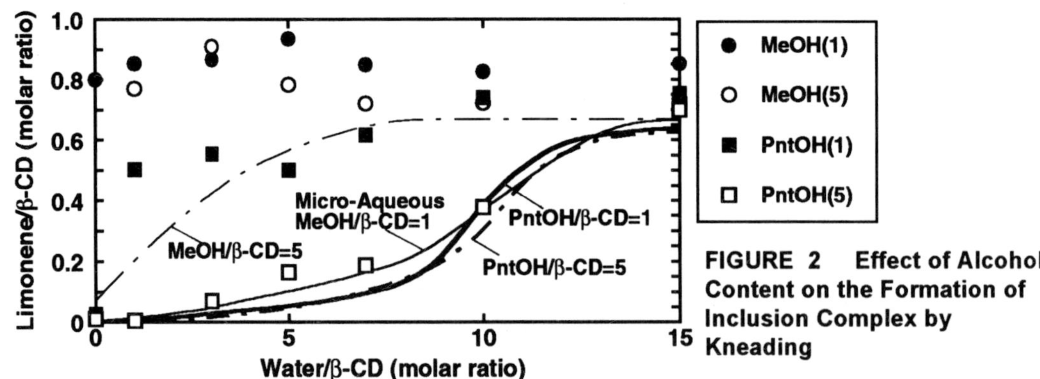

FIGURE 2 Effect of Alcohol Content on the Formation of Inclusion Complex by Kneading

length of alcohol added the inclusion becomes less formed particularly at lower water content. These results suggest that the shear stress created by kneading at these conditions may promote the formation of the inclusion complex. Figure 3 shows the relationship between the kneading torque (shear stress) and the inclusion fraction of d-limonene. The inclusion fraction increases sharply with the increase in the kneading torque.

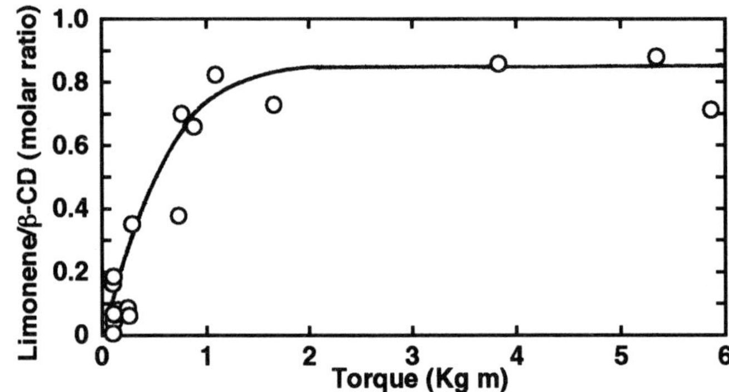

FIGURE 3 Correlation of Inclusion Fraction with Kneading Torque

α-CD/d-Limonene System

The effect of ethanol on the formation of the inclusion complex was examined at a constant water content of 15. When the ethanol content was larger than 2 times the moles of α-CD, the inclusion was inhibited by ethanol and no inclusion complex could be formed. This was similar to the result observed in the micro-aqueous method.

Acknowledgement

This work was supported by the IIjima Memorial Foundation for the Production of Food Science and Technology, and a Grant-in Aid for Scientific Research from the Ministry of Education, Science and Culture. We also appreciate Ensuiko Sugar Refining Co. for the gift of cyclodextrins.

REFERENCES

1. Szejtli, J. Szente, L. and Banky-Elod, E., Molecular Encapsulation of Volatile, Easy Oxidizable Labile Flavour Substances by Cyclodextrin, *Acta Chimica Scientiarum Hungaricae*, 1979, 101, pp.27-46
2. Reineccius, G.A. and Rish, S.J., Encapsulation of Artificial Flavors by β-Cyclodextrin, *Perfumer and Flavorist*, 1986, 11, pp.3-6
3. Furuta, T., Yoshii, H., Miyamoto, A. Yasunishi, A. and Hirano, H., Effect of water and alcohol on the formation of inclusion complexes of d-limonene and cyclodextrin, *Supramolecular Chemistry*, 1993, 1, pp.321-325

ENTRAPMENT OF LIQUID LIPIDS INTO POWDERY MATRIXES OF SACCHARIDES AND PROTEINS

RYUICHI MATSUNO, JUN IMAGI, and SHUJI ADACHI

Department of Food Science and Technology, Faculty of Agriculture, Kyoto University, Sakyo-ku, Kyoto 606-01, Japan

SUMMARY

The emulsifying activity (1), the high stabilizing activity of emulsion (2) and the formation of a fine dense skin layer during drying (3) were the properties of agents that effectively entrap liquid lipids. Gum arabic and gelatin having the three properties were effective. Addition of an agent having a property to a base agent lacking the property improved the entrapment.

Oxidation of entrapped liquid lipid was retarded. However, extent of retardation depended on kinds of lipids and of entrapping agents. Oxidation processes of some combinations of lipids and entrapping agents were expressed by a kinetic model including oxygen diffusion through dehydrated entrapping agents, but most were not. Ethyl eicosapentaenoate was also stabilized by the entrapment.

INTRODUCTION

The most popular method for encapsulation is emulsification of lipids in a solution of entrapping agents followed by spray drying. By this encapsulation, the lipids might gain various advantages, that is, the autoxidation of lipids is retarded, the enhanced stability allows full utilization of the lipids and so on.[1)] To determine the best conditions under which to encapsulate a lipid for a particular purpose, a large number of entrapping agents and combinations needs to be tested under an enormous range of conditions. A convenient method for assessing encapsulation has been introduced, in which a single lipid emulsion of about 5 $\mu\ell$ was dried and a resultant dried powder was subjected to the analyses for the amount of lipids emerged at the surface and the oxidation process of lipids.

METHODS

Drying method for a single droplet of lipid emulsion

A 5 $\mu\ell$ droplet of emulsion containing 15% lipid and 15% entrapping agent(s) was suspended on a glass globule 1 mm in diameter fixed on a chromel–alumel thermocouple junction (100 μm diameter) or on a glass filament, and dehydrated by a single droplet drying apparatus described elsewhere.[2)]

Measurement of lipid appeared at the surface of dehydrated droplet

A droplet was usually dehydrated for 15 minutes. Encapsulation efficiency was then tested by soaking in 0.5 mℓ *n*-hexane for 3 s, and the amount of extracted lipid was determined by gas chromatography. It was assumed that the lipid extracted during the first 3 s was lipid not covered by the entrapping agent.

Method for following the oxidation process of lipids encapsulated

Methyl linoleate (ML), linoleic acid (LA) and ethyl eicosapentaenoate (EE) were used as model lipids susceptible to oxidation. Methyl oleate (MO) was selected as the internal standard. Equal amounts of the model lipids and MO were emulsified in a solution of entrapping material(s), then a 5 μℓ droplet of each emulsion was suspended at the lower end of a glass filament and dehydrated. The dehydrated droplets on the glass filaments were placed upright in a desiccator with controlled humidity and temperature. Periodically a dried droplet was removed and analysed to determine the ratio of the model lipid to MO.

RESULTS AND DISCUSSION

Screening of entrapping agents

Encapsulating agents suitable for use should have the following properties: 1) high emulsifying activity, 2) high stabilizing activity of emulsion, and 3) a tendency to form a fine dense skin layer during drying. The third property might be assessed by the rate of isothermal drying. Maltodextrin, pullulan, gum arabic and gelatin have the third property. Various entrapping agents have been used to encapsulate ML, and then the encapsulation was evaluated. Table 1 summarizes the measurement of the exposed lipid. We define 'optimal' encapsulation as corresponding to a surface lipid level of less than 1%. As shown in Table 1, excellent encapsulation was achieved using gum arabic or gelatin alone, both of which were excellent emulsifiers and had the third property. Maltodextrin and pullulan, which might have the third property, but had no emulsifying activity, were not suitable when used alone. Egg albumin and sodium caseinate, which had high emulsifying activity but not the third property, were also not effective. Glucose, maltose and mannitol lacked all three properties and were not useful at all.

To improve emulsifying activity, a surfactant of low or high molecular weight was added. In general, the addition of a high molecular weight surfactant had positive effects; sodium caseinate gave the best results. The amount of surface lipid with maltodextrin and lecithin was 25.9%. However, when xanthan gum, which is highly viscous in solution, was added to this system, the amount of surface lipid was nearly zero. Thus, this combination (15% ML, 14% maltodextrin, 0.5% lecithin and 0.5% xanthan gum) was the optimal combination for encapsulation.

Retardation of lipid oxidation by encapsulation

ML encapsulated with α–cyclodextrin (α–CD), maltodextrin and pullulan was extremely resistant to autoxidation. ML encapsulated with sodium caseinate was rapidly oxidized after a week. Although gum arabic and gelatin were good encapsulating agents, their abilities to retard oxidation were not good. Albumin was not a good entrapping agent with respect to the retardation of ML oxidation.

α–CD was a good material for retarding LA oxidation. LA encapsulated in maltodextrin and pullulan matrixes was rapidly oxidized, unlike ML. The reason for this remains unclear. The oxidation of LA encapsulated with albumin was relatively rapid, as for ML, whereas LA encapsulated with sodium caseinate oxidized more slowly than ML. Gum arabic and gelatin exhibited a similar retardation effect on the oxidation of both ML and LA.

EE is very susceptible to oxidation due to its high degree of unsaturation. When EE was encapsulated in α–CD, which greatly retarded the oxidation of both ML and LA, 65% was lost by oxidation within two weeks. However, the remaining 35% was unoxidized thereafter. Pullulan significantly retarded the oxidation of EE. Maltodextrin and gum arabic did not prevent EE oxidation.

To elucidate theoretically the reason why the oxidation is retarded, the oxgen diffusion equation and lipid oxidation equation were solved simultaneously, and calculated and experimental average concentrations of unoxidized lipid were compared. Some experimental results correlated well with the calculations, but most of the experimental results indicated that lipids were more stable than predicted by the simulation. Therefore, other mechanisms must be investigated.

REFERENCES

1) Matsuno, R. and Adachi, S., Lipid encapsulation technology – techniques and application to food. *Trends in Food Sci. & Technol.*, 1993, **4**(8), 256–261.
2) J. Imagi, Kako, N., Nakanishi, K., and Matsuno, R., Entrapment of liquid lipids in matrixes of saccharides. *J. Food Eng.*, 1990, **12**, 207–222.

Table 1 Amount of Methyl Linoleate Exposed at the Surface of Dehydrated Samples of the Emulsion Solution of Entrapping Agent with 0.5% Surfactant, 0.5% Stabilizer, or Neither.

	Lipid exposed as percentage of total lipid									
		Surfactant or stabilizer (B)								
Entrapping agents (A)	None	LE (+XG*)	Sugar ester	Tween 80	CMC	Gum arabic 0.5%	Gum arabic 5.0%	Casein	XG	α–CD
Glucose	22.9	3.3	0.3	39.1	5.1	3.1	0.5	0.7	19.0	47.3
Mannitol	34.7	6.9	0.4	38.2	24.4	26.1	1.9	0.5	28.6	40.9
Maltose	32.2	3.9	1.0	59.0	4.9	5.2	0.8	2.2	21.4	36.4
Maltodextrin	1.8	25.9 (≃ 0)	8.0	30.5	1.2	2.0	0.7	1.1	4.3	16.2
α–CD	15.9	13.0	13.5	12.0	7.4	16.3	0.2	13.6	11.1	
Pullulan	2.5	0.8	0.4	48.9	0.2	1.1	0.9	0.4	16.7	28.5
Gum arabic	0.3	0.9	42.0	52.3	0.4			1.0	0.6	1.1
Egg albumin	5.6	8.7	9.4	53.3	3.9	3.0		3.5	3.5	6.7
Casein	3.7	4.8	7.3	6.2	0.8	2.1			1.5	1.5
Gelatin	0.3	2.9	22.6	35.1	0.5	0.9		0.2	0.6	0.8

LE, lecithin; CMC, carboxymethylcellulose; casein, sodium caseinate; XG, xanthan gum; α–CD, α–cyclodextrin; non, neither surfactant or stabilizer. The concentration of methyl linoleate in the original emulsion solution was 15%. Total concentration of entrapping agent was 15%. The concentrations of A and B were 14.5% and 0.5%, respectively, unless otherwise noted.
* The concentration of XG was 0.5%.

AUTHOR INDEX

SUBJECT INDEX